Theory of Unimolecular Reactions

Wendell Forst

ERRATA

Page	*Correction*
68	line 3, for "ethene" read "ethane"
69	penultimate line of Exercise 2, for "(*54*)" read "(*55*)"
117	line 9, for "density (or sum) of states" read "density function"
126	line 4 of Exercise 4, for "(18, . . .)" read "(8, . . .)"
142	line above Eq. (ii), the expression for E_t should read $$E_t \approx [(\mathscr{J}_m^{f})^2 - (\mathscr{J}^{f})^2]\hbar^2/2I_B.$$
164	Eq. (8-68), in the expression for $f(z)$ the middle factor should read $$\left(\frac{n}{n-2}\, bz^{2/(n-2)} - 1\right)$$
172	Eq. (8-96), second line, the middle factor should read $$\left(\frac{n}{n-2}\, bz^{2/(n-2)} - 1\right)$$
196	line 1, for "Guggenhein" read "Guggenheim"
246	Eq. (9-94) should read $$P(E - E_\lambda)_e = N(E - E_\lambda)e^{-(E-E_\lambda)/kT}/Q \qquad (9\text{-}94)$$
263	Eq. (10-8), in the denominator, for "e—$(E)t$" read "e—$k(E)t$"
267	Replace Fig. 10-1 with that shown below

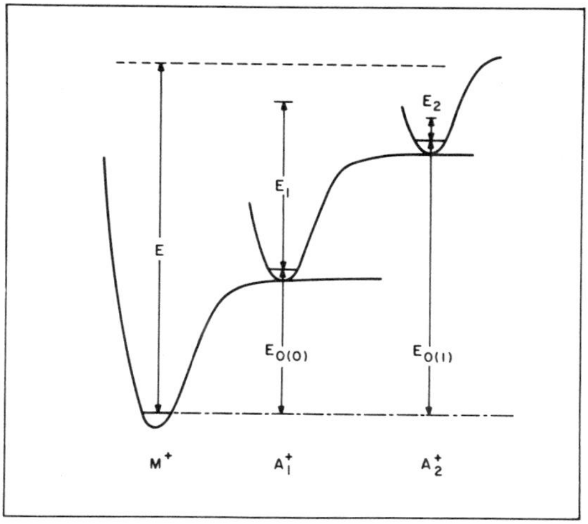

269	line above Eq. (10-25), for "$E_{0(1)}$" read "$E_{0(0)}$"
312	line 12, for "disallowed" read "allowed"
315	two lines below Eq. (10-71a), for "photoelectron spectra" read "total photoionization current"
	five lines above Section 7, for "about" read "above"
321	third paragraph, item 3, for "A^{∞}_{1j}" read "A^{+}_{1j}" and for "$x = E_{0(0j)}$ read "$x = 0$"
329	line above Eq. (10-84), for "translation" read "translational"
380	line 3, the exponent in the exponential term should be preceded by a minus sign
	three lines below Eq. (11-30), for "Table 11-3" read "Table 11-4"
395	line 2, for "Chapter 7" read "Chapter 6"

Theory of Unimolecular Reactions

This is Volume 30 of
PHYSICAL CHEMISTRY
A series of monographs
Edited by ERNEST M. LOEBL, *Polytechnic Institute of Brooklyn*

A complete list of the books in this series appears at the end of the volume.

THEORY OF UNIMOLECULAR REACTIONS

WENDELL FORST

Department of Chemistry
Université Laval
Québec, Canada

1973

ACADEMIC PRESS New York and London
A Subsidiary of Harcourt Brace Jovanovich, Publishers

ACADEMIC PRESS, INC.
111 Fifth Avenue, New York, New York 10003

United Kingdom Edition published by
ACADEMIC PRESS, INC. (LONDON) LTD.
24/28 Oval Road, London NW1

Library of Congress Cataloging in Publication Data

Forst, Wendell.
Theory of unimolecular reactions.

(Physical Chemistry)
Includes bibliographical references.
1. Chemical reaction, Rate of. 2. Chemical reaction, Conditions and laws of. 3. Statistical mechanics. I. Title. II. Series.
QD502.F67 541'.394 72-7686
ISBN 0-12-262350-9

PRINTED IN THE UNITED STATES OF AMERICA

To A. C. F.

Contents

Foreword

It has now been almost exactly fifty years since Lindemann, at a Discussion of the Faraday Society, made his famous suggestion of a time lapse after activation before reaction could take place, thus making it possible to understand how collisional activation could be involved in reactions which were nevertheless first order. Two interesting observations might be made by a curious reader who takes the trouble to refer to the original account of this event. His first observation would be that Lindemann's statement apparently had little impact upon those who heard it. All the discussion centered upon the radiation theory of thermal reactions, upon its difficulties, and upon its alleged triumphs. Probably this was because most of those present had already discarded a collision theory, not only on the grounds that it might be expected to result in a second-order reaction, but also because they believed it could not account for the high rate at which unimolecular reactions proceed. This was a natural belief if one did not take into account the fact that the internal degrees of freedom greatly increase the equilibrium number of molecules in the higher energy states. Indeed, that same year, in an article in *Zeitschrift für physikalische Chemie*, Christiansen expressed essentially the same ideas and discarded them for the reasons indicated. The second observation one might make would be that the thoughts which were obviously going through Lindemann's mind differed to a considerable extent from the presently accepted ideas, and, indeed, he did not do much to clarify the difficulty in accounting for the rate of decomposition. Nevertheless, one can note hints that a probability of reaction is involved and one sees very definitely the prediction that unimolecular gas reactions will no longer be first order at sufficiently low pressures. Only a year after the Faraday Discussion, Rodebush, in a paper in the *Journal of the American Chemical Society*, approached modern ideas more closely.

My own interest in unimolecular reactions was aroused some years later by Ramsperger's experiments on the decomposition of azomethane, especially by the realization that Hinshelwood's hypothesis, that the probability of reaction of an activated molecule was independent of its energy (an idea that Ramsperger and I were considering independently at almost the same time), was inadequate to explain the details of the falling-off of the rate at low pressures. Thus, we came to the conclusion that the energy had to be localized and that the probability of its being so localized could be calculated by statistical mechanics. Illustrating the intensity of interest in this subject in the twenties, this same idea was arrived at by Kassel, independently and only slightly later.

The present working theory of unimolecular reactions is based upon the assumption that it is possible to apply statistical methods to calculate the distribution and rate of redistribution of energy within the molecule; and connection has been made with absolute reaction-rate theory. Quantization was introduced by Kassel, using a crude but reasonable model. Much later, Marcus and I broached the idea of using the actual energy-level densities of molecules to calculate the rates of reaction, and this approach was pursued in detail by Marcus. In recent years many people have contributed to the further understanding of unimolecular reactions. The trend has been toward the elucidation of detailed processes, involving particular energy levels. While it would be futile to attempt to list all the recent contributions, reference may be made to the work of Rabinovitch and his collaborators on chemical activation which penetrates into the details of the level distributions, and also to their work on the transfer of energy at collisions.

Wendell Forst, my valued one-time co-worker, and the author of the present volume, has contributed to the methods of calculating energy level densities, to the understanding of the role of rotational energy, and to the general theory, in addition to his excellent experimental work. He now presents a book which is at once an introduction to the theory, a general overview of what has been accomplished to date, and a manual to assist those who are interested in making the actual calculations necessary to correlate theory and experiment. It is my hope and belief that this book will be a stimulus to further research, and I am very pleased to have had the opportunity to write a foreword to it.

OSCAR K. RICE
Chapel Hill, North Carolina
April 1972

Preface

This book is an account of a theory of unimolecular reactions known generally to kineticists as the Rice–Marcus or the Rice–Ramsperger–Kassel–Marcus (RRKM) theory, and to those working in mass spectrometry and related fields as the quasi-equilibrium theory or the theory of mass spectra. The theory, under any one of its four names, has been successful in accounting for a large body of experimental data in a variety of systems, and is today in wide—if not always correct—use.

It is basically a statistical rate theory couched in transition state formulation. The success of the theory is due mainly to two factors: first, the use of statistics avoids the spelling out of detailed molecular dynamics, and the transition state formulation makes do with very limited information about molecular potentials; second, the various theoretical parameters of the theory can be related fairly directly to experimental observables.

On the debit side, the transition-state approach has certain inherent conceptual difficulties, as well as some vagueness, which does not make for easy or meaningful application of the theory. In addition, the many refinements of the theory that have been worked out in recent years are scattered throughout the literature, so that it takes a very considerable effort just to review the literature, let alone integrate the refinements properly into the framework of the theory.

It is also apparent that gas kineticists and mass spectrometrists, though using the same basic theory, seem to work separately in airtight compartments, with the result that interesting theoretical developments in one field are sometimes ignored in the other, and vice versa.

The present work is intended as an integrated up to date treatment of the theory and its applications, sufficiently detailed and complete to be immediately useful to active workers in any one of the fields where unimolecular phenomena are studied. The objectives are to impart a good understanding of the basic theory, to show how theoretical parameters are related to experimental observables, and to discuss in some detail the methods that are used to obtain useful numerical answers. Chapter 6, and the Appendix are devoted to a discussion of the methods for calculation of energy level densities, the most important computational problem in unimolecular rate theory.

It is assumed that the reader has had fairly good grounding in quantum mechanics and in statistical mechanics, roughly on the senior or first year graduate level, but not necessarily in reaction rate theory. Some of the theoretical parts are actually an outgrowth of a course in the applications of statistical mechanics that the author has been teaching to second- and third-year graduate students in physical chemistry. One hopes that after reading this book, a reader familiar with the rudiments of computer programming, but otherwise heretofore only superficially acquainted with unimolecular rate theory, should be able to apply the theory to his specific system of interest and write his own computer program.

Most chapters end with a few problems which offer useful sidelights on the developments of the main text, and which the reader may use to test his mettle. Although a majority of these problems are simple enough to be worked out on the back of an envelope, they do require some thought.

In a simple statistical theory, like the one discussed here, the excitation process that imparts to a molecule sufficient energy to react, can, to a good approximation, be separated from the actual process of reaction. The various systems of excitation—thermal systems, chemical activation systems, mass spectra, and so on—then differ essentially only by producing different populations of excited molecules. In other words, the observable rate in all systems is the same basic rate averaged over an energy distribution function appropriate to each specific system.

Accordingly, the book is divided into two parts. Part I discusses the derivation of the expression for the basic rate $k(E)$. This part contains most of the theory. Part II deals with the energy distribution functions appropriate to each system, the averaging of $k(E)$, and the relations between theoretical and experimental parameters. This part contains most of the applications of the theory.

The last chapter is an attempt to deal in a rational and systematic way with the ambiguities of the transition state, and provides a guide to the specification of transition state parameters for the various types of reaction.

Acknowledgments

I take particular pleasure to acknowledge my indebtedness to Professor O. K. Rice who has been instrumental in showing me what the theory of unimolecular reactions is all about, and who has read most of the manuscript and offered valuable advice. I am also indebted to the following colleagues who were good enough to send me their comments on various portions of the manuscript: Jacques E. Collin (Liège), Jean-Claude Lorquet (Liège), Zdeněk Prášil (Prague), Pierre C. Roberge (Quebec), Donald W. Setser (Manhattan, Kansas), Herbert L. Strauss (Berkeley), Philip Empedocles (Ann Arbor), Henryk Wincel (Warsaw), and an anonymous referee of Academic Press. It is understood that I alone bear responsibility for errors and omissions remaining in the manuscript. I also wish to thank Jean-Claude Lorquet, Jürgen Troe (Lausanne) and Donald W. Sester for sending me preprints of their work.

Thanks are due to the National Research Council of Canada for support during the preparation of the manuscript, and the publisher's staff for care and attention to my wishes in the production of this volume.

I am no less indebted to Miss D. Lacasse who did the difficult and seemingly endless typing with imperturbable cheerfulness, to Mr. A. Laverdière for the figures, to Mr. G. Plante for photographic work, and last, but not least, to my wife for patience and understanding.

PART I

Introduction

Part I of this text deals with the derivation of an expression for $k(E, \mathscr{J})$, the rate constant for the unimolecular decomposition of a molecule of specified energy E and angular momentum quantum number $\mathscr{J}$. Although $k(E, \mathscr{J})$ does not represent an observable property of a real system, it is the cornerstone of the theoretical treatment because, to a good approximation, observable rates can be considered merely as averages of $k(E, \mathscr{J})$ over distributions of energies and angular momenta particular to each real system (cf. the introduction to Part II).

Chapter 1 examines the by no means trivial problem of finding the conditions under which $k(E)$ can in fact be defined. The formulation of the rate constant as $k(E)$ assumes implicitly that just one parameter, the total energy E, is sufficient for a description of the unimolecular process. The implications of the energy criterion for reaction are discussed, and experimental evidence for or against it are reviewed in Chapter 2. The next chapter deals with the information that can be deduced from the almost pictorial representation of the unimolecular process as the movement of a mass point over a constant-energy surface. Actual derivation of the rate constant $k(E)$ is undertaken in Chapter 4 from two different points of view, both essentially classical. Nonclassical effects and some symmetry properties are also discussed here. Chapter 5 considers the role of the various degrees of freedom in the unimolecular process, in particular, which degrees of freedom are "pertinent" and which are not because of conservation requirements. Introduced here is a simplified model for the dissociating particle which helps to make the rate problem solvable. Chapter 6 develops the mathematical apparatus required for the evaluation of energy-level densities which are necessary for actual rate calculations. Finally, Chapter 7 considers the effect of angular momentum and leads to the derivation of an expression for the more general rate constant $k(E, \mathscr{J})$.

CHAPTER I

The Unimolecular Rate Constant

In this chapter we shall discuss what is meant by the unimolecular rate constant and examine under what conditions a unimolecular process can be described by a single time-independent rate constant; such a description is the central assumption of most unimolecular rate theories. The unimolecular process on which the discussion is focused is the time evolution of an undisturbed molecular system prepared in a very special way, which is largely unrealizable experimentally; nevertheless, the properties of such a system are fundamental to the understanding of the time behavior of real, more complicated systems.

The problem is basically quantum mechanical; while it is presently the subject of active research interest, it has not really been completely worked out as yet. (If it had, we would not need a statistical theory of unimolecular reactions.) Hence the crux of the problem is only touched upon and the discussion is limited for the most part to exploring the implication of the premise, assumed to be true in the absence of any incontrovertible evidence to the contrary, that a single time-independent rate constant does represent an adequate description of the time evolution of such a specially prepared and undisturbed, i.e., isolated, system.

This chapter is not only the shortest, but probably also the most difficult and abstract part of the book. If the material here makes for heavy reading on the first try, the reader should not get discouraged because subsequent chapters progressively come down to earth, in that most of the basic concepts introduced here are reintroduced from another point of view throughout Part I, and then are discussed again in connection with actual experimental systems in Part II.

The basic problem is stated in Section 1, and in Section 2 is expressed in terms of the detailed rate constant k_{ij}. Both theoretical and practical considerations make this an unsatisfactory formalism, and therefore in Section 3 a less detailed rate constant $k(X)$ is considered together with the averaging it implies. The connection with average lifetime is noted in Section 4, and nonrandom lifetime is briefly mentioned here. Finally, the basic assumption of unimolecular rate theory is restated in Section 5.

1. The unimolecular process

Suppose we have a reactant species[1] A that under appropriate conditions decomposes[2] according to

$$\mathsf{A} \longrightarrow \text{Product(s)} \tag{1-1}$$

(for example $C_2H_6 \rightarrow 2CH_3$). The basic problem of unimolecular rate theory is to calculate, and ultimately compare with an experimental observable, the time history of the undisturbed system in which reactant A in state i yields products in state j. By state we mean, ideally, a single quantum mechanical state characterized by a set of quantum numbers; for brevity, let i and j represent such a set for the reactant and product(s), respectively. The reactant state i which decomposes is necessarily some excited state, since the species A is normally stable and does not yield products unless suitably excited; the product state j, on the other hand, can be either a ground state or an excited state. A possible experimental arrangement might consist, for example, of a device producing a beam of particles A of a given flux, another device that would raise all beam particles into state i, and a counter that would count those products which are in state j.

Implicit in the statement of the basic process are the assumptions that the excitation process which produces reactant in state i is much more rapid than the unimolecular process forming products in state j, since otherwise the decomposition rate cannot be defined, and that the inverse reaction product(s) $\rightarrow$ A can be neglected. With these restrictions in mind, the time

[1] The "species" can be molecules or radicals, neutral or charged (i.e., ions). We shall use the terms "species" and "molecule" interchangeably.

[2] The term "decomposition" is to be understood in its general sense: reaction (1-1) will be termed a decomposition whenever the product(s) are chemically distinguishable from reactant A and are separated in space. In this sense, "decomposition" includes isomerization, for example.

history of the unimolecular process can be described in two ways: by the number of molecules in state(s) i decaying per unit time into product(s) in state(s) j (decay rate into channel j, for short), or by the average lifetime, for decomposition into channel j, of a reactant molecule in state(s) i. The information content of each description is different and they are, in principle, not equivalent, although, depending on the complexity of the system being studied, one or the other may be the easier parameter to compute or measure.

2. Decay of a pure bound state into a single channel

Suppose for the moment that i is a pure bound state and j a well-defined dissociated state, and that we choose to describe the process of dissociation as a transition between these two only lightly coupled states. If the decay rate of state i into channel j is characterized by a *single, time-independent* constant, we have the phenomenological relation [cf. Fig. 1-1(a)]

$$-\left\{\frac{1}{[\mathrm{A}(i)]}\frac{d[\mathrm{A}(i)]}{dt}\right\}_j = k_{ij} \tag{1-2}$$

where $[\mathrm{A}(i)]$ represents the number or concentration[3] of molecules in state i and the subscript j on the left-hand side of Eq. (1-2) is there to remind us that the concentrations refer to molecules that decay only into channel j. The parameter of interest is k_{ij} (units, second^{-1}), the detailed rate constant, which depends of course on both the initial (i) and final (j) states. Integration of Eq. (1-2) leads to the familiar exponential relation

$$[\mathrm{A}(i)]_j = [\mathrm{A}(i)]_j^{\circ} e^{-k_{ij}t} \tag{1-3}$$

where $[\mathrm{A}(i)]_j$ is the number or concentration of molecules in state i which decompose into channel j, and the superscript $^\circ$ refers to initial conditions.

Now this representation is a rather unsatisfactory formalism from both the theoretical and practical points of view. Strictly speaking, a bound state is a stationary state with infinite lifetime, and as such it can undergo only radiative transitions (*1*).[4] An excited molecule that undergoes dissociation must be in a metastable state (actually a resonant scattering state) having a finite lifetime; as a result, because of the uncertainty principle, all such metastable states are broadened and cannot be precisely specified, although they do decay exponentially into a given channel, at least initially.[5]

[3] Equation (1-2) makes it obvious that concentration units of $[\mathrm{A}(i)]$ cancel out so that it is sufficient to use for $[\mathrm{A}(i)]$ any quantity merely proportional to concentration.

[4] The reference list for this chapter is to be found on p. 12.

[5] Basically, the detailed description of a decay process cannot be separated from the process of excitation. It turns out, however, that the decay is likely to be nonexponential only if the excitation process produces a very unusual distribution of excited states. Cf. (*2*).

From the practical point of view, one cannot hope ever to prepare a severely state-selected beam of particles, if for no other reason than that the intensity of such a beam would be virtually zero. Even if such a state selection were possible, k_{ij} could not be obtained from Eq. (1-3) because it is not possible to measure, among all the molecules in state *i*, the concentration of only those molecules that decay into channel *j*. We therefore have to give up some detail in the description of the process on this ground alone, and as a compromise we may consider only the total yield of products, i.e., we may consider the decay rate into *all* channels from *one* initial state *i*:

$$-\frac{1}{[\mathrm{A}(i)]}\frac{d[\mathrm{A}(i)]}{dt}=\sum_j k_{ij} \tag{1-4}$$

This is shown schematically in Fig. 1-1(b).

3. Average rate

A more substantial compromise involves the less ambitious specification of initial state *i* imposed by the uncertainty principle. We shall suppose that instead of a single pure bound state it is only possible to prepare A in a perhaps large, but nevertheless limited,[6] collection of bound states, characterized by some parameter[7] X; for example, X might be energy. We should then be able to measure the decay into all channels (or lifetime) of A in states with a specified value of X, and if this decay is characterized by a single, time-independent constant, we have, in analogy with Eq. (1-2),

$$-\frac{1}{[\mathrm{A}(X)]}\frac{d[\mathrm{A}(X)]}{dt}=k(X) \tag{1-5}$$

where $[\mathrm{A}(X)]$ is now the instantaneous concentration of A in states characterized by X, decomposing into all channels, and $k(X)$ is the appropriate rate constant. Note that the collection of states described by X is taken to be a collection of bound states, our comments in Section 2 notwithstanding. We may regard this as a sort of zeroth-order approximation, justified by the fact that our ultimate objective is a statistical, not quantum-mechanical, treatment of unimolecular decay.

Since $[\mathrm{A}(X)]=\sum_i[\mathrm{A}(i)]$, where the summation is over all states *i* specified by X, Eq. (1-5) implies

$$-\sum_i d[\mathrm{A}(i)]/dt=k(X)\sum_i[\mathrm{A}(i)] \tag{1-6}$$

[6] We specifically exclude for the moment the case of an infinite collection of states present in a system about which the only available information is its temperature.

[7] X stands for any convenient means of describing and identifying the states of a collection, and therefore need not be just one parameter.

i.e., each state i of the collection X decomposes into all channels available to it with the same rate constant $k(X)$ [cf. Fig. 1-1(c)]. Equation (1-4) summed over the same i gives

$$-\sum_{i} d[\mathrm{A}(i)]/dt = \sum_{i} \left\{ \sum_{j} k_{ij}[\mathrm{A}(i)] \right\} \tag{1-7}$$

Comparing (1-6) and (1-7), we see that the only way all [A(i)]'s of Eq. (1-7) can have a common factor, as in (1-6), is when

$$k(X) = \left\langle \sum_{j} k_{ij} \right\rangle_{X} \tag{1-8}$$

where $\langle \cdots \rangle_X$ represents an appropriate average over states specified by X, i.e., average over i; thus $k(X)$ is the average rate *per state i* of the collection X.

In the most general case, the result of the summation over j will be some function of i, say $f(i)$: $\sum_j k_{ij} = f(i)$. If $P(i)$ is the (normalized) probability of state i, then in general we have

$$k(X) = \sum_{i} [P(i)f(i)] \tag{1-9}$$

Note that $P(i)$ is assumed implicitly to be independent of time, i.e., the "time zero" at which we choose to start the observation of the decay is left unspecified since it (hopefully) does not matter, although k_{ij} depends in principle on i. Such time independence characterizes decay that is termed "random"; this is explored more fully in Section 4. The implication of a time-independent $P(i)$ is that the individual initial states composing the collection X must interchange and must do so more rapidly than the time it would take for any one individual state to decay into a final state j. Clearly it makes sense to speak of a rate constant $k(X)$ that is the same for all members of the collection X of initial states only if, on the relevant time scale, the initial states have no clearly definable individuality with respect to the decay process. This point touches the very essence of the statistical theory of unimolecular reactions and is taken up again at some length in Chapter 2, Section 3, where the interchange among initial states is discussed in terms of intramolecular energy transfer when the parameter X is total energy.

The immediate problem is to simplify Eq. (1-9). If the collection X is composed of b initial states i, each equally probable, $P(i) = 1/b$ for every i, and (1-9) becomes

$$k(X) = (1/b) \sum_{i} f(i) \tag{1-10}$$

In order to evaluate $\sum_i f(i)$, the simplest possible case is to have every k_{ij} independent of i and j; then if a is the total number of final states which can be

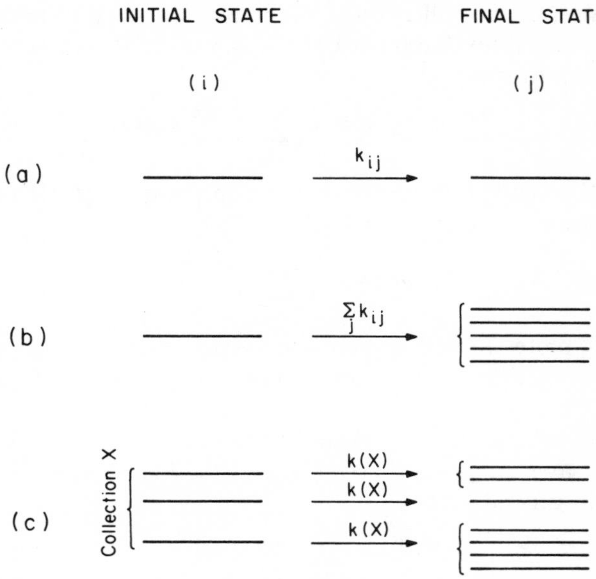

Fig. 1-1. *Transition from initial to final state(s) at various levels of detail (schematic).*

(a) One initial state i, one final state j. There is only one decay channel, and it is described by the rate constant k_{ij}. (b) One initial state i, several final states j. There are now several decay channels but the rate constant $\sum_j k_{ij}$ describes only overall decay into *all* channels (five in the example shown). (c) Parameter X specifies, in this case, a collection consisting of three initial states i, each of which has open to it one or more decay channels j. The overall decay of each initial state i into all channels j that are available to it is described by the rate constant $k(X)$ which is the *same for every initial state i* of the collection X. In the specific case shown, Eq. (1-11) yields $k(X) = \frac{7}{3}k_{ij}$.

reached from the collection X of initial states, we have simply $\sum f(i) = ak_{ij}$ and (1-10) yields

$$k(X) = (a/b)k_{ij}, \qquad k_{ij} = \text{const} \tag{1-11}$$

A schematic representation is given in Fig. 1-1(c) and is explained further in the legend to the figure. We shall recover $P(i) = 1/b$ and $k(X)$ of Eq. (1-11) in Chapter 4, Section 3 on the basis of a more complicated argument which will render perfectly respectable what at this point may seem a very crude simplification.

4. Average lifetime

Equation (1-5) has an interesting implication[8] for the lifetime of a decomposing molecule. Let us write Eq. (1-5) in the form

$$P = -\Delta[\text{A}]/[\text{A}] = k\,\Delta t \tag{1-12}$$

[8] Or indeed any other equation giving exponential decay with a single time-independent constant, such as Eq. (1-2) or Eq. (1-4).

where the dependence of [A] and k on X is not indicated explicitly, to simplify the notation. The ratio $\Delta[A]/[A]$ is the fraction of molecules dissociating, i.e., the probability of dissociation (P), and the rate constant k may therefore be interpreted as the probability of dissociation per unit time. It is noteworthy that P is independent of time, so that the probability of A dissociating is independent of its past or future history and depends only on Δt, the length of the interval during which A is under observation; such incidence of dissociation is said to be random (*3*, p. 15 ff.) (cf. Section 3). The probability of A surviving an interval Δt is

$$1 - P = 1 - k\,\Delta t \tag{1-13}$$

and the probability of A surviving n intervals Δt is $(1 - k\,\Delta t)^n$, since in a random dissociation these probabilities are independent. Let $n\,\Delta t = \tau$; then the probability of A surviving in the interval $(0, \tau)$, i.e., the probability of lifetime τ, is

$$P(\tau) = \lim_{n\to\infty} (1 - k\,\Delta t)^n = \lim_{n\to\infty} [1 - k(\tau/n)]^n = e^{-k\tau} \tag{1-14}$$

If we let $n\,\Delta t = \tau_2 - \tau_1$, we obtain the slightly more general result

$$P(\tau_2 - \tau_1) = e^{-k(\tau_2 - \tau_1)} \tag{1-15}$$

where $P(\tau_2 - \tau_1)$ is the probability of survival in the interval (τ_1, τ_2), $\tau_2 > \tau_1$. We shall make use of Eq. (1-15) in Chapter 10, Section 2.

Lifetimes with a distribution given by Eq. (1-14) are said to be random. The average lifetime is then

$$\langle\tau\rangle = \frac{\int_0^\infty \tau e^{-k\tau}\,d\tau}{\int_0^\infty e^{-k\tau}\,d\tau} = 1/k \tag{1-16}$$

In terms of experimental observables, we can define $\langle\tau\rangle$ as

$$\langle\tau\rangle = (1/[A]^\circ)\int_0^\infty [A]\,dt \tag{1-17}$$

where $[A]^\circ$ is the number of molecules present at $t = 0$. It can be easily verified that if

$$[A] = [A]^\circ e^{-kt} = [A]^\circ P(t) \tag{1-18}$$

Eq. (1-17) yields again $\langle\tau\rangle = k^{-1}$.

We thus have the interesting result that if dissociation is random, the decay rate constant k and average lifetime $\langle\tau\rangle$ are equivalent. The rate constant k is generally the easiest of the two to measure experimentally by means of Eq. (1-5).

The definitions (1-5) and (1-12) refer the rate constant to [A], the number of molecules actually present at t. However, we can also refer the rate to $[A]^\circ$, the number of molecules present originally, and then (1-5) becomes, using (1-18),

$$-\frac{1}{[A]^{\circ}}\frac{d[A]}{dt} = ke^{-kt} \tag{1-19}$$

If we normalize $P(t)$ of Eq. (1-14) to unity in $(0, \infty)$, the result is

$$P_{\text{norm}}(t) = ke^{-kt} \tag{1-20}$$

so that (1-19) can be written

$$-\frac{d[\text{A}]}{dt} = [\text{A}]^{\circ}P_{\text{norm}}(t) \tag{1-21}$$

Note that whereas $P(t)$ is dimensionless, $P_{\text{norm}}(t)$ has the dimension second^{-1}. Equation (1-21) is obviously the derivative of Eq. (1-18), so that

$$P_{\text{norm}}(t) = -\frac{dP(t)}{dt} \tag{1-22}$$

We shall have occasion to make use of these relations in Chapter 8, Section 1.

There is one other interesting aspect concerning Eq. (1-16). The integration with respect to time comprises the time interval $(0, \infty)$ but the infinite upper limit can be looked upon merely as a mathematical convenience that represents macroscopic sampling of events over a time that is long compared with the microscopic time scale on which individual decay events take place, which ensures that the sample is sufficiently large to contain all lifetimes, as implied by the integral (1-16).

Physically, as Eq. (1-17) shows, this means that the system is under observation for a time that is long compared to the microscopic time scale; mathematically, this is again given as an infinite time of observation, but in practice "infinite" works out to a time interval of the order of minutes which is usual in conventional thermal systems. If the time of observation is very much less than this, e.g., of the order of milliseconds or less, we shall eventually reach the point where $\langle\tau\rangle$ decreases (see Exercise 1), for as the time of observation of the decay gets shorter and shorter, only short lifetimes are sampled. This point is particularly important in connection with not-so-short times of observation on real systems prepared with a whole spectrum of values of the parameter X, and therefore this subject is developed in more detail in the brief introduction to Part II, and again in Chapter 10, Section 1.

If the decay cannot be described by a single time-independent rate constant, Eq. (1-16) is not valid (i.e., $\langle\tau\rangle \neq k^{-1}$) and dissociation is said to be nonrandom (see Exercise 2). The average lifetime defined by Eq. (1-17) is then usually referred to as the mean first passage time (*4*), and is the parameter which is easier to obtain from experimental data; however, it does not offer as much detailed information about the system as would the individual (perhaps time-dependent) constants that characterize the decay. Nonrandom dissociation requires a model for the decay process that must of necessity be more detailed than for the ultrasimple random case. We shall therefore defer further discussion of nonrandom decay until the end of Section 5 in Chapter 2.

5. Basic assumption

An experiment has not yet been performed in which the decay or the lifetime of an undisturbed and reasonably well-resolved $\mathsf{A}(X)$ has been measured,[9] so that we do not really know if the decay strictly obeys Eqs. (1-5) and (1-16); presumably this would depend on the preparation of $\mathsf{A}(X)$. However, available experimental evidence from both the radiative and nonradiative decay of less well-resolved systems shows the decay to be essentially exponential, and suggests therefore that Eq. (1-5) represents a useful approximation to the time behavior of real systems, at least until experimental methods and our detailed understanding of the properties of dissociating systems become more refined.

We shall therefore concern ourselves exclusively with the exponential decay derivable from Eq. (1-5) and its implied random incidence of dissociation, and this will of course have the felicitous consequence that we shall have just one value of a time-independent rate constant to worry about for each value of X. The assumption that dissociation is random is made implicitly in all unimolecular rate theories where only one time-independent rate constant is used to characterize the process.

Exercises

1. Show that $\langle \tau \rangle$ decreases when the infinite upper limit of the integration in Eq. (1-16) is replaced by a finite upper limit (assuming k remains constant).

2. The following is a more general form of $P_{\text{norm}}(t)$, based on a suggestion of Slater [(*3*, p. 200)]:

$$ {}_uP_{\text{norm}}(t) = [k/\Gamma(u)] \int_{ukt}^{\infty} x^{u-1} e^{-x}\, dx, \qquad 1 \leqslant u \leqslant \infty $$

where u is a parameter (cf. (*8*)). Show that this function yields the average lifetime $\langle \tau \rangle \sim 1/uk$; in other words, the distribution of lifetimes ${}_uP_{\text{norm}}(t)$ is random only if $u = 1$, in which case it reduces to Eq. (1-20).

[9] An experimental technique that comes closest to this is laser-induced photodissociation in crossed molecular and laser beams ("photofragment spectroscopy"). See (*5–7*).

Hint: The integral in ${}_uP_{\mathrm{norm}}(t)$ is the incomplete gamma function $\Gamma(a, z) = \int_z^\infty x^{a-1}e^{-x}\,dx$. For $a/z \ll 1$, which will be a good approximation when u is not very large and t not too small, we can approximate $\Gamma(a, z) \approx z^{a-1}e^{-z}$. See (*9*).

References

1. J. Jortner, S. A. Rice, and R. M. Hochstrasser, *Advan. Photochem.* **7**, 149 (1969).
2. M. L. Goldberger and K. M. Watson, *Phys. Rev.* **136B**, 1472 (1964).
3. N. B. Slater, "Theory of Unimolecular Reactions." Cornell Univ. Press, Ithaca, New York, 1959.
4. B. Widom, *Advan. Chem. Phys.* **1**, 353 (1958).
5. R. W. Diesen, J. C. Wahr, and S. E. Adler, *J. Chem. Phys.* **55**, 2812 (1971).
6. S. R. Riley and K. R. Wilson, *Discuss. Faraday Soc.* **53** (1972).
7. G. E. Busch and K. R. Wilson, *J. Chem. Phys.* **56**, 3626, 3638, 3655 (1972).
8. D. L. Bunker, *J. Chem. Phys.* **40**, 1946 (1964).
9. M. Abramowitz and I. A. Stegun, eds., "Handbook of Mathematical Functions." Nat. Bur. Std. Appl. Math. Ser. No. 55, p. 263, item 6.5.32 (1965).

CHAPTER 2

Intramolecular Energy Transfer

Having decided to characterize the decay rate by one time-independent rate constant, which is a function of X, we now have to specify X and the states it describes. Obviously, we are not interested in a collection of states for which $k(X)$ is zero, so that X must necessarily describe reactive states, i.e., states of the reactant which ultimately yield products. As a first step, we therefore have to find a criterion which will permit reactive states to be distinguished from nonreactive states; this information will be necessary for calculating $k(X)$, but need not be sufficient.

One such criterion, due to Slater, is briefly mentioned in Section 1, and another, due to Rice, Ramsperger, and Kassel, is presented in Section 2. This criterion is simply total internal energy; it has one important implication discussed in Section 3, namely that there must be intramolecular energy transfer. Experimental evidence for such energy transfer is discussed in Section 4, and theoretical evidence is summarized in Section 5. Finally the essentials of the Rice–Marcus theory, which is the basis of modern statistical theory of unimolecular reactions, are outlined in Section 6.

It is assumed throughout that only one (ground) electronic state is involved in the reaction. The question of randomization of electronic energy in cases where two or more electronic states might be involved is deferred to Chapter 10, which deals with the theory of mass spectra. Section 3 of Chapter 10 also discusses experimental evidence related to the randomization of internal (vibrational) energy in the ground electronic state of ionic (positively charged) systems, and therefore represents an adjunct to Section 4 of this chapter.

1. Theory of Slater

It turns out that the specification of reactive states, or of a criterion for reaction expressed in terms of X and the subsequent calculation of $k(X)$, is a very difficult problem in molecular dynamics which at present defies rigorous solution unless the reaction model is drastically simplified. One ambitious approach to the problem has been made by Slater (*1*)[1] who took as his molecular model a collection of harmonic oscillators whose internal configuration is described in terms of normal mode coordinates. The criterion for reaction is the extension of a bond, seen as a linear combination of normal modes, to some critical value; this depends on energy and vibrational phase. The rate of dissociation is given by the average frequency with which the critical extension is attained and depends on the time behavior of all the normal modes. We are thus given a fairly clear picture of molecular dynamics but the difficulty is that a decomposing molecule is in general very highly excited (as will be discussed) and therefore its molecular vibrations are highly anharmonic. Under these conditions the concept of normal modes loses its meaning (see the discussion by Pitzer *et al.* in (*2*)) and the model is only remotely related to the behavior of real *reacting* molecules. The theory cannot be easily amended by introducing anharmonicity as a perturbation, because anharmonicity is a large perturbation for which perturbation theory is invalid.

2. Theory of Rice, Ramsperger, and Kassel

Another approach to the same problem goes back to *a.* 1927 when Rice and Ramsperger (*3*) and Kassel (*4*) (RRK) made a very shrewd guess that the only criterion for reaction is energy: A species will not decompose [as in Eq. (1-1)] unless it contains a certain minimum internal energy, which we will call henceforth the critical energy E_0, and will react at a rate that increases as a function of the energy it contains in excess of E_0. Symbolically, we have

$$\begin{aligned} k(E) &= 0 && \text{if} \quad E < E_0 \\ &= f(E - E_0) && \text{if} \quad E > E_0 \end{aligned} \tag{2-1}$$

where f is an as yet unspecified function of $(E - E_0)$. Note that energy is considered to be the necessary *and sufficient* information for calculating the rate constant, so that $X = E$. The implication for Eq. (1-8) is that the states involved in the averaging that leads to $k(E)$ must be *all* the quantum states

[1] The reference list for this chapter can be found on p. 28.

having energy in the narrow range E to $E + \delta E$, since no additional[2] information is available as to which (if any) of the usually large number of states are to be excluded from the average.

The problem as posed by RRK is then: Given the energy E, what is the value of $k(E)$? To answer this question, they construct a molecular model which consists of a collection of s independent oscillators, generally harmonic[3] but which, in contrast with Slater's treatment, are assumed to be strongly coupled so as to be able to exchange energy freely. The form of the function f can then be discovered by a statistical argument to within an energy-independent constant factor, which we may call A. The rate constant, or the specific dissociation probability $k(E)$, is taken as proportional to the probability that one particular (critical) oscillator[4] has at least energy E_0 while the remaining $s - 1$ oscillators have energy less than $E - E_0$. The result is[5]

$$k(E) = A[(E - E_0)/E]^{s-1} \tag{2-2}$$

In applications of the theory to thermal reactions, A and E_0 must be obtained from experimental data, while s is used as an adjustable parameter. E_0 can be considered as more or less related to the dissociation energy of the breaking bond, and s as being the number of "effective" oscillators in the decomposing molecule. The RRK theory does not connect A with any observable property of the molecule, although interpretations of A have been made (*6–11*). Equation (2-2) predicts that, at a given excitation energy, $k(E)$ should decrease with increasing s, i.e., roughly with increasing molecular complexity.

3. Intramolecular energy transfer

We are not concerned here with the explicit expressions obtained by RRK, which are now slightly dated, but with one important implication of their criterion for reaction. Without loss of generality, we can use RRK's model of a reacting molecule and think for the moment of a unimolecular reaction as consisting of the dissociation of a critical bond or oscillator whose dissociation energy is E_0. If *all* states at some energy E ($E > E_0$) are considered as reactive, then included among the reactive states will be a state where (for example) the critical oscillator has no energy[6] and the entire energy E is shared only among the other remaining oscillators. The point is that with

[2] Actually, in a more detailed theory such additional information is available from momentum conservation requirements. See Chapter 5, Section 1.

[3] The requirement that oscillators be harmonic is by no means essential but it helps to keep the mathematics simple ("zeroth-order approximation"). See Chapter 6.

[4] We neglect here the historical fact that RR of RRK actually considered accumulation of E_0 in one squared term rather than in one oscillator.

[5] (*5*, pp. 35, 96 ff). Equation (2-2) gives the classical form of $k(E)$; Kassel also derived a quantum form (*4*).

[6] We mean energy in excess of zero-point energy.

energy as the only criterion for reaction, reactive states perforce include states where the critical oscillator has insufficient energy to dissociate, so that energy at least equal to E_0 must somehow reach the critical oscillator if the molecule is to dissociate eventually; the lifetime of a dissociating species can be interpreted as the time it takes, on the average, for an amount of energy E_0 to accumulate in the critical oscillator. Therefore if energy is the necessary and sufficient criterion for reaction, it is thereby implied that energy flows or is exchanged among oscillators, or, in other words, that there is intramolecular energy transfer. RRK then go on to assume in effect that this transfer is a random scrambling process in which energy is shuttled about freely, and this permits them to use the statistical argument mentioned above to find the probability that the critical oscillator has accumulated energy $E > E_0$.

The assumption of energy randomization has the further important consequence that the molecule in a reactive state has no memory and its "remembrance of things past" is completely obliterated as soon as it is formed. As a result, activation, i.e., the process by means of which molecules are raised into reactive states, is immaterial[7] in calculating $k(E)$ as long as all states of energy E are generated, so that under these conditions all unimolecular processes are basically similar, regardless of the mode of activation. If the assumption of energy randomization were not true, only *some* of the many states belonging to E would contribute to $k(E)$, and we would then have to specify which states they are. Thus energy would be a necessary, though not sufficient, criterion for reaction, and consequently $k(E)$, at a given E, would in principle have a different value for different modes of activation, since different states would be involved.[8]

We should note in this context that the parameter $\mathscr{s}$, i.e., the number of degrees of freedom among which the internal energy is assumed to be randomized, is left unspecified in the original RRK theory. By manipulating $\mathscr{s}$ we therefore have a very crude way to allow for some nonrandomness of energy distribution. In the early days, $\mathscr{s}$ always seemed to come out smaller than the actual number of internal degrees of freedom of the molecule studied, so that it looked as if energy were randomized only among a part of the available degrees of freedom.[9] It has since been shown that this is largely an artifice due

[7] In a truly dynamical theory, the mode of activation, even if of short duration, could not be so neatly separated from the unimolecular decomposition process itself since, in principle, the decay of reactive states would be a function of their mechanism of formation. See footnote 5 in Chapter 1.

[8] For calculations involving random and nonrandom activation by collisions, see Baetzold and Wilson (*12*).

[9] See, for example, Kondratev (*13*, p. 315 ff. and Table 23). It was realized even in the old days that at least part of the discrepancy was due to quantum effects.

to the crude approximations that had to be made then in the calculations.

Note also that beyond the statement that oscillators are coupled, the mechanism responsible for intramolecular energy transfer is left unspecified, and, in fact, need not be specified explicitly.

The assumption of rapid intramolecular energy transfer, leading to energy randomization among internal degrees of freedom of the reacting molecule, is central to the RRK argument. What is the evidence in its favor?

4. Randomization of energy and experimental evidence

It took over thirty years before an experimental test[10] was performed of RRK's original surmise about internal energy redistribution. The first such experiment[11] was done by Butler and Kistiakowsky (*15*), who prepared hot methylcyclopropane by reacting (a) methylene with propylene

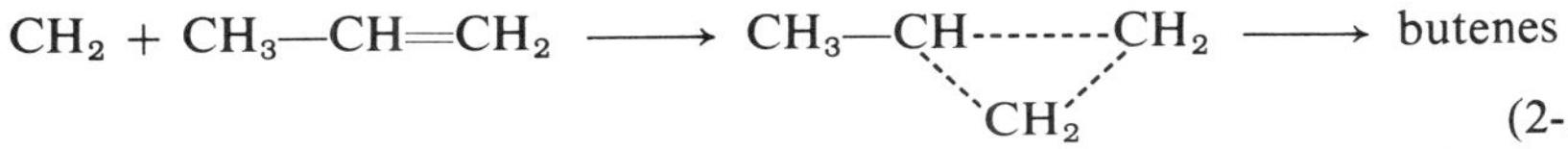

$$CH_2 + CH_3-CH=CH_2 \longrightarrow CH_3-CH\text{-------}CH_2 \longrightarrow \text{butenes}$$
$$CH_2 \qquad\qquad (2\text{-}3a)$$

and (b) methylene with cyclopropane

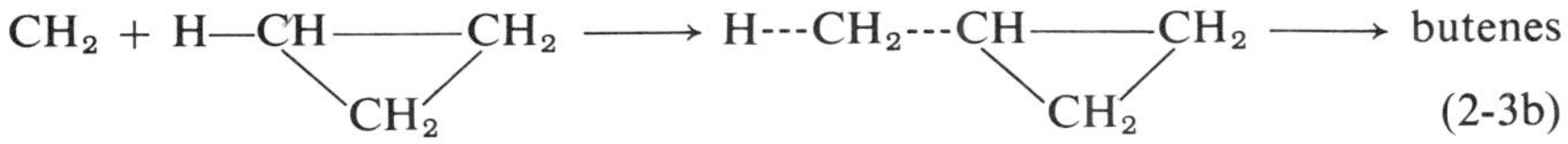

$$CH_2 + H-CH\text{------}CH_2 \longrightarrow H\text{---}CH_2\text{---}CH\text{------}CH_2 \longrightarrow \text{butenes}$$
$$CH_2 \qquad\qquad CH_2 \qquad\qquad (2\text{-}3b)$$

Dashed lines indicate the newly formed bonds. In case (a) the excitation energy is originally located in the carbon ring vibrations, whereas in case (b) it is located in the methyl vibrations, yet both kinds of hot methylcyclopropane give rise to the same reactions, i.e., to the same mixture of butenes. The implication is that by the time the hot cyclopropane molecule has reacted, it has "forgotten" its initial state as a result of rapid internal energy redistribution.

Harrington *et al.* (*17*), in a somewhat similar experiment, prepared excited *sec*-butyl radicals by reacting hydrogen atoms with (a) butene-1,

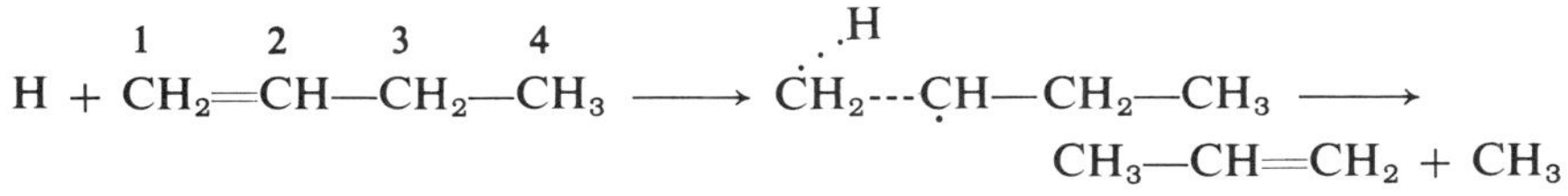

$$\overset{1}{C}H_2=\overset{2}{C}H-\overset{3}{C}H_2-\overset{4}{C}H_3$$
$$H + CH_2=CH-CH_2-CH_3 \longrightarrow \overset{\cdot}{C}H_2\text{---}\overset{\cdot}{C}H-CH_2-CH_3 \longrightarrow$$
$$CH_3-CH=CH_2 + CH_3$$
$$(2\text{-}4a)$$

and (b) *cis*-butene-2,

[10] Actually, indirect tests were available earlier. For example, Douglas *et al.* (*14*) studied the cis–trans isomerization of 1,2-ethylene-d_2 and found by calculation that the parameter s [Eq. (2-2)] is about six, although we would expect s much lower than this since the reaction involves only one degree of freedom, the torsion about a double bond. The fact that $s = 6$ represents half the total number of degrees of freedom in the molecule suggests that the C=C torsion must be coupled to other degrees of freedom.

[11] This type of experiment was simultaneously suggested by Frey (*16*). See also (*61*).

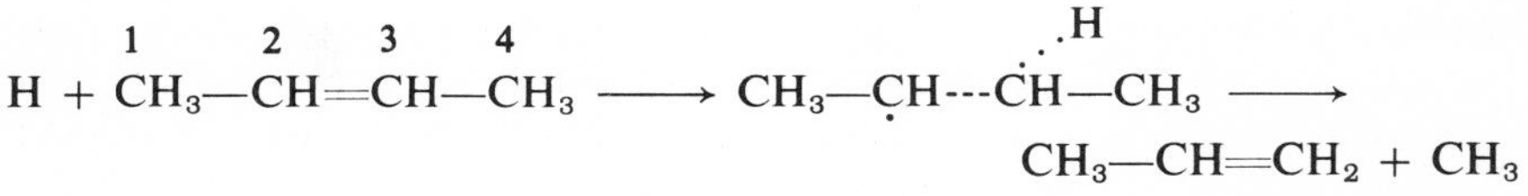

$$\begin{array}{cccc} 1 & 2 & 3 & 4 \end{array}$$
$$H + CH_3-CH=CH-CH_3 \longrightarrow CH_3-\overset{\cdot}{C}H\text{---}\overset{\cdot\cdot H}{C}H-CH_3 \longrightarrow$$
$$CH_3-CH=CH_2 + CH_3$$
$$(2\text{-}4b)$$

Dashed lines again indicate the newly formed bonds. In case (a) the excitation energy is originally located in the C—H bond on carbon 1, and in case (b) in the C—H bond on carbon 3, yet decomposition in both cases must occur by rupture of the C—C bond between carbons 3 and 4, since the products are found to be the same and are formed with the same rate in both cases. Hence migration of internal energy preceding decomposition must have obliterated all difference between the two modes of activation.

In a related system, Rodgers (*18*) has measured the decomposition of *sec*-2,3-dichloroperfluorobutyl radical

$$CH_3-\overset{\overset{\textstyle F}{\downarrow}}{\underset{\underset{\textstyle Cl}{\downarrow}}{C}}-\overset{\cdot}{C}Cl-CF_3 \longrightarrow CF_3-CF=CCl-CF_3 + Cl \qquad (2\text{-}5a)$$

and of *sec*-1,4-dichloroperfluorobutyl radical

$$CF_2Cl-\overset{\overset{\textstyle F}{\downarrow}}{C}F-\overset{\cdot}{C}F-\overset{\underset{\textstyle Cl}{\downarrow}}{C}F_2 \longrightarrow CF_2Cl-CF_2-CF=CF_2 + Cl \qquad (2\text{-}5b)$$

both prepared by fluorine atom addition to the respective dichloroperfluoro-butene-2. The inward arrow indicates the point of attack of the fluorine atom, i.e., the point at which the radical has received its excitation energy, and the outward arrow indicates where excitation energy must become localized for the reaction to occur. The points of excitation and of reaction are different in the two cases, but the two reactions occur at essentially the same rate, again pointing to rapid intramolecular transfer of energy.

A technique which is in many ways similar, except that the level of activation is much higher, involves reactions of recoil tritium. Energetic tritium atoms produced in a nuclear reaction substitute easily for hydrogen in organic molecules and in the process deposit substantial energy (~ 5 eV in a broad distribution) in the target molecule. The random or nonrandom character of the unimolecular decomposition of such an excited, suitably labeled target molecule is reflected in the isotopic composition of the products. Thus Tang and Rowland (*19*) prepared 2-chloropropane labeled with tritium at carbon 1. The unimolecular decomposition proceeds by elimination of HCl, which can

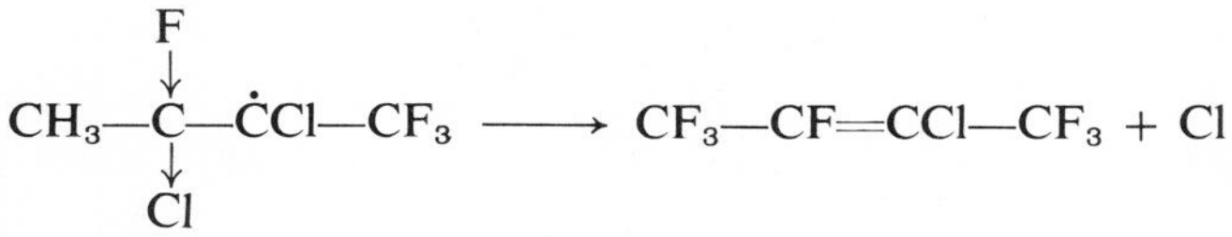

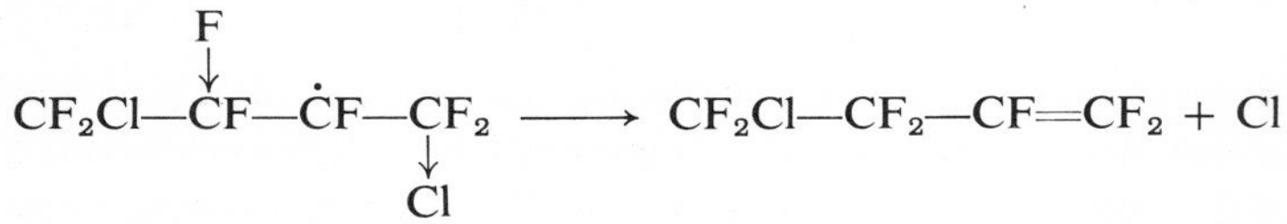

involve, in principle, either the hydrogen on carbon 1 (case a), or the hydrogen on carbon 3 (case b):

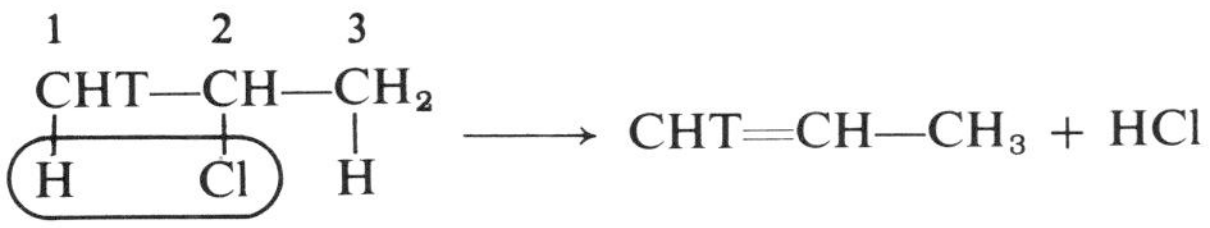

$$\text{CHT}-\text{CH}-\text{CH}_2 \longrightarrow \text{CHT}=\text{CH}-\text{CH}_3 + \text{HCl} \qquad (2\text{-}6a)$$

$$\text{CHT}-\text{CH}-\text{CH}_2 \longrightarrow \text{CH}_2\text{T}-\text{CH}=\text{CH}_2 + \text{HCl} \qquad (2\text{-}6b)$$

On being formed, the chloropropane originally contains all the excitation energy in the vicinity of carbon 1, and if the energy remains there before decomposition, we would expect the decomposition to follow mostly scheme (a), i.e., the tritiated propylene product would be olefinically labeled. If, on the other hand, energy becomes randomized inside the molecule before decomposition, schemes (a) and (b) would both be equally likely, and we would expect the products to consist of a 1:1 mixture of olefinically labeled and alkyl-labeled propylene. Tang and Rowland found 50% $\text{CH}_2\text{TCH}=\text{CH}_2$ in the gas phase and 45% $\text{CH}_2\text{TCH}=\text{CH}_2$ in the liquid phase. Thus energy randomization is complete in the gase phase and is very nearly so in the liquid phase.

Deuterium labeling was used in an interesting experiment by Doering and collaborators (20) in which methylene cyclopropane (I) was reacted with dideutero methylene (CD_2) to produce excited spiropentane[12] (II), which then decays to methylene cyclobutane (III):

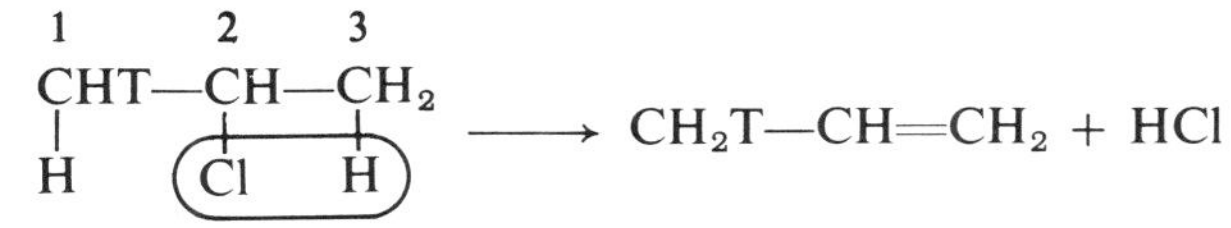

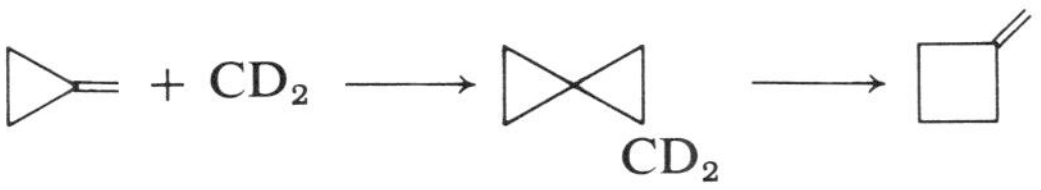

$$\triangleright= + \text{CD}_2 \longrightarrow \bowtie_{\text{CD}_2} \longrightarrow \square \qquad (2\text{-}7)$$

(I)   (II)   (III)

If there is complete internal scrambling of energy in (II), any one of the four methylene groups in (II) can become attached to the central carbon of (II) by a double bond to yield the lateral $=\text{C}\big\langle$ group in (III), so that the products should contain (III) with a random distribution of deuterium, although only one of the methylene groups in (II) is deuterium labeled. This is what has been found by actual analysis of the products, confirming that under the conditions of the experiment the energy randomization hypothesis holds.

[12] Excited spiropentane was also prepared with 5-eV excitation energy by recoil-tritium substitution reaction, and it also decayed in a random fashion below 1500 torr. See Su and Tang (21).

Compression of the time scale All these experiments compare the composition of products formed from a molecule in which the excitation energy was initially deposited in two different sites, the idea being that for different excitation sites the composition of the products would be different if the energy did not become redistributed in the interval between excitation and reaction. The time scale on which this interval is accessible to measurement in the above type of experiment is determined by the ambient pressure.[13] As we shall see in Chapter 8, almost every collision suffered by a molecule in a reactive state causes it to lose enough energy to knock it down to a nonreactive state, so that the molecule in a reactive state has a chance to react, i.e., form a product, only during the time that precedes its first collision. As the pressure is made higher, the frequency of collisions increases and the interval between excitation and reaction becomes shorter, and therefore the time available for the energy to become randomized before the molecule decomposes becomes shorter also. Clearly, if the pressure is made high enough, a point must inevitably be reached when there is insufficient time for the energy to become randomized, which should be detectable either as a change in the composition of the products or as a change in the rate of formation of one specific product.

The gas-phase experiments cited earlier were done generally at pressures well below 1 atm, which obviously is not "high enough." In the case of the recoil-tritium experiments, the collision rate in the liquid was 10^{12} sec^{-1}, and still there was no significant change in the composition of the products. It was with the purpose of finding out what pressure is "high enough" that the addition of H atoms to *cis*-butene-2, producing excited *sec*-butyl radicals, was later studied by Rabinovitch and co-workers (*23,24*) at H_2 pressures from 0.036 to 203 atm. Within experimental error, the average rate constant for decomposition to propylene was found to be invariant over the whole range of pressures studied. About three collisions with H_2 are necessary to remove sufficient energy from the butyl radical to render it incapable of decomposing into propylene. At 203 atm and at the experimental temperature, the collision rate between H_2 and the butyl radical is $\sim 5 \times 10^{12}$ collisions sec^{-1}, which means that the butyl radical must have decomposed in 10^{-13} sec or less. At the other end of the pressure range, at 0.036 atm, the collision rate is $\sim 10^9$ collisions sec^{-1}, and therefore $\sim 3 \times 10^{-9}$ sec was available for the radical to decompose after the initial activation. Yet the rate constant, i.e., the probability of dissociation per unit time, has remained constant over this range of pressures. This means that intramolecular energy transfer was fast

[13] In a collision-free system, the interval depends on the experimental technique. A technique exists for the direct observation of decay in ionic systems 10^{-11} sec after excitation. See Chapter 10, Section 1. Another technique is laser-induced vibrational fluorescence, by means of which extremely rapid equilibration among vibrational modes of CH_3F has been demonstrated (*22*).

enough[14] to randomize the internal energy throughout the pressure range, i.e., for times as short as 10^{-13} sec.

In a similar experiment, Wilson *et al.* (*25*) studied the unimolecular decomposition of ethyl cyclobutane in the presence of N_2 pressures from 7 to 170 atm. They found that the unimolecular rate constant was a slowly decreasing function of pressure, while Dutton and Bunker (*26*), who measured the rate of the unimolecular decomposition of NO_2Cl in the presence of 50–300 atm of N_2, found no significant effect on the rate constant.

The interpretation of experiments designed to detect a change in the unimolecular rate constant at extreme pressure conditions depends on what one expects to find when redistribution of energy ceases to be random. A case can be made for the rate constant to *increase* when nonrandomization of energy sets in. On each increase of pressure, molecules of increasingly smaller lifetime are sampled, and if the energy has not become randomized among all internal degrees of freedom during the lifetime of the molecule, the molecule would behave as if it were a progressively smaller molecule than it actually was (*27,28*), and this would lead to a progressively larger rate constant, as can be seen qualitatively from Eq. (2-2). The tacit assumption is that the molecule receives excitation energy near the breaking bond. On the other hand, it can also be argued that nonrandomization is due to the activation process having deposited internal energy "in the wrong place" of the molecule, so that the energy does not have time to reach the bond to be broken before the next collision, which it must if dissociation is to take place; hence the rate constant should *decrease* when nonrandomization of energy sets in. It should also be noted that the rate constant measured in these high-pressure experiments is not $k(E)$ but an average of $k(E)$ over the equilibrium thermal distribution of energies (cf. Chapter 8, Section 2). The implicit assumption is therefore that if a change is observed in the measured rate constant at very high pressure, the change is due to the effect of high collision frequency on $k(E)$ but not on the energy distribution function. Thus the meaning of the result obtained by Wilson *et al.* is not clear insofar as possible failure of randomization is concerned; the authors themselves thought the result could be due to the effect of pressure on the free energy of activation.

In another series of experiments, Pearson and Rabinovitch (*29,30*) have prepared, by H-atom addition to the appropriate α-olefin, chemically

[14] There is a possibility, mentioned by the authors of ref. *23*, that the molecular collisions themselves (between excited reactant and inert gas) cause a perturbation of the internal states of the reactant molecule sufficient to promote randomization of internal energy. Therefore the cited experiments do not necessarily mean that in an *undisturbed* excited reactant molecule internal energy randomization would also take as little as 10^{-13} sec.

activated alkyl radicals from 2-pentyl to 2-octyl; these radicals then decompose to yield propylene and a smaller radical:

$$H + CH_3(CH_2)_nCH_2CH\text{=}CH_2 \longrightarrow CH_3(CH_2)_nCH_2\overset{\cdot}{C}HCH_3 \longrightarrow$$

$$CH_3(CH_2)_n^{\cdot} + CH_2\text{=}CHCH_3$$

$$n = 1, 2, 3, 4 \tag{2-8}$$

In this homologous series, the mode of decomposition is the same and the level of activation is roughly constant: The average internal excitation energy for 2-pentyl radical is 14 kcal mole^{-1} above threshold, for 2-octyl it is 16.6 kcal mole^{-1} above threshold. In contrast with the experiments mentioned above, here the collision interval is constant but the size of the reactant is varied. It is found that the logarithm of the average rate constant for decomposition to propylene is a linear, decreasing function of the number of carbon atoms in the activated radical. For radicals with a relatively narrow distribution of internal energies, such as in the present case, the logarithmic dependence of the rate constant on the effective number of degrees of freedom (σ) can be expected from Eq. (2-2). Since the number of carbon atoms is related to the *total* number of vibrational degrees of freedom, the experimental results thus point to all vibrational degrees of freedom as effective, and this suggests that internal excitation energy is randomized among essentially all internal degrees of freedom of the decomposing radical. It is notable that this happens even for as large a radical as 2-octyl. For an extension to C_{16} radicals see (*62*).

A nonrandom case   Recently, however, Rynbrandt and Rabinovitch (*31,32*) have found a system, related to reaction (2-7), that exhibits nonrandom behavior. The molecule in question is excited hexafluorobicyclopropyl (II), which is prepared by a reaction between CD_2 and hexafluorovinyl cyclopropane (I):

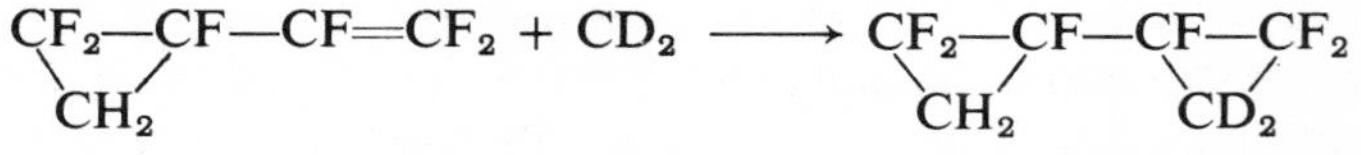

$$\underset{\text{(I)}}{\underset{CH_2}{\overset{CF_2-CF-CF=CF_2}{\diagup\diagdown}}} + CD_2 \longrightarrow \underset{\text{(II)}}{\underset{CH_2\qquad CD_2}{\overset{CF_2-CF-CF-CF_2}{\diagup\diagdown\diagup\diagdown}}} \tag{2-9}$$

The deuterium thus serves to label the newly formed cyclopropane ring. The exited bicyclopropyl (II) can then decompose in two ways:

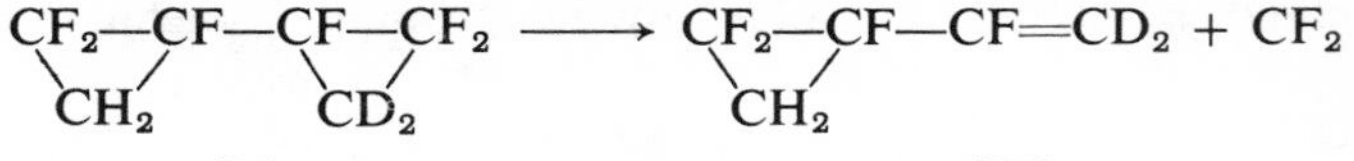

$$\underset{\text{(II)}}{\underset{CH_2\qquad CD_2}{\overset{CF_2-CF-CF-CF_2}{\diagup\diagdown\diagup\diagdown}}} \longrightarrow \underset{\text{(III)}}{\underset{CH_2}{\overset{CF_2-CF-CF=CD_2}{\diagup\diagdown}}} + CF_2 \tag{2-10a}$$

$$\longrightarrow CF_2 + \underset{\text{(IV)}}{\underset{CD_2}{\overset{CH_2\text{=}CF-CF-CF_2}{\diagdown\diagup}}} \tag{2-10b}$$

Compound (III) arises by rupture of the newly formed ring, while compound

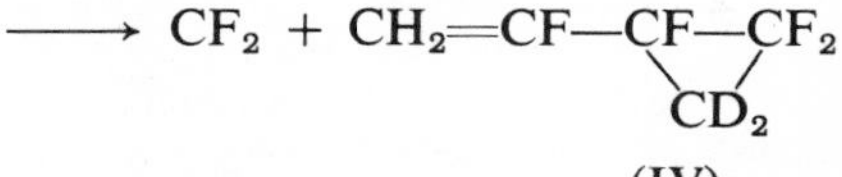

(IV) arises by rupture of the original ring. At pressures below 1 atm, products (III) and (IV) are formed in equal amounts (*31*), as one would expect if internal energy of (II) is randomized before decomposition, in agreement with the conclusions reached in the study of the similar reaction (2-7) at comparable pressures.

However, when the pressure is increased above 1 atm, i.e., when the time available for randomization of energy is decreased, the ratio (III)/(IV) in the products *increases* (*32*), showing that the newly formed ring is preferentially ruptured. When the isotopic labeling is reversed, by forming (II) from

$$CF_2—CF—CF=CF_2 + CH_2$$
$$\diagdown\diagup$$
$$CD_2$$

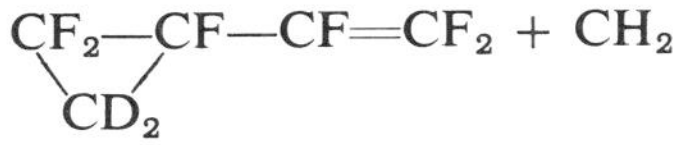

the ratio (III)/(IV) in the products *decreases*, indicating that the rupture of the original ring, which now gives rise to compound (III), is more difficult, thus further confirming the result of the original deuterium labeling. The conclusion is therefore that energy finds its way from the newly formed ring to the original ring only with some difficulty, so that when the pressure is "high enough," the rupture of the newly formed ring becomes sufficiently preponderant to alter the composition of the products. In this case, "high enough" means a pressure of only a few atmospheres. Similarly, the decomposition of hot spiropentane-t (footnote 12) becomes nonrandom at a pressure of about 6 atm (*59*).

The hexafluorobicyclopropyl molecule is somewhat special in that the two cyclopropane rings are connected only by one C—C bond, which apparently acts as an effective bottleneck, preventing easy transfer of energy from one ring to another, and this seems to be the reason why in this case the failure of randomization could be detected under relatively mild conditions. However, when the *total* available experimental evidence from all sources is considered, it can be said, in summary, that there is strong circumstantial experimental evidence that intramolecular energy transfer exists, is very rapid, and results in the randomization of energy among many, if not all, internal degrees of freedom in every molecule that has been studied so far, except for one rather special case. The hypothesis of the randomization of energy is thus borne out by experiment as a useful concept applicable to most real-life situations. It is of interest now if this intramolecular energy transfer can be justified on theoretical grounds.

5. Theoretical treatment of energy randomization

Although the mechanism responsible for the intramolecular energy transfer and for the subsequent randomization of energy need not be specified in the RRK theory, it was then, as now, generally understood to be due essentially to vibrational anharmonicity. Its existence is amply demonstrated in vibrational spectroscopy,[15] where it accounts for a good deal of the complications in vibrational spectra, especially at high resolution. However, vibrational spectroscopy is concerned mainly with transitions confined to the vicinity of the ground state, whereas a molecule in a reactive state is more likely to be in an excited state near its dissociation limit. If the effect of anharmonicity, as observed in vibrational spectroscopy, is extrapolated to highly excited reactive molecular states (cf. also Chapter 5, Section 3), the argument for energy randomization among reactive states can be put, in a general way,[16] as follows.

Molecular vibrations in a real molecule are basically anharmonic, with the result that, except in the vicinity of the potential miminum (low quantum states, i.e., small bond stretching), they cannot be represented by independent normal vibrations because of interaction or coupling among them. However, if the concept of normal mode vibrations is formally retained, it is found[17] that the introduction of anharmonic coupling leads to a time dependence of the vibrational amplitude of the quasinormal modes. Since amplitude is proportional to $(\text{energy})^{1/2}$, this time dependence can be formally interpreted as energy exchange or redistribution among the individual normal modes.

A molecule undergoing decomposition is highly excited, and during the time lag between activation and decomposition most of its vibrations will pass repeatedly through high quantum states, where the departure from a harmonic potential (anharmonicity) is considerable; the resultant coupling among the vibrations can therefore be expected to give rise to rapid redistribution of energy, at least among some of the vibrations (*39*).

A more rigorous demonstration that anharmonicity does indeed lead to such energy transfer requires a rather detailed knowledge of molecular vibrations which is generally not available at the present time, and even if it were, would still present a formidable problem. It is therefore necessary to work with simplified models in this sort of calculations.

Using classical mechanics, Tredgold (*36*) obtained an exact solution for the case of three mass points constrained to move in a straight line and pairwise subject to a potential with a quartic (anharmonic) term. Such a system may be considered to be a crude model of a linear triatomic molecule.[18] He found rapid periodic interchange of energy between the two quasi-normal modes, such that ultimately all possible ways of sharing a given total energy between the two oscillators were realized; in other words, the energy become randomly distributed.

[15] For a recent review, see Strauss (*33*) and Overend (*34*).

[16] For an interesting qualitative discussion of anharmonicity, see Stevens (*35*).

[17] See refs. *36* and *37*. For a different approach, see Bunker (*38*).

[18] Cf. footnote 2 in the following chapter.

Computer studies A somewhat similar classical model was used by Thiele and Wilson (*37*), except that they used Morse potentials and unequal masses, as in CO_2. By numerical integration of the equations of motion, they found very rapid energy flow between the two vibrational modes. Later others (*40,41*) extended the calculations to a linear tetraatomic molecule and found that when the atomic masses were made very unequal, as in acetylene, there was very inefficient vibrational energy transfer between different C—H bonds undergoing stretching, and also between carbon skeleton vibration and C—H stretching.

Bunker (*38*, *42–44*) used the Monte Carlo technique to simulate on a computer the (classical) motions on a potential energy surface of various models of rotating triatomic molecules with Morse potentials, until the molecules "dissociated." He found that randomization of internal energy was virtually complete in 10^{-11} sec except for triatomic molecules with highly unequal masses, such as H_2O, a finding similar to Wilson's (*40*). Bunker also found —and this is of interest in connection with RRK's criterion for reaction—that all molecules with energy $E > E_0$ "decomposed" (E_0 is the critical energy for decomposition).

When it comes to more complicated model systems, mention must be made of an early study by Fermi *et al.* (*45*), who made a computer study of a one-dimensional dynamical system of 64 particles with fixed end points and with nonlinear (cubic) forces acting between neighbors; there was little, if any, tendency for redistribution of energy among the degrees of freedom of this model system. This is the first study in which randomization of energy could not be substantiated. It seems that this result is peculiar to the chosen model, which does not resemble any real molecule known to decompose unimolecularly. In fact, the model does not allow much internal resonance among the "oscillators," and it can be shown (*46*) that under such conditions a system of weakly coupled oscillators does not achieve a redistribution of energy if originally excited at one end of the chain.

It is difficult to assess the significance of the computer results mentioned in the previous two paragraphs. For one thing, the dimensionality of the problem has usually been drastically reduced by eliminating nonessential degrees of freedom in order to keep machine time within reasonable bounds, so that the models bear only a superficial resemblance to real molecules, although this is not quite the case with Bunker's results. Furthermore, extrapolation of these results to more complex molecules of greater chemical interest is uncertain.

Preliminary results have now been reported by Harris and Bunker (*47*) of classical trajectory calculations on the polyatomic molecule CH_3NC, which is known to isomerize to CH_3CN in a thermal reaction that has been extensively studied experimentally and found to be an excellent example of a "clean" unimolecular reaction exhibiting perfect statistical behavior (see Chapters 8 and 11). The theoretical model for the molecule and for the reaction used in this computer study is based on the model discussed in Chapter 3, Section 1 and shown in Fig. 3-3(a), where the critical coordinate is the angle θ between the methyl group and the N—C bond. All seven atoms were explicitly considered in the calculations; one hydrogen of the methyl was taken to be tritium.[19] It turns out that this model molecule refuses to "react" for the most part. At 150 kcal mole^{-1} excitation energy, essentially no "reaction" occurs (i.e., only very few trajectories lead to a chemically distinct product) when the energy is originally deposited in the hydrogens of the methyl; if the C—N—C skeleton is excited, one-tenth of the trajectories lead to CH_2TCN, one-tenth lead to the fragmentation $CH_2T + NC$, and full four-fifths of the trajectories lead to no reaction. Even if up to 300 kcal mole^{-1} is put in the entire molecule, to simulate the random excitation in a thermal reaction, some bond rupture is observed but little isomerization, although experimentally the isomerization under such conditions is quite rapid.

It must be noted, however, that the model for the reaction used in this computer study, while reasonably realistic, does not seem to be borne out by the thermal data, and that the discovery of unsuspected experimental complications has now brought the data themselves under suspicion (see Chapter 11, end of Section 4). For the moment it is not clear if what is measured in the laboratory and what is calculated on the computer are one and the same thing in this particular case.

A more fundamental general objection comes from Levine (*50*) who argues that one cannot accept a limited number of classical trajectories because of the position–momentum uncertainty principle, so that a classical model cannot represent the behavior of an individual system. The consensus of opinion seems to be at present that quantum effects are in general small (*51*, p. 372) but an unequivocal demonstration is not available because the quantum treatment is very difficult (*51*, p. 407). In a comparison of classical and quantum-mechanical trajectory calculations for the *bimolecular* reaction $H + Cl_2 \rightarrow HCl + Cl$, it has been found (*60*) that classical calculations yield good results if the total energy is not too low and the zero-point energy is not neglected. As total energy decreases, however, tunneling becomes important, which no classical scheme can handle adequately. See also (*63*).

Stochastic approach A different theoretical approach to the intramolecular energy transfer problem can be based on a stochastic approach, borrowed from the theory of collisional activation in thermal reactions (cf. Chapter 8,

[19] The reason was to investigate some unusual recoil-tritium results obtained on CH_2TNC by Ting and Rowland (*48*). It now appears that some of the unusual results can be ascribed to rotational effects (*49*).

Section 8 and Chapter 9, Section 5). In one version (*52*), which addresses itself specifically to excitation in thermal reactions, the reactant molecule is seen as if composed of oscillators that exchange energy as a result of interoscillator "collisions." The mechanism of these "collisions" need not be specified; it is simply some effect that effaces all memory on a given time scale. The theory starts from the assumption that the energy in the critical oscillator is not initially the one corresponding to a random distribution throughout the molecule, but is governed by a collision-type interaction between oscillators in pairs. The theory eventually leads to a lifetime distribution (Chapter 1, Section 4) that is nonrandom, but only very slightly so. The theory predicts, in fact, that the condition of detailed balance in thermal reactions [see Eq. (8-123)] places such strict limits on the allowable transitions probabilities for the critical oscillator that pronounced nonrandom behavior in thermal reactions is unlikely.

Another stochastic treatment, due to Gelbart *et al.* (*53*), employs a different model that is more easily interpreted in terms of observable properties of real molecules. A polyatomic molecule is seen as a collection of "cells" (oscillators, not necessarily harmonic), each of which is populated by any number of "particles" (vibrational quanta). The system is assumed to interact with an infinite reservoir of particles. Two random variables are defined; one measures the number of particles in cell i at time t, and the other determines whether at time t cell i is intact or "broken" (dissociated). The model contains full off-diagonal interaction, i.e., every "cell" is able to exchange "particles" with every other "cell" and any number of "cells" can be broken at any one time. The model can thus answer the question, "Which bond is most likely to break first and why?"

The answer will depend to a degree on the choice of transition probabilities which determine the extent of coupling among the cells. These probabilities are not known, and therefore the authors chose simple linear probabilities which allow for an exact treatment of the stochastic process. The treatment is rather involved and we shall give only the results, which in any event are only qualitative because the molecular parameters required for a quantitative evaluation in a specific case are not known. It is found, in the decay of an isolated molecule excited in some nonrandom fashion, that the more weakly coupled "cell" is least likely to be broken at all times, and that the population (energy) of the more weakly coupled "cell" rises most slowly to reach the value corresponding to equipartition. If the system is made linear, with near-neighbor interactions only, it is conversely the weakly coupled cell that is most likely to break first, since it "traps" energy flowing through it longest. Actual molecules are probably somewhere between the full off-diagonal and linear systems.

In summary, then, full theoretical treatment of the intramolecular redistribution of energy is not yet available, either because the model is not realistic enough or because the requisite molecular parameters and probabilities are not known. On balance, however, the doubt that these theoretical studies may create concerns only the extent to which internal energy redistribution is efficient and rapid in a given specific case, and not the existence of the redistribution itself.

6. The Rice–Marcus (RRKM) theory

The difficulties outlined in the preceding section, which attend a rigorous demonstration of rapid intramolecular energy transfer, demonstrate the nature of the problem that would have to be solved first before a truly dynamical unimolecular rate theory can be developed. It was therefore astute on the part of RRK to sidestep the whole problem and proceed instead by means of a statistical argument, thereby avoiding all reference to molecular dynamics; much detail is of course lost, but the essence of behavior of real molecules is preserved.

In later years, the RRK theory was cast into transition-state formulation by Marcus and Rice (*54*), Eyring and collaborators (*55*), and particularly Marcus (*56–58*), who provided most of the apparatus for evaluating the RRK parameters A and s. This version of the theory is often referred to as the RRKM, or Rice–Marcus, theory. It has proved very successful in interpreting a variety of unimolecular processes, despite some often (needlessly) formidable-looking equations and a certain lingering ambiguity, which we shall point out in due course.

The following chapters are essentially a somewhat updated and expanded account of the transition-state formulation of Rice and Marcus, and also include the various refinements that have been contributed by numerous workers in the field.

References

1. N. B. Slater, "Theory of Unimolecular Reactions." Cornell Univ. Press, Ithaca, New York, 1959.

2. K. S. Pitzer, F. T. Smith, and H. Eyring, *in* "The Transition State," p. 53. Chem. Soc. Special Publ. No. 16, 1962.

3. O. K. Rice and H. C. Ramsperger, *J. Amer. Chem. Soc.* **49,** 1617 (1927).

4. L. S. Kassel, *J. Phys. Chem.* **32,** 225, 1065 (1928).

5. L. S. Kassel, "Kinetics of Homogeneous Gas Reactions." Chemical Catalog Co., New York, 1932.

6. M. Polanyi and E. Wigner, *Z. Phys. Chem.* **139A**, 439 (1928).
7. W. H. Rodebush, *J. Chem. Phys.* **1**, 440 (1933).
8. H. Gershinowitz and O. K. Rice, *J. Chem. Phys.* **2**, 275 (1934).
9. H. Eyring, *J. Chem. Phys.* **3**, 107 (1935).
10. S. W. Benson, *J. Chem. Phys.* **19**, 802 (1951).
11. M. Šolc, *Z. Phys. Chem. (Leipzig)* **234**, 185 (1967).
12. R. C. Baetzold and D. J. Wilson, *J. Chem. Phys.* **43**, 4299 (1965).
13. V. A. Kondratev, "Chemical Kinetics of Gas Reactions." Pergamon, Oxford, 1964.
14. J. E. Douglas, B. S. Rabinovitch, and F. S. Looney, *J. Chem. Phys.* **23**, 315 (1955).
15. J. N. Butler and G. B. Kistiakowsky, *J. Amer. Chem. Soc.* **82**, 759 (1960).
16. H. M. Frey, *Trans. Faraday Soc.* **56**, 51 (1960).
17. R. E. Harrington, B. S. Rabinovitch, and H. M. Frey, *J. Chem. Phys.* **33**, 1271 (1960).
18. A. S. Rodgers, *J. Phys. Chem.* **72**, 3400, 3407 (1968).
19. Y.-N. Tang and F. S. Rowland, *J. Phys. Chem.* **72**, 707 (1968).
20. W. E. von Doering, J. C. Gilbert, and P. A. Leermakers, *Tetrahedron* **24**, 6863 (1968).
21. Y. Y. Su and Y. N. Tang, *J. Phys. Chem.* **76**, 2187 (1972).
22. E. Weitz and G. Flynn, *J. Chem. Phys.* **58**, 2781 (1973).
23. D. W. Placzek, B. S. Rabinovitch, and F. H. Dorer, *J. Chem. Phys.* **44**, 279 (1966).
24. I. Oref, D. Schuetzle, and B. S. Rabinovitch, *J. Chem. Phys.* **54**, 575 (1971).
25. J. Aspden, N. A. Khawaja, J. Readon, and D. J. Wilson, *J. Amer. Chem. Soc.* **91**, 7850 (1969).
26. M. L. Dutton, D. L. Bunker, and H. H. Harris, *J. Phys. Chem.* **76**, 2614 (1972).
27. O. K. Rice, *Z. Phys. Chem.* **B7**, 226 (1930).
28. M. Šolc, *Z. Phys. Chem. (Leipzig)* **236**, 213 (1967).
29. M. J. Pearson and B. S. Rabinovitch, *J. Chem. Phys.* **41**, 280 (1964).
30. M. J. Pearson and B. S. Rabinovitch *J. Chem. Phys.* **42**, 1624 (1965).
31. J. D. Rynbrandt and B. S. Rabinovitch, *J. Phys. Chem.* **74**, 4175 (1970).
32. J. D. Rynbrandt and B. S. Rabinovitch, *J. Phys. Chem.* **75**, 2164 (1971).
33. H. L. Strauss, *Ann. Rev. Phys. Chem.* **19**, 419 (1968).
34. J. Overend, *Ann. Rev. Phys. Chem.* **21**, 265 (1970).
35. B. Stevens, *Can. J. Chem.* **36**, 96 (1958).
36. R. H. Tredgold, *Proc. Phys. Soc. (London)* **A68**, 920 (1955).
37. E. Thiele and D. J. Wilson, *J. Chem. Phys.* **35**, 1256 (1961).
38. D. L. Bunker, *J. Chem. Phys.* **40**, 1946 (1964).
39. O. K. Rice, *J. Phys. Chem.* **65**, 1588 (1961).
40. R. J. Harter, E. B. Alterman, and D. J. Wilson, *J. Chem. Phys.* **40**, 2137 (1964).
41. T. H. Latta, Ph.D. Thesis, Univ. of Illinois, Urbana (1964).
42. D. L. Bunker, *J. Chem. Phys.* **37**, 393 (1962).
43. D. L. Bunker, "Theory of Elementary Gas Reaction Rates." Pergamon, Oxford, 1966, p. 70ff.
44. D. L. Bunker, *in* "Molecular Beams and Reaction Kinetics," Course XLIV, Italian Phys. Soc. p. 355. Academic Press, New York and London, 1970.
45. E. Fermi, J. Pasta, and S. Ulam, Los Alamos Scientific Lab. Rep. LA-1940 (1955).
46. J. Ford, *J. Math. Phys.* **2**, 387 (1961).
47. H. H. Harris and D. L. Bunker, *Chem. Phys. Lett.* **11**, 433 (1971).
48. C. T. Ting and F. S. Rowland, *J. Phys. Chem.* **74**, 763 (1970).
49. D. L. Bunker, *J. Chem. Phys.* **57**, 332 (1972).
50. R. D. Levine, *J. Chem. Phys.* **44**, 3597 (1966).

51. M. Karplus, *in* "Molecular Beams and Reaction Kinetics," Course XLIV, Italian Physical Soc. Academic Press, New York and London, 1970.
52. M. R. Hoare and E. Thiele, *Discuss. Faraday Soc.* **44**, 30 (1967).
53. W. M. Gelbart, S. A. Rice, and K. F. Freed, *J. Chem. Phys.* **52**, 5718 (1970).
54. R. A. Marcus and O. K. Rice, *J. Phys. Coll. Chem.* **55**, 894 (1951).
55. H. M. Rosenstock, M. B. Wallenstein, A. L. Wahrhaftig, and H. Eyring, *Proc. Nat. Acad. Sci. U. S.* **38**, 667 (1952).
56. R. A. Marcus, *J. Chem. Phys.* **20**, 359 (1952).
57. G. M. Wieder and R. A. Marcus, *J. Chem. Phys.* **37**, 1835 (1962).
58. R. A. Marcus, *J. Chem. Phys.* **43**, 2658 (1965).
59. Y. N. Tang and Y. Y. Su, *J. Chem. Phys.* **57**, 4048 (1972).
60. D. Russell and J. C. Light, *J. Chem. Phys.* **51**, 1720 (1969).
61. G. B. Kistiakowsky and B. B. Saunders, *J. Phys. Chem.* **77**, 427 (1973).
62. E. A. Hardwidge, B. S. Rabinovitch, and R. C. Ireton, *J. Chem. Phys.* **58**, 340 (1973).
63. K. P. Fong and D. J. Diestler, *J. Chem. Phys.* **56**, 3200 (1972).

CHAPTER 3

Potential Energy Surfaces in Unimolecular Reactions

An interesting representation of a unimolecular process (or, in fact, of any rate process) can be obtained by observing the motion of a "system point" on a potential energy surface. Such a surface shows the dependence of the molecular potential energy on all the internal coordinates of the molecule, and the system point, which represents the instantaneous configuration of all the atoms in the molecule, can then be thought of as moving[1] in coordinate space along a trajectory. Although calculations of rate can be done in this way with the help of a computer (cf. Section 5 of Chapter 2), here we shall focus attention only on the general topography of the surface. Such general survey will be a useful preliminary to actual rate calculations.

When the molecule of interest contains more than two atoms the potential energy surface is obviously a multidimensional hypersurface. The largest molecule whose potential energy surface can be conveniently represented on

[1] If a purely classical representation of the movements of the atoms in the course of a reaction is desired, such as would be represented by a frictionless mass point rolling on the potential energy surface, cross-products in the kinetic energy expression must be eliminated by appropriate skewing and scaling of the axes. See H. S. Johnston (*1*). For a comprehensive discussion see Hirschfelder (*2*).

(The reference list for this chapter can be found on p. 44.)

paper is the linear triatomic molecule A—B—C with two[2] internal bond-stretching coordinates, while the third dimension, energy, can be shown as contour lines of constant energy. This may seem disappointingly restrictive, but even triatomic potential energy surfaces can be useful for our purpose.

If it is assumed, to a first-order approximation, that the motion of electrons in a molecule can be separated from the motion of its nuclei (Born–Oppenheimer approximation), the stationary state Schrödinger equation can be solved (in principle) for electronic motion in the field of a given (clamped) nuclear configuration, yielding a manifold of electronic states. As the relative position of the nuclei is changed, the energy of each given electronic state is changed, and if the calculations are repeated for all possible interatomic distances, a potential energy surface belonging to the given electronic state is mapped out. Such a surface corresponds to vibrational potential and is discussed in Section 1. When the condition of fixed nuclear configuration is

---

[2] A linear triatomic molecule actually has four vibrational modes, two stretching and two bending modes, shown schematically as follows (arrows indicate displacement of atoms during vibration):

$$
\begin{array}{ccc}
\text{A} & \text{B} & \text{C} \\
\leftarrow\cdot & \cdot & \cdot\rightarrow \\
\leftarrow\cdot & \cdot\rightarrow & \leftarrow\cdot \\
\uparrow & & \uparrow \\
\cdot & & \cdot \\
& \downarrow &
\end{array}
$$

symmetric stretch (ν_1)

antisymmetric stretch (ν_2)

bending (ν_3 and ν_4)

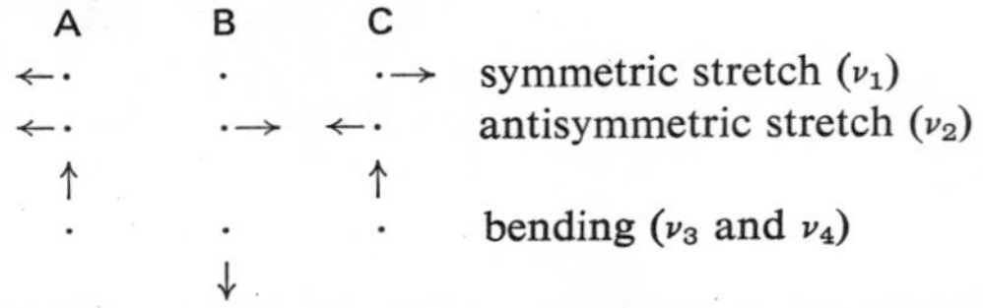

If we are interested in dissociation upon bond stretching, the two bending modes may be ignored (i.e., assumed to be constant). [For potential energy surface for bending (in CO_2) see Herzberg (*3*, pp. 435, 436).] A vibrational mode ν is the so-called normal mode, which represents a characteristic internal motion of the molecule, and could be used, in principle, as a coordinate for describing the relative motion of the nuclei, usable for constructing the potential energy surface. The simple diagrams of ν_1, ν_2, ν_3, and ν_4 for the linear ABC case show, however, that a normal mode coordinate represents in general a complicated motion of the system, one in which all the nuclei move. It is therefore somewhat easier on the imagination if the potential energy surface shows energy as a function of simple relative positional coordinates of the atoms, rather than the normal mode coordinates. In principle, there are three positional coordinates for a triatomic ABC, but in a *linear* ABC two are sufficient, say r_1 and r_2 (as defined in the legends of Figs. 3-1 and 3-2), since the third, which would be the distance A—C, must be necessarily equal to $r_1 + r_2$. Thus molecular symmetry reduces the number of independent positional coordinates, and therefore reduces also the dimensionality of the potential energy surface needed for an adequate representation of the dissociation process on simple bond stretching. There exists a simple relationship between these positional coordinates r_1 and r_2 and the normal mode motions of the system for bond stretching: The arrow in Fig. 3-2 shows displacement of the atoms in the antisymmetric stretch normal mode motion ν_2, and an arrow roughly at right angle to the first would show the displacement of atoms in the course of the symmetric stretch normal mode motion ν_1. It is perhaps worthwhile noting that symmetry reduces the number of independent coordinates only if symmetry is maintained throughout the process being considered (e.g., ABC remaining linear throughout the dissociation ABC → AB + C).

relaxed somewhat by allowing the molecule to rotate, we obtain a potential energy surface for an effective potential, which may be taken to represent a second-order approximation; this is considered in Section 2. Only the potential energy surface belonging to the lowest (ground) electronic state is taken into account in Sections 1 and 2, and it is assumed that this surface does not come into close contact with the potential energy surface of a higher electronic state; such an eventuality is deferred until Section 3. Finally in Section 4 the general features of triatomic potential energy surfaces are extrapolated to polyatomic potential energy surfaces.

It would be outside the scope of this treatise to discuss methods for calculating a potential energy surface.[3] Suffice to say that the task is not an easy (or inexpensive) one, and there are very few checks on the calculations that can be made. Generally the only direct information available is the equilibrium internuclear distances in reactant and product, and perhaps also their dissociation energies. This constitutes a check on possibly four or five points, out of the hundred or so necessary to trace out a surface. In the past, when available methods were crude, it was common to draw more or less fanciful "schematic" or "approximate" potential energy surfaces, but recent *ab initio* methods are much better, and even if, as a rule, the energies are not yet accurate enough for rate calculations, these surfaces are not likely to be too far off insofar as the general topography is concerned, which is the feature of interest in the present context. The potential energy surfaces given in the present chapter are all based on work published in the last few years.

1. Triatomic potential energy surface for a vibrational potential

We shall first consider the potential energy surface for a linear triatomic molecule ABC with a purely vibrational potential. Figures 3-1–3-3 illustrate three different important examples of such a surface. The zero of energy is always taken to be the energy of the completely dissociated system A + B + C, so that bound systems have negative energies.

Figure 3-1 shows a case which we shall henceforth call a type 1 surface; an actual case of a linear triatomic molecule with this sort of surface would be HCN, for example (*3*, pp. 438, 441). The principal feature of this type of surface is revealed by taking a cut along the bottom of the "valley" leading from reactant ABC to products A + BC [Fig. 3-1(b)]: Energy increases asymptotically as A pulls away from BC, so that under this potential there is no easily identifiable feature that could be used to decide at which point the system ABC has dissociated into A + BC. The dissociation is endothermic (as

[3] A brief general description of methods for calculating potential energy surfaces can be found in Bunker (*4*, pp. 20–29). A more comprehensive account is given by Karplus (*5*, p. 320).

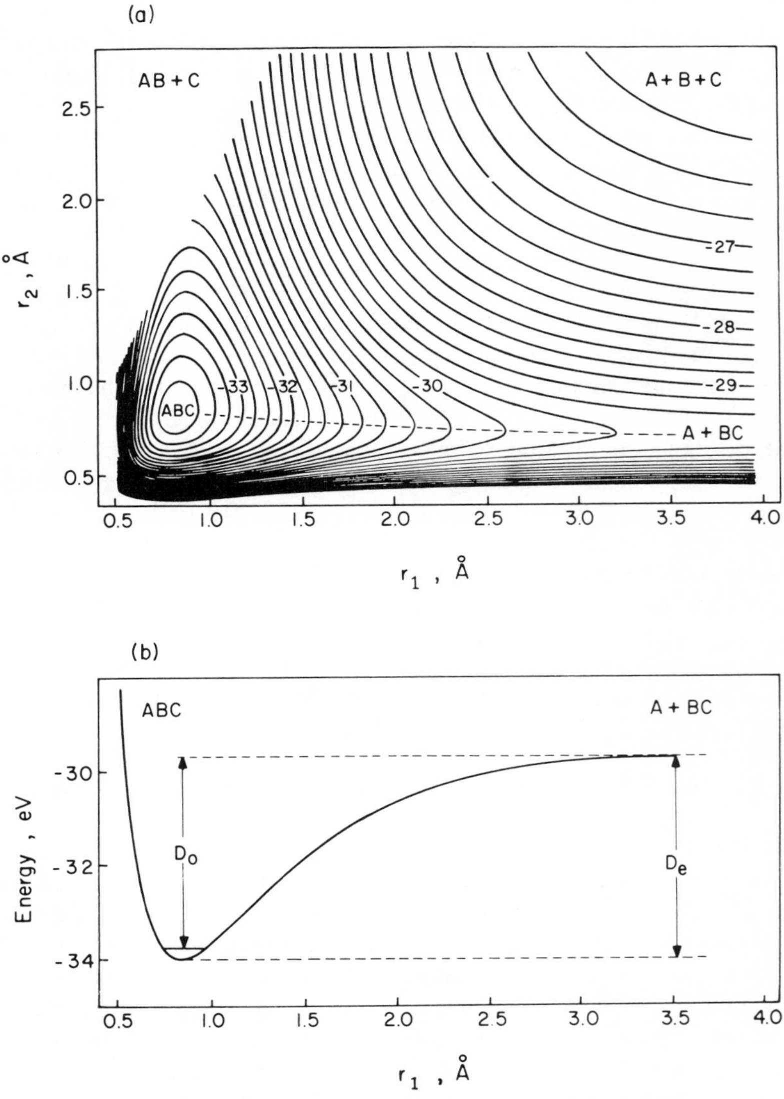

Fig. 3-1. *Potential energy surface for type 1 process.*

The process is ABC → A + BC, with no activation energy for the reverse recombination A + BC → ABC. Energy in eV, distance in Å. (a) Potential energy surface with contours of constant energy at 0.25-eV intervals. The coordinates are $A \overset{r_1}{\cdots} B \overset{r_2}{\cdots} C$. (b) Section through the potential energy surface along the dotted line in (a). D_e is the potential energy of dissociation; D_0 is the dissociation energy at absolute zero. Drawings are based on potential energy surface for triangular isosceles H_3^+ ground ($^1A_1$) state calculated by H. Conroy, *J. Chem. Phys.* **51**, 3979 (1969).

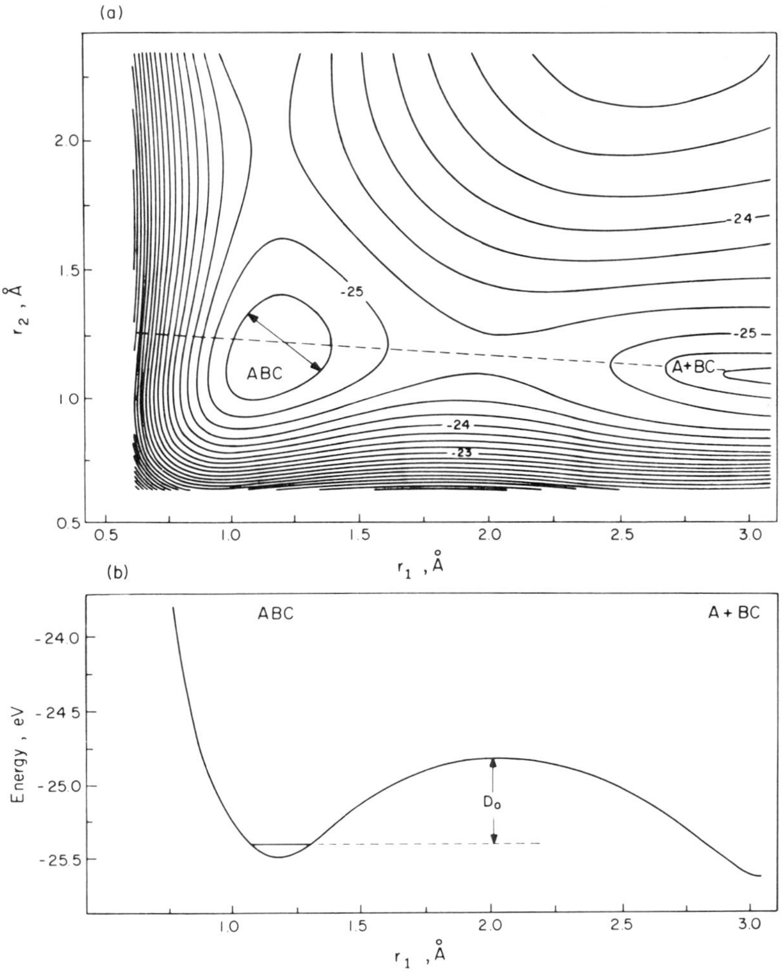

Fig. 3-2. *Potential energy surface for type 2 process.*

The process is ABC → A + BC, as in Fig. 1, but with activation energy for the reverse recombination A + BC → ABC. Energy in eV, distance in Å. (a) Potential energy surface with contour lines of constant energy at 0.25-eV intervals. The coordinates are $A \overset{r_1}{\cdots} B \overset{r_2}{\cdots} C$. Arrow indicates motion during the antisymmetric vibration ν_2. (b) Section through the potential energy surface along the broken line in (a). D_0 is the activation energy for the reaction ABC → A + BC at absolute zero. Drawings are based on potential energy surface for square H_4^{++} ($^1A_{1g}$ state), calculated by H. Conroy and G. Malli, *J. Chem. phys.* **50**, 5049 (1969). The part of the surface beyond $r_1 = 2.25$ Å is an added extrapolation.

it must be if ABC is to be a stable structure), but the reverse reaction of association A + BC → ABC does not require any expenditure of energy since along the path of Fig. 3-1(b) energy is monotonically decreasing. Note that dissociation into A + BC (or the reverse) takes place while the distance B–C remains very nearly constant; therefore the potential energy curve in Fig. 3-1(b) is essentially the vibrational potential for the pseudodiatomic molecule A–(BC).

The qualitative features of the surface in Fig. 3-1(a) are symmetric, so that arguments concerning dissociation into A + BC apply equally well to dissociation into AB + C, except that the AB and BC valleys may have different depths and slopes.

Figure 3-2 represents a type 2 surface; an actual case of a linear triatomic molecule with this type of surface would be CO_2.[4] The surface here is similar to Fig. 3-1(a), but the essential difference is that a cut along the bottom of the valley leading from reactant to products [Fig. 3-2(b)] shows a *maximum* ("hump") separating A + BC from ABC. As a result, the reverse recombination A + BC now also requires an expenditure of energy to get the system over the top of the hump, i.e., there is now an activation energy for reaction in both directions. In three dimensions the hump represents a "col," i.e., a *minimum* in the direction of increasing r_2 at a fixed r_1. The hump (in two dimensions) or the col (in three dimensions) thus represent a convenient landmark.

Figures 3-1 and 3-2 show surfaces for unimolecular processes involving fragmentation, i.e., bond breaking, which constitute a large class of unimolecular reactions. A second large class consists of isomerization reactions, in which there is just one product; we shall call these type 3. If no bonds are broken, the product contains only a different geometric arrangement of the same bonds as in the reactant, and the isomerization is said to be geometrical. Cis–trans isomerizations are an example. If bonds are broken in the isomerization, the product contains atoms of the reactant bonded together differently, and the isomerization is said to be structural.

Figure 3-3 shows the potential energy surface for the structural isomerization hydrogen isocyanide–hydrogen cyanide, which is a prototype of several isocyanide–cyanide unimolecular isomerizations which have been extensively studied.[5] Since kinetic evidence from the isomerization of methyl and higher alkyl isocyanides shows that the alkyl group does not become fully detached from the CN group during the reaction, the surface in Fig. 3-3 explores the

[4] Herzberg (*3*, p. 430). See also Machida and Overend (*6*, Fig. 3).

[5] Computer studies using the potential energy surface of Fig. 3-3 are mentioned in Chapter 2, Section 5; experimental results are discussed in Chapter 11, Section 4. There are a number of as yet unexplained discrepancies between theory and experiment in the isocyanide isomerization.

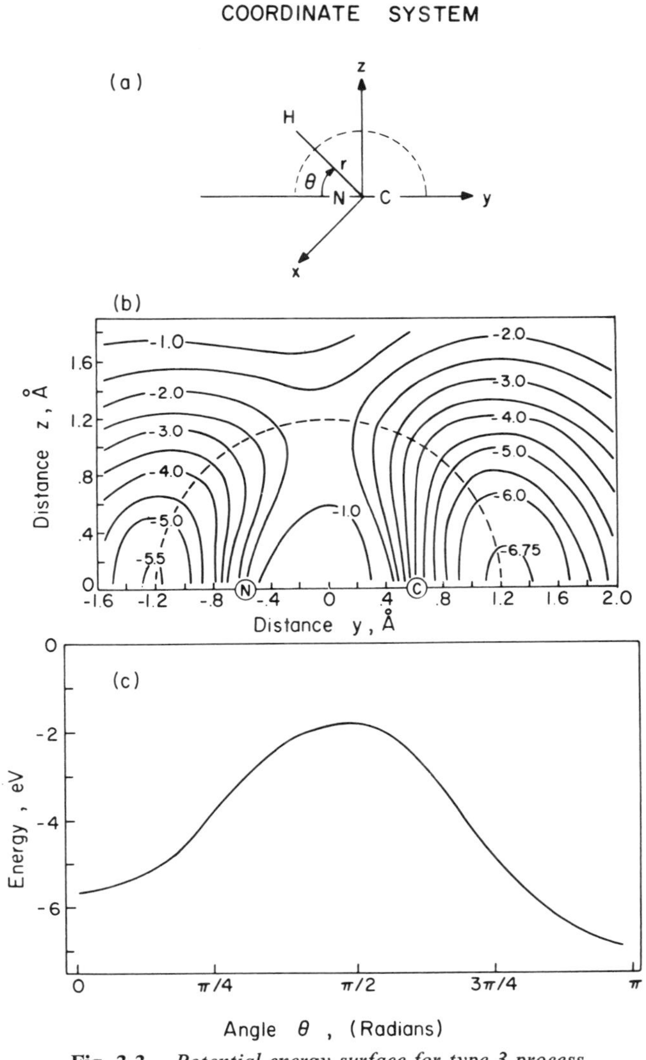

Fig. 3-3. *Potential energy surface for type 3 process.*

The process is an isomerization; the case shown is HNC → HCN, after calculations of G. W. Van Dine and R. Hoffmann, *J. Amer. Chem. Soc.* **90** 3227 (1968). (a) System of coordinate axes. Origin is at midpoint between carbon and nitrogen atoms. The molecular axis is designated as the *y* axis. (b) Potential energy surface for motion of a hydrogen atom in the vicinity of the CN group at a fixed CN distance of 1.2 Å. Contour lines of constant energy at 0.5-eV intervals. The surface has cylindrical symmetry around the C—N axis. (c) Section through the surface along the dashed semicircle in (b). Distance *r* assumed constant at 1.2 Å. See also D. Booth and J. N. Murrell, *Mol. Phys.* **24,** 1117 (1972).

potential energy associated with the motion of the hydrogen in the more or less immediate vicinity of the CN group. The distance C—N is assumed to be constant, because it is virtually the same in cyanide and isocyanide. The surface has cylindrical symmetry around the C—N axis. Thus two coordinates, y and z, as defined in Fig. 3-3(a), are sufficient to describe the surface. It shows two minima, one corresponding to HNC, and the other, somewhat deeper, to NCH, which are 180° apart since both isomers are linear. The two minima are separated by a saddle point, located roughly at the midpoint between C and N, and connected to the minima by two valleys which straddle an arc of almost constant radius. Thus the angle θ becomes a natural variable, and if a cut is taken through the surface along the bottom of the valleys, the energy profile, as a function of θ, has the form shown in Fig. 3-3(c).

There is therefore an activation energy for reaction in both directions, as of course there must be if both isomers are to be stable, and in this respect isomerization could be considered a special case of type 2 processes.

2. Triatomic potential energy surfaces for an effective potential

We have not said anything yet about rotational motions. Real molecules have degrees of freedom corresponding to overall rotation which give rise to additional centrifugal and Coriolis forces, and a potential energy surface should take these into account. A rigorous treatment of the coupling of rotational degrees of freedom with internal degrees of freedom is a difficult proposition, but a simple and tractable treatment can be set up by ignoring the Coriolis coupling entirely and approximating the effect of rotations through the addition of a rotational potential to the electronic energy for each nuclear configuration. The neglect of the various couplings has the result that the quantum mechanical rotational quantum number $\mathscr{J}$ becomes independent of internuclear distance. An *effective* potential (*7*, p. 425) V_{eff} is thus obtained, which in the case of a diatomic molecule AB has the form shown in Fig. 3-4. The main feature is that the potential no longer rises asymptotically to approach the energy of the dissociated system A + B but shows a hump separating the bound and dissociated system.

To visualize the effect of the addition of rotational potential on the potential energy surface of a triatomic molecule (Figs. 3-1–3-4) would require an additional dimension to represent the dependence on $\mathscr{J}$. However, it is not too difficult to see that the potential energy profile for type 1 reaction [Fig. 3-1(b)] will acquire a hump similar to that of Fig. 3-4; in other words there will now be a saddle point separating the reactant ABC from the products A + BC. Note that the location of the saddle point will depend on the rotational quantum number $\mathscr{J}$. By contrast, the addition of rotational potential in the case of type 2 and type 3 reactions affects only the *height* of the col

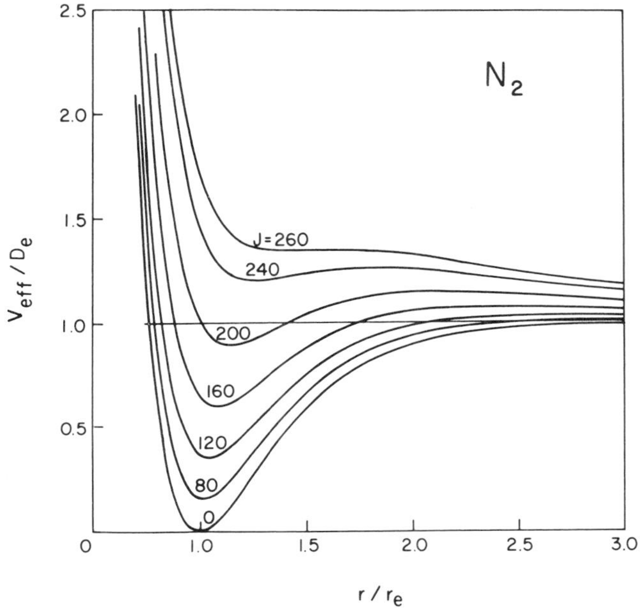

Fig. 3-4. *The reduced effective potential V_{eff}/D_e for ground state* N_2.

It is shown as a function of reduced interatomic distance r/r_e (r_e is the equilibrium internuclear distance) for several values of the rotational quantum number $\mathscr{J}$. The state at $\mathscr{J} = 260$ is not stable and represents a rotationally dissociated N_2. The zero of energy is at the potential minimum of the curve for $\mathscr{J} = 0$. Based on computations of S. H. Bauer and S. C. Tsang, *Phys. Fluids* **6**, 182 (1963) using a "best-fit" Morse function.

separating reactants and products but not its *location*, as shown in Chapter 7, where the problem of effective potential is treated in more detail.

The important consideration for our present purpose is that the addition of a rotational potential qualitatively modifies the features of potential energy surface for type 1 reactions: It removes the arbitrariness[6] noted in Section 1 arising from the absence of a landmark useful for specifying the intermolecular distance at which the system is to be considered as dissociated, and makes it qualitatively similar to potential energy surfaces for type 2 and type 3 reactions.

If we examine the potential energy surfaces for the more realistic effective potential, we see that they all have certain topological features in common: a dip or bowl representing the stable initial state (reactant), and another bowl or valley corresponding to the stable final state (product), the two states

[6] For this reason rotational potential cannot be ignored in type 1 reactions. Cf. (*8*).

separated by an energy col or saddle point. The progress of reaction can be conveniently followed by means of a progress variable, or reaction coordinate, suitably defined with respect to these topological features of the potential energy surface. For instance, if we assume that the reaction path follows the valley separating reactants and products in the direction of minimum curvature, the reaction coordinate may be defined as the distance along this path.[7] The structure located at the top of the col or saddle point is usually referred to as the transition state[8] since it obviously refers to a configuration intermediate between that of reactants and products. In "transition-state theory" (*9*) the motion of the system along the reaction coordinate is assumed to be separable from the motions of the system in all other possible modes, at least in the vicinity of the saddle point, and the calculation of the rate of reaction is then reduced to calculating the rate of passage of systems across the col or saddle point *in one dimension*.

This is a quite remarkable reduction of the dimensionality of the problem; in other words, transition-state theory makes do with information about the potential energy surface that is limited to the knowledge of the change in effective potential energy associated with the degree of freedom directly involved in the chemical transformation. The other degrees of freedom not involved in the chemical transformation are of course not discarded; as we shall see in Chapter 4, Section 3, they act merely as a sort of reservoir of energy and contribute to the rate only through their density of states. We might say, in the terminology of Eq. (1-11), that, over a range of energies $\delta E\ (\equiv \delta X)$ in the vicinity of some fixed total energy E, the one-dimensional passage of the system across the energy col defines k_{ij}, while the other degrees of freedom not directly involved in the reaction contribute the factor a/b.

3. Triatomic potential energy surfaces for nonadiabatic reactions

In some instances the potential energy surface for the first excited electronic state crosses[9] that for the ground state, as, for example, in the case of isoelec-

[7] In a simple dissociation, as in type 1 or type 2 surfaces, the reaction coordinate may be directly related to one of the normal mode coordinates, e.g., the antisymmetric stretch. In an isomerization (type 3), the reaction coordinate may be also a simple coordinate, as the angle θ in Fig. 3-3(c), for example, but it usually cannot be related in a simple way to one of the normal mode coordinates.

[8] We shall studiously avoid the term "complex" since it is likely to lead eventually to confusion, as there are all kinds of complexes and complex molecules. Its usage in unimolecular rate theory is a carry-over of nomenclature from bimolecular reactions, where the transition state is a "complex" or supermolecule formed from the two reactants.

[9] For the described effects to be observable, it is sufficient that the two potential energy surfaces merely approach each other. The rotational potential is ignored throughout the discussion of Section 3.

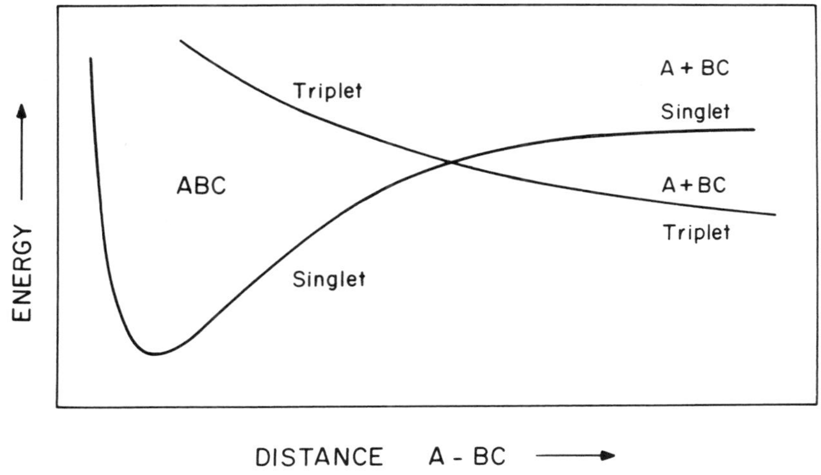

Fig. 3-5. *Crossing of potential energy surfaces (schematic).*

Potential energy surface profile of Fig. 3-1(b) redrawn to show schematically the crossing of surfaces for ground-state singlet and repulsive triplet states of *reactant*. As a result of the crossing, ground state of *products* is a triplet, and hence dissociation of ground-state reactant to yield ground-state products involves a switchover from singlet to triplet potential energy surface. The essential point is that the multiplicities of the reactant and product(s) are different. Hence similar curve crossing is involved when reactant is decomposing from an *excited* state, as in photoactivation (Chapter 9, Section 6), to yield products in some other electronic state.

tronic molecules CO_2 and N_2O; similar examples[10] are the dissociation of CS_2 and COS. The potential energy surface profile looks roughly as in Fig. 3-5 [cf. Fig. 3-1(b)], i.e., the reactant ground state (singlet) correlates with products in an *excited* state,[11] while the reactant excited state correlates with products in the *ground* state (triplet). Since it is known experimentally that ground-state CO_2 or N_2O yields ground-state products, the reaction must proceed through a switchover from a singlet potential energy surface to a triplet surface. Dissociation resulting from such a switchover is often referred to as predissociation, and reactions where the potential energy surface of more than one electronic state is involved are called nonadiabatic. A potential energy surface switchover is a "forbidden" transition, and as such has a

[10] A compilation of experimental results on spin-forbidden unimolecular dissociations of triatomic molecules is given in Troe and Wagner (*10*).

[11] The products in the two cases cited are CO + O and N_2 + O. Their ground state is $^1\Sigma + {}^3P$, respectively, which is a triplet; a singlet requires the oxygen to be in the ^1D state, which is an excited state.

probability of considerably less than unity (about 10^{-2} in the case of N_2O), so that this must be taken into account in calculating the rate. The usual procedure is to calculate the rate as if the reaction took place entirely on one single potential energy surface and then multiply the rate by the probability of transition from one surface to another. The calculation of the transition probability is a purely quantum mechanical problem and will not be treated here; for a recent account of the calculations see Nikitin (*11,12*); see also (*13*).

Predissociation is a type of radiationless transition known in spectroscopy as the Auger effect. These transitions have in common the fact that certain discrete levels of a system (e.g., singlet reactant levels above the crossover point in Fig. 3-5) have the same energy as the continuum of translational levels of another system (e.g., of the triplet products in Fig. 3-5); as a result, a mixing of the wave function occurs and the systems in discrete levels will, after a while, find itself in the continuum of levels, i.e., it will dissociate.

It is tempting to visualize a unimolecular process, even when reactant and continuum levels belong to the same electronic state, as an Auger-type transition between discrete and continuous levels of a system[12]; that such a connection exists between predissociation and unimolecular reactions was recognized very early by Rice (*14*). However, this approach to unimolecular rate theory has not so far proved very fruitful because, as early work has shown (*15*), unimolecular reactions represent a large perturbation of energy levels for which the usual theory of predissociation, based on first-order time-dependent perturbation treatment, is not applicable. A new attack on the problem has been made by Mies (*16–18*) [this approach has been further developed by Knewstubb (*19*)] and, more recently, by Rice (*20*). A general discussion of the problem and its relation to intersystem crossing in photochemical reactions has been given by Jortner *et al.* (*21*); see also Nikitin (*22*) and ref. (*28*).

4. Extrapolation to polyatomic potential energy surfaces

There is very little reliable information in existence about potential energy surfaces, or even segments thereof, for molecules containing more than three atoms. The most ambitious effort to date is the *ab initio* calculation of the potential energy surface done by Clementi (*23,24*) for NH_4Cl and its dissociation into NH_4 + Cl and NH_3 + HCl. The reaction involves both charge transfer and hydrogen bonding, which makes it a rather special case of a unimolecular dissociation; nevertheless, the overall features of the surface

[12] For example, vibrational levels of ν_2, the antisymmetric stretching vibration in Fig. 3-2(a), do not converge to the dissociation energy for A + BC, but continue above this limit, at which point they coincide with translational levels of the system A + BC; dissociation then takes place by predissociation, the whole process taking place on the same potential energy surface.

closely parallel those of Fig. 3-1. In general, however, we are reduced to deducing the properties of potential energy surfaces of polyatomic systems by extrapolating the knowledge obtained on triatomic systems.

In a polyatomic system it seems reasonable to assume that atoms not directly involved in the chemical transformation will move in a force field that remains basically the same throughout the reaction, so that the dimensionality of the reaction rate problem can be reduced. For the remaining degrees of freedom that cannot be so easily disposed of, it can be assumed, by analogy with triatomic molecules, that they define a reaction path which can be described by a single progress variable, i.e., by the *one-dimensional* passage of the system point across an energy barrier.[13] The reaction coordinate is then assumed to involve only those degrees of freedom directly concerned in the chemical transformation, mostly just one or two.[14]

Since, as was mentioned in Section 2, the reaction coordinate represents distance along the path of least curvature, i.e., along the path of least energy, between reactant and product(s), it may be sufficient in the case polyatomic reactants to calculate only segments of the potential energy surface, and then by an appropriate optimization technique deduce the minimum-energy path. As an example of such an approach, we may cite calculations by Hsu *et al.* (*27*) on the isomerization of cyclobutene to *cis*-butadiene, a rather intriguing process because of its stereospecificity (cf. Chapter 11, Section 3).

The reduction of the problem of unimolecular reaction rates to the one-dimensional passage of a system point over an energy barrier has the consequence that sometimes, on account of the symmetry of the complete multidimensional potential energy hypersurface, there may exist several equivalent reaction paths over this hypersurface leading from reactant to products. Since the transition-state theory calculates the rate for only one path, the total rate will be a multiple of the single-path rate, The multiplying factor is called the reaction path degeneracy; its calculation will be discussed in Chapter 4, Section 5.

Assuming the validity of extrapolation from triatomic to polyatomic potential energy surfaces, we may conveniently classify unimolecular reactions according to the basic features of their potential energy surface as typified by the potential energy surfaces of Figs. 3-1–3-3 as follows.

Type 1 These reactions have a potential energy surface similar to that of Fig. 3-1. The reaction involves simple bond breaking and yields two radicals,

[13] A method for calculating the reaction path in a multidimensional potential surface, not requiring the calculation of the entire surface, has been proposed by Empedocles (*25*).

[14] In the simple treatment used here and in most of the literature the reaction coordinate is assumed to be a straight line. For a generalization to curvilinear reaction coordinates see Marcus (*26*).

so that the reverse reaction of association requires no formal activation energy (but does require the clearing of a rotational barrier). An example is the dissociation of ethane, $C_2H_6 \rightarrow CH_3 + CH_3$.

Type 2 These are unimolecular dissociations which require an activation energy for the reverse reaction of association. The dissociation may involve (i) simple bond breaking, as in $C_2H_5 \rightarrow H + CH_2{=}CH_2$, in which case the potential energy surface is similar to that of Fig. 3-2, or (ii) complex bond breaking, as in $C_2H_5Cl \rightarrow HCl + CH_2{=}CH_2$, which is a four-center reaction. More complicated five- and six-center reactions are known but have not been studied quite as extensively.

Type 3 These are unimolecular isomerizations which have a potential energy surface similar to that of Fig. 3-3, although perhaps with a more complicated reaction coordinate. The one feature they have in common with type 2 reactions is an activation energy for the reverse isomerization. The simplest cases are cis–trans isomerizations, e.g., *cis*-butene-2 → *trans*-butene-2, where the reaction coordinate is simply the angle of twist about the C=C bond; more complicated cases that have been studied extensively are the isomerization cyclopropane → propylene and variants thereof, and the already mentioned isomerizations isocyanide → cyanide.

The calculation of the unimolecular rate is essentially the same for types 1, 2, and 3; the basic rate expression will be derived in the next chapter. An adequate consideration of type 1 reactions requires, in addition, special attention to centrifugal effects, which will be left for Chapter 7.

References

1. H. S. Johnston, "Gas Phase Reaction Rate Theory," Appendix C. Ronald Press, New York, 1966.
2. J. O. Hirschfelder, *Int. J. Quantum Chem.* **3s**, 17 (1969).
3. G. Herzberg, "Molecular Spectra and Molecular Structure," Vol. III, Electronic Spectra and Electron Structure of Polyatomic Molecules. Van Nostrand Reinhold, Princeton, New Jersey, 1966.
4. D. L. Bunker, "Theory of Elementary Gas Reaction Rates." Pergamon, Oxford, 1966.
5. M. Karplus, *in* "Molecular Beams and Reaction Kinetics," Course LXIV, Italian Physical Soc., p. 320. Academic Press, New York and London, 1970.
6. K. Machida and J. Overend, *J. Chem. Phys.* **50**, 4429 (1969).
7. G. Herzberg, "Molecular Spectra and Molecular Structure," Vol. I, Spectra of Diatomic Molecules. Van Nostrand Reinhold, Princeton, New Jersey, 1957.
8. F. P. Buff and D. J. Wilson, *J. Amer. Chem. Soc.* **84**, 4063 (1962).
9. S. Glasstone, K. J. Laidler, and H. Eyring, "The Theory of Rate Processes." McGraw-Hill, New York, 1941.
10. J. Troe and H. Gg. Wagner, *Ber. Bunsenges.* **71**, 937 (1967), Table 2b.

11. E. E. Nikitin, *in* "Chemische Elementarprozesse" (H. Hartmann, ed.), p. 43. Springer-Verlag, Berlin, 1968.
12. E. E. Nikitin, *Advan. Quantum Chem.* **5**, 135 (1970).
13. B. R. Henry and M. Kasha, *Ann. Rev. Phys. Chem.* **19**, 161 (1968).
14. O. K. Rice, *Phys. Rev.* **35**, 1538 (1930).
15. N. Rosen, *J. Chem. Phys.* **1**, 319 (1933).
16. F. H. Mies and M. Krauss, *J. Chem. Phys.* **45**, 4455 (1966).
17. F. H. Mies, *J. Chem. Phys.* **51**, 787 (1969).
18. F. H. Mies, *J. Chem. Phys.* **51**, 798 (1969).
19. P. F. Knewstubb, *Int. J. Mass Spectrom. Ion Phys.* **6**, 229 (1971).
20. O. K. Rice, *J. Chem. Phys.* **55**, 439 (1971).
21. J. Jortner, S. A. Rice, and R. M. Hochstrasser, *Advan. Photochem.* **7**, 149 (1969).
22. E. E. Nikitin, *Mol. Phys.* **8**, 473 (1964).
23. E. Clementi, *J. Chem. Phys.* **46**, 3851 (1967); **47**, 2323 (1967).
24. E. Clementi and J. N. Gayles, *J. Chem. Phys.* **47**, 3837 (1967).
25. P. Empedocles, *Int. J. Quantum Chem.* **3s**, 47 (1969).
26. R. A. Marcus, *J. Chem. Phys.* **41**, 2614, 2624 (1964).
27. K. Hsu, R. J. Buenker, and S. D. Peyerimhoff, *J. Amer. Chem. Soc.* **93**, 2117 (1971).
28. R. G. Gilbert and I. G. Ross, *Aust. J. Chem.* **24**, 1541 (1971).

CHAPTER 4

Statistical Calculation of Unimolecular Rate under Vibrational Potential

In this chapter we shall elaborate the statistical argument by means of which is obtained $k(E)$, the unimolecular rate constant for molecules of energy E ($E \geqslant E_0$, where E_0 is the critical or threshold energy for reaction). We proceed on the basis of conclusions reached in the previous three chapters: (1) one time-independent rate constant describes the process; (2) energy is the necessary and sufficient criterion for reaction; (3) rate is given by the one-dimensional passage of a system point over an energy barrier.

The reaction is assumed to take place under the influence of the vibrational potential, i.e., the rotational part of the effective potential is here neglected. It will be shown more fully in Chapter 7 that the rotational potential influences only the height of the energy barrier the system has to cross (i.e., E_0) so that the statistical argument is not affected and therefore can be presented here uncluttered with rotational considerations; these are deferred until Chapter 7.

Although our objective is to calculate the quantum-mechanical rate constant, in Section 1 we shall first develop the main theme in classical terms and then in Section 2 amend it, as necessary, in quantum-mechanical terms. This hybrid semiclassical approach conserves most of the intuitive imagery of classical mechanics. The basic quantum-mechanical equation is then rederived in Section 3 from a different point of view which is useful for illustrating another facet of the problem. Pure quantum effects associated with the passage of a system over an energy barrier are discussed in Section 4, and in Section 5 the reaction path degeneracy is considered in some detail.

1. Unimolecular process in classical phase space

Suppose that the degrees of freedom of the decomposing molecule which are involved in the chemical transformation are n in number. In principle, n will be less than the total number of degrees of freedom in the molecule, since the three translational degrees of freedom are excluded by the necessity to conserve linear momentum, and some rotational degrees of freedom may have to be excluded in order to conserve angular momentum (cf. Chapter 5). We shall defer the actual determination of n also to Chapter 5 and shall, for the moment, merely consider n to represent the number of "pertinent" degrees of freedom, with the understanding that these are essentially internal. This procedure does not in any way affect the generality of the argument that follows.

Let us construct a $2n$-dimensional phase space consisting of n generalized coordinates q_i ($i = 1, 2, \ldots, n$) and n conjugate momenta p_i. If the decomposing molecule (henceforth referred to as "particle," for short) moves in a time-independent potential with no external forces (the usual case in a chemical reaction), we may take the q_i as Cartesian positional coordinates and the p_i as corresponding linear momenta; the total energy (kinetic plus potential), i.e., the Hamiltonian of the system $\mathscr{H}(p_i; q_i)$, is then a constant of motion and as such is independent of time. The instantaneous dynamical state of the particle will be given by a point ("representative point") in phase space and its time evolution by a trajectory.[1] We can imagine a $(2n - 1)$-dimensional (hyper)surface, which we may call the critical surface, separating the total phase space into two regions, one corresponding to bound particles (positive part) and the other to product particles (negative part). In general, this critical surface will be defined by a function of all the p_i and q_i.

We are interested in the rate $k(E)$ at a specified E ($E \geqslant E_0$). Particles in reactive configurations[2] of energy E, i.e., those for which $\mathscr{H}(p_i; q_i) = E$, will be represented by that part of a $(2n - 1)$-dimensional hypersurface of constant energy which lies on the bound particle (positive) side of the critical surface. For the purpose of this chapter, and in fact throughout Part 1, energy E is taken to be constant.

If the critical surface is defined in such a way as to intersect all constant-energy surfaces of the reactive configurations, i.e., those for which $E \geqslant E_0$,

[1] For a fuller account see, for example, Lichtenberg (*1*) (the reference list for this chapter can be found on p. 69).

[2] By "configuration" we mean a particular set of p_i and q_i which locates a given volume element in phase space. A reactive configuration is one from which dissociation can occur, i.e., one with $E \geqslant E_0$.

the intersections will constitute a $(2n - 2)$-dimensional hypersurface which will define particles belonging to a "critical" configuration with the Hamiltonian $\mathscr{H}'(p_i; q_i) = E - E_0$. In terms of the discussion of the previous chapter, the Hamiltonian $\mathscr{H}'$ describes therefore the internal configurations of particles located at the top of the barrier separating reactants and products, and E_0 is then the difference in potential energy between the reactant particles and particles in the critical configuration. See Fig. A-1 (p. 404).

We shall make the assumption that once the representative point of a bound particle reaches a critical configuration, its trajectory in phase space[3] is by definition such that it will *always* cross the critical surface once and dissociate. The rate constant $k(E)$ is then simply the total flux of representative points through the critical surface in the direction of decomposition, the points approaching along the subsurface $\mathscr{H}' = E - E_0$.

The equation for $k(E)$ can be set up as for any other problem involving the flux of a fluid [see e.g. (*4*, p. 19)], the "fluid" in this case being the movement of representative points. Consider the element $d\mathscr{S}$ of the critical "surface" [Fig. 4-1(a)]; quotation marks are used here to underline the fact that common terms like surface, volume, velocity, etc. actually refer to multidimensional hyperspaces. If the "velocity" of representative points near $d\mathscr{S}$ is $\boldsymbol{u}$ (a vector), representative points crossing $d\mathscr{S}$ in the time interval dt will sweep out the parallelepiped $|\boldsymbol{u}|\, d\mathscr{S} \cos\theta\, dt$. Recall now that $k(E)$ is defined as the *fractional* number of molecules decomposing per unit time [cf. Eq. (1-5) or (1-12)]; therefore if N_f is the fractional number of representative points per unit "volume" in the vicinity of $d\mathscr{S}$, the contribution of $d\mathscr{S}$ to the rate constant is the fractional number of representative points passing through the element of critical surface $d\mathscr{S}$ per unit time, which is

$$N_f|\boldsymbol{u}|\, d\mathscr{S} \cos\theta = N_f \boldsymbol{u}\cdot\boldsymbol{k}\, d\mathscr{S} \tag{4-1}$$

i.e., a scalar product of two vectors, $\boldsymbol{k}$ being interpreted as a unit vector normal to $d\mathscr{S}$ in the outward direction. The total fractional number of representative points passing through the entire critical "surface" is the total flux, i.e., $k(E)$, and is obtained by integrating the right-hand side of (4-1) over $\mathscr{H}'$:

$$k(E) = \underset{\mathscr{H}' = E - E_0}{\int \cdots \int} N_f\, \boldsymbol{u}\cdot\boldsymbol{k}\, d\mathscr{S} \tag{4-2}$$

The integration here is with respect to a surface element but we shall find it more convenient to interpret the equation in terms of an integral with respect

[3] The concept of phase-space trajectories can be developed into an alternative method for rate calculation. See Keck (*2*) and Keck and Kalelkar (*3*).

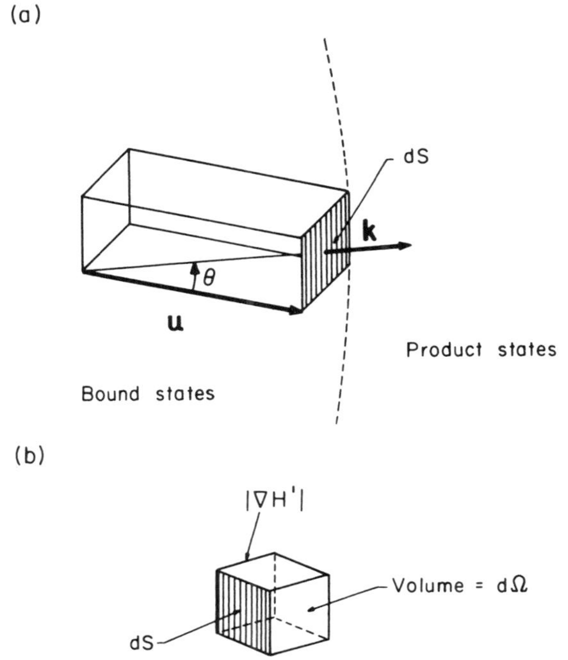

Fig. 4-1. (a) *Flux of representative points through element $d\mathscr{S}$ of critical surface.*

$d\mathscr{S}$ is the element of the critical "surface" represented by $\mathscr{H}' = E - E_0$, $\boldsymbol{k}$ is a unit vector normal to $d\mathscr{S}$ and pointing in an outward direction, i.e., from the bound-state side of the critical surface to the product side, and $\boldsymbol{u}$ is the velocity of representative points, i.e., "distance" traveled in unit time near $d\mathscr{S}$. Therefore in unit time all the representative points inside the parallelepiped shown will pass through $d\mathscr{S}$. The "volume" of the parallelepiped is $\boldsymbol{u}\cdot\boldsymbol{k}\, d\mathscr{S}$. If there are N_{f} representative points per unit volume near $d\mathscr{S}$, their number passing through $d\mathscr{S}$ in unit time is therefore $N_{\mathrm{f}}\boldsymbol{u}\cdot\boldsymbol{k}\, d\mathscr{S}$.

(b) *Relation between surface and volume elements in the vicinity of the critical surface.*

$|\nabla\mathscr{H}'|$ is the magnitude of the gradient of $\mathscr{H}'$, i.e., the magnitude of the derivative of $\mathscr{H}'$ in the direction normal to $\mathscr{H}'$ (i.e., normal to $\mathscr{S}$) at $d\mathscr{S}$. The critical surface is therefore considered as if of "thickness" $|\nabla\mathscr{H}'|$.

to a volume element because N_{f} is defined per unit "volume." We have by the theorem of the gradient (*5*, Chapter 5, p. 159)

$$d\mathscr{S} = d\Omega/|\nabla\mathscr{H}'| \tag{4-3}$$

where [Fig. 4-1(b)] $\nabla\mathscr{H}' = \mathrm{grad}\,\mathscr{H}'$, an element of "distance" perpendicular to $d\mathscr{S}$, and $d\Omega$ is an element of "volume." We can then write

$$\boldsymbol{u}\cdot\boldsymbol{k}/|\nabla\mathscr{H}'| = \Phi \tag{4-4}$$

where Φ is "velocity"/"distance," so that the dimension of Φ is second^{-1}; Φ is therefore sometimes referred to, rather misleadingly, as the frequency of crossing the critical energy surface, although no back-and-forth movement is involved, as the term frequency would seem to imply. Equation (4-4) makes it clear that Φ refers to velocity per unit distance in an outward direction normal to the critical surface.

Using Eqs. (4-3) and (4-4), and indicating the dependence on $(p_i; q_i)$ explicitly, Eq. (4-2) can be transformed to Eq. (4-5) below in which N_f is expressed as the number of representative points per unit volume at $(p_i; q_i)$, designated by $P(p_i; q_i)$, divided by the total number of representative points on the bound particle side of the hypersurface E = const, designated by DEN:

$$k(E) = \underset{\mathscr{H}' = E - E_0}{\int \cdots \int} P(p_i; q_i)\Phi(p_i; q_i)\, d\Omega \Big/ \underset{\mathscr{H} = E}{\int \cdots \int} P(p_i; q_i)\, d\Omega \tag{4-5}$$

where $d\Omega = \prod_{i=1}^{n} dp_i\, dq_i$, $P(p_i; q_i)/\text{DEN} = N_f$, DEN representing the denominator of (4-5), i.e., the total "area" on the positive side of hypersurface E = const; thus $P(p_i; q_i)\, d\Omega/\text{DEN}$ can be interpreted as the probability of finding the representative point of a bound molecule within a volume element $d\Omega$ centered at a specified $(p_i; q_i)$; DEN therefore normalizes $P(p_i; q_i)$ to unity at a specified E. The integrals in (4-5) are $(2n - 1)$-fold and the domain of integration extends over all p_i and q_i for which $\mathscr{H}' = E - E_0$ and $\mathscr{H} = E$ in the numerator and denominator, respectively.

Critical surface function of coordinates only To evaluate (4-5), we must first specify the critical surface. Following previous discussion, we shall assume that it depends on coordinates only; in this case, we can always transform to a new set of coordinates q_i' by taking a suitable linear combination of the original coordinates q_i and then write the equation for the critical surface in the simple form[4] $q_1' = q_1^{0\prime}$ ($q_1^{0\prime}$ = const), where for convenience we may put the constant equal to zero. We will henceforth drop the prime on q_i' with the understanding that q_i shall refer to the new, transformed coordinates. We have bound particles for $q_1 < q_1{}^0$ and dissociated particles for $q_1 > q_1{}^0$, so that dissociation amounts to a unidimensional translation along q_1. Let the conjugate momentum be p_1; the corresponding Hamiltonian is then $\mathscr{H}_1(p_1; q_1{}^0)$. As a further result of the conclusion reached at the end of Chapter 3 that the rate depends largely on the properties of the system in the one degree of freedom that involves the crossing of the critical energy surface (to use the present terminology), we shall now assume that the degree

[4] The transformation that allows us to write $q_1' = q_1^{0\prime}$ is called a point transformation. Cf. Rice (*6*, p. 18).

of freedom involving p_1 and q_1 separates from the total Hamiltonian, at least within some small distance δq_1 in the vicinity of $q_1{}^0$:

$$\mathscr{H}'(p_1, \ldots, p_n; q_1, \ldots, q_n) = \mathscr{H}^*(p_2, \ldots, p_n; q_2, \ldots, q_n) + \mathscr{H}_1(p_1; q_1{}^0) \quad (4\text{-}6)$$

where $\mathscr{H}^*$ is the Hamiltonian for the system of $n - 1$ degrees of freedom, not involving p_1 and q_1, located in δq_1 and moving across the critical surface; such a system is called the transition state (we shall henceforth use an asterisk to denote quantities that refer to the transition state). As a result of the separability in the Hamiltonian $\mathscr{H}'$, we have quite similarly for the probability $P\,d\Omega$

$$P(p_1, \ldots, p_n; q_1, \ldots, q_n)\,d\Omega = P^*(p_2, \ldots, p_n; q_2, \ldots, q_n)\,d\Omega^* \times P_1(p_1; q_1{}^0)\,dp_1\,dq_1 \quad (4\text{-}7)$$

where $d\Omega^* = \prod_{i=2}^{n} dp_i\,dq_i$. We will also assume that the potential energy surface has no curvature along q_0; we then have, for a given p_1, $\mathscr{H}_1(p_1; q_1{}^0) = \varepsilon_t = p_1{}^2/2\mu^*$, since in such case $\mathscr{H}_1$ represents the energy of simple rectilinear motion in one dimension. Here μ^* is the effective mass[5] of atoms moving along q_1. Clearly, for a critical surface so defined, Φ of Eq. (4-5) depends only on p_1: $\Phi(p_i; q_i) = \Phi(p_1)$.

It is now convenient to introduce a rate constant for molecules that not only have a specified total energy E, but also a specified energy ε_t ($\varepsilon_t \leqslant E - E_0$) in the unidimensional translation along q_1. Defined per unit interval in ε_t, this rate constant will be called $k(E, \varepsilon_t)$ [units: energy^{-1}]. For a given ε_t, the Hamiltonian $\mathscr{H}^*$ is given by

$$\mathscr{H}^* = E - E_0 - \varepsilon_t \quad (4\text{-}8)$$

It follows from (4-5)–(4-8) that

$$k(E, \varepsilon_t) = \mathrm{DEN}^{-1} \int_{\mathscr{H}^* = E - E_0 - \varepsilon_t} \!\!\cdots\! \int P^*(p_i; q_i)\,d\Omega^* \times \iint_{\mathscr{H}_1 = \varepsilon_t} P_1(p_1; q_1{}^0)\Phi(p_1)\,dp_1\,dq_1 \quad (4\text{-}9)$$

Since decomposition corresponds to systems moving only in one (positive) direction across the critical surface, integration with respect to p_1 is performed only over *positive* momenta. The rate constant, when only the total energy is specified (which is the form we want), is obtained by integrating (4-9) over all values of ε_t, i.e., between 0 and $E - E_0$:

$$k(E) = \int_0^{E - E_0} k(E, \varepsilon_t)\,d\varepsilon_t \quad (4\text{-}10)$$

[5] See, for example, Johnston (*7*, Chapter 3).

Microcanonical ensemble Up to this point, the argument is quite general and the only assumptions involved concern the separability in the total Hamiltonian [Eq. (4-6)] and the unit probability of crossing the critical surface (= unit transmission coefficient) for all particles whose representative points have reached the critical configuration. Let us now assume that the representative points of the bound particles are uniformly distributed over the positive part of the hypersurface of constant energy, i.e., that the bound-state part of the system is in equilibrium; such a representation corresponds to a microcanonical ensemble.[6] The meaning of equilibrium is that *all* bound state configurations corresponding to E = const are represented, and it is thus implicitly assumed that whatever the mechanism of activation, it is such as to generate each and every bound state configuration of energy E with equal probability, leading to a uniform (= random) distribution of representative points. Since the configurations of the transition state are part of the bound-state configurations, the representative points of the transition state will be also uniformly distributed over its subsurface $\mathscr{H}^* =$ const. Note that the equilibrium assumption *excludes* the product states, and in fact no information is available about the negative part of the hypersurface of constant energy.

The consequence of the uniform distribution of representative points is that, on the hypersurface E = const, the probability of finding a representative point in a volume element $d\Omega$ is simply the ratio of the volume element to the total volume, so that we have $P(p_i; q_i)\, d\Omega/\text{DEN} = d\Omega/\text{DEN}$, from which it follows that

$$\begin{aligned} P(p_i; q_i) &= 1 \quad \text{for} \quad \mathscr{H}(p_i; q_i) = E \\ &= 0 \quad \text{otherwise} \end{aligned} \tag{4-11}$$

Since the integrations in (4-5) or (4-9) are precisely over those $(p_i; q_i)$ for which $P(p_i; q_i)$ is constant, $P(p_i; q_i)$ drops out and its specification in the case of an equilibrium (microcanonical) distribution is a redundant formalism that may be dispensed with. Using (4-7) and (4-11), we can therefore rewrite (4-9) and (4-10) for the case of the microcanonical rate constant in the form

$$k(E) = \frac{\displaystyle\int_0^{E-E_0} \Bigg(\underset{\mathscr{H}^* = E - E_0 - \varepsilon_t}{\int \cdots \int} d\Omega^* \times \underset{\mathscr{H}_1 = \varepsilon_t}{\iint} \Phi(p_1)\, dp_1\, dq_1 \Bigg)\, d\varepsilon_t}{\displaystyle\underset{\mathscr{H} = E}{\int \cdots \int} d\Omega} \tag{4-12}$$

[6] For a good discussion on this point see Andrews (*8*, p. 22 ff.). It is worth stating emphatically here that the equilibrium alluded to in Sections 1 and 2 of this chapter is *not thermal* equilibrium.

The difficult part here is the evaluation of the integrals with respect to $d\Omega$ and $d\Omega^*$. We shall return to this subject briefly in Chapter 6, Section 3.

2. The quantum-mechanical unimolecular rate constant

Inasmuch as we really want the quantum-mechanical $k(E)$, it is necessary at this point to present the foregoing argument in quantum-mechanical language.

Because of the position–momentum uncertainty $\delta p\ \delta q \sim h$, where δp is the uncertainty in momentum and δq the uncertainty in positional coordinate, phase space for a system of n degrees of freedom must be considered as having a cellular structure, with each cell of volume h^n (h is Planck's constant) containing one quantum state. Therefore we cannot speak of definite trajectories in phase space, and likewise we cannot speak of "volume" in phase space but only of the number of cells (i.e., of the number of states) contained in a given region of phase space. Thus the term "configurations" used in Section 1 must be interpreted as meaning quantum states. By the same token, we have to acknowledge that energy can be specified only to within an allowance δE, so that a hypersurface of constant energy must really be considered as a "shell" of thickness δE. Consequently when evaluating the area[7] of a given region on a hypersurface of constant energy we actually count the number of states contained in a shell of thickness δE. Let $W(E)$ denote the number of states contained with δE on the positive side of hypersurface $E =$ const. It is also convenient to introduce the density of states $N(E)$ [units: energy^{-1}], i.e., the number of states per unit energy interval at E; we then have $W(E) = N(E)\ \delta E$. (Cf. Chapter 6, Section 1.)

We now return to the microcanonical rate constant $k(E)$. Transcribed in quantum-mechanical terms, the ratio of multiple integrals in (4-12) becomes a fraction where the numerator refers to transition state, i.e., $W^*(E - E_0 - \varepsilon_t) = N^*(E - E_0 - \varepsilon_t)\ \delta E$ refers to the number of those bound states within δE on the subsurface $E - E_0 - \varepsilon_t =$ const.:

$$\frac{\displaystyle\int \cdots \int_{\mathscr{H}^* = E - E_0 - \varepsilon_t} d\Omega^*}{\displaystyle\int \cdots \int_{\mathscr{H} = E} d\Omega} \approx W^*(E - E_0 - \varepsilon_t)/W(E) = N^*(E - E_0 - \varepsilon_t)/N(E) \tag{4-13}$$

[7] In general, if integration is over $\mathscr{H} = E$ or $E \leqslant \mathscr{H} \leqslant E + \delta E$, we shall refer to it as a surface integral, and if integration is over $0 \leqslant \mathscr{H} \leqslant E$, we shall refer to it as a volume integral.

while $W(E)$ and $N(E)$ refer to *all* bound states of energy E. Equation (4-13) therefore represents, as it should, the fraction of all states of energy E which belong to the critical configuration of specified ε_t. The normalization factor is seen to be $W(E) = N(E)\,\delta E$. The double integral in Eq. (4-12) is in fact $\langle\Phi\rangle$, the average value of Φ. Only translation is involved, which may be treated semiclassically, i.e., the classical integral can be corrected for nonclassical effects merely by division with h and by taking the integral over an interval $\delta\varepsilon_t$ in ε_t:

$$\langle\Phi\rangle = (1/h) \iint_{\mathscr{H}_1 = \varepsilon_t}^{\mathscr{H}_1 = \varepsilon_t + \delta\varepsilon_t} \Phi(p_1)\, dp_1\, dq_1 \tag{4-14}$$

Integration with respect to q_1 yields only δq_1, the length[8] in the vicinity of $q_1{}^0$ where the total Hamiltonian is separable. The significance of $\Phi(p_1)$ is velocity u (now only one-dimensional) divided by δq_1, where

$$\tfrac{1}{2}\mu^* u^2 = p_1{}^2/2\mu^* \tag{4-15}$$

since all the energy is kinetic. Hence

$$\Phi(p_1) = u/\delta q_1 = p_1/\mu^*\, \delta q_1 \tag{4-16}$$

Substituting into (4-14) and integrating over positive momenta yields

$$\langle\Phi\rangle = (1/\mu^* h) \int_{\mathscr{H}_1 = \varepsilon_t}^{\mathscr{H}_1 = \varepsilon_t + \delta\varepsilon_t} p_1\, dp_1 = \delta\varepsilon_t/h \tag{4-17}$$

Equation (4-9) becomes, using (4-13) and (4-17),

$$k(E, \varepsilon_t)\, \delta\varepsilon_t = N^*(E - E_0 - \varepsilon_t)\, \delta\varepsilon_t/hN(E) \tag{4-18}$$

Writing $d\varepsilon_t$ for $\delta\varepsilon_t$ in (4-18), Eq. (4-10) gives then

$$k(E) = [1/hN(E)] \int_0^{E - E_0} N^*(E - E_0 - \varepsilon_t)\, d\varepsilon_t \tag{4-19}$$

an equation which was first derived almost simultaneously by Marcus (*10*) and Rosenstock *et al.* (*11*).[9] Let us write for the integral with respect to ε_t

$$\int_0^{E - E_0} N^*(E - E_0 - \varepsilon_t)\, d\varepsilon_t = G^*(E - E_0) \tag{4-20}$$

[8] Bell (*9*) has pointed out that δq_1 cannot be taken too small lest the principle of uncertainty come into operation and the linear movement cease to be classical.

[9] Actually, Eq. (4-19) was derived essentially in the above form several years earlier by Bohr and Wheeler (*12*) in connection with the statistical model of nuclear fission. As pointed out by Rosenstock *et al.*, there are many analogies between nuclear and unimolecular fissions.

where $G^*(E - E_0)$ is the integrated density (cf. Chapter 6, Section 1) for the transition state. In this notation we have simply

$$k(E) = G^*(E - E_0)/hN(E) \tag{4-21}$$

The transition state It is important to realize that $G^*(E - E_0)$ involves states of a critical configuration from which the states of the translational degree of freedom along q_1 have been eliminated, so that it refers to a fictitious species having one degree of freedom less than the original molecule. This species—the transition state—is merely a convenient device for defining that part of the total phase space of bound states which is relevant to the description of the decomposition process as a flux of phase points across a critical boundary. The virtue of this approach is that this relevant part of phase space may be treated as if belonging to a molecular system which behaves in all respects as a normal molecule, except in the one degree of freedom corresponding to the reaction coordinate, and thus could be computed, at least in principle, by standard methods if the potential energy surface for the reaction were known in sufficient detail. This aspect of the problem will be treated more fully in Chapter 11; suffice to say at this point that in practice there is not much detailed information about the transition state and many of its parameters can only be surmised.

As we have pointed out before, the representative points of the transition state are also uniformly distributed over the subsurface $\mathscr{H}^* =$ const. Therefore *every* quantum state of the transition state of energy $E - E_0$ is represented, so that $G^*(E - E_0)$ can be evaluated by the same methods of equilibrium[10] statistical mechanics as $N(E)$.

For more precise interpretation of the various energy quantities, we have to take into account the existence of the quantum-mechanical zero-point energy. Thus if classically E_0 was defined as the difference between the potential energy of the reactant and the transition state, quantum mechanically we have to interpret E_0 as the difference in energy between the ground state of the reactant and the ground state of the transition state. Similarly, the excitation energy E must be interpreted as energy in excess of zero-point energy. Figure 4-2 shows the relation of the various energy quantities in the case of a type 1 potential energy surface. Thus D_0 is the dissociation energy of the bond that is breaking, and E_0 is related to it by $E_0 = D_0 + E_z^*$, where E_z^* is the zero-point energy of the transition state. The energy E_0 is unavailable to the transition state as internal energy because it was used up, so to speak, in breaking the

[10] For this reason some workers, particularly in the field of mass spectrometry, refer to this theory as the "quasi-equilibrium" theory. The qualification "quasi" is meant to bring attention to the equilibrium hypothesis as a subterfuge for using equilibrium methods in what is basically a nonequilibrium situation.

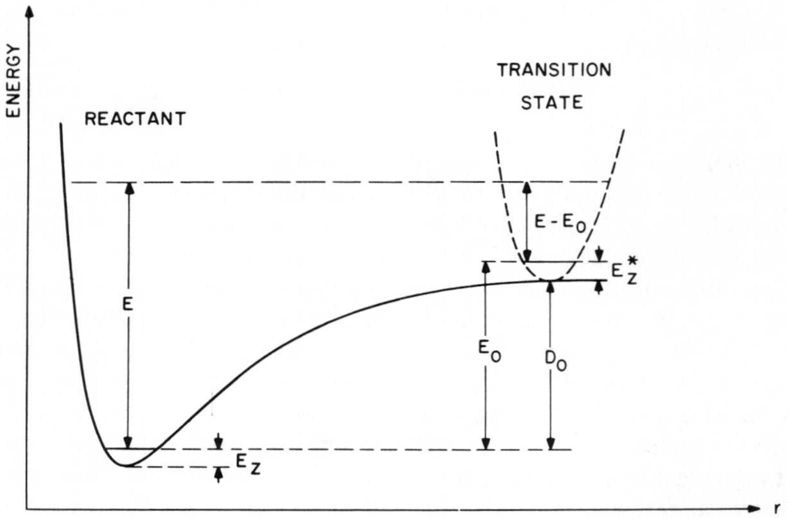

Fig. 4-2. *Energy relations between reactant and transition state for type 1 potential energy surface (schematic).*

Rotational potential assumed to be zero ($\mathscr{J} = 0$). E_0 is the quantum mechanical critical energy, D_0 is the dissociation energy of reactant, E_z is the zero-point energy of reactant, $E_z{}^*$ is the zero-point energy of transition state, E is the internal excitation energy of reactant, and $E - E_0$ is the excitation energy in transition state; this energy is partitioned between the internal excitation energy of the transition state ($E - E_0 - \varepsilon_t$), and the relative translational energy of fragments (ε_t).

bond. Note that for a type 1 potential energy surface, E_0 is in fact the endothermicity of the reaction. For type 2 potential energy surfaces the various relations are obviously quite similar, except that E_0 is no longer equal to the heat of reaction. A simple application of Eq. (4-21) is the subject of Exercise 1.

3. An alternative derivation

The derivation which will be now given has the advantage of bringing out certain aspects which are somewhat obscured in the treatment of Sections 1 and 2. It is originally due to Rice (*13*; *6*, p. 495 ff.) and is here slightly modified.

Let us consider first the dissociation of a diatomic molecule AB. The molecule will be regarded as dissociated when the internuclear distance r just exceeds some value r_m, and the equilibrium rate constant for dissociation will ge given by the frequency with which, at equilibrium, molecules cross the boundary at

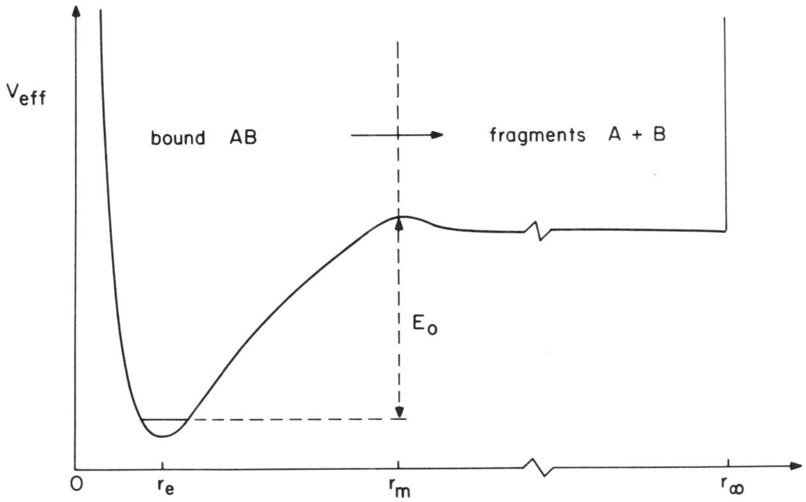

Fig. 4-3. *Potential used by Rice for dissociation of diatomic molecule* AB.

The potential is in essence an effective potential V_{eff} (cf. Chapter 3, Section 2) which is made to become infinite at $r = r_\infty$. The vertical broken line represents the boundary separating bound AB from fragments A + B; it is located at the maximum of the hump where the A–B distance is r_m. In the present case the zero-point energy of the transition state is zero ($E_z^* = 0$), and therefore $D_0 = E_0$.

$r = r_m$ while the fragments A and B are moving *away* from each other. It is then assumed, as is usual in rate theory, that rate of dissociation at equilibrium represents also the rate of dissociation in a nonequilibrium situation. (Cf. footnote 4 in Chapter 8.) The term equilibrium is used here in reaction kinetic sense, i.e., that there exists a fixed relation between the concentrations of bound and dissociated molecules.

The potential used by Rice for the interaction of atoms A and B is given in Fig. 4-3. It is essentially the effective potential for the molecule AB (cf. Fig. 3-4) which is made infinite at $r = r_\infty$; it is assumed that $r_\infty \gg r_m$. This rise to infinity[11] is a device to convert the continuum of translational states of A + B into discrete states and to prevent atoms A and B from going off to infinity and bring them back together eventually, so that we may have equilibrium between bound molecules and dissociated molecules. The actual value of r_∞ is immaterial as it will drop out in the final expressions.

The quantized energy of relative motion of A + B in the region $0 \leqslant r \leqslant \infty$

[11] See also Macomber and Colvin (*14*).

is given by the usual formula for the energy of a particle in a one-dimensional box:

$$\varepsilon_t = \tilde{n}^2 h^2 / 8 r_\infty^2 \mu \tag{4-22}$$

where μ is the reduced mass of A and B, $\tilde{n}$ is the translational quantum number. The zero of ε_t is the asymptote at r_∞ in Fig. 4-3. Obviously, no relative motion of A + B can occur unless $\varepsilon_t > 0$, that is, unless the molecule AB has at least energy E_0 in excess of ground-state energy. Thus the energy criterion for reaction is preserved. The density of translational states at ε_t is (Chapter 6, Exercise 1)

$$N(\varepsilon_t) = d\tilde{n}/d\varepsilon_t = (r_\infty/h)(2\mu/\varepsilon_t)^{1/2} \tag{4-23}$$

Now the energy corresponding to ε_t is all kinetic, $\varepsilon_t = \frac{1}{2}\mu u^2$, so that the velocity u of relative motion of A + B is

$$u = (2\varepsilon_t/\mu)^{1/2} \tag{4-24}$$

The time required for a (dissociating) pair A + B to increase their relative distance from r_m to r_∞ and back is $2(r_\infty - r_m)/u$, during which time we shall find AB (i.e., A + B with $r = r_m$) once while their mutual distance is increasing. Therefore the reciprocal of this time is the rate constant (k_{ε_t}) for dissociation of a molecule AB with excitation energy $E = E_0 + \varepsilon_t$, yielding products A + B in a state with relative translational energy ε_t:

$$k_{\varepsilon_t} = u/2r_\infty = (2\varepsilon_t/\mu)^{1/2}/2r_\infty \tag{4-25}$$

Because of the energy–time uncertainty relation, the energy of the decomposing molecule AB can be specified only within an interval δE; therefore the rate constant for dissociation of AB is meaningful only over a range of energies δE. The number of translational states in δE is $(d\tilde{n}/d\varepsilon_t)\,\delta E$, and for each of these the rate is k_{ε_t}; thus the contribution to the rate over the range δE is, from (4-23) and (4-25),

$$k(E) = (d\tilde{n}/d\varepsilon_t)\,\delta E\, k_{\varepsilon_t} = \delta E/h \tag{4-26}$$

Equation (4-26) refers to one translational quasicontinuum of the fragments A and B (the only continuum there is), i.e., to one product channel. The states of AB are nondegenerate, and therefore δE contains just one state[12] of AB; hence in terms of the discussion of Chapter 1, Section 3, we recognize (4-26) as the equation for the "elementary" rate constant k_{ij}. Thus $k_{ij} = \delta E/h$, which is a constant independent of E, i, and j, and so bears out the assumption made in Chapter 1.

[12] Since we mean the state above the dissociation energy of AB, the state is only "virtual," i.e., the state that would exist if the molecule did not dissociate. One consequence of its being virtual is that in the reverse process of association, A + B will never form AB in the absence of a third body.

In quantum-scattering theory (*15*, p. 450, Eq. 117) the probability of dissociation from one state into one channel is given by $\Gamma/\hbar$, where Γ is the energy width of the state, i.e., δE in our notation. Thus (4-26) is essentially the analog of the quantum result.[13]

If AB is represented by an oscillator of frequency ν, its vibrational states are spaced by the amount $h\nu$. Hence one state is contained in the interval $h\nu$, i.e., $\delta E = h\nu$. Thus

$$k(E) = \nu \tag{4-27}$$

If the oscillator is harmonic, ν is constant, so that the (hypothetical) rate constant for unimolecular dissociation of one harmonic oscillator is independent of energy.

The polyatomic case Suppose now that AB represents a polyatomic molecule which dissociates into fragments A and B, one or both of which are polyatomic and, as a result, can exist in internal energy states. Whether the reaction is of type 1, 2, or 3, the potential between A and B will again look very much like in Fig. 4-3 [cf. Figs. 3-2(b) and 3-3(c)]. The basic argument thus remains the same, except that there is no longer just one state of AB in δE, but $N(E)\,\delta E$ states, where $N(E)$ is the density of states of AB at E. Consider now δE in the vicinity of E_0, i.e., when the internal energy of AB just exceeds its effective dissociation energy E_0. Equation (4-26) then becomes, remembering that the rate constant expresses rate *per* initial state (Chapter 1, Section 3)

$$k(E_0) = \delta E/[hN(E_0)\,\delta E] = 1/hN(E_0) \tag{4-28}$$

Equation (4-28) represents the rate of dissociation of a polyatomic molecule at threshold, i.e., into one (the only one) continuum or channel available, as there is not enough energy for internal excitation of fragments. When $E > E_0$ the relation

$$k^1(E) = 1/hN(E) \tag{4-29}$$

represents the rate constant with which a polyatomic AB with internal excitation energy E would dissociate into one channel; it is thus the equivalent of k_{ij}/b [Eq. (1-11)].

Since in general $N(E)\,\delta E > 1$, we see that Eq. (4-26) gives the *maximum* rate of dissociation into one product channel. For a harmonic oscillator potential between A and B this maximum rate is the oscillator frequency.

However, when the energy E of a polyatomic AB exceeds E_0 by a finite amount the difference $E - E_0$ is in general large enough to raise the products

[13] Actually $\Gamma/\hbar$ differs from (4-26) by the factor 2π. The origin of this factor has been discussed by Rice (*16,17*).

(A + B) into one or more excited internal states. The energy $E - E_0$ is thus shared between the internal states of the fragments (A + B) and the continuum of translational states representing the relative motion of A with respect to B, i.e., there will be one continuum (or channel) associated with each internal state of the products. Equation (4-29) gives the rate for one channel, and the rate into all channels, which is the information of interest, is obtained by multiplying $k^1(E)$ of Eq. (4-29) by the number of available channels. This is given by the number of states of the products in the interval 0, $E - E_0$, or, what amounts to the same thing, by the number of ways the energy $E - E_0$ can be distributed between the internal states of (A + B) and the continuum.

The difficulty is, of course, that the number, or even the nature, of the product states is not known. The transition state is a device by means of which the enumeration of the largely unknown product states is replaced by the equivalent process of enumeration of states of the transition state. Once the transition state is set up, the nature of its states is known, and since these are equilibrium states, *every*[14] state in the interval 0, $E - E_0$ will be present with equal probability, so that enumeration can be done by the usual methods of equilibrium statistical mechanics. If there are $G^*(E - E_0)$ states of the transition state in the interval 0, $E - E_0$, this is also the number of channels available for dissociation, and the rate constant into all channels is

$$k(E) = k^1(E) \times G^*(E - E_0) = G^*(E - E_0)/hN(E) \qquad (4\text{-}30)$$

a result obtained before [Eq. (4-21)]. For ramifications, see Exercise 2.

It is instructive to take Eq. (4-30) apart and interpret it in terms of the averaging process described in Chapter 1, Section 3. The "elementary" rate k_{ij} is $\delta E/h$, the summation of $f(i)$ over i is $\sum f(i) = ak_{ij} = (\delta E/h) \times G^*(E - E_0)$. Finally, the probability $P(i) = 1/[N(E)\,\delta E] = \text{const}$, as of course it must for a microcanonical ensemble, where every member (= state) is realized with equal probability.[15]

We have seen in Chapter 1, Section 4 that a unique unimolecular rate

[14] Note that every *product* state in the interval 0, $E - E_0$ will be present only when the reaction has come to equilibrium, at which point we are usually no longer interested in the rate. The discussion is not meant to suggest that states of the transition state and product states are necessarily similar in number and kind; however, if the transition state is "loose" (Chapter 11, Section 3), such a similarity exists. In such a case we can actually calculate the rate by counting all those product states not disallowed by momentum conservation requirements (Chapter 5); we have then a unimolecular version of the phase-space theory [cf. Klots (*18*) and Chapter 10, Section 9].

[15] The equation $P(i) = [N(E)\,\delta E]^{-1}$ gives the probability of state i as the reciprocal of the total number of states contained within a thickness δE of the hypersurface $E = \text{const}$; it is therefore the quantum version of $d\Omega/\text{DEN}$ [cf. Eq. (4-11)].

constant is the reciprocal of the average lifetime of a molecule. On writing $1/k^1(E) = \tau^1(E)$, Eq. (4-29) becomes

$$\tau^1(E) \times 1/N(E) = h \tag{4-31}$$

where $\tau^1(E)$ is the average lifetime of a molecule with energy E decomposing into one channel. Rice has observed that Eq. (4-31) has the form of the usual uncertainty relation, whereby the product of the time of observation [$\tau^1(E)$ in this instance] and the uncertainty in energy is of the order of h. However, the uncertainty in energy is, in Eq. (4-31), the inverse of the energy level density at E, i.e., the average spacing of energy levels at E. Therefore the energy levels are just enough broadened to overlap slightly. This means that in the reverse reaction of association $\mathrm{A} + \mathrm{B} \rightarrow \mathrm{AB}$ there is no difficulty of matching the energy levels of $(\mathrm{A} + \mathrm{B})$ with those of AB, and the fragments will recombine at every collision if there is no activation energy for association.

Finally, we can make a connection between the developments of Sections 2 and 3 by observing that $\delta\varepsilon_t$ in (4-17) is in fact δE in disguise, so that Eqs. (4-17) and (4-26) are analogs of each other. We can see that $\langle\Phi\rangle$ is in fact an elementary flux, i.e., k_{ij}.

4. Nonclassical effects near potential barrier maximum

The quantum mechanical considerations that were introduced into the derivation of $k(E)$ in Section 2 of this chapter consist principally of replacing volumes in phase space by the density of states. However, the essentially classical criterion of reaction, i.e., that all particles with $E \leqslant E_0$ have zero probability of reaction and those with $E \geqslant E_0$ have unit probability of crossing the critical surface and ultimately yielding products, was left untouched. This probability will now be examined.

It is known from elementary quantum mechanics of one-dimensional systems [e.g. (*19*, pp. 102 ff.)] that there is a finite probability for a particle to pass (tunnel) *through* a potential barrier even if it has energy insufficient to surmount the barrier classically, and also that there is a probability of less than unity for a particle to pass *over* a barrier even if classically it has sufficient energy to do so. These nonclassical effects depend on the mass of the particle and on the relative change of potential within one de Broglie wavelength ("steepness" of potential), and therefore are most pronounced for a discontinuous potential such as the familiar rectangular barrier.

Insofar as the passage *over* a barrier is concerned, the quantum effect amounts to a partial reflection of an incoming wave packet; in the present context of a particle crossing the critical surface located at the top of the potential energy barrier separating reactants from products, the result is that the particle reverts to reactant, although it has (classically) sufficient energy to react. The passage *through* a barrier corresponds to formation of product from reactant having (classically) insufficient energy to react. Consequently the probability of reaction in a quantum system becomes a function of energy[16] such that it goes from zero to one over a range of energies, rather than changing abruptly from zero to one at E_0, as in the classical system; the criterion for reaction therefore becomes blurred.

To correct for these nonclassical effects, the rate constant $k(E)$ [Eq. (4-21)] should be multiplied by an energy-dependent correction factor. For $E \geqslant E_0$, this factor is commonly referred to as the "transmission coefficient" and for $E \leqslant E_0$ as the "tunneling correction." The reason that two different names are used for what is essentially the same effect seems to be that some authors [e.g., Laidler (*20*)] tend to interpret the transmission coefficient in a wider sense as including also a correction for any basic defect of the theory that might make the calculated rate too large, although usually no recipe is given how such a correction might be calculated. Lumped together in a transmission coefficient is sometimes also a correction for the crossing of potential energy surfaces in nonadiabatic reactions (Chapter 3, Section 3). The tunneling correction,[17] on the other hand, is usually interpreted to mean exactly what the name implies.

The transmission coefficient We shall use a single term "transmission coefficient" in the more restricted sense as correction factor for quantum effects associated with passage both *through* and *over* a potential barrier, a definition which has the virtue that a transmission coefficient so defined can be calculated (*22–24*). Let us denote this transmission coefficient as $\kappa(E)$, and recall that it refers to the unidimensional passage, along the reaction coordinate, of a particle of effective mass μ^* through or over a barrier of height E_0. If the barrier is assumed to be a truncated parabola of width $2d$ at the base, the transmission coefficient is approximately (*25,26*)

$$\kappa(E) \sim \{1 + \exp[\mathrm{D}(1 - E/E_0)]\}^{-1} \tag{4-32}$$

where $\mathrm{D} = \pi d(2\mu^* E_0)^{1/2}/\hbar$. Figure 4-4 shows $\kappa(E)$ of Eq. (4-32) as a function of E/E_0 for several values of D; note that at $E = E_0$ (reaction threshold), $\kappa(E) = \frac{1}{2}$. Expressions are also available giving $\kappa(E)$ for a symmetric (*27*) and asymmetric (*28*)[18] Eckart barrier (*30*), and the dependence of $\kappa(E)$ on

[16] Because of resonances, it is actually an oscillating function of energy. Cf. Merzbacher (*19*, pp. 102 ff.).

[17] For a review, see Caldin (*21*).

[18] For recent calculations see Truhlar and Kuppermann (*29*).

E/E_0 in these cases is similar to that shown in Fig. 4-4. On introducing $\kappa(E)$ into the expression for the microcanonical rate constant, we have

$$k(E) = \kappa(E)[G^*(E - E_0)/hN(E)] \tag{4-33}$$

where E now may take values below, as well as above, E_0. In the former case, $E - E_0$ is negative and we have to adopt the convention that $G^*(-y) = 1$ for all y; such tunneling calculations have been done (*31*). However, using $G^*(-y) = 1$ is an unsatisfactory formalism because Eq. (4-21), on which (4-33) is based, is really a semi classical expression that makes no allowance for the possibility of reaction at energies below E_0. Therefore (4-33) is usable only for $E \geqslant E_0$, i.e., for calculating a possible reflection *above* the barrier. Reflection reduces the rate just above threshold, and may account for some aspects of the reaction $CH_4^+ \rightarrow CH_3^+ + H$ observed in the mass spectrum of methane (*32*). To account properly for nonclassical effects associated with the crossing of the potential energy barrier at energies *below* as well as above the barrier height, we would have to use a nonclassical treatment of the reaction coordinate, but this seems hardly worthwhile, because in most cases of interest the constant D in (4-32) is so large as to make the correction factor $\kappa(E)$ unnecessary, as the following calculation shows.

If we take $\mu^* = 30$ amu, $E_0 = 85$ kcal mole^{-1}, $d = 3$ Å, which are numbers that would roughly apply to the reaction $C_2H_6 \rightarrow 2CH_3$, it turns out that D $\sim 2 \times 10^3$, and from Fig. 4-4 it may be inferred that $\kappa(E)$ for such a large D would behave for all practical purposes as the classical step function.

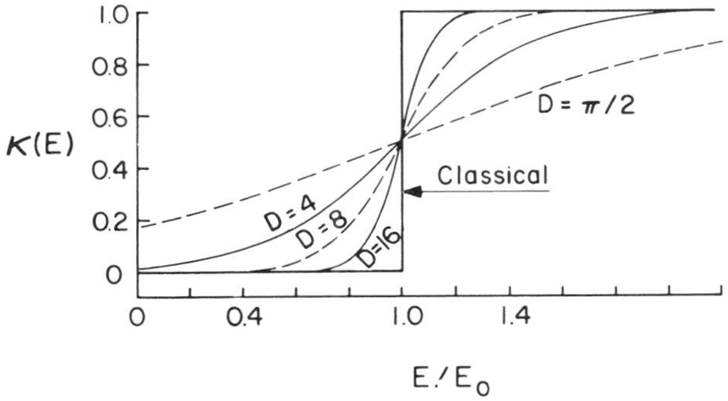

Fig. 4-4. *Transmission coefficient $\kappa(E)$ [Eq. (4-32)] as a function of E/E_0 for a parabolic potential barrier.*

E is the energy of the incident particle of reduced mass μ^*, E_0 is the height of the barrier, and $\text{D} = \pi d(2\mu^* E_0)^{1/2}/\hbar$, where d is essentially the thickness of the barrier. The larger the constant D, the more classical the behavior of the system. For a fully classical system, $\kappa(E)$ becomes a step function: $\kappa(E) = 0$ for $E \leqslant E_0$, $\kappa(E) = 1$ for $E \geqslant E_0$. After H. S. Johnston and D. Rapp, *J. Amer. Chem. Soc.* **83**, 1 (1961).

Even if $2\mu^* E_0$ were ten times smaller than in the case considered, D would still be $\sim 7 \times 10^2$, which is large enough to make again $\kappa(E)$ very nearly classical.

It is therefore fairly obvious that the importance of $\kappa(E)$ is limited to the immediate vicinity of threshold and only for reactions where μ^* or E_0 or both are small.[19] However even in such cases, calculations of $\kappa(E)$ make sense only if E_0 is known fairly accurately and if the curvature of the potential energy surface, which determines d (or similar constants in the Eckart potential), is also known. This is rarely, if ever, the case, and therefore we shall henceforth assume that $\kappa(E) = 1$ for $E \geqslant E_0$, which seems preferable to a dubious manipulation of d or μ^*.

One special case where calculation of $\kappa(E)$ might have some merit is where *relative* rates of isotopic species are involved, since it can be supposed that the only variables are μ^* and the part of E_0 attributable to the difference in zero-point energies.

5. Reaction path degeneracy

As mentioned in Chapter 3, Section 4, the full multidimensional potential energy surface may have a symmetry such that there are α equivalent paths leading from reactant to products, and therefore the rate constant $k(E)$ in Eq. (4-21), which represents the single-path rate, must be multiplied by α. The factor α, which arises ultimately from the symmetry properties of the reacting species, is variously referred to as "statistical factor" (*34*) or "reaction path degeneracy or multiplicity" (*35*). We shall use the term "reaction path degeneracy," which seems to have greater currency.

In transition-state theory, rate is determined only by the rate of forming the transition state, so that reaction path degeneracy α refers to the number of equivalent reaction paths leading from reactant to transition state and therefore depends on the symmetry properties of *both* the reactant and the transition state. "Reaction path" is not meant here in the literal sense of an actual trajectory over the potential energy surface but refers merely to one particular (abstract) way of producing a given transition state from the atoms of the reactant.

As a simple example, let us take methane and the reaction $CH_4 \rightarrow CH_3 + H$. Because the four hydrogens of methane are all equivalent and indistinguishable,[20] the potential energy surface for the reaction will exhibit one potential

[19] Hisatsune (*33*), using an Eckart barrier, finds that in the unimolecular cis-trans isomerization of nitrous acid ($E_0 \sim 10$ kcal mole^{-1}), $\kappa(E)$ goes from zero to one between $0.9E_0$ and $1.1E_0$.

[20] Unless isotopic substitution makes them distinguishable; cf. Chapter 11, Section 2.

minimum on the reactant side (for CH_4) and four equivalent minima on the product side, one for each of the four equivalent sets of products $CH_3 + H$, since in principle any one of the four identical hydrogens can become detached. Each of the product minima will be separated from the reactant minimum by an energy barrier identical to the other three barriers, and at the top of each barrier will be located a transition state, indistinguishable from the other three transition states. There are thus four identical reaction paths leading from reactant to products, or, what amounts to the same thing, there are four ways of forming the transition state; thus $\alpha = 4$.

In more complicated reactants, however, the determination of α is not nearly as simple as in the H-atom elimination from methane. In general the simplest and surest procedure is to obtain α by a direct enumeration of the number of ways of forming a given transition state. A simple rule for such an enumeration, based on group-theoretical arguments and adapted here for unimolecular reactions, has been given by Schlag and Haller (*36*).

(1) Number all atoms in reactant and count the number of transition states (n) that can be produced as a consequence of reaction and that are not related to one another by simple rotation. All such transition states must be superimposable[21] if the numbering on the atoms is removed.

(2) Determine whether reactant or transition state, or both, are asymmetric (can exist in two enantiomeric forms, i.e., have distinguishable mirror images). (i) If reactant and transition state are both either symmetric or asymmetric, $\alpha = n$. (ii) If reactant is symmetric and transition state is asymmetric, $\alpha = 2n$, since in (1) only one enantiomer was counted and both can be formed; (iii) If reactant is asymmetric and transition state is symmetric, $\alpha = n/2$, since both were counted in (1).

As an example, we might consider the reaction $C_2H_5Cl \rightarrow C_2H_4 + HCl$ which is believed (*38*) to proceed through a four-center transition state:

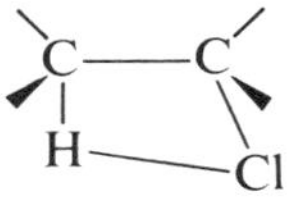

If the internal torsion of C_2H_5Cl is treated as torsional vibration, then assuming a staggered configuration for the molecule, there are two equivalent hydrogens on the CH_3 group that can be placed in the C—C—Cl—H ring, and therefore $n = 2$. If the torsion is treated as internal rotation, all three hydrogens of the CH_3 group become equivalent, and therefore $n = 3$. In

[21] Marcus (*37*) considers the possibility that, though superimposable, such transition states might have different $G^*(E - E_0)$ values, and therefore writes his total rate as a *sum* over geometrically isomeric paths rather than as a multiple of the single-path rate. It seems doubtful that it is possible in practice to make such a fine distinction among geometrically isomeric transition states as to assign them distinct integrated densities of states.

addition, if the transition state is assumed to be planar, then both reactant and transition state are symmetric, so that $\alpha = 2$ or 3, depending on the treatment of internal torsion. If the transition state is assumed to be nonplanar due to the ring H atom being out of plane, two mirror images of the transition state can exist, one with the ring hydrogen above the plane of the ring, and the other below the ring, so that $\alpha = 2n$.

The symmetry number Although the reaction path degeneracy α is related to the so-called symmetry numbers σ and σ^* for the reactant and transition state, respectively, it must not be confused with them. The symmetry number σ may be defined (*39*) as the number of indistinguishable positions into which a molecule can be turned by simple rigid rotations, and therefore its inverse appears as a factor in the partition function for overall rotations, since it affects the available phase space for these rotations. The ultimate origin of σ is the requirement that the total molecular wave function, including nuclear spin, must be symmetric or antisymmetric with respect to exchange of identical nuclei, depending on whether the number of elementary particles in the nuclei is even or odd (*40*, p. 174). A tabulation exists (*41*, p. 508, Table 140) giving σ for each of the various point groups. The reaction path degeneracy α used to be written[22] as $\alpha = \sigma/\sigma^*$ until it was realized that α is not given by the ratio σ/σ^* except as a special case. Schlag (*44*) recognized two cases when the relation $\alpha = \sigma/\sigma^*$ fails: (a) when the transition state has several paths open for return to reactant ["type (a) failing"]; (b) when either reactant or transition state (but not both) contains a symmetry element other than rotation ["type (b) failing"].

As an example (*44,45*) we might consider the formation of cyclopentadiene by dehydrogenation of cyclopentene:

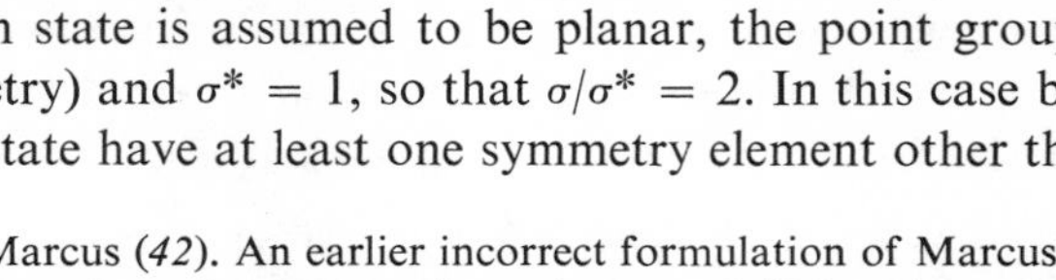

The point group of reactant cyclopentene is C_{2v} for which $\sigma = 2$; the symmetry elements are a twofold axis of rotation and two planes of symmetry. If the transition state is assumed to be planar, the point group is C_s (one plane of symmetry) and $\sigma^* = 1$, so that $\sigma/\sigma^* = 2$. In this case both reactant and transition state have at least one symmetry element other than rotation.

[22] Wieder and Marcus (*42*). An earlier incorrect formulation of Marcus (*43*) has been perpetuated by a number of authors, who write $\alpha = Q_r^*/Q_r$, where Q_r^* and Q_r are partition functions for adiabatic rotations (Chapter 5, Section 1) of transition state and reactant, respectively. The α so defined involves, in addition to the ratio σ/σ^*, also the ratio of moments of inertia of transition state and reactant.

Presumably the H_2 bridge between carbons 3 and 4 is formed from the two hydrogens that are closest to each other. Which two depends on how we imagine the hydrogens of the cyclopentene to go planar; several ways can be imagined, none of which really looks very plausible, showing that the H_2 bridge could be formed in one, or two, or perhaps even four different ways, the latter presenting complete scrambling of the hydrogens; direct count thus gives $\alpha = 2$, 4, or 8, since the second double bond in cyclopentene may be formed not only across carbons 3 and 4 but also across carbons 4 and 5, with one, two, or four ways of forming the H_2 bridge at each location. When we suppose that there is more than one way of forming the H_2 bridge at a given location, there is more than one path for the transition state to return to reactants, so that we have a "type (a) failing" and the direct count then disagrees with σ/σ^*.

If we assume that the transition state is nonplanar, with the H_2 bridge bent out of the plane of the carbon ring, the point group is C_1 (no symmetry) and $\sigma^* = 1$, so that $\sigma/\sigma^* = 2$, as before. However, this time only the reactant contains a symmetry element other than rotation, and therefore we have a "type (b) failing." In fact, direct count gives $\alpha = 4$, on the assumption that the H_2 bridge is formed from *cis*-hydrogens only; there is then just one path for the transition state to return to reactant.

The cited example $CH_4 \rightarrow CH_3 + H$ is a case with neither type (a) or type (b) failing. The reactant CH_4 belongs to point group T_d, for which $\sigma = 12$. If the transition state is assumed to have the structure $CH_3 \ldots H$, its point group is C_{3v} and $\sigma^* = 3$, so that $\alpha = \sigma/\sigma^* = 4$, as obtained by direct count.

It must be noted that because α depends on reaction path, it cannot appear in the equilibrium constant, whereas σ for reactant (and products) does since it represents an intrinsic property. This has led to a controversy (*46–48*) whether or not α must be given by the ratio of symmetry numbers after all, the consequence being that this would impose restrictions on the allowed symmetries of transition states.

It seemed for a while that such restrictions could be derived from the Woodward–Hoffmann symmetry rules (*49*). However, it takes only a very slight deviation from a given symmetry to destroy the symmetry argument, and it has now been realized that, except in special cases, it is unlikely that a given symmetry will be strictly conserved in the course of a chemical reaction, for even the vibrational motion of the nuclei may be sufficient to destroy the symmetry of the transition state (*50*).

Pearson (*51*) has recently given a somewhat similar symmetry rule based on perturbation theory. In this formulation, the only reactions where symmetry considerations are of any help in determining the symmetry of the transition state are type 2 reactions involving complex bond-breaking in a concerted process (cf. Chapter 11, Section 3). It then turns out that, according to Pearson's rule, the dehydrogenation of cyclopentene mentioned above, a process entirely analogous to the dehydrogenation of ethene to form ethylene, which he discusses in some detail, would actually be forbidden as a gas-phase reaction.

One should bear in mind that the purpose of symmetry rules is merely to exploit the symmetry properties of reactants and products for the discovery of the lowest-energy reaction path from reactants to products. The ultimate criterion for the feasibility of a given reaction remains the energy expenditure along the given reaction path, i.e. the critical energy E_0, and its associated transition state. There is some doubt whether symmetry rules can always pick out the transition state with the lowest E_0, for in at least one case *(52)* orbital symmetry alone appears to be insufficient to determine whether, among several possible transition states, less symmetrical geometries have lower energy. In any event, highly symmetric transition states are suspect *(54)*.

The original problem, whether α must be equal to σ/σ^* or not, therefore appears to be unresolved at present. We shall always use α exclusively for reaction path degeneracy.

The most general formulation of $k(E)$ [Eq. (4-21)], which includes α but neglects factors discussed in Section 4, is thus

$$k(E) = \alpha G^*(E - E_0)/hN(E) \tag{4-34}$$

It should be realized that the need for the reaction path degeneracy factor arises from the neglect of part of the nuclear dynamics, so that if our microcanonical rate constant expressed the nuclear dynamics properly, we would not have to correct it by multiplication with α. The need for α would disappear, for example, if we used a potential energy function that adequately expressed the symmetry properties of the reactant molecule,[23] but it is then almost certain that the microcanonical rate constant derived on such a basis would not have the simple form of (4-21).

Our immediate concern now is how to calculate $G^*(E - E_0)$ and $N(E)$ in expression (4-34). This is the subject of the following two chapters.

Exercises

1. It can be shown (Chapter 6, Section 7) that for a collection of v classical oscillators, $G(E)$ and $N(E)$ are proportional to E^{v} and $E^{\mathsf{v}-1}$, respectively. Use Eq. (6-55) to show that Eq. (4-21) then yields

$$k(E) = \frac{\prod_{i=1}^{\mathsf{v}} \nu_i}{\prod_{i=1}^{\mathsf{v}-1} \nu_i^*} \left(\frac{E - E_0}{E}\right)^{\mathsf{v}-1}$$

[23] See, for example, Pickett and Strauss *(53)*. The author is grateful to Prof. Strauss for correspondence on this matter.

where ν_i are the vibrational frequencies of the reactant and ν_i^* are those of the transition state. This is the classical Kassel form [Eq. (2-2)].

2. In the spirit of Section 2, $\langle\Phi\rangle$ of Eq. (4-17) can be interpreted as $[(u/\delta q_1)$ = unidimensional flux along δq_1 per translational state at $\varepsilon_t] \times [N(\varepsilon_t)\, d\varepsilon_t$ = number of translational states in δq_1 within $d\varepsilon_t$ at $\varepsilon_t]$. The total flux over the range $d\varepsilon_t$ at E' is then $\alpha \int_0^{E'} N^*(E' - \varepsilon_t)\, d\varepsilon_t \langle\Phi\rangle = \alpha G^*(E')\, d\varepsilon_t/h$, where $E' = E - E_0$. Use this approach to calculate the total flux through the critical surface in the *reverse* direction, i.e. for the *association of fragments*. Show that if V = reaction volume, μ = reduced mass of fragments, v = their relative velocity, $E_t = \mu v^2/2$, and $\sigma(E', E_t)$ = cross section for association of fragments with E_t at total energy E', the total reverse flux over the range dE_t at E' is

$$\int_0^{E'} N^f(E' - E_t)\, dE_t \times (v/V)\, \sigma(E', E_t) \times N(E_t)\, dE_t, \qquad \text{(i)}$$

i.e. integral of [number of internal states of separated fragments (index f) within dE_t at $E' - E_t$] $\times$ [reverse flux from volume V at relative velocity v] $\times$ [number of translational states in V within dE_t at E_t]. Over the same range $d\varepsilon_t = dE_t$ at E' the forward and reverse fluxes must be equal by virtue of microscopic reversibility, and therefore

$$G^*(E') = h(2/\mu)^{1/2} V \int_0^{E'} N^f(E' - E_t)\, (E_t)^{1/2}\, \sigma(E', E_t)\, N(E_t)\, dE_t \qquad \text{(ii)}$$

This is an expression for $G^*(E')$ in terms of the properties of the separated fragments, valid if the transition state is "loose" [cf. (*18*) and (*54*)]. For further evaluation of (ii) see Exercise 4, Chapter 6.

References

1. A. J. Lichtenberg, "Phase-Space Dynamics of Particles." Wiley, New York, 1969.
2. J. C. Keck, *Advan. Chem. Phys.* **13**, 85 (1966).
3. J. Keck and A. Kalelkar, *J. Chem. Phys.* **49**, 3211 (1968).
4. E. Butkov, "Mathematical Physics." Addison-Wesley, Reading, Massachusetts, 1968.
5. G. A. Korn and T. M. Korn, "Mathematical Handbook for Scientists and Engineers." McGraw–Hill, New York, 1961.
6. O. K. Rice, "Statistical Mechanics, Thermodynamics, and Kinetics." Freeman, San Francisco, California, 1967.
7. H. S. Johnston, "Gas Phase Reaction Rate Theory." Ronald Press, New York, 1966.
8. F. C. Andrews, "Equilibrium Statistical Mechanics." Wiley, New York, 1963.
9. R. P. Bell, *Trans. Faraday Soc.* **66**, 2770 (1970).
10. R. A. Marcus, *J. Chem. Phys.* **20**, 359 (1952).
11. H. M. Rosenstock, M. B. Wallenstein, A. L. Wahrhaftig, and H. Eyring, *Proc. Nat. Acad. Sci. U. S.* **38**, 667 (1952).
12. N. Bohr and J. A. Wheeler, *Phys. Rev.* **56**, 426 (1939).
13. O. K. Rice and H. Gershinowitz, *J. Chem. Phys.* **2**, 853 (1934).
14. J. D. Macomber and C. Colvin, *Int. J. Chem. Kinet.* **1**, 483 (1969).
15. M. Goldberger and K. M. Watson, "Collision Theory." Wiley, New York, 1964.
16. O. K. Rice, *J. Phys. Chem.* **65**, 1588 (1961).
17. O. K. Rice, *J. Chem. Phys.* **55**, 439 (1971).

18. C. E. Klots, *J. Phys. Chem.* **75**, 1526 (1971).
19. E. Merzbacher, "Quantum Mechanics." Wiley, New York, 1961.
20. K. J. Laidler, "Theories of Chemical Reaction Rates." McGraw–Hill, New York, 1969, p. 56.
21. E. F. Caldin, *Chem. Rev.* **69**, 135 (1969).
22. J. O. Hirschfelder and E. Wigner, *J. Chem. Phys.* **7**, 616 (1939).
23. H. M. Hulburt and J. O. Hirschfelder, *J. Chem. Phys.* **11**, 276 (1943).
24. R. A. Marcus, *J. Chem. Phys.* **43**, 1598 (1965).
25. R. P. Bell, *Trans. Faraday Soc.* **55**, 1 (1959).
26. K. W. Ford, D. L. Hill, M. Wakano, and J. A. Wheeler, *Ann. Phys.* **7**, 239 (1959).
27. I. Shavitt, *J. Chem. Phys.* **31**, 1359 (1959).
28. H. S. Johnston and D. Rapp, *J. Amer. Chem. Soc.* **83**, 1 (1961).
29. D. G. Truhlar and A. Kuppermann, *J. Amer. Chem. Soc.* **93**, 1840 (1971).
30. C. Eckart, *Phys. Rev.* **35**, 1303 (1930).
31. H. Heydtmann, *Ber. Bunsenges.* **72**, 1009 (1968).
32. L. P. Hills, M. L. Vestal, and J. H. Futrell, *J. Chem. Phys.* **54**, 3834 (1971).
33. I. C. Hisatsune, *J. Phys. Chem.* **72**, 269 (1968).
34. D. M. Bishop and K. J. Laidler, *J. Chem. Phys.* **42**, 1688 (1965).
35. E. W. Schlag, *J. Chem. Phys.* **38**, 2480 (1963).
36. E. W. Schlag and G. L. Haller, *J. Chem. Phys.* **42**, 584 (1965).
37. R. A. Marcus, *J. Chem. Phys.* **43**, 2658 (1965).
38. J. C. Hassler and D. W. Setser, *J. Chem. Phys.* **45**, 3246 (1966).
39. E. B. Wilson, Jr., *Chem. Rev.* **27**, 17 (1940).
40. J. E. Mayer and M. G. Mayer, "Statistical Mechanics." Wiley, New York, 1940.
41. G. Herzberg, "Molecular Spectra and Molecular Structure," Vol. II, Infrared and Raman Spectra of Polyatomic Molecules. Van Nostrand Reinhold, Princeton, New Jersey, 1945.
42. G. M. Wieder and R. A. Marcus, *J. Chem. Phys.* **37**, 1835 (1962).
43. R. A. Marcus, *J. Chem. Phys.* **20**, 359 (1952).
44. E. W. Schlag, *J. Chem. Phys.* **38**, 2480 (1963).
45. D. M. Bishop and K. J. Laidler, *J. Chem. Phys.* **42**, 1688 (1965).
46. J. N. Murrell and K. J. Laidler, *Trans. Faraday Soc.* **64**, 371 (1967).
47. J. N. Murrell and G. L. Pratt, *Trans. Faraday Soc.* **66**, 1680 (1970).
48. D. M. Bishop and K. J. Laidler, *Trans. Faraday Soc.* **66**, 1685 (1970).
49. R. B. Woodward and R. Hoffmann, "The Conservation of Orbital Symmetry." Verlag Chemie, GmbH, Weinheim, and Academic Press, New York and London, 1970.
50. T. F. George and J. Ross, *J. Chem. Phys.* **55**, 3851 (1971).
51. R. G. Pearson, *J. Amer. Chem. Soc.* **94**, 8287 (1972).
52. L. M. Raff, D. L. Thompson, L. B. Sims, and R. N. Porter, *J. Chem. Phys.* **56**, 5998 (1972).
53. H. M. Pickett and H. L. Strauss, *J. Amer. Chem. Soc.* **92**, 728 (1970).
54. K. J. Laidler, *in* "Reaction Transition States," (J. E. Dubois, ed.) p. 23. Gordon and Breach, London, 1972.
55. R. A. Marcus, *J. Chem. Phys.* **45**, 2138 (1966).

CHAPTER 5

Pertinent Degrees of Freedom

A general expression has been derived in the preceding chapter for the unimolecular rate constant $k(E)$ [Eq. (4-34)] with the understanding that $G^*(E - E_0)$ and $N(E)$ are to be computed for "pertinent" degrees of freedom. The specification of these degrees of freedom will now be considered.

In essence, this chapter elaborates the general argument presented in Chapter 2, particularly Chapter 2, Section 5. Let us recall that "pertinent" degrees of freedom are those that are involved in reactive states, and therefore they may be defined as those degrees of freedom whose energy is available for randomization. In other words, these degrees of freedom are actively participating in intramolecular energy transfer, and for this reason Marcus *(1)*[1] has called them "active."

It is convenient to start, in Section 1, by eliminating first those degrees of freedom that are specifically excluded by conservation requirements. Internal degrees of freedom are considered in Section 2, where the assumption is justified that all internal degrees of freedom are "pertinent." The pertinence of external degrees of freedom is considered next in Section 3. Finally, on the basis of arguments presented in Sections 1–3, a model for the decomposing reactant molecule and its transition state is presented in Section 4.

[1] The reference list for this chapter can be found on p. 81.

1. Degrees of freedom excluded by conservation requirements

In transition-state theory, the process of reaction depends only on the formation of the transition state: Once it is formed, reaction is assured and subsequent events (e.g., the actual formation of fragments) are immaterial. Thus insofar as the kinematics of the process is concerned, there is just one initial "particle" (reactant) and just one final "particle" (transition state). The conditions assumed in the previous chapter, that is, time-independent potential and absence of external (or dissipative) forces, define a conservative system, in which the constants of motion are the total energy, linear momentum, and total angular momentum and its component along a space-fixed axis [see, e.g., (*2*, pp. 28–31, 242–243)]; these four must therefore be conserved in the course of forming the transition state from the reactant.

Conservation of total energy is implicit in the derivation of the basic equation (4-21) of the preceding chapter and is shown schematically in Fig. 4-1.

Linear momentum Conservation of linear momentum requires that $M\boldsymbol{v} = M^*\boldsymbol{v}^*$, where M and $\boldsymbol{v}$ are molecular mass and velocity, respectively, and quantities with asterisks refer as usual to the transition state. Since the molecular mass of the transition state is the same as that of the reactant, we have $M = M^*$, and therefore necessarily also $\boldsymbol{v} = \boldsymbol{v}^*$. The velocities $\boldsymbol{v}$ and $\boldsymbol{v}^*$ are vector quantities that refer to the translation in three dimensions of the center of gravity of the reactant and the transition state, i.e., represent their translation *as a whole*, and therefore $\boldsymbol{v}^*$ must not be confused with the velocity $\boldsymbol{u}$ of Chapter 4.

Overall translation Equality of masses and of translational velocities implies equality of translational energies E_t and E_t^* of the reactant and transition state, respectively; again E_t^* must be distinguished from ε_t, the one-dimensional translational energy in the reaction coordinate. The condition $E_t = E_t^*$ means that the three translational degrees of freedom, which we may call "overall" or "external," to distinguish them from the "internal" translation with energy ε_t, do not in any way partake in the reaction process and the energy E_t, whatever its value, therefore does not describe reactive states. Hence overall translations are not "pertinent" degrees of freedom.

Overall rotation We now turn to rotations of the "particle" (reactant or transition state) as a whole; such rotations shall be called "external" or "overall," to distinguish them from internal rotations to be considered later.

Quantum mechanically, the square of the total angular momentum $\mathscr{M}$ for overall rotations is given by

$$\mathscr{M}^2 = \mathscr{J}(\mathscr{J} + 1)\hbar^2 \tag{5-1}$$

where $\mathscr{J}$ is the rotational quantum number. The component of $\mathscr{M}$ along a space-fixed axis (say, c axis) is

$$\mathscr{M}_c = m\hbar^2 \tag{5-2}$$

where the quantum number m is restricted to $-\mathscr{J} \leqslant m \leqslant \mathscr{J}$. Conservation of angular momentum and its component requires that $\mathscr{M}$ and $\mathscr{M}_c$, and therefore $\mathscr{J}$ and m, remain the same throughout the process of forming the transition state from reactant. Since the rotational energy E_r also depends on $\mathscr{J}$ (but not on m), there is a restriction on E_r, and consequently also a restriction on the participation of degrees of freedom for overall rotation in intramolecular energy transfer. However, the expression for E_r is a function of the symmetry of the particle, so that each symmetry case must be discussed separately, although Eqs. (5-1) and (5-2) are valid regardless of symmetry (*3*).

Rigid rotor approximation In discussing the individual symmetry cases, we shall make the simplifying assumption that the particle behaves as a rigid rotor (negligible centrifugal distortion, no coupling with vibrations). Three principal axes of rotation may be defined, designated a, b, and c, and associated with these will be three principal moments of inertia,[2] I_a, I_b, and I_c, and the rotational constants $\mathbf{A} = h/8\pi^2 I_a$, $\mathbf{B} = h/8\pi^2 I_b$, $\mathbf{C} = h/8\pi^2 I_c$. A convenient point of departure is the symmetric top, from which most of the other symmetry cases can be obtained as special cases.

The symmetric top is a particle having two equal moments of inertia, with the third different from the other two. If the axis of the smallest moment of inertia (I_a) is also the symmetry axis, $I_a < I_b = I_c$ and the top is prolate; the energy is (*3*)

$$E_r = \mathscr{J}(\mathscr{J} + 1)\mathbf{B}h + \mathscr{K}^2(\mathbf{A} - \mathbf{B})h \tag{5-3}$$

If the symmetry axis is the c axis, $I_a = I_b < I_c$, the top is oblate and (*3*)

$$E_r = \mathscr{J}(\mathscr{J} + 1)\mathbf{B}h + \mathscr{K}^2(\mathbf{C} - \mathbf{B})h \tag{5-4}$$

$\mathscr{K}$ is a quantum number for component of angular momentum along a molecule-fixed axis, and is limited to the range $-\mathscr{J} \leqslant \mathscr{K} \leqslant \mathscr{J}$. A special

[2] See, for example, Slater and Frank (*4*, p. 100). The convention is that I_a represents the smallest moment of inertia, I_b the next larger, and so on.

case is the spherical top, with $I_a = I_b = I_c$, in which case (5-3) and (5-4) reduce to (*3*)

$$E_r = \mathscr{J}(\mathscr{J} + 1)\mathbf{B}h \tag{5-5}$$

The asymmetric top has all three moments of inertia different, $I_a < I_b < I_c$, and its energy is a complicated function of $\mathscr{J}$, **A**, **B**, and **C**, but lies somewhere between the energies of the corresponding prolate [Eq. (5-3)] and oblate [Eq. (5-4)] symmetric tops, depending on the degree of asymmetry (*5*, p. 86). The rotational energy of a linear (or diatomic) particle is given by an equation similar to Eq. (5-3), except that in this case $\mathscr{K}$ represents the component of *electronic* angular momentum along the molecular symmetry axis. If the particle is in $^1\Sigma$ electronic state (the most frequent case), $\mathscr{K} = 0$ and the rotational energy is again given by Eq. (5-5).

The process of interest is where a reactant molecule of known symmetry starts the decomposition process with a specified $\mathscr{J}$ and m, to yield a transition state of the same $\mathscr{J}$ and m but of a symmetry that can generally only be surmised because of incomplete information concerning the transition state. Fortunately, rotational energies are independent of m, and the $\mathscr{J}$-dependent part of the rotational energy in all symmetry cases has the same form, as can be easily seen from Eqs. (5-3)–(5-4). Thus the $\mathscr{J}$-dependent part of the rotational energy in the reactant is

$$E_r(\mathscr{J}) = \mathscr{J}(\mathscr{J} + 1)\mathbf{B}h \tag{5-6}$$

and in the transition state is

$$E_r^*(\mathscr{J}) = \mathscr{J}(\mathscr{J} + 1)\mathbf{B}^*h \tag{5-7}$$

where $\mathbf{B}^*$ is the rotational constant of the transition state. Recall that the rotational constant **B** is associated with moment of inertia I_b, and $I_b = I_c$ in the prolate top, $I_b = I_a$ in the oblate top, and $I_a = I_b = I_c$ in the spherical top. We can therefore state as a general rule that conservation of angular momentum affects only rotational degrees of freedom associated with two equal (or most nearly equal, if the top is asymmetric) moments of inertia of the reactant or transition state, except that in the unlikely case when they both are spherical tops, all three rotational degrees of freedom are involved.

If $\mathbf{B} = \mathbf{B}^*$, we have $E_r(\mathscr{J}) = E_r^*(\mathscr{J})$, so that the rotational energy is fixed, i.e., is unavailable for randomization and the two (three for spherical tops) rotational degrees of freedom are excluded from participating in intramolecular energy transfer. In general, as shown in Chapter 3, the transition state is a structure where at least one bond is stretched beyond its equilibrium value, so that we may expect $I_b^* > I_b$. Hence $\mathbf{B}^* < \mathbf{B}$ and therefore $E_r^*(\mathscr{J}) < E_r(\mathscr{J})$, and the energy difference

$$\Delta E_r(\mathscr{J}) = E_r(\mathscr{J}) - E_r^*(\mathscr{J}) = \mathscr{J}(\mathscr{J} + 1)h(\mathbf{B} - \mathbf{B}^*) \tag{5-8}$$

is available to drive the reaction,[3] not by increasing the energy (or the number of degrees of freedom) available for intramolecular energy transfer, but by decreasing the critical energy. This is explored more fully in Chapter 7, where a quasidiatomic model is used for the dissociation process to circumvent the problem that **B*** (like most transition-state parameters) is difficult to estimate.

The point concerning the quantum number $\mathscr{m}$ is somewhat more subtle (*7*). There are $2\mathscr{J} + 1$ values of $\mathscr{m}$ for a given $\mathscr{J}$, which accounts for $(2\mathscr{J} + 1)$-fold degeneracy[4] of each $E_r(\mathscr{J})$ and $E_r^*(\mathscr{J})$. If $\mathscr{m}$ is a "good" quantum number, in the sense that $\mathscr{m}$ is strictly conserved in the process reactant $\rightarrow$ transition state, *none* of the rotational states involving **B** or **B*** can participate in intramolecular energy transfer. If $\mathscr{m}$ is not a "good" quantum number and therefore is not conserved, there are $2\mathscr{J} + 1$ states available for randomizing $E_r(\mathscr{J})$ or $E_r^*(\mathscr{J})$ in reactant and transition state, respectively. Note that this does not make the two overall rotations "pertinent" degrees of freedom in the usual sense. It can be shown (*7*) that the consequence is that the same factor $2\mathscr{J} + 1$ appears in both $G^*(E - E_0)$ and $N(E)$, and since the rate constant $k(E)$ involves the ratio $G^*(E - E_0)/N(E)$, the factors $2\mathscr{J} + 1$ cancel out, so that the rate is the same as when $\mathscr{m}$ is a "good" quantum number.

Conclusions Unless mentioned otherwise, it will always be assumed for simplicity that $\mathscr{m}$ is a "good" quantum number. With this proviso, the principal conclusions may be stated as follows: Conservation requirements exclude from participation in intramolecular energy transfer all states belonging to the three degrees of freedom for overall translation, and all states of degrees of freedom for overall rotation associated with moments of inertia that are equal; this means two degrees of freedom for overall rotation in all cases except in the rather unlikely case that both reactant and transition state are spherical tops, in which case all three degrees of freedom for overall rotation are excluded. Rotations that are constrained to remain in the same quantum state throughout the decomposition have been termed "adiabatic" by Marcus (*1*).

It is useful to emphasize that the discussion in this chapter concerns only the question of whether or not states for overall rotations are to be included in $G^*(E - E_0)$ or $N(E)$. The answer, as we have seen, is a qualified no, but this does not mean that overall rotations are without influence on the *rate*.

[3] Computer results of Hung and Wilson (*6*) obtained on a classical model triatomic system with Morse potentials show that even a small $\Delta E_r(\mathscr{J})$ (our notation) has an appreciable influence on $k(E)$ near threshold.

[4] The degeneracy of $E_r(\mathscr{J})$ for a spherical top is $(2\mathscr{J} + 1)^2$ since any space-fixed axis is also a molecule-fixed symmetry axis and each contributes a degeneracy factor $(2\mathscr{J} + 1)$.

In fact, it turns out (Chapter 7) that the rate constant does depend on the quantum number $\mathcal{J}$ for adiabatic rotations, but not through $G^*(E - E_0)$ or $N(E)$ [cf. comments following Eq. (5-8)].

2. Internal degrees of freedom

A particle with N atoms has $3N - 6$ internal degrees of freedom if it is nonlinear and $3N - 5$ internal degrees of freedom if it is linear; in principle, these degrees of freedom are vibrational. The evidence, both experimental and theoretical, in favor of vibrations participating in intramolecular energy transfer has been discussed in Chapter 2, Sections 4 and 5. The conclusion was that experiment provides strong circumstantial evidence for the randomization of energy among most, if not all, vibrations, and that such a randomization can be reasonably justified by invoking anharmonic coupling among the vibrations.

The present state of the art is such, however, that even if the coupling problem were to be solved, it runs afoul of the present lack of knowledge of the properties of excited molecules (*8*), and an even greater lack of knowledge of the properties of the transition state. We are therefore forced to make an assumption, and the simplest is not only to assume that coupling is strong, i.e., that there is free intramolecular transfer of energy, as assumed originally by Rice and Kassel, but also that it is free among *all* the internal degrees of freedom of our "particle," meaning the reactant as well as the transition state.

The obvious advantage of the assumption of free intramolecular transfer of energy is that we do not have to specify the (largely unknown) numerical values of parameters that would be required for describing the transfer of energy from one degree of freedom to the next; as a result, the model is a very crude one, where internal energy randomization is treated as an all-or-nothing affair: Either a degree of freedom participates freely in the randomization process or it does not participate at all. The meaning of *free* participation is that equal probability is assigned to every state of specified energy belonging to a participating degree of freedom. But this is precisely the probability assignment that characterizes a microcanonical ensemble (cf. Chapter 4, Section 1), so that the assumption of free intramolecular energy transfer necessarily follows from the statistical representation of the reacting system as a microcanonical ensemble, which in turn is a consequence of the still more basic assumption that energy is the necessary and sufficient criterion for reaction. Thus free randomization of energy is not in any way a new assumption, merely a consequence of the fundamental assumption made in Chapter 2.

However, the assumption that *all* internal degrees of freedom are involved in the randomization of energy is a supplementary assumption, although one that is strongly suggested by experimental data. If this assumption is not made, at best we would have to specify which internal degrees of freedom are to be excluded, and on what grounds, and at worst specify some sort of "effective" number of degrees of freedom which would in practice become an adjustable parameter as in the old RRK theory. The former possibility requires a qualitative solution of the coupling problem, and the latter is a step backward since it has been shown *(9,10)*[5] that the empirical fit-parameter s (cf. Chapter 2, Section 2) is a complicated function of several variables of which the molecular complexity (number of degrees of freedom) is only one.

One internal degree of freedom that requires a special mention is internal rotation (e.g., *12*, p. 438), which is usually associated with a hindering potential barrier. If the hindering potential is very large compared with kT, then, in molecules containing average thermal energy, the internal rotation is a torsional vibration, which is just like any other vibration. If the barrier is low, the motion is a free rotation. The coupling of internal rotation to vibrations is a well-known phenomenon in infrared spectroscopy (*13,14*); in accordance with previous discussion, we will assume that whether it is as torsional vibration or as free rotation, internal rotation exchanges energy freely with other internal degrees of freedom.

3. External degrees of freedom

We now consider external degrees of freedom not specifically excluded by conservation requirements. This means essentially the external rotation associated with the smallest (or largest) moment of inertia in the prolate (or oblate) symmetric top. In order that an external degree of freedom may be involved in intramolecular energy transfer, it must be able to interact with internal degrees of freedom. The absence or presence of such an interaction, for one external degree of freedom, is not likely to have a very large effect on the calculated rate constant, especially after averaging over a distribution of energies for the purpose of comparison with thermal or similar experimental rate data. Given the fairly small magnitude of the effect, presently available experimental rate data do not furnish decisive evidence either for or against the participation of an external degree of freedom in intramolecular

[5] In a thermal system, the parameter s can be variously interpreted (*9*) as $s = \langle E_v \rangle / RT$, or (*10*) as $s = C_v/R$, where $\langle E_v \rangle$ is the average vibrational energy and C_v the vibrational heat capacity of the reactant, assuming that only vibrational degrees of freedom are "pertinent." See however, Skinner and Rabinovitch (*11*).

randomization of energy, because, on the one hand, experimental rate data are generally not accurate enough, and on the other hand, calculated rate constants are subject to some uncertainty in the specification of transition-state parameters. It is therefore necessary to consider information from another source.

Vibration–rotation interaction Almost all of the information on the subject of vibration–rotation interaction comes from rotational or vibrational spectroscopy. As mentioned in Chapter 2, Section 5, spectroscopic information is mostly concerned with events occurring in relatively unexcited molecules, and therefore such information is only indirectly useful for describing highly excited reactive molecules. We may observe, however, that while molecules of low level of excitation are subject to restrictive spectroscopic selection rules, arising from harmonic potentials and symmetry requirements due to the rigidity of the molecular framework, these rules become considerably relaxed as the amplitude of vibrational motion increases and as the rigidity of the molecule decreases, i.e., with rising level of excitation. Therefore, as in the case of anharmonicity, we may suppose that whatever is true of vibration–rotation interaction in a fairly unexcited molecule observed in spectroscopy will be even more true of a highly excited molecule that is reacting. Hence spectroscopic information may be used as a sort of lower limit on the behavior of reacting molecules.

From rotational spectroscopy, it is known (*15,16*) that there are three ways in which rotation–vibration interaction can arise.

The first is due to the change of moment of inertia during vibrational motion. In quasilinear molecules, which have a very small moment of inertia (I_a) along the molecular axis, such interaction is particularly likely since vibrational skeletal bending motion produces a very large relative change in I_a. There are several examples (*17,18*) of this sort of coupling between bending and rotation observed spectroscopically in quasilinear molecules of low level of excitation, but, as an example of what to expect in a reactive molecule, the most interesting case[6] is the highly excited molecule C_3 produced by flash photolysis (*19*). There is very large amplitude of bending vibrations in the excited C_3 and therefore large anharmonicity and considerable vibration–rotation interaction, which makes an appreciable contribution to the thermodynamic functions of C_3 at high temperature (*20*).

The second source of rotation–vibration interaction is Coriolis coupling (*21*), which arises from the interaction of vibrational and rotational angular momenta. The vibrations that are subject to Coriolis coupling with overall rotations can be identified by group-theoretical considerations: The condition

[6] Actually C_3 in the ground state is linear but in excited states spends most of its time in a bent configuration.

is that the product of the symmetry species of the interacting vibrations contains the symmetry species of a rotation (*22*). In practice, this interaction will be appreciable only if the vibrations are degenerate, or very nearly so, and even in the case of a linear molecule the Coriolis coupling between its doubly degenerate vibrations and overall rotations is not very large (*23*, p. 374). This type of coupling, on the other hand, will be much stronger in the case of a symmetric top and will involve coupling between the angular momentum due to the doubly degenerate vibrations of the symmetric top and the angular momentum (quantum number $\mathcal{K}$) due to component of overall rotation about the top axis (*23*, p. 402).

The third source of rotation–vibration interaction is the so-called Fermi resonance which causes two neighboring vibrational levels to interact; this in turn affects the rotational constants and through them produces a perturbation of rotational levels with the same $\mathcal{K}$. This effect does not result in large rotation–vibration coupling in unexcited molecules, but is likely to increase in importance as the excitation level rises.

The net result of these considerations is that the part of the rotational energy of a symmetric top corresponding to the $\mathcal{K}$-dependent term in Eqs. (5-3) and (5-4), i.e.,

$$E_r(\mathcal{K}) = \mathcal{K}^2(\mathbf{A} - \mathbf{B})h \qquad \text{(prolate top)} \tag{5-9}$$

$$E_r(\mathcal{K}) = \mathcal{K}^2(\mathbf{C} - \mathbf{B})h \qquad \text{(oblate top)} \tag{5-10}$$

is in principle available for randomization, subject to the restriction $-\mathcal{J} \leqslant \mathcal{K} \leqslant \mathcal{J}$. This means that in both the reactant and transition states the external rotation about the particle symmetry axis must be considered as a "pertinent," though slightly restricted, degree of freedom, unless reactant or transition state is linear or a spherical top. Because of the difficulty of specifying the exact degree to which $E_r(\mathcal{K})$ is coupled to the internal degrees of freedom in the excited particle, we shall assume, as in Section 2, that the coupling is strong, i.e., that $E_r(\mathcal{K})$ exchanges energy freely with the other pertinent degrees of freedom, and later, in Chapter 6, we shall also drop the restriction on $\mathcal{K}$ to simplify matters further.

This is a fairly crude approximation, forced on us by the fact that the treatment of the conservation of angular momentum in Section 1 of this chapter is really only very summary. In a more detailed treatment, such an approximation should not be necessary; it turns out, however, that the general case is difficult to solve, although detailed treatments of some of the simpler case (generally type 1 reactions with "loose" transition states) have been presented (*24–26*).[7] See Exercise 1 in Chapter 7 (p. 141).

[7] A treatment of angular momentum applicable to the calculation of average relative kinetic energy of recoiling fragments has been given by Safron *et al.* (*27*).

4. Model for particle

On the basis of arguments presented in Sections 1–3 of this chapter, we may now proceed to construct a model for our particle (= reactant or transition state). Its internal degrees of freedom (and one external rotation in symmetric tops) will be all strongly coupled together, but its external rotations will be totally decoupled from internal degrees of freedom and will be subject to strict conservation of angular momentum. Beyond this point it is quite arbitrary how we proceed, as long as we find ways to reasonably approximate the density of states for the pertinent degrees of freedom.

The most convenient way to proceed is to consider the internal degrees of freedom formally as a collection of *independent* normal modes. Normal modes are based on small-vibration theory; if we use the harmonic oscillator representation, we may look at this as a sort of zeroth-order approximation. The advantage is that usually all the parameters (normal or fundamental frequencies, equilibrium moments of inertia and the like) are known from vibrational and rotational spectroscopy.[8] A first-order approximation would consist of replacing the harmonic oscillators with independent anharmonic oscillators, which of course would require the knowledge of the corresponding anharmonicity constants, also generally known from vibrational spectroscopy. Strong coupling among the normal modes is introduced by ignoring all the symmetry relations and by evaluating the density of states by the simple device of counting the number of ways a given energy can be shared among the pertinent degrees of freedom and assuming that any one way of distributing this energy is just as likely as another; this is of course just what is meant by the microcanonical density of states.

The procedure just outlined couples, for example, all the overtones of a vibration not only to its own ground state, but to the ground states and overtones of all other vibrations as well; furthermore, since the degrees of freedom are assumed to be independent, the energy contained in one oscillator is quite independent of whatever energy is contained in the other oscillators, subject only to the requirement that the total energy have the prescribed value, so that none of the usual selection rules apply. Thus the normal mode concept serves merely to delimit the size of energy quanta, which are then freely shuffled about among energy levels, like so many billiard balls among identical boxes, at least in the harmonic oscillator approximation. Therefore the density of states can be discovered by a purely statistical argument. This will be considered in the next chapter.

[8] Strictly speaking, this refers only to reactant. In the transition state corresponding information is obtained by a circuitous route (cf. Chapter 11) from reactant parameters or from parameters of stable molecules resembling the transition state, and so in the final analysis is also (indirectly) based on spectroscopic information.

References

1. R. A. Marcus, *J. Chem. Phys.* **20**, 359 (1952).
2. R. C. Tolman, "The Principles of Statistical Mechanics." Oxford Univ. Press, London and New York, 1938.
3. I. N. Levine, "Quantum Chemistry," Vol. II. Allyn and Bacon, Boston, Massachusetts, 1970, p. 240 ff.
4. J. C. Slater and N. H. Frank, "Mechanics." McGraw–Hill, New York, 1947.
5. C. H. Townes and A. L. Schawlow, "Microwave Spectroscopy." McGraw–Hill, New York, 1955.
6. N. C. Hung and D. J. Wilson, *J. Chem. Phys.* **38**, 828 (1963).
7. W. Forst and Z. Prášil, *J. Chem. Phys.* **53**, 3065 (1970).
8. W. M. Gelbart, S. A. Rice, and K. F. Freed, *J. Chem. Phys.* **52**, 5718 (1970).
9. J. Troe and H. Gg. Wagner, *Ber. Bunsenges.* **71**, 937 (1967).
10. D. M. Golden, R. K. Solly, and S. W. Benson, *J. Phys. Chem.* **75**, 1333 (1971).
11. G. B. Skinner and B. S. Rabinovitch, *J. Phys. Chem.* **76**, 2418 (1972).
12. G. N. Lewis and M. Randall, "Thermodynamics," 2nd ed., revised by K. S. Pitzer and L. Brewer. McGraw–Hill, New York, 1961.
13. M. L. Unland, J. R. van Wazer, and J. H. Letcher, *J. Amer. Chem. Soc.* **91**, 1045 (1969).
14. R. Meyer and H. H. Günthard, *J. Chem. Phys.* **50**, 353 (1969).
15. Y. Morino and E. Hirota, *Ann. Rev. Phys. Chem.* **20**, 139 (1969).
16. G. Amat and H. H. Nielsen, *J. Mol. Spectrosc.* **23**, 359 (1967).
17. S. R. Polo, *J. Mol. Spectrosc.* **23**, 117 (1967).
18. G. O. Neely, *J. Mol. Spectrosc.* **27**, 177 (1968).
19. L. Gausset, G. Herzberg, A. Lagerquist, and B. Rosen, *Discuss. Faraday Soc.* **35**, 113 (1963).
20. H. L. Strauss and E. Thiele, *J. Chem. Phys.* **46**, 2473 (1967).
21. W. Forst and P. St. Laurent, *Can. J. Chem.* **45**, 3169 (1967).
22. H. A. Jahn, *Phys. Rev.* **56**, 680 (1939).
23. G. Herzberg, "Molecular Spectra and Molecular Structure," Vol. II, Infrared and Raman Spectra of Polyatomic Molecules. Van Nostrand Reinhold, Princeton, New Jersey, 1960.
24. E. E. Nikitin, *Theoret. Exp. Chem.* **1**, 83, 90 (1965) (English transl.).
25. C. E. Klots, *J. Phys. Chem.* **75**, 1526 (1971).
26. C. E. Klots, *Z. Naturforsch.* **27a**, 553 (1972).
27. S. A. Safron, N. D. Weinstein, D. R. Herschbach, and J. C. Tully, *Chem. Phys. Lett.* **12**, 564 (1972).

CHAPTER 6

Calculation of Energy-Level Densities

We have seen that $k(E)$, the microcanonical rate constant for unimolecular decomposition of species with internal excitation energy E, is given by the remarkably simple expression [Eq. (4-34)]

$$k(E) = \alpha G^*(E - E_0)/hN(E) \qquad (E \geqslant E_0)$$

where α is the reaction path degeneracy and E_0 is the critical energy for reaction. Before this expression can be used for actual calculations of $k(E)$, ways must be found for calculating the density of states $N(E)$ for the reactant and the integrated density $G^*(E - E_0)$ for the transition state. Note that $G^*(E - E_0)$ refers to a system having one degree of freedom *less* than the reactant (Chapter 4, Section 1) and having, in principle, molecular parameters (vibrational frequencies, moments of inertia, etc.) that are different from those of the reactant. In addition, of course, $G^*(E - E_0)$ is evaluated at energy $E - E_0$, whereas $N(E)$ is evaluated at energy E.

We now propose to examine the quite general problem of calculating $G^*(E - E_0)$ and $N(E)$ at any energy and for any set of specified molecular parameters. When the problem is stated in these terms, the fact that $G^*(E - E_0)$ and $N(E)$ actually refer to different molecular systems and are evaluated at different energies becomes immaterial insofar as the subject matter of this chapter is concerned.[1] We shall therefore refer henceforth, for the purpose

[1] This chapter is adapted from Forst (*1*) (the reference list for this chapter can be found on p. 127).

of this chapter, only to the general functions $G(E)$ and $N(E)$. Since they are related in a simple way, $G(E)$ being the integral of $N(E)$ [Eq. (4-20)], we can deal with both at the same time; in fact, it will turn out that a trivial modification of the expression for $G(E)$ transforms it into the expression for $N(E)$ and vice versa. Therefore the term "density of states function" will be understood to refer to both $N(E)$ and $G(E)$.

The model molecular system for which we wish to find ways of calculating $N(E)$ and $G(E)$ will be the model of Chapter 4, Section 4 with degrees of freedom assumed to be coupled strongly enough to permit free exchange of energy, but for convenience assumed to be, at the same time, independent, i.e., separable. In particular, vibrational degrees of freedom will be assumed to be adequately represented as a collection of normal modes. This sort of simplification is necessary because, on the one hand, the energy-level density problem was solvable until recently only for the simplified molecular model, and on the other hand, not enough is known about interactions among internal degrees of freedom in polyatomic molecules to make calculations of $N(E)$ or $G(E)$ for interacting degrees of freedom worthwhile at the present time.

Plan of chapter After an introductory section (Section 1) concerned mainly with the definition of symbols, the combinatorial nature of the problem is shown in Sections 2 and 6. It is pointed out in Section 3 that, while the calculation of $N(E)$ and/or $G(E)$ does not present any difficulty in principle, the bookkeeping gets quite out of hand for all but the simplest systems and the lowest energies. Practical considerations[2] therefore dictate the use of some suitable approximation not afflicted with the bookkeeping disability.

A general method for obtaining such an approximation can be based on the inversion of the partition function, as is shown in Section 4; all the various approximations considered in the subsequent sections are discussed from this point of view. Inversion of the classical partition function is discussed in Section 5 for rotational states and in Section 7 for vibrational states; inversion of the quantum mechanical vibrational partition function is treated in Section 8. Section 9 deals with some refinements, notably

[2] Since the ultimate object of rate calculations is usually to compare observed and calculated rate constants and since observed rates are always averages over some distribution of energies (see Part II), the calculated $k(E)$ must therefore be (numerically) integrated over a more or less extensive energy range to make it comparable with experiment. Hence for the purposes of numerical integration $k(E)$, and therefore also $G^*(E)$ and $N(E)$, must be generated over a large number of equally spaced energy intervals, from zero upward for $G^*(E)$ and from E_0 upward for $N(E)$. Thus practical considerations also include reasonable speed of computation of $G(E)$ and $N(E)$ and reasonable accuracy over the entire energy range, two requirements that are difficult to reconcile.

anharmonicity. Finally Section 10 discusses the pros and cons of the various approximations and gives some practical hints for those wishing to undertake such calculations.

1. Terminology and basic concepts

To avoid any confusion of symbols and meaning, define for *positive* E:

$N(E)$ is the number of states *per unit energy range*, i.e., the density of states at energy E [units: energy^{-1}];

$W(E)$ is the number of states *at* energy E, i.e., the quantum mechanical degeneracy of state of energy E; this quantity is dimensionless. Note that

$$W(E) = N(E)\,\delta E \qquad (\delta E = \text{allowance in } E) \tag{6-1}$$

so that $N(E)$ may be considered a smooth-function version of $W(E)$.

$G(E)$ is the *total* number of states between zero and E, i.e., the integrated density; this quantity is likewise dimensionless. Note that

$$G(E) = \sum_{E=0}^{E} W(E) \quad \text{or} \quad \int_0^E N(E)\,dE; \qquad N(E) = dG(E)/dE \tag{6-2}$$

Since the quantum form of $G(E)$ involves a summation, $G(E)$ is sometimes also referred to as the "sum of states." We shall avoid this term because some authors use it to mean the partition function Q (the German word for Q is *Zustandsumme*, which translates to "sum of states").

It is useful to establish a connection between the density (or number) of states of a system and its partition function. If Q is the partition function for the specified system, we have, by definition,

$$Q = \sum_E W(E)e^{-E/kT} \quad \text{or} \quad \int_0^\infty N(E)e^{-E/kT}\,dE \tag{6-3}$$

(quantum mechanically) (semiclassically)

E shall always be taken to mean energy in excess of zero-point energy, so that the first term of the quantum mechanical partition function will be unity, and the lower limit of the integral in the semiclassical expression for Q will be zero. Usually Q for bound states is desired, in which case the summation or integration should end at some large but finite value of E. However, it is preferable to keep the definition of Q as an integral with limits $(0, \infty)$, with the understanding that above some large and finite E, when there are no more bound states, $N(E)$ is zero.

The absence of a subscript in these definitions is to indicate that they apply to a collection of states belonging to any type of degree of freedom.

A subscript "v" will be used for vibrational states, subscript "r" for rotational states, and subscript "vr" for vibration–rotation[3] states; for example. $N_r(E)$, $W_v(E)$, and $G_{vr}(E)$ are, respectively, the density of rotational states, the number of vibrational states, and the total number of vibration–rotation states, all at energy E. Similarly, Q_v, Q_r, and Q_{vr} represent, respectively, the vibrational, rotational, and vibration–rotation partition function.

The subscripts are also meant to have a quantitative significance, in the sense that, unless assigned an explicit value (r = 2, v = 3, etc.), it is understood the formula in question refers to a total of r rotors [r as defined by Eq. (6-42)] and v oscillators.

Quantum mechanically, only discrete values of E are allowed, so that $W(E)$ will be zero everywhere except at those values of E that are allowed. Consequently, $W(E)$ is, in principle, a string of δ-functions, and $G(E) = \sum_{E=0}^{E} W(E)$ is a series of step functions (see Fig. 6-1). When E is sufficiently high and the number of degrees of freedom not too small there will be an allowed state at almost every energy, so that $W(E)$ becomes practically a smooth function of E; under these conditions the density of states $N(E)$ is the more useful and convenient representation. Then $G(E) = \int_0^E N(E)\,dE$ is obviously also a smooth function of E. Conversely, when the allowed values of E are widely spaced, as when E is low and the number of degrees of freedom small, $N(E)$ and $G(E) = \int_0^E N(E)\,dE$ will not be a useful representation.

It should be obvious that $N(E)$, $W(E)$, or $G(E)$ cannot be calculated unless the allowed energy levels of the system under consideration are known. The calculation of these levels is a purely quantum mechanical problem which is not dealt with here, because for separable degrees of freedom it is sufficient to consider only energy levels for a system of independent oscillators and/or rotors, and these can be found in any standard treatise on quantum mechanics [e.g., (*2*, Chapter V, p. 68 ff.)].

2. Direct evaluation of $W(E)$ and $G(E)$ in simple systems

For the purpose of illustrating the general principles involved in the enumeration of vibrational states, it is useful to consider a few simple systems at low energies where $W_v(E)$ and $G_v(E)$ can be easily calculated by hand exactly. Generalization to more complex oscillator systems and higher energies should be obvious. The practical interest of these simple calculations is

[3] The term means that the available energy is shared between vibrational and rotational states, but does *not* imply that vibrational and rotational states are coupled, i.e., that the expression for total energy has crossterms involving both vibrations and rotations.

that, as we shall see, direct enumeration of states is the only accurate method for calculating $W_v(E)$ [or $N_v(E)$] and $G_v(E)$ at low energies.

Direct enumeration of $N_{vr}(E)$ and $G_{vr}(E)$ for vibration–rotation systems will be considered in Section 6.

Independent harmonic oscillators In the case of a single harmonic oscillator of frequency ν, the energy levels are nondegenerate and are given by

$$E = nh\nu, \qquad n = 0, 1, 2, \ldots \tag{6-4}$$

where n is the vibrational quantum number and E is energy in excess of zero-point energy. Hence $W(E)$ is unity when $E = nh\nu$ and is zero otherwise, $G(E) = \sum_n 1$, and $N(E) = 1/h\nu =$ const. These relations are illustrated graphically in Fig. 6-1.

When there are several independent oscillators, $W(E)$ is the number of ways a given total energy E can be distributed among the oscillators. The problem is particularly simple when the oscillators are all of the same

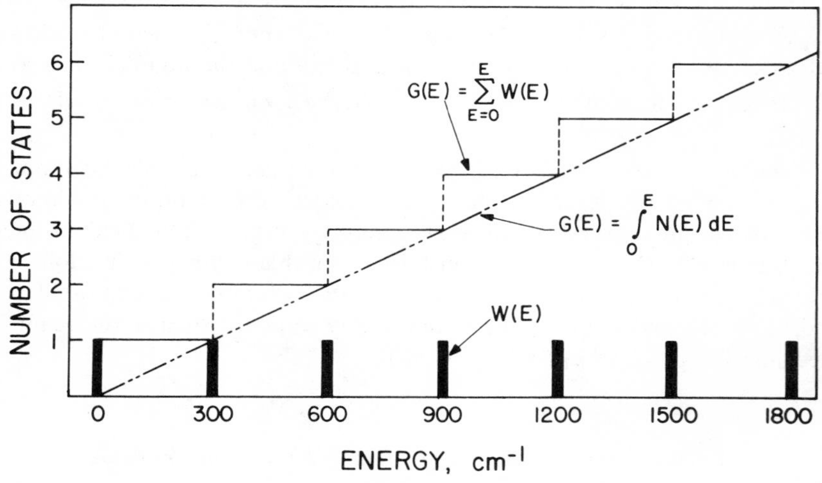

Fig. 6-1. *Number of states for one harmonic oscillator of frequency 300 cm^{-1}.*

Heavy vertical line is the quantum mechanical degeneracy $W(E)$. The step function represents the quantum-mechanical integrated density $G(E)$; line (— - - —), the continuous-function approximation to $G(E)$. The density $N(E)$ is the slope of this line, which in the present instance is constant at 1/300 (cm^{-1})$^{-1}$. Quantum mechanically, $G(0) = 1$. The continuous-function approximation is drawn to give the classical result $G(0) = 0$. The thickness of the line for $W(E)$ represents δE, the allowance in E. Graphs of harmonic $G(E)$ versus E for a number of polyatomic molecules can be found in G. Z. Whitten and B. S. Rabinovitch, *J. Chem. Phys.* **38**, 2466 (1963).

frequency. Suppose their number is $\mathfrak{q}$; if the total energy E is regarded as n quanta ($E = nh\nu$), $W(E)$ will be the number of ways n quanta can be distributed among $\mathfrak{q}$ oscillators, an elementary combinatorial problem which has the solution[4]

$$W_{\mathrm{v}=\mathfrak{q}}(E) = \binom{n + \mathfrak{q} - 1}{n} = \frac{(n + \mathfrak{q} - 1)!}{n!\,(\mathfrak{q} - 1)!} \tag{6-5}$$

and

$$G_{\mathrm{v}=\mathfrak{q}}(E) = \sum_{n=0}^{n} \frac{(n + \mathfrak{q} - 1)!}{n!\,(\mathfrak{q} - 1)!} = \frac{(n + \mathfrak{q})!}{n!\,\mathfrak{q}!} \tag{6-6}$$

It is convenient to regard the $\mathfrak{q}$ oscillators, all of the same frequency ν, as *one* $\mathfrak{q}$-fold degenerate oscillator. If we now have a total of v oscillators belonging to v' groups of frequencies, some oscillators will be degenerate if $\mathrm{v} > \mathrm{v}'$; the problem then reduces to the equivalent, and somewhat simpler, problem of v' oscillators, the ith being $\mathfrak{q}_i$-fold degenerate ($\mathrm{v} = \sum_{i=1}^{\mathrm{v}'} \mathfrak{q}_i$). The total energy is $E = \sum_{i=1}^{\mathrm{v}'} n_i h\nu_i$, where n_i and ν_i are the total quantum number and frequency, respectively, of the ith oscillator. The number of ways of distributing $\sum_{i=1}^{\mathrm{v}'} n_i$ quanta among the v' oscillators, with n_i quanta in the ith,

$$\prod_{i=1}^{\mathrm{v}'} \frac{(n_i + \mathfrak{q}_i - 1)!}{n_i!\,(\mathfrak{q}_i - 1)!} \tag{6-7}$$

is a simple product, since the oscillators are assumed to be independent. If some of the frequencies are commensurable, a given value of E can be realized by several sets of n_i, so that in general

$$W_{\mathrm{v}}(E) = \sum_{\Sigma n_i h\nu_i = E} \prod_{i=1}^{\mathrm{v}'} \frac{(n_i + \mathfrak{q}_i - 1)!}{n_i!\,(\mathfrak{q}_i - 1)!} \tag{6-8}$$

where the summation is taken over all sets of positive, integral (zero included) values of n_i satisfying $\sum_{i=1}^{\mathrm{v}'} n_i h\nu_i = E$. Similarly

$$G_{\mathrm{v}}(E) = \sum_{\Sigma n_i h\nu_i \leqslant E} \prod_{i=1}^{\mathrm{v}'} \frac{(n_i + \mathfrak{q}_i - 1)!}{n_i!\,(\mathfrak{q}_i - 1)!} \tag{6-9}$$

[4] Mayer and Mayer (*3*, Appendix VII). The problem amounts to counting the number of ways of placing n objects into $\mathfrak{q}$ compartments. We can imagine the n objects to be laid out in a linear array; to delimit the $\mathfrak{q}$ compartments, we require $\mathfrak{q} - 1$ partitions, which we can imagine as vertical strokes within the array. The number of permutations of n objects and $\mathfrak{q} - 1$ partitions is $(n + \mathfrak{q} - 1)!$ if both objects and partitions are distinguishable; since they are not, we have to divide by $n!\,(\mathfrak{q} - 1)!$, and Eq. (6-5) follows. When the problem is to determine the total number of ways of placing 0, 1, 2, 3, . . . , n objects in the same $\mathfrak{q}$ compartments, we can simply add a partition, thus creating an additional compartment to accommodate the $n - y$ objects left over after y of them ($0 \leqslant y \leqslant n$) had been placed in the first $\mathfrak{q}$ compartments; thus Eq. (6-6) is obtained.

where the summation is taken over all sets of n_i such that all values of $\sum_{i=1}^{v'} n_i h\nu_i$ between zero and E are generated.

A simple example of seven oscillators with four different frequencies is worked out explicitly in Table (6-1). In this case the overtones are partially commensurable, so that the energy of 1000 cm^{-1}, for example, is realized by four sets of n_i.

Independent anharmonic oscillators When the oscillators are independent and anharmonic, the procedure for determining $W_v(E)$ and $G_v(E)$ by direct counting is more tedious. A specific example is worked out in Table 6-1 for the case of Morse oscillators, whose individual energy levels are given by

$$E = h\nu[n(1 - x) - n^2 x] \qquad (6\text{-}10)$$

where x is the anharmonicity constant. The basic frequency ν of each oscillator is assumed to be the same as that of its harmonic counterpart. Energy levels of an anharmonic oscillator are no longer integral multiples of the basic frequency, and therefore oscillators of the same basic frequency cannot be usefully combined into degenerate sets, since even if the total quantum number is the same, some combinations of levels have slightly different energies. For example, the oscillator with $\nu = 300$ cm^{-1} has energy levels at 0, 294, 582,... cm^{-1}. This oscillator is triply degenerate; let us label the three members a, b, and c. If the total quantum number $n_i = n_a + n_b + n_c = 2$, we can have, for example, $n_a = 2$, $n_b = 0$, $n_c = 0$, in which case the energy is 582 cm^{-1}; by permutation of the labels there are two other ways of realizing this energy, making this a total of three states at 582 cm^{-1}. We can also have $n_i = 2$ with $n_a = 1$, $n_b = 1$, $n_c = 0$, in which case the energy is $2 \times 294 = 588$ cm^{-1}, and by permutation of the labels we obtain two other states at this energy, for a total of three. In the harmonic case, all such states have energy 600 cm^{-1}, and their number (six) follows immediately from Eq. (6-5).

For the purpose of enumerating states by computer in more complex systems, the procedure indicated in Table 6-1 can be systematized (*4*).

Independent rigid free rotors The energy levels of a one-dimensional rigid free rotor are given by

$$E = m^2\hbar^2/2I, \qquad m = 0, 1, 2, \ldots \qquad (6\text{-}11)$$

where I is the moment of inertia and m the rotational quantum number. Except for the nondegenerate level at $m = 0$, all other levels are doubly degenerate. Therefore $W_{r=1}(E) = 2$ except that $W_{r=1}(0) = 1$, and $G_{r=1}(E) = 2m^\star + 1$, where $m^\star = (2IE/\hbar^2)^{1/2}$, i.e., $m^\star$ is the quantum number corresponding to the (allowed) energy E at which $G(E)$ is to be calculated. If for

TABLE 6-1

$W_v(E)$ AND $G_v(E)$ BY EXACT ENUMERATION OF STATES FOR A SYSTEM OF INDEPENDENT OSCILLATORS[a] AT $E < 1000\ \mathrm{cm}^{-1}$

n_i				Morse			Harmonic		
$i = 1$	$i = 2$	$i = 3$	$i = 4$	E	$W_v(E)$	$G_v(E)$	E	$W_v(E)$	$G_v(E)$
0	0	0	0	0	1	1	0	1	1
1	0	0	0	196	1	2	200	1	2
0	1	0	0	294	3	5	300	3	5
2	0	0	0	388	1	6	400	1	6
1	1	0	0	490	3	9	500	3	9
3	0	0	0	576	1	10			
0	2	0	0	582	3	13	600	7	16
0	2	0	0	588	3	16			
2	1	0	0	682	3	19	700	5	21
0	0	1	0	686	2	21			
4	0	0	0	760	1	22			
1	2	0	0	778	3	25	800	7	28
1	2	0	0	784	3	28			
0	3	0	0	864	3	31			
3	1	0	0	870	3	34			
0	3	0	0	876	6	40	900	15	43
0	3	0	0	882	1	43			
1	0	1	0		2				
5	0	0	0	940	1	44			
2	2	0	0	970	3	47			
2	2	0	0	976	3	50	1000	14	57
0	1	1	0	980	6	57			
0	0	0	1		1				

[a] The frequencies in cm^{-1} are (degeneracies in brackets): 200 (1), 300 (3), 700 (2), 1000 (1). If the oscillators are harmonic, total energy E in cm^{-1} is given by $E = 200n_1 + 300n_2 + 700n_3 + 1000n_4$. Here n_i [cf. Eq. (6-7)] is the total quantum number of the ith q_i-fold degenerate oscillator, defined as $n_i = \sum_{j=1}^{q_i} n_j$, where the n_j are the quantum numbers of the individual oscillators in the degenerate set. If the oscillators are Morse oscillators, the anharmonicity constant is assumed for simplicity to be $x = 0.01$ for all oscillators, and the total energy E in cm^{-1} is then given by $E = 200n_1' + 300n_2' + 700n_3' + 1000n_4'$, where, since Eq. (6-7) no longer applies, $n_i' = \sum (0.99n_j - 0.01n_j^2)$, the n_j being the quantum numbers of the individual oscillators of the q_i-fold degenerate set, as before. The sum is taken over all sets of n_j such that $\sum_{j=1}^{q_i} n_j = n_i$.

At values of E not listed, there are no states, so that $W_v(E)$ is zero and $G_v(E)$ is the same as for the next lower listed E.

reasons of symmetry[5] only each σth value of m is allowed, then $G_{r=1}(E) = 2(m^\star/\sigma) + 1$.

The energy levels of a two-dimensional rigid free rotor are given by

$$E = \mathscr{J}(\mathscr{J} + 1)(\hbar^2/2I), \qquad \mathscr{J} = 0, 1, 2, \ldots \tag{6-12}$$

where I is again the moment of inertia and $\mathscr{J}$ the rotational quantum number. Every level is $(2\mathscr{J} + 1)$-fold degenerate. Hence $W_{r=2}(E) = 2\mathscr{J}^\star + 1$, $G_{r=2}(E) = \sum_{\mathscr{J}=0}^{\mathscr{J}=\mathscr{J}^\star} (2\mathscr{J} + 1) = (\mathscr{J}^\star + 1)^2$, where $\mathscr{J}^\star$ is, as before, the rotational quantum number corresponding to the energy at which $G_{r=2}(E)$ is to be calculated. If for reasons of symmetry only every σth value of $\mathscr{J}$ is allowed, then $G_{r=2}(E) = [(\mathscr{J}^\star/\sigma) + 1](\mathscr{J}^\star + 1)$.

Observe that for a typical value of $I \sim 5 \times 10^{-40}$ g cm^2, the ratio $\hbar^2/2I \sim 5$ cm^{-1}. If $\sigma = 1$, the rotational energy corresponding, to, say, $\mathscr{J} = 10$ is ~ 500 cm^{-1} and so $G_{r=2}(500) = 121$, i.e., roughly one state every 5 cm^{-1} on the average. Contrast this with a (nondegenerate) harmonic oscillator having a frequency of 500 cm^{-1} (this is in fact a rather low frequency for a typical molecular oscillator); therefore $G_{v=1}(500) = 1$, i.e., one state every 500 cm^{-1}. Thus $G_r(E)$ for rotors is almost a continuous function of E and $N_r(E)$, rather than $W_r(E)$, would be the more convenient representation.

Coupled rigid free rotors We shall now consider a three-dimensional rotor of the symmetric-top type discussed previously in Chapter 5. We have seen that its energy may be written as the sum

$$E = E_r(\mathscr{J}) + E_r(\mathscr{K}) \tag{6-13}$$

where $E_r(\mathscr{J})$ is the $\mathscr{J}$-dependent part [Eq. (5-6)], and $E_r(\mathscr{K})$ is the $\mathscr{K}$-dependent part [Eqs. (5-9) and (5-10)] of the rotational energy; the two parts are coupled by the requirement $-\mathscr{J} \leqslant \mathscr{K} \leqslant \mathscr{J}$.

To simplify matters, we shall first assume that the top is very prolate; i.e., that the rotational constant $\mathbf{B} = h/8\pi^2 I_b$ is smaller than $\mathbf{A}$; then

$$E_r(\mathscr{K}) \sim \mathscr{K}^2\hbar^2/I_a \qquad \text{(very prolate top)} \tag{6-14}$$

We shall next drop the restriction on $\mathscr{K}$, which has the result of uncoupling the rotors, so that the energy [Eq. (6-13)] is merely the sum of energies of a two-dimensional rigid free rotor [Eq. (6-12)] and of a one-dimensional rigid free rotor [Eq. (6-11)]. The neglect of the term $-\mathscr{K}^2\hbar^2/I_b$ in the energy of the one-dimensional rotor has the effect of diminishing the total number of

[5] Mayer and Mayer (*3*, p. 174 ff.). It is understood then that only integral values of $m^\star/\sigma$ appear in $G_{r=1}(E)$ and only integral values of $\mathscr{J}^\star/\sigma$ in $G_{r=2}(E)$. We then get for the exact densities $N_{r=1}(E) = \Delta G_{r=1}(E)/\Delta E = 4I/\hbar^2(2m^\star\sigma + \sigma^2)$ and $N_{r=2}(E) = \Delta G_{r=2}(E)/\Delta E = (2I/\hbar^2) \times (2\mathscr{J}^\star + 2\sigma + 1)/(2\mathscr{J}^\star\sigma + \sigma + \sigma^2)$.

states within a specified energy interval, but the absence of a restriction on $\mathscr{K}$ results in the inclusion of normally disallowed states having $\mathscr{K} > \mathscr{J}$, which has the effect of increasing the number of states within a specified energy interval. Thus the two errors tend to compensate each other, with the result that the separated rotor approximation for the density of states is not a bad one.[6]

Calculation of the density or sum of states for a collection of independent one- and two-dimensional rigid free rotors could be done by a direct enumeration procedure similar to that used for oscillators in Table 6-1; however, such a tedious process is not really necessary because an analytic function expression, sufficiently accurate for most applications, can be easily obtained, as shown in Section 5. It must be emphasized that calculation of the density or sum of states for the $\mathscr{J}$-dependent (two-dimensional rotor) part of the symmetric top representing overall (external) rotations of the reactant or transition state implies randomization of $E_r(\mathscr{J})$, which is normally prohibited by angular momentum conservation (cf. Chapter 5, Section 1), and therefore these rotational states are normally not considered when calculating reactant and transition-state density or sum of states, respectively. The calculation of the density of states for two-dimensional rotors is included in this chapter for completeness for possible use in the case of a "loose" transition state. (Cf. Chapter 11, Section 2.)

Symmetry numbers for rotations Mention is made, in connection with Eqs. (6-11) and (6-12), of the symmetry numbers σ (say, σ_i) for individual one- and two-dimensional rotors. In the present context, these rotors represent some or all of the rotational degrees of freedom of the reactant or transition state, and their symmetry numbers are therefore factors[7] that appear in the overall symmetry number σ for the species in question. This is the same σ which was mentioned in Chapter 4, Section 5, where it was noted that σ is connected with the reaction path degeneracy α.

As a result, the symmetry numbers for individual rotations are taken into account in computing the rotational density or sum of states only if the symmetry numbers have not been already included in the reaction path degeneracy.

For example, in the cited case of the reaction $C_2H_5Cl \rightarrow C_2H_4 + HCl$

[6] Current and Rabinovitch (*5*). However, if the top is very oblate, $\mathbf{C} \ll \mathbf{B} = \mathbf{A}$, and then $E_r(\mathscr{K}) \sim -\mathscr{K}^2\hbar^2/I_b$, and the separated unrestricted rotor approximation becomes very poor.

[7] If the symmetry number of the *i*th rotation is σ_i, then $\sigma = \mathbf{\Pi}_i\, \sigma_i$, where the product is taken over all rotations in the molecule. Cf. Herzberg (*6*, p. 508 ff.). It is these factors σ_i that are discussed in the text (distinguished, when necessary, by subscripts *i* and *j*; then $\sigma_{ij} = \sigma_i\sigma_j$). The symbol σ_i^* refers to symmetry number of an individual rotor in the transition state.

and its four-center transition state (Chapter 4, Section 5), treating the three hydrogens of the CH_3 group in the reactant as equivalent for the purpose of determining the reaction path degeneracy is tantamount to taking the internal rotation in C_2H_5Cl as a one-dimensional rotor with $\sigma_i = 3$; therefore in calculating $N(E)$ for reactant, the one-dimensional rotor contribution is introduced with $\sigma_i = 1$, so as not to allow for this symmetry element twice.

An example of a case where rotational symmetry is not necessarily counted in α is the dissociation of ethane, $C_2H_6 \rightarrow 2CH_3$, with a transition state that may be represented as essentially an ethane molecule in which the two methyls are moving away from each other. In this reaction $\alpha = 1$ by direct count. If the internal rotation in both reactant and transition state is taken as a free one-dimensional rotation, in computing $N(E)$ and $G^*(E - E_0)$ we may take either $\sigma_i = 3$, $\sigma_i^* = 3$ or $\sigma_i = 1$, $\sigma_i^* = 1$; the reason is that the rate constant $k(E)$ is proportional to $G_i^*(E - E_0)/N(E)$, and therefore σ_i and σ_i^* cancel out since $1/\sigma_i$ is a factor in $N(E)$ and $1/\sigma_i^*$ is a factor in $G^*(E - E_0)$ (cf. Section 5). If, in addition, the transition state is assumed to be "loose" with both methyls freely tumbling (cf. Chapter 11, Section 2), there are two additional rotational degrees of freedom per methyl. For planar methyl (point group D_{3h}, $\sigma_{ij}^* = 6$), one (one-dimensional) rotation is about a threefold axis ($\sigma_i^* = 3$), and the other about a twofold axis ($\sigma_j^* = 2$) at right angle to the preceding. Symmetry numbers for both of these rotations (actually their square since there are two methyls) will appear in $G^*(E - E_0)$ since these symmetry elements are without influence on the reaction path degeneracy.

In general, every reaction needs individual scrutiny to determine which, if any, rotations have their symmetry properties taken into account in the reaction path degeneracy.

3. Approximation to $N(E)$ and $G(E)$: general considerations

Need for an approximation The fairly complex bookkeeping required even in the almost trivial examples of counting of states worked out explicitly in the previous section suggests that, at energies of chemical interest, where 10^{10} states or more might be involved, direct enumeration of states is not a feasible procedure, except by computer, and even there machine time becomes quite appreciable at high energies.[8] A vast expenditure of machine time on exact sums or densities of states is quite unwarranted in the course of a

[8] As a very rough guide, $G(E)$ in steps of 100 cm^{-1} by actual counting of states in a moderately complex molecular system would require machine time (IBM 360) of the order of seconds up to 5000 cm^{-1}, of the order of minutes up to about 15,000 cm^{-1}, and of the order of hours above about 30,000 cm^{-1}.

unimolecular rate calculation since the number and nature of the approximations that have to be made in the theory are such that these densities or sums of states are substituted into expressions that are at best only approximate.

Ideally, we would like to replace the time-consuming enumeration of individual states by an analytical function expression for $G(E)$ [as a minimum, the function should at least have a first derivative so that $N(E)$ can be obtained] and in the process we would be willing to trade some accuracy for a significantly shortened machine time. As a reasonable compromise, one might wish to find an approximate expression for $G(E)$ giving 10% accuracy or so, in exchange for a reasonable machine time, i.e., a few minutes for the 50 or 100 different energy values at which $G(E)$ or $N(E)$ might be required.

Number of states as volume in phase space A closed-form approximation to $N(E)$ or $G(E)$ can be in principle obtained from statistical ensemble theory. We have considered this in classical terms in Chapter 4, Sections 1 and 2 [see, in particular, Eq. (4-13)], where the classical equivalent of the density of states was called DEN. Semiclassically, one quantum state is considered as occupying the volume h, and therefore there are $dp\,dq/h$ quantum states in the volume element $dp\,dq$. The semiclassical density of states is the number of quantum states within a shell of thickness dE on the hypersurface $E = \text{const}$, which is given by

$$N(E)\,dE = \underset{E \leqslant \mathscr{H} \leqslant E+dE}{\int \cdots \int} d\Omega/h^n \tag{6-15}$$

for a system of n degrees of freedom. This is merely the expression for DEN slightly modified. Similarly, for a system of n degrees of freedom,

$$G(E) = \underset{0 \leqslant \mathscr{H} \leqslant E}{\int \cdots \int} d\Omega/h^n \tag{6-16}$$

Since the statistical representation corresponds to a microcanonical ensemble, $N(E)$ and $G(E)$ are sometimes called the microcanonical density and integrated density, respectively. A simple application of these formulas is the subject of Exercise 1. For fully classical calculations see Šolc (*67*).

The evaluation of the multiple integrals in (6-15) and (6-16) is difficult and tedious; more convenient is another semiclassical equivalent of (6-16). We can take E as a (continuous) function of the quantum numbers $n_1, n_2, \ldots, n_n$:

$$E = f(n_1, n_2, \ldots, n_n) \tag{6-17}$$

where n_i is the quantum number for the ith degree of freedom. If g_i is the degeneracy (i.e., the number of states) associated with n_i, then

$$G(E) = \int \cdots \int_{f(n_1, n_2, \ldots, n_n) \leqslant E} g_1\, dn_1, g_2\, dn_2, \ldots, g_n\, dn_n \tag{6-18}$$

where the integration is over a range of quantum numbers such that the total energy does not exceed the specified value. The integral in (6-18) is difficult to evaluate for a general case, but if the degrees of freedom are separable (i.e., independent), there are no cross-terms in the energy expression and the function f can often be written in the form of a sum,

$$f(n_1, n_2, \ldots, n_n) = \sum_{i=1}^{n} \left(\frac{n_i}{c_i}\right)^{c_i'} \tag{6-19}$$

where c_i and c_i' are constants. With this form of f, the integral (6-18) is known as Dirichlet's integral, which has a simple solution.[9] Then $N(E)$ is easily obtained by differentiation with respect to E.

As an example, $G_r(E)$ and $N_r(E)$ are calculated for a collection of n independent two-dimensional rigid free rotors ($r = 2n$). The total energy is clearly

$$E = \sum_{i=1}^{n} \mathcal{J}_i(\mathcal{J}_i + 1)(\hbar^2/2I_i) \tag{6-20}$$

where $\mathcal{J}_i$ and I_i are the quantum number and moment of inertia, respectively, of the ith rotor. Assuming that I_i is not too small, so that the rotors are sufficiently semiclassical and integration may be performed, it follows that

$$G_{r=2k}(E) = \int \cdots \int \prod_{i=1}^{n} (2\mathcal{J}_i + 1)(d\mathcal{J}_i/\sigma_i) \tag{6-21}$$

subject to

$$\sum_{i=1}^{n} \frac{\mathcal{J}_i(\mathcal{J}_i + 1)\hbar^2}{2I_i} \leqslant E$$

The result is

$$G_{r=2k}(E) = \prod_{i=1}^{n} \left(\frac{2I_i}{\hbar^2 \sigma_i}\right) \frac{E^n}{\Gamma(n+1)} \tag{6-22}$$

where σ_i is the symmetry number of the ith rotor and $\Gamma(n+1) = n!$. Differentiation with respect to E yields

$$N_{r=2k}(E) = \prod_{i=1}^{n} \left(\frac{2I_i}{\hbar^2 \sigma_i}\right) \frac{E^{n-1}}{\Gamma(n)} \tag{6-23}$$

[9] Tolman (*7*, Appendix II). For another derivation see Rice (*8*, p. 570 ff.).

Equations (6-22) and (6-23) represent, as mentioned in Section 5, a very good approximation to $N_r(E)$ and $G_r(E)$. Unfortunately, the classical approach described here is unsuitable for calculating the vibrational $N_v(E)$ and $G_v(E)$ with any decent accuracy because of the much larger quantization of vibrational states. Even though the resulting inaccurate expressions can be corrected empirically to yield quite accurate harmonic $N_v(E)$ and $G_v(E)$, there is another, much more general approach to the problem of calculating $N(E)$ and $G(E)$. It is based on the inversion of the partition function and has the virtue of being very simple mathematically and of yielding approximations of high accuracy; all the semiclassical-type expressions can be easily obtained in this context as special cases.

Because of the inherent simplicity of the method, further treatment and development of approximations to $N(E)$ and $G(E)$ will be based entirely on the inversion of the partition function.

4. Inversion of the partition function

In this section are derived a few general relations which will be useful later.

Partition function as a Laplace transform *(9,10)* The partition function Q can be written, with $1/kT = s$, in the form

$$Q = \int_0^\infty N(E)e^{-Es}\,dE \tag{6-24}$$

This integral is a function of s, or, symbolically,

$$Q = \ell(s) = \mathfrak{L}\{N(E)\} \tag{6-25}$$

where $\mathfrak{L}\{\ \}$ denotes the operation to be performed on the function inside the compound bracket, i.e., multiplication by e^{-Es} and integration with respect to E between zero and infinity. $\mathfrak{L}\{N(E)\}$ is called the Laplace transform of $N(E)$ and s is the transform parameter. The interest of this representation for our purpose is that there exist standard ways for reversing the operation, i.e., $N(E)$ can be obtained from Q:

$$N(E) = \mathfrak{L}^{-1}\{\ell(s)\} \tag{6-26}$$

where $\mathfrak{L}^{-1}\{\ \}$ denotes the inverse Laplace transform. The reciprocal correspondence between a function and its transform is the subject of operational calculus which has its own theorems and rules for operations on, and relations between, $N(E)$ and $\ell(s)$. These may be found in standard treatises[10] on the Laplace transformation; there are also extensive tables [e.g., (*12*)] of transform pairs, so that simple problems may be solved by merely finding the appropriate entry in the tables.

The integration theorem of the Laplace transformation applied to (6-25) asserts that

$$\ell(s)/s = \mathfrak{L}\left\{\int_0^E N(x)\,dx\right\} = \mathfrak{L}\{G(E)\} \tag{6-27}$$

since by Eq. (6-2) the integral of density is $G(E)$. Inverting (6-27) gives $G(E) = \mathfrak{L}^{-1}\{\ell(s)/s\}$, which can be combined with (6-26) in the more compact expression

$$\left.\begin{matrix} N(E) \\ {\scriptstyle (k=0)} \\ \\ G(E) \\ {\scriptstyle (k=1)} \end{matrix}\right\} = \mathfrak{L}^{-1}\left\{\frac{\ell(s)}{s^k}\right\} \tag{6-28}$$

Sections 5–7 involve essentially trivial forms of $\ell(s)$, so that the inverse transform (6-28) may be found in tables and the result easily verified by very unsophisticated calculation. The actual evaluation of (6-28) for more complicated forms of $\ell(s)$ is given in Sections 8 and 9.

Convolution Suppose there are two independent systems (indexed 1 and 2), for which

$$\ell_1(s) = \mathfrak{L}\{N_1(E)\}, \qquad \ell_2(s) = \mathfrak{L}\{N_2(E)\} \tag{6-29}$$

Then according to the convolution theorem

$$\mathfrak{L}^{-1}\{\ell_1(s)\ell_2(s)\} = \int_0^E N_1(x)N_2(E-x)\,dx \quad \text{or} \tag{6-30a}$$

$$\int_0^E N_1(E-x)N_2(x)\,dx \tag{6-30b}$$

Since by definition $\ell_1(s) \equiv Q_1$, $\ell_2(s) \equiv Q_2$, then

$$\ell_1(s)\ell_2(s) \equiv Q_1Q_2 = Q_{1,2} \equiv \ell_{1,2}(s) \tag{6-31}$$

where $Q_{1,2}$ is the partition function of the combined system, assumed to factorize in Q_1 and Q_2. The inverse transform of $Q_{1,2}$ is $N_{1,2}(E)$, the density

[10] See, for example, Doetsch (*11*). Our notation using the symbol s for the transform parameter is consistent with most of the mathematical literature. In physical problems, where the transform parameter is $(kT)^{-1}$, as here, the notation $\beta = -1/kT$ is common. We do not use this notation here because β is used in another sense.

of states of the combined system, and from (6-28) and 6-29) there follows immediately

$$N_{1,2}(E) = \int_0^E N_1(E - x)N_2(x)\,dx \quad \text{or} \quad \int_0^E N_1(x)N_2(E - x)\,dx \tag{6-32}$$

In other words, if the partition function factorizes into two factors, the densities of separated and combined systems are related by convolution.[11] The quantum mechanical analog of Eq. (6-32) is

$$N_{1,2}(E) = \sum_{x=0}^{E} N_1(E - x)N_2(x)\,\delta x = \sum_{x=0}^{E} N_1(E - x)W_2(x), \quad \text{etc.} \tag{6-33a}$$

or, after multiplying both sides by δE,

$$W_{1,2}(E) = \sum_{x=0}^{E} W_1(E - x)W_2(x), \quad \text{etc.} \tag{6-33b}$$

where $W_1(E)$ and $W_2(E)$ are the number of states of constituent systems and $W_{1,2}(E)$ is the number of states of the combined system.[12] Writing $L(s) = \ell(s)/s$, there follows from (6-27) for a combined system consisting of parts 1 and 2,

$$\begin{aligned} L_{1,2}(s) = \mathfrak{L}\{G_{1,2}(E)\} &= \ell_1(s)\ell_2(s)/s \\ &= L_1(s)\ell_2(s) \quad \text{or} \quad \ell_1(s)L_2(s) \end{aligned} \tag{6-34}$$

Applying the convolution theorem to the product $L(s)\ell(s)$,

$$\begin{aligned} G_{1,2}(E) = \mathfrak{L}^{-1}\{L_1(s)\ell_2(s)\} &= \int_0^E G_1(E - x)N_2(x)\,dx \\ &\quad \text{or} \quad \int_0^E G_1(x)N_2(E - x)\,dx \qquad \text{(6-35a)} \\ = \mathfrak{L}^{-1}\{\ell_1(s)L_2(s)\} &= \int_0^E N_1(x)G_2(E - x)\,dx \\ &\quad \text{or} \quad \int_0^E N_1(E - x)G_2(x)\,dx \qquad \text{(6-35b)} \end{aligned}$$

The quantum-mechanical equivalent of these relations is

$$G_{1,2}(E) = \sum_{x=0}^{E} G_1(E - x)W_2(x), \quad \text{etc.} \tag{6-36}$$

[11] Kubo (*13*, p. 103). The principle of convolution was first used by Rice and Ramsperger (*14*).

[12] For two degenerate independent harmonic oscillators, $W_1(E)$ and $W_2(E)$ are each given by the appropriate form of Eq. (6-5); when these are substituted into Eq. (6-33b,) Eq. (6-8) is recovered.

Therefore the total sum of states of a combined system whose partition function factorizes into two factors is given by the convolution of the *density* (or number of states) of one part and the *integrated density* of the other part.

The result (6-35) or (6-36), like the previous results (6-32) or (6-33), depends on the partition function for the total system $Q_{1,2}$ being factorizable into two parts, Q_1 and Q_2. Such factorizability arises if there are no cross-terms in the total energy, i.e., if the degrees of freedom involved in Q_1 are independent of, and therefore separable from, those in Q_2.

Therefore the general problem of calculating $N(E)$, $W(E)$, or $G(E)$ for a complex system of many degrees of freedom can be sometimes simplified by calculating these functions separately for each of two (or more) groups of degrees of freedom which are separable, and then by obtaining the density and related functions for the total system by convolution. Such two groups of degrees of freedom that can be usefully treated separately are the vibrational and rotational degrees of freedom because of their very different quantization.

5. Approximation to $N(E)$ and $G(E)$ for rotational states

A good approximation to $N_r(E)$ and $G_r(E)$ can be obtained very easily and therefore will be considered first.

Separability of rotations It shall be assumed that in *every* case the rotational degrees of freedom of a molecule can be represented as a collection of free independent one- and two-dimensional rigid rotors. The one-dimensional rotors (partition function $Q_{r=1}$), characterized by one axis of rotation and one moment of inertia, will be useful as a representation of internal rotations, and the two-dimensional rotor (partition function $Q_{r=2}$), characterized by two axes of rotation and two equal moments of inertia, will be useful as a representation of rotations in a "loose" transition state. If the three external rotations of a very prolate symmetric top are involved, it shall be assumed that the rotation having the smallest moment of inertia is separable from the other two, so that the partition function for the three rotations of a symmetric top ($Q_{r=3}$) can be written

$$Q_{r=3} = Q_{r=1} Q_{r=2} \tag{6-37}$$

Similarly, if the three external rotations of an asymmetric (or similar) top are involved, with three unequal moments of inertia, it shall be assumed that the total partition function $Q_{r=3}$ is the product of three one-rotor partition functions

$$Q_{r=3} = Q^{I}_{r=1}\, Q^{II}_{r=1}\, Q^{III}_{r=1} \tag{6-38}$$

where the superscripts denote that the moments of inertia are different.

Hence it will suffice to consider only $Q_{r=1}$ and $Q_{r=2}$.

Classical rotational partition function Q_r ***and its inversion*** The energy levels for a one-dimensional rotor are given by Eq. (6-11), and therefore the partition function is

$$\begin{aligned} Q_{r=1} &= (2/\sigma)\int_0^\infty \exp\,(-m^2\,\hbar^2/2IkT])\,dm \\ &= (8\pi^3 IkT)^{\frac{1}{2}}/h\sigma \end{aligned} \tag{6-39}$$

The energy levels of a two-dimensional rotor are given by Eq. (6-12), and hence

$$\begin{aligned} Q_{r=2} &= (1/\sigma)\int_0^\infty (2\mathcal{J} + 1)\exp\,[-\mathcal{J}(\mathcal{J} + 1)\hbar^2/2IkT]\,d\mathcal{J} \\ &= 8\pi^2 IkT/h^2\sigma \end{aligned} \tag{6-40}$$

Division by σ has been performed in both cases to allow for symmetry considerations.

The results (6-39) and (6-40) assume that the rotations behave classically, i.e., that the spacing between energy levels is small (moment of inertia I not too small, i.e., larger than that for H_2 or D_2), and that integration is permissible (temperature above $\sim 20°K$). These conditions are usually satisfied in systems of chemical interest.

In the general case when there are n_1 one-dimensional rotors and n_2 two-dimensional rotors, all separable, the total partition function is

$$Q_r = Q_r'(kT)^{r/2} \tag{6-41}$$

where

$$Q_r' = \left(\frac{8\pi^2}{h^2}\right)^{r/2} \pi^{n_1/2} \prod_{i=1}^{n_1}\left(\frac{I_i}{\sigma_i^2}\right)^{1/2} \prod_{j=1}^{n_2}\left(\frac{I_j}{\sigma_j}\right) \tag{6-42}$$

and $r = n_1 + 2n_2$. The exponent r may be considered to be the total number of rotations; note that a two-dimensional rotor counts for two rotations.

In Laplace transform notation,

$$Q_r \equiv \ell_r(s) = Q_r' s^{-r/2} \qquad (s = 1/kT) \tag{6-43}$$

From tables of Laplace transforms

$$\mathcal{L}^{-1}\{s^{-\mathrm{r}/2}\} = \frac{E^{(\mathrm{r}/2)-1}}{\Gamma(\mathrm{r}/2)}; \qquad \mathcal{L}^{-1}\{s^{-(1+\mathrm{r}/2)}\} = \frac{E^{\mathrm{r}/2}}{\Gamma(1+\mathrm{r}/2)} \tag{6-44}$$

so that there follows immediately the result

$$N_{\mathrm{r}}(E) = \frac{Q_{\mathrm{r}}'E^{(\mathrm{r}/2)-1}}{\Gamma(\mathrm{r}/2)}; \qquad G_{\mathrm{r}}(E) = \frac{Q_{\mathrm{r}}'E^{\mathrm{r}/2}}{\Gamma(1+\mathrm{r}/2)} \tag{6-45}$$

With $\mathrm{r} = 2n$, this is the same result obtained more laboriously from Dirichlet's integral in Section 3.

For two rotations, $\mathrm{r} = 2$, and noting that $\Gamma(1) = \Gamma(2) = 1$, there follows from (6-45)

$$N_{\mathrm{r}=2}(E) = Q'_{\mathrm{r}=2}, \qquad G_{\mathrm{r}=2}(E) = Q'_{\mathrm{r}=2}E \tag{6-46}$$

Thus the approximate density of states for two rotations is a constant. Note that the constant $Q'_{\mathrm{r}=2}$ is equal to $8\pi^3(I_1I_2)^{1/2}/h^2\sigma_1\sigma_2$ if the two rotations consist of two (independent) one-dimensional rotors (indexed 1 and 2), and equal to $8\pi^2I/h^2\sigma$ if they consist of one two-dimentional rotor.

The approximation (6-45) is compared in (*1*, Tables II and III) with the exact $N_{\mathrm{r}}(E)$ and $G_{\mathrm{r}}(E)$ for one- and two-dimensional rotors, and the result is that (6-45) gives an acceptable accuracy at energies above about 500 cm^{-1}; for a chemical system this is a quite low energy. Therefore the approximation is suitable for use at all energies in rate calculations where accurate behavior near reaction threshold is not desired, e.g., when averaging over some distribution of energies is involved. Note, however, that rate calculations near threshold involve the lowest energy states of the transition state (but not of the molecule). Given the usual ambiguity about the structure of the transition state, the error introduced by the classical approximation in the rotational part of $G^*_{\mathrm{vr}}(E)$ at low E is insignificant compared with the error in the vibrational part of $G^*_{\mathrm{vr}}(E)$ arising from uncertain assignment of vibrational states of the transition state.

For the same reason, there is little advantage in considering a *restricted* internal rotor: The transition state is not known well enough to make worthwhile explicit consideration of energy levels inside the potential barrier hindering rotation, while in the molecule, if it has n degrees of freedom, the average excitation energy per degree of freedom is at least E_0/n, and this is usually larger than barrier height, so that the rotor is basically a free rotor for all practical purposes.

The good low-energy behavior of the classical approximation to $N_{\mathrm{r}}(E)$ and $G_{\mathrm{r}}(E)$ is consistent with the inverse relationship between energy and temperature involved in the Laplace transformation: If the partition function being inverted is valid down to low temperatures, $N(E)$ and $G(E)$ so obtained will be valid down to low energies. A slight nuisance is that when $r = 1$, $N_r(E)$ of Eq. (6-45) goes off to infinity as $E \to 0$, and this must be taken into account in programming.

6. Direct evaluation of $N_{vr}(E)$ and $G_{vr}(E)$ in simple systems

In the previous section is presented a very simple approximation to $N_r(E)$ and $G_r(E)$ which, with a certain proviso, is valid essentially at all energies. We now anticipate the result of the following sections, where it is concluded that for vibrational states below about 3000 or 4000 cm^{-1} no sufficiently accurate approximation to $N_v(E)$ and $G_v(E)$ is available. Therefore $N_v(E)$ and $G_v(E)$ below these energies must be obtained by direct count, while this is not necessary for $N_r(E)$ and $G_r(E)$.

Convolution for vibrational–rotational states To obtain $N_{vr}(E)$ and $G_{vr}(E)$ for vibrational–rotational states of a system where vibrations and rotations are independent and therefore separable, one may usefully put to profit relations (6-33) and (6-36) obtained by convolution. These general relations, applied to the specific problem on hand, show how to combine *exact* $N_v(E)$ with approximate (but very good) $N_r(E)$ [or $G_r(E)$] to yield an essentially exact $N_{vr}(E)$ and $G_{vr}(E)$. Because of the simple form of $N_r(E)$ the expenditure of mathematical labor (and of machine time) for $N_{vr}(E)$ and $G_{vr}(E)$ is thus no more than necessary for good accuracy. Direct and exact counting of vibrational–rotational states not involving an approximation for the rotational part is of course possible, but the much greater machine time does not make the modest increase in accuracy (mostly below 500 cm^{-1}) worthwhile.

Substituting (6-45) for $N_r(E)$ and $G_r(E)$, (6-33a) yields

$$N_{vr}(E) = \frac{Q_r'}{\Gamma(r/2)} \sum_{x=0}^{E} W_v(x)(E - x)^{(r/2)-1} \tag{6-47}$$

and (6-36) yields

$$G_{vr}(E) = \frac{Q_r'}{\Gamma(1 + r/2)} \sum_{x=0}^{E} W_v(x)(E - x)^{r/2} \tag{6-48}$$

The quantum form of the convolution integral was deliberately chosen since (6-47) and (6-48) are meant to be used at low energies where vibrational states are widely spaced; for the same reason, expressions are used that involve $W_v(x)$ rather than the less appropriate $N_v(x)$. Note that x in (6-47) and (6-58) means vibrational energy, and to underline that fact E_v is sometimes written[13] for x.

[13] Equations (6-47) and (6-48) were first derived by Marcus (*15*), who used the notation $x \equiv E_v$, $W_v(x) = P(E_v)$.

Direct enumeration of vibrational–rotational states The only nontrivial part of (6-47) and (6-48) is $W_v(x) \equiv W_v(E_v)$, and it is given by Eq. (6-8). Thus the direct[14] enumeration of vibrational–rotational states may be accomplished by a procedure similar to that shown in Section 2. The major difference is that as a consequence of the smooth-function approximation to the rotational part, there is now a state or states at every value of E, and that all allowed values of $E - E_v$ must be generated and raised to the proper power. For purposes of subsequent numerical integration, it is usually sufficient to generate $N_{vr}(E)$ or $G_{vr}(E)$ in 100-cm^{-1} intervals.

In Table 6-2 is worked out the explicit calculation, at 100-cm^{-1} intervals, of the first few values of $N'_{vr}(E) = (Q_r')^{-1}N_{vr}(E)$ and of $G'_{vr}(E) = (Q_r')^{-1}G_{vr}(E)$ for a collection of seven oscillators, harmonic or anharmonic, and two rigid free rotors; the purpose of the division by Q_r' is to make the results independent of the properties of the individual rotors. This is a particularly simple case since the terms $(E - E_v)$ appear only to unit power in $G'_{vr}(E)$, and contribute merely a factor of one in $N'_{vr}(E)$. To render the calculation immediately verifiable, the collection of oscillators is the same as that used in Table 6-1, so that E_v and $W_v(E_v)$ may be found there.

It may be noted that the numbers in the column under $N'_{vr}(E)$ in Table 6-2 are the same as the numbers under $G_v(E)$ in Table 6-1, i.e., the so-called reduced density $N'_{vr}(E)$ for r = 2 (two rotors) is the same as $G_v(E)$ for the oscillators alone. This result may be easily generalized (*16*) as follows.

Write the convolution integral (6-30) for v vibrations and r + 2 rotations as

$$N'_{v(r+2)}(E) = \int_0^E N'_{vr}(x)N'_{r=2}(E - x)\,dx \tag{6-49}$$

Since by Eq. (6-46) $N'_{r=2}(E - x) = 1$, it follows that

$$N'_{v(r+2)}(E) = \int_0^E N'_{vr}(x)\,dx = G'_{vr}(E) \tag{6-50}$$

When r = 0, the reduced densities (integrated or not) become the densities (integrated or not) of pure vibrational states, $N_v(E)$ and $G_v(E)$, respectively, i.e.,

$$[G'_{vr}(E)]_{r=0} = G_v(E); \qquad [N'_{vr}(E)]_{r=0} = N_v(E) \tag{6-51}$$

Equation (6-50) remains valid for r = 0, so that

$$[N'_{vr}(E)]_{r=2} = G_v(E) \tag{6-52}$$

Relations (6-50)–(6-52) are useful for reducing the amount of computation

[14] In the literature Eqs. (6-47) and (6-48) are usually represented as the "exact" enumeration of rotational–vibrational states, although they actually involve an approximation for the rotational part.

TABLE 6-2

$N'_{vr}(E)$ AND $G'_{vr}(E)$ BY DIRECT ENUMERATION OF STATES[a]

E (cm^{-1})	$N'_{vr}(E)$	$G'_{vr}(E)$
		Harmonic[b]
0	1	0
100	1	$1 \times 100 = 100$
200	2	$1 \times 200 = 200$
300	5	$1 \times 300 + 1 \times (300 - 200) = 400$
400	6	$1 \times 400 + 1 \times (400 - 200) + 3 \times (400 - 300) = 900$
500	9	$1 \times 500 + 1 \times (500 - 200) + 3 \times (500 - 300) + 1 \times (500 - 400) = 1500$
		Anharmonic[b]
0	1	0
100	1	$1 \times 100 = 100$
200	2	$1 \times 200 + 1 \times (200 - 196) = 204$
300	5	$1 \times 300 + 1 \times (300 - 196) + 3 \times (300 - 294) = 422$
400	6	$1 \times 400 + 1 \times (400 - 196) + 3 \times (400 - 294) + 1 \times (400 - 388) = 934$
500	9	$1 \times 500 + 1 \times (500 - 196) + 3 \times (500 - 294) + 1 \times (500 - 388) + 3 \times (500 - 490) = 1564$

[a] For a system consisting of two rigid free rotors and seven oscillators (same oscillators as those of Table 6-1). $N'_{vr}(E) = \sum W_v(E_v)$ [cf. Eq. (6-47)]. $G'_{vr}(E) = \sum \{W_v(E_v) \times (E - E_v)\}$ (cm^{-1}) [cf. Eq. (6-48)]. Note that $\Gamma(1) = \Gamma(2) = 1$.

[b] All terms of sum shown.

since they show that except for the definition of r, $N'_{vr}(E)$ and $G'_{vr}(E)$ are basically equivalent. This equivalence arises from the use of approximation (6-45) for the rotations, but does not otherwise depend on the use of any particular expression for the vibrations; in fact, relations (6-50)–6-52) could also have been derived using the quantum form (6-33a) of the convolution integral, for the vibrational part.

7. Approximations based on inversion of classical partition function

We now follow essentially the historical development of the subject, even though the techniques originally used were somewhat different. Thus Eqs. (6-55) were first obtained from Dirichlet's integral (*7*, p. 492, Eq. 111.2) and Eqs. (6-62) by convolution (*17*).

Vibrational states The classical vibrational partition function for one harmonic oscillator is $Q_{v=1} = kT/h\nu$, and for v independent oscillators is therefore $Q_v^{\text{class}} = (kT)^v/\prod_{i=1}^{v} h\nu_i$, which in Laplace transform notation becomes

$$Q_v^{\text{class}} = 1\Big/s^v \prod_{i=1}^{v} h\nu_i \qquad (s = 1/kT) \tag{6-53}$$

Since

$$\mathcal{L}^{-1}\{1/s^v\} = E^{v-1}/\Gamma(v); \qquad \mathcal{L}^{-1}\{1/s^{v+1}\} = E^v/\Gamma(v+1) \tag{6-54}$$

then the *classical* density and sum of states are, respectively,

$$N_v(E) = E^{v-1}\Big/\left[\Gamma(v)\prod_{i=1}^{v} h\nu_i\right]; \quad G_v(E) = E^v\Big/\left[\Gamma(v+1)\prod_{i=1}^{v} h\nu_i\right] \tag{6-55}$$

As is well known, the true (quantum-mechanical) vibrational partition function has the form (6-53) only at high temperature, which means that (6-55) will approach the true (quantum-mechanical) count only at high energies, given the reciprocal relation between temperature and energy of the Laplace transformation. At low energies, (6-53) will give results that are too low [cf. (*1*, Table V)].

One obvious source of error is the zero-point energy of the v oscillators, $E_z = \frac{1}{2}\sum_{i=1}^{v} h\nu_i$, which is of course ignored in the classical expression (6-55). E_z is fixed and only energy $E \geqslant E_z$ of the classical oscillators should be available for distribution among the oscillators. Since E in $N_v(E)$ and $G_v(E)$ should refer only to this distributable part of total energy (cf. definition of E in Section 1), we can expect that (6-55) will perform better if E on the right-hand side of (6-55) is not allowed to drop too much below E_z. Taking advantage of the fact that $N_v(E)$ and $G_v(E)$ are defined only for positive E, this can be accomplished by shifting the zero of energy to the vicinity of E_z, say aE_z ($a \leqslant 1$), so that energies below aE_z will be negative, and therefore excluded from $N_v(E)$. The result of such energy zero shift is that Eqs. (6-55) become

$$N_v(E) = (E + aE_z)^{v-1} \Big/ \left[\Gamma(v) \prod_{i=1}^{v} h\nu_i\right] \tag{6-56}$$

and

$$G_v(E) = (E + aE_z)^{v} \Big/ \left[\Gamma(v + 1) \prod_{i=1}^{v} h\nu_i\right] \tag{6-57}$$

which for $a = 1$ is referred to as the semiclassical approximation and was used in that form for the first time by Marcus and Rice (*18*).

The factor a appears in front of E_z in the expectation that appropriate manipulation of E_z may further improve the approximation (*19–21*). In expression (6-57) for $G_v(E)$ Whitten and Rabinovitch (*22*) have used with good results the empirical relations

$$\begin{aligned} a &= 1 - \beta w \\ \log_{10} w &= -1.0506(E/E_z)^{1/4} \qquad (E > E_z) \\ w^{-1} &= 5(E/E_z) + 2.73(E/E_z)^{1/2} + 3.51 \qquad (E < E_z) \end{aligned} \tag{6-58}$$

where

$$\beta = \nu_d(v - 1)/v \tag{6-59}$$

ν_d is the frequency dispersion parameter,

$$\nu_d = \langle\nu^2\rangle/\langle\nu\rangle^2, \qquad \langle\nu^2\rangle = (1/v)\sum_{i=1}^{v} \nu_i^2, \qquad \langle\nu\rangle = (1/v)\sum_{i=1}^{v} \nu_i \tag{6-60}$$

Since the relation between $G_v(E)$ and $N_v(E)$ should be that of a function and its derivative, note that $G_v(E)$ with an energy-dependent a does not yield upon differentiation $N_v(E)$ of the simple form given by Eq. (6-56).

Vibrational–rotational states The classical vibrational–rotational partition function for independent vibrations and rotations is, from (6-41) and (6-53),

$$Q_{vr}^{class} = Q_r' \Big/ s^{v+(r/2)} \prod_{i=1}^{v} h\nu_i \tag{6-61}$$

which, apart from a constant factor and different exponent, is identical with (6-53), so that there follow immediately

$$\begin{aligned} N_{vr}(E) &= Q_r' E^{v-1+(r/2)} \Big/ \left[\Gamma(v + r/2) \prod_{i=1}^{v} h\nu_i\right] \\ G_{vr}(E) &= Q_r' E^{v+(r/2)} \Big/ \left[\Gamma(v + 1 + r/2) \prod_{i=1}^{v} h\nu_i\right] \end{aligned} \tag{6-62}$$

For a derivation of these formulas by convolution see Exercise 2.

The problem with vibrational zero-point energy persists, of course, and we may therefore correct, as in the case of pure vibrational states, by writing

$$\left.\begin{array}{l} N_{\mathrm{vr}}(E) \\ {\scriptstyle (k=0)} \\ \\ G_{\mathrm{vr}}(E) \\ {\scriptstyle (k=1)} \end{array}\right\} = \frac{Q_{\mathrm{r}}'(E + a_k E_z)^{\mathrm{v}-1+k+(\mathrm{r}/2)}}{\Gamma(\mathrm{v} + k + \mathrm{r}/2) \prod_{i=1}^{\mathrm{v}} h\nu_i}; \qquad a_k = 1 - \beta_k w \qquad (6\text{-}63)$$

The w of Eq. (6-58) may still be used (*22,23*) but β needs to be redefined. A definition which was found (*16*) to give very good results is

$$\beta_k = \nu_{\mathrm{d}}(\mathrm{v} - 1)[\mathrm{v} - 1 + k + (\mathrm{r}/2)]/\mathrm{v}^2 \qquad (6\text{-}64)$$

where ν_{d} is the same as in Eq. (6-60). This definition, while slightly different from the one proposed (*23*) originally, has the virtue that it preserves the rule of (6-50), whereby $G'_{\mathrm{vr}}(E)$ can be obtained from the formula for $N'_{\mathrm{vr}}(E)$ by replacing everywhere r with r + 2, and whereby one and the same formula yields $G_{\mathrm{v}}(E)$ ($k = 1$) and $N_{\mathrm{v}}(E)$ ($k = 0$) by letting $\mathrm{r} = 0$ [note that with $\mathrm{r} = 0$, $k = 1$, Eq. (6-64) reduces to Eq. (6-59)].

The performance of the classical approximation to $N_{\mathrm{vr}}(E)$, as well as of the two semiclassical approximations, one with $a = 1$ and the other with energy-dependent a, is compared in Table V of (*1*). The classical approximation gives a gross underestimate, and the semiclassical approximation with $a = 1$ gives a gross overestimate that improves at higher energies, just what one would expect. As always when there is an energy-dependent factor in front of E_z, differentiation of $G_{\mathrm{vr}}(E)$ [Eq. (6-61), $k = 1$] with respect to E gives $N_{\mathrm{vr}}(E)$ of a form that is different from that given by (6-61) with $k = 0$; nevertheless, $N_{\mathrm{vr}}(E)$ or $G_{\mathrm{vr}}(E)$ of (6-61) represents a very good approximation [Table V, (*1*)].

A different semiclassical formula for the density of states was derived by Kinsey[15] using the WKB approximation, but it has not been extensively tested.

8. Approximations based on the inversion of the quantum mechanical partition function

The quantum mechanical vibrational partition function, unlike its classical counterpart, is valid at all temperatures, so that it may be expected that its inversion will yield a smooth-function approximation to $N_{\mathrm{vr}}(E)$ or $G_{\mathrm{vr}}(E)$

[15] Kinsey (*24*). For applications of the WKB method to vibration–rotation states of diatomics see Dickinson and Bernstein (*25*) and Mahan (*26*).

valid to much lower energies than was the case with the inversion of the classical Q_{vr} [Eq. (6-62)]. A complicating factor, however, is that the quantum Q_v is no longer simply proportional to a power of s, the transform parameter. Thus instead of (6-61), we will have $Q_{vr} = \ell(s)$, where $\ell(s)$ is now sufficiently complicated to require the actual evaluation of the inverse transform (6-28).

By the Fourier–Mellin integration theorem, we have quite generally for any $\ell(s)$

$$\mathcal{L}^{-1}\left\{\frac{\ell(s)}{s^k}\right\} = \frac{1}{2\pi i} \int_{c-i\infty}^{c+i\infty} \frac{\ell(s)e^{sE}\,ds}{s^k} = \Upsilon \qquad (6\text{-}65)$$

The integral Υ involves s as the complex variable $s = x + iy$, and the path of integration is along a straight line parallel to the imaginary axis with abscissa c. There are essentially two methods[16] that have been used for the approximate evaluation of Υ: the method of residues, which yields a polynomial related to the generalized Bernoulli polynomial, and the method of steepest descents. They will be discussed in turn.

We shall assume familiarity with standard procedures of integration in the complex plane and therefore will mention them, when necessary, only in the barest outline.

Evaluation of inversion integral by Cauchy's residue theorem The path of integration in (6-65) can be closed (Jordan's lemma), and then by Cauchy's theorem the value of the integral Υ is given by the sum of residues at poles [see, e.g., Korn (*29*, Section 7.7)]. A residue at pole of order (n + 1), located at $s = s_1$, is the coefficient of s^n in Taylor-series expansion of ($s^{n+1} \times$ integrand) about $s = s_1$. Thus the first, and major, part of the problem is to find a suitable expansion for the integrand of Υ.

For one harmonic oscillator of frequency ν, the quantum partition function is

$$Q_{v=1} = \frac{1}{(1 - e^{-h\nu/kT})} = \frac{e^{h\nu/2kT}}{2\sinh(h\nu/2kT)} \qquad (6\text{-}66)$$

if the ground vibrational level is taken as the zero of energy. We know therefore that for a system comprising a collection of v quantum harmonic oscillators, the vibrational part of the function $\ell(s)$ in (6-65) will involve a factor of the form

$$\ell_v(s) = \frac{1}{2^v} \prod_{j=1}^{v} \frac{e^{h\nu_j s/2}}{\sinh(h\nu_j s/2)} \qquad (6\text{-}67)$$

and this product of hyperbolic sines must now be expanded in powers of s.

---

[16] A third, less common method is based on Khinchin's central limit theorem [see Khinchin (*27*)]. It was used in connection with the energy-level density problem by Lurçat and Mazur (*28*).

This is the difficult part of the expansion of the total integrand, since the rotational part contributes only $s^{-r/2}$ [Eq. (6-43)].

A ready-made expansion, useful for the present purpose, may be obtained from relations involving generalized Bernoulli polynomials (*30*, Vol. I, p 40, Eq. (38); *31–33*; *34*, Chapter 6, Sections 2 and 4):

$$\prod_{j=1}^{n} \frac{(\alpha_j z)}{\sinh(\alpha_j z)} = \sum_{\substack{i=0 \\ (\text{even } i)}}^{\infty} D_i^{(n)}(\alpha_1, \ldots, \alpha_n) \frac{z^i}{i!} \tag{6-68}$$

which is interesting in that it consists of even powers of z only; $D_i^{(n)}(\alpha_1, \ldots, \alpha_n)$ is a coefficient which will be discussed later. Comparing the left-hand side of (6-68) with (6-67), we note that the product $\prod_{j=1}^{\mathsf{v}} \exp(h\nu_j s/2) = \exp(\frac{1}{2} \sum_j h\nu_j s)$ spoils the identity. However $\frac{1}{2} \sum_{j=1}^{\mathsf{v}} h\nu_j$ is E_z, the zero-point energy of the v oscillators, which suggests that $\ell_{\mathsf{v}}(s)$ of Eq. (6-67) would have the right form if we used instead a quantum vibrational partition function with energy zero at the potential minimum. With this energy zero, the partition function for one oscillator is obtained by multiplying (6-66) by $e^{-h\nu/2kT}$; for v oscillators the factor which multiplies (6-67) is therefore $\prod_{j=1}^{\mathsf{v}} e^{-h\nu_j s/2}$. Hence the complete partition function for r rotations and v vibrations, with vibrational energy zero at the potential minimum, is, in Laplace transform notation,

$$\ell_{\mathrm{vr}}(s) = \frac{Q_{\mathrm{r}}'}{2^{\mathsf{v}} s^{\mathsf{r}/2} \prod_{j=1}^{\mathsf{v}} \sinh(h\nu_j s/2)} \tag{6-69}$$

The assumption implicit in (6-69) is that vibrations and rotations are independent, and that rotations behave classically as discussed in Section 5. The integrand of Υ is therefore

$$\text{integrand} = \frac{Q_{\mathrm{r}}' e^{s\varepsilon}}{2^{\mathsf{v}} s^{k+(\mathsf{r}/2)} \prod_{j=1}^{\mathsf{v}} \sinh(h\nu_j s/2)}$$

$$= \frac{Q_{\mathrm{r}}' e^{s\varepsilon}}{s^{\mathsf{v}+k+(\mathsf{r}/2)} \prod_j h\nu_j} \prod_{j=1}^{\mathsf{v}} \frac{(h\nu_j s/2)}{\sinh(h\nu_j s/2)} \tag{6-70}$$

To underline the fact that total energy now includes vibrational energy referred to vibrational potential minimum as zero, write ε for E in (6-70). Then $\varepsilon = E + E_z$. From (6-68) it follows that

$$\text{integrand} = \frac{Q_{\mathrm{r}}' e^{s\varepsilon}}{s^{\mathsf{v}+k+(\mathsf{r}/2)} \prod_j h\nu_j} \sum_{\substack{i=0 \\ (\text{even } i)}}^{\infty} D_i^{(\mathsf{v})}\left(\frac{h\nu_1}{2}, \ldots, \frac{h\nu_{\mathsf{v}}}{2}\right) \frac{s^i}{i!} \tag{6-71}$$

The coefficient $D_i^{(\mathsf{v})}(h\nu_1/2,\ldots, h\nu_{\mathsf{v}}/2)$ is defined by (32,35)

$$D_i^{(\mathsf{v})}\left(\frac{h\nu_1}{2},\ldots,\frac{h\nu_{\mathsf{v}}}{2}\right) = \sum\left\{\frac{i!}{m_1!\,m_2!\ldots m_{\mathsf{v}}!}\right.$$

$$\times\left.\left[\left(\frac{h\nu_1}{2}\right)^{m_1}\cdots\left(\frac{h\nu_{\mathsf{v}}}{2}\right)^{m_{\mathsf{v}}}\right]D_{m_1}\cdots D_{m_{\mathsf{v}}}\right\} \quad (6\text{-}72)$$

where the sum[17] is taken over all positive even values (zero included) of $m_1, m_2,\ldots, m_{\mathsf{v}}$ such that

$$m_1 + m_2 + \cdots + m_{\mathsf{v}} = i \quad (6\text{-}73)$$

D_m is related to the Bernoulli numbers B_m through

$$D_{2m} = 2(-1)^m(2^{2m-1} - 1)B_m; \qquad m > 1; \qquad D_0 = 1 \quad (6\text{-}74)$$

The first three values of $D_i^{(\mathsf{v})}$ [this shortened notation will be used henceforth for $D_i^{(\mathsf{v})}(h\nu_1/2,\ldots, h\nu_{\mathsf{v}}/2)$] are (33,35)

$$D_0^{(\mathsf{v})} = 1 \quad (6\text{-}75)$$

$$D_2^{(\mathsf{v})} = -\tfrac{1}{3}\sum_j (\tfrac{1}{2}h\nu_j)^2 \quad (6\text{-}76)$$

$$D_4^{(\mathsf{v})} = \tfrac{1}{3}\left[\sum_j (\tfrac{1}{2}h\nu_j)^2\right]^2 + \tfrac{2}{15}\sum_j (\tfrac{1}{2}h\nu_j)^4 \quad (6\text{-}77)$$

Since both the expansions of $e^{s\varepsilon}$ and $\ell_{\mathsf{v}}(s)$ have unity as the leading term [$D_0^{(\mathsf{v})} = 1$], we see immediately from (6-71) that the integrand has a pole of order ($\mathsf{v} + k + \mathsf{r}/2$) at $s = 0$. There are lower order poles on the imaginary axis, but the approximation is now made that only the residue at the pole at origin ($s = 0$) is significant, the other poles merely contributing oscillating terms which are assumed to approximately cancel out.

Now,

$$(s^{\mathsf{v}+k+(\mathsf{r}/2)} \times \text{integrand}) = \frac{Q_{\mathsf{r}}'}{\prod_j h\nu_j}\left\{e^{s\varepsilon}\sum_{\substack{i=0\\(\text{even } i)}}^{\infty} D_i^{(\mathsf{v})}\frac{s^i}{i!}\right\} \quad (6\text{-}78)$$

To obtain the residue at $s = 0$, we need the coefficient of $s^{\mathsf{v}+k-1+(\mathsf{r}/2)}$ in (6-78). Expanding $e^{s\varepsilon}$ in powers of s and multiplying together the two series inside the compound bracket of (6-78), it is found that the coefficient of s^{n} is

$$\sum_{\substack{i=0\\(\text{even } i)}}^{\mathsf{n} \text{ or } \mathsf{n}-1} \frac{D_i^{(\mathsf{v})}\varepsilon^{\mathsf{n}-i}}{(\mathsf{n} - i)!\,i!} \quad (6\text{-}79)$$

[17] The factorial term in the sum will be recognized as the multinomial coefficient.

where the summation is taken to $i = \mathsf{n}$ if n is even and to $i = \mathsf{n} - 1$ if n is odd; therefore the series (6-79) has $(\mathsf{n}/2) + 1$ terms for n even and $(\mathsf{n} + 1)/2$ terms for n odd. Writing $\varepsilon = E + E_z$, there finally follows the result

$$\left.\begin{array}{l} N_{vr}(E) \\ {\scriptstyle (k=0)} \\ \\ G_{vr}(E) \\ {\scriptstyle (k=1)} \end{array}\right\} \simeq (\text{residue at } s = 0) = \frac{Q_r'}{\prod_j h\nu_j} \sum_{\substack{i=0 \\ (\text{even } i)}}^{\mathsf{n} \text{ or } \mathsf{n}-1} \frac{D_i^{(\mathsf{v})}(E + E_z)^{\mathsf{n}-i}}{(\mathsf{n} - i)!\, i!} \tag{6-80}$$

where $\mathsf{n} = \mathsf{v} + k - 1 + (\mathsf{r}/2)$.

Equation (6-80) was first derived for $k = 1$, $\mathsf{r} = 0$ by Thiele (*35*), and in essentially the above general form by Forst *et al.* (*36,37*); see also (*38*). A truncated form of (6-80) was obtained by Haarhoff (*39*) and also by Schlag and Sandsmark (*40*). Equation (6-80) was extensively tested (*16, 36–38*) against direct count (Section 6) and found to represent an excellent approximation to harmonic $N_{vr}(E)$ and $G_{vr}(E)$, valid to very low energies ($<3000\ \text{cm}^{-1}$).

It will be noted that the leading term of (6-80) is the semiclassical approximation (6-56) or (6-57) with $a = 1$, so that (6-80) may be considered a generalization of the semiclassical approximation. The importance of terms beyond the first increases as the energy decreases (the terms are alternately positive and negative), so that at low energies (below $\sim 5000\ \text{cm}^{-1}$) all the $(\mathsf{n}/2) + 1$ or $(\mathsf{n} + 1)/2$ terms, as the case may be, must be computed for good accuracy, whereas at high energies the first few terms are usually sufficient. For this reason the truncated expression of Haarhoff (*39*) is also quite adequate at higher energies. Since his formula is relatively easy to program, it is quite useful for some purposes and is therefore given here, transcribed into our notation:

$$\left.\begin{array}{l} N_{vr}(E) \\ {\scriptstyle (k=0)} \\ \\ G_{vr}(E) \\ {\scriptstyle (k=1)} \end{array}\right\} = Q_r' \left\{ \left(\frac{2}{\pi \mathsf{v}}\right)^{1/2} \left(\frac{\mathsf{v}}{\mathsf{n} + 1}\right)^{\mathsf{n}+(1/2)} \left[1 - \frac{1}{12(\mathsf{n} + 1)}\right] \frac{\lambda(h\langle\nu\rangle)^{\mathsf{n}-\mathsf{v}}}{1 + \eta} \right\}$$

$$\times \left[\left(1 + \frac{\eta}{2}\right)\left(1 + \frac{2}{\eta}\right)^{\eta/2}\right]^{\mathsf{n}+1} \left(1 - \frac{1}{(1 + \eta)^2}\right)^{\beta_H} \tag{6-81}$$

where $\mathsf{n} = \mathsf{v} + k - 1 + (\mathsf{r}/2)$, $\beta_H = [\mathsf{n}(\mathsf{n} - 1)\nu_d - \mathsf{v}(\mathsf{n} + 1)]/6\mathsf{v}$, $\eta = E/E_z$, $1/\lambda = \prod_i (\nu_i/\langle\nu\rangle)$, and ν_d is given by Eq. (6-60). The β_H should not be confused with β of Eq. (6-59) or β_k of Eq. (6-64).

Comparison of Eqs. (6-80) and (6-81) [(*1*, Table V)] at various energies and various values of r shows that both perform quite well, but at higher values of r Eq. (6-80) is better.

Evaluation of inversion integral by the method of steepest descents The method of steepest descents (*41*, p. 503) is based on the proposition that at some value of s, say $s = s^\star$, the integrand will have a saddle point, i.e., a minimum with respect to the real plane, and a steep maximum with respect to the imaginary plane; the maximum is assumed to be so steep that contributions to the integral from values of s other than $s = s^\star$ can be neglected. The accuracy of the approximation to Υ so obtained depends on the steepness of the maximum.

It is convenient to define (*13,16*) a function $\phi(s)$,

$$\phi(s) = \ln[\ell(s)] - k \ln s + sE \tag{6-82}$$

Then the integrand of Υ is simply $\exp[\phi(s)]$. The integrand will have a saddle point at $s = s^\star$, so that $\phi'(s^\star) = 0$, i.e., $s^\star$ is the solution of

$$\frac{1}{\ell(s^\star)}\left[\frac{d\ell(s)}{ds}\right]_{s=s^\star} - \frac{k}{s^\star} + E = 0 \tag{6-83}$$

The existence of a saddle point at $s^\star$ suggests that the path of integration in (6-65) be so chosen that $\mathrm{c} = s^\star$, then

$$\Upsilon = (1/2\pi \mathrm{i}) \int_{s^\star - \mathrm{i}\infty}^{s^\star + \mathrm{i}\infty} \exp[\phi(s)]\, ds = (1/2\pi) \int_{-\infty}^{+\infty} \exp[\phi(s^\star + \mathrm{i}y)]\, dy \tag{6-84}$$

Expanding $\phi(s)$ around $s^\star$ [and keeping in mind that $\phi'(s^\star) = 0$] yields

$$\phi(s) = \phi(s^\star) + \tfrac{1}{2}(s - s^\star)^2 \phi''(s^\star) + \cdots \tag{6-85}$$

If the series is cut off after the second derivative,[18] then, since $s - s^\star = \mathrm{i}y$, the integral (6-84) becomes simply

$$\Upsilon = \frac{\exp[\phi(s^\star)]}{2\pi} \int_{-\infty}^{+\infty} \exp\left[-\frac{1}{2}\, \phi''(s^\star) y^2\right] dy = \frac{\exp[\phi(s^\star)]}{[2\pi\phi''(s^\star)]^{1/2}} \tag{6-86}$$

To avoid the nuisance of handling exponentials, put $e^{-s} = z$. Then $\ell(s)$ becomes a functions of z, $\ell(z)$, which we will write $Q(z)$ to underline the fact that $\ell(s)$ or $\ell(z)$ is actually the partition function Q in disguise. The function $\phi(s)$ then becomes

$$\phi(z) = \ln[Q(z)] - k \ln[\ln z^{-1}] - E \ln z \tag{6-87}$$

The saddle point of this function is now at $\theta = e^{-s^\star}$, where θ is the solution of $\phi'(s^\star) = -\theta\phi'(\theta) = 0$;

$$\theta \left(\frac{d \ln Q(z)}{dz}\right)_{z=\theta} + \frac{k}{\ln \theta^{-1}} - E = 0 \tag{6-88}$$

[18] For an account of the "full" steepest-descent method which includes higher derivatives, see Hoare (*42*).

Thus Eq. (6-83), which is a transcendental equation in $s^\star$, becomes an algebraic equation in θ. Since $\phi''(s^\star) = \theta^2\phi''(\theta)$, Eqs. (6-65) and 6-86) become

$$\left.\begin{array}{c} \underset{(k=0)}{N(E)} \\ \\ \underset{(k=1)}{G(E)} \end{array}\right\} = \frac{Q(\theta)}{[\ln \theta^{-1}]^k \theta^E [2\pi\theta^2\phi''(\theta)]^{1/2}} \tag{6-89}$$

$Q(\theta)$ and θ^E are dimensionless, and the units of $\ln \theta^{-1}$ are energy^{-1} and those of $\theta^2\phi''(\theta)$ are (energy)^2. Equation (6-89) was first derived by Forst and Prášil (*16*) and in a slightly different form by Hoare and Ruijgrok (*43*). In the case of v independent harmonic oscillators and r classical rigid free rotors,

$$Q_{\mathrm{vr}}(z) = Q_{\mathrm{v}}(z)Q_{\mathrm{r}}(z), \quad Q_{\mathrm{v}}(z) = \prod_{i=1}^{\mathrm{v}} (1 - z^{h\nu_i})^{-1}, \quad Q_{\mathrm{r}}(z) = Q_{\mathrm{r}}'(\ln z^{-1})^{-\mathrm{r}/2} \tag{6-90}$$

where ν_i is the frequency of the ith oscillator; hence (*16*)

$$\left.\begin{array}{c} \underset{(k=0)}{N_{\mathrm{vr}}(E)} \\ \\ \underset{(k=1)}{G_{\mathrm{vr}}(E)} \end{array}\right\} = \frac{Q_{\mathrm{r}}' \prod_{i=1}^{\mathrm{v}} (1 - \theta^{h\nu_i})^{-1}}{\theta^E (\ln \theta^{-1})^{k+(\mathrm{r}/2)} [2\pi\theta^2\phi''(\theta)]^{1/2}} \tag{6-91}$$

where

$$\theta^2\phi''(\theta) = \sum_{i=1}^{\mathrm{v}} \frac{(h\nu_i)^2\theta^{h\nu_i}}{(1 - \theta^{h\nu_i})^2} + \frac{k + (\mathrm{r}/2)}{(\ln \theta^{-1})^2} \tag{6-92}$$

and θ is the solution of

$$\sum_{i=1}^{\mathrm{v}} \frac{h\nu_i\theta^{h\nu_i}}{1 - \theta^{h\nu_i}} + \frac{k + (\mathrm{r}/2)}{\ln \theta^{-1}} - E = 0 \tag{6-93}$$

Note that because k and r/2 always appear together, Eq. (6-91) satisfies relations (6-50)–(6-52). When tested against the direct count (*1*, Table V), Eq. (6-91) gave an excellent approximation, though not quite as good as the formula (6-80) based on Bernoulli polynomials. For a different version of the method of steepest descents making use of a redundant mean frequency see Lin and Eyring (*44,45*) and Lin *et al.* (*46–48*). Thermodynamic implications of the steepest-descent method are discussed in (*43*). Exercise 3 shows the connection between Eqs. (6-89) and (6-63).

While the evaluation of the inversion integral (6-65) by the method of residues or by the method of steepest descents already gives $N(E)$ and $G(E)$

with sufficient accuracy for most rate calculations at all but the lowest energies, still better results can be obtained by numerical inversion (*49*).

9. Approximation to $N_{vr}(E)$ and $G_{vr}(E)$ for more realistic systems

Methods shown in Sections 7 and 8 give $N_{vr}(E)$ and $G_{vr}(E)$ for a system comprising a collection of independent harmonic oscillators. The limitation of such a $N_{vr}(E)$ or $G_{vr}(E)$, no matter how accurately determined, is that it neglects several important features of real molecular vibrations.

Foremost among these is vibrational anharmonicity. In the present context, the important consequences of anharmonicity are twofold:

(1) Vibrational level spacing of an anharmonic oscillator decreases as vibrational energy increases, while for the harmonic oscillator it remains constant (Section 2); hence compared with the anharmonic case at a given E, the harmonic $G_v(E)$ is smaller;

(2) As a consequence of (1), the anharmonic oscillator eventually dissociates when E is sufficiently high, while the harmonic oscillator does not, no matter how large the supplied energy may be. This means that at high E the harmonic $G_v(E)$ includes so-called "virtual states," i.e., states that would be dissociated in a real molecule, and therefore should be excluded; consequently this effect tends to make the harmonic $G_{vr}(E)$ larger than it should be.

Effects (1) and (2) both become appreciable at higher energies, and since they work in opposite directions, we may expect they will somewhat neutralize each other, although the extent to which this happens cannot be accurately predicted in the absence of an actual calculation.

Another feature of molecular vibrations (and rotations) that has been neglected is the coupling among the vibrations, and among vibrations and rotations. For the moment the state of the art is such that this sort of coupling in polyatomic molecules is not well known and is difficult to describe, but eventually the assumption of total separability among all the internal degrees of freedom of a molecule may have to be abandoned if it turns out that the effect is important.

Finally, the third feature, though not really due to molecular vibrations or rotations themselves, is nevertheless important in the context of unimolecular rate theory. It concerns certain states that must be excluded from the density functions on account of conservation requirements, e.g., certain rotational states that are disallowed on account of conservation of angular momentum.

Each of the three complicating features mentioned can be easily taken into account when states are enumerated directly at low energies; in fact, we have considered direct evaluation of anharmonic $W_v(E)$ and $G_v(E)$ in Section 2, and of anharmonic $N_{vr}(E)$ and $G_{vr}(E)$ in Section 6. However, the same sort of high-energy bookkeeping problem mentioned previously makes it imperative to find a smooth-function approximation incorporating some or all of the complicating features mentioned, particularly since their importance is likely to increase with energy.

The generalized method of steepest descents A method in which all the various complicating features can be easily accommodated is the method of steepest descents. Note that in the derivation that leads to Eq. (6-89) no assumptions were made as to the form of $\ell(s)$ or $Q(z)$, and therefore we are in no way confined to considering $N_{vr}(E)$ or $G_{vr}(E)$ for harmonic oscillators only. It is not even necessary that the various degrees of freedom be separable, since it was not assumed that the partition function factorizes; in fact, all that is required is that we should be able to write down the partition function for the system in question.

It turns out (*50*), however, that when $Q(z)$ is a polynomial (rather than an infinite series) Eq. (6-88) does not have a solution for all E when $k = 0$ but always has one when $k = 1$. A more general formulation of (6-89) is therefore

$$G(E) = \frac{Q(\theta)}{(\ln \theta^{-1})\theta^E[2\pi\theta^2\phi''(\theta)]^{1/2}} \tag{6-94}$$

$$N(E) = \frac{dG(E)}{d\theta}\frac{d\theta}{dE} = G(E)\left\{\ln \theta^{-1} - \frac{1}{\theta^2\phi''(\theta)} - \frac{\theta^3\phi'''(\theta)}{2[\theta^2\phi''(\theta)]^2}\right\} \tag{6-95}$$

Define the various logarithmic derivatives of $Q(z)$ as

$$\mathrm{F} = \theta\left(\frac{d\ln Q(z)}{dz}\right)_{z=\theta}; \quad \mathrm{S} = \theta^2\left(\frac{d^2\ln Q(z)}{dz^2}\right)_{z=\theta}; \quad \mathrm{T} = \theta^3\left(\frac{d^3\ln Q(z)}{dz^3}\right)_{z=\theta} \tag{6-96}$$

Also let

$$\mathrm{C}^{-1} = \ln \theta^{-1}$$

Then the various derivatives of the function $\phi(\theta)$ to be used in conjunction with (6-94) and (6-95) are

$$\theta\phi'(\theta) = \mathrm{F} + \mathrm{C} - E = 0 \qquad \text{(this defines } \theta\text{)} \tag{6-97a}$$

$$\theta^2\phi''(\theta) = \mathrm{S} + \mathrm{F} + \mathrm{C}^2 \tag{6-97b}$$

$$\theta^3\phi'''(\theta) = \mathrm{T} + \mathrm{S} - \mathrm{F} + 2(\mathrm{C}^3 - \mathrm{C}^2) \tag{6-97c}$$

Equations (6-94) and (6-95) have been used to calculate $N_{vr}(E)$ and $G_{vr}(E)$ in various systems, harmonic and anharmonic, involving the exclusion of disallowed states. However, the following shall be limited to considerations of anharmonicity in a system of separable degrees of freedom, which is of more immediate interest. The model used is the Morse oscillator, which has been used by all workers in the field. It is certainly quite adequate for the purpose on hand.

Correction for anharmonicity in a system of classical Morse oscillators We might start with one of the approximations obtained for the harmonic case and attempt to modify it so as to include anharmonicity. This has been done by Haarhoff (*39*), who has so modified the semiclassical expression (6-56). What follows is an outline of his argument.

The energy levels of a classical Morse oscillator, with energy referred to potential minimum as zero, are given by

$$\varepsilon = (n + \tfrac{1}{2})h\nu - (n + \tfrac{1}{2})^2(h\nu)^2/4D_e \qquad (6\text{-}98)$$

where D_e is the dissociation energy of the oscillator (cf. Fig. 3-1) and $\varepsilon = E + E_z$. Solving for the quantum number n,

$$(n + \tfrac{1}{2}) = (2D_e/h\nu)\{1 - [1 - (\varepsilon/D_e)]^{1/2}\} \qquad (6\text{-}99)$$

so that the density of levels at E is

$$\frac{dn}{dE} = \frac{1}{h\nu[1 - (\varepsilon/D_e)]^{1/2}} = \frac{1}{h\nu}\left[1 + \frac{\varepsilon}{2D_e} + \frac{3}{8}\left(\frac{\varepsilon}{D_e}\right)^2 + \cdots\right] \qquad (6\text{-}100)$$

The Laplace transform of dn/dE gives an expansion for $\ell(s)$, the corresponding classical partition function for the Morse oscillator. Suppose the oscillator is the ith; then (writing D_i for $D_{e(i)}$)

$$\mathfrak{L}\left\{\frac{dn}{dE}\right\} = \frac{1}{h\nu_i}\left[\frac{1}{s} + \frac{1}{2D_i s^2} + \frac{3}{4D_i^2 s^3} + \cdots\right] = \ell_i(s) \qquad (6\text{-}101)$$

For a collection of v such oscillators,

$$\ell(s) = \prod_{i=1}^{\mathrm{v}} \ell_i(s) \qquad (6\text{-}102)$$

and then

$$N_{\mathrm{v}}(E) = \mathfrak{L}^{-1}\{\ell(s)\} = \mathfrak{L}^{-1}\left\{\prod_{i=1}^{\mathrm{v}} \ell_i(s)\right\} \qquad (6\text{-}103)$$

Straightforward multiplication in (6-102) and term by term inversion in (6-103) yields

$$N_{\mathrm{v}}(E) = \frac{\varepsilon^{\mathrm{v}-1}}{\Gamma(\mathrm{v})\prod_i h\nu_i}\left\{1 + \frac{\varepsilon}{\mathrm{v}}\sum_i \frac{1}{2D_i} + \frac{\varepsilon^2}{\mathrm{v}(\mathrm{v}+1)}\left[\frac{5}{2}\sum_i\left(\frac{1}{2D_i}\right)^2 + \frac{1}{2}\left(\sum_i \frac{1}{2D_i}\right)^2\right] + \cdots\right\} \qquad (6\text{-}104)$$

Since $\varepsilon = E + E_z$, we recognize the factor in front of the brace as the semi-classical approximation (6-56). Hence the series inside the brace may be considered as a correction factor for anharmonicity, which we shall write as $C_0(E)$, to emphasize that it depends on E; subscript 0 anticipates Eq. (6-105). Equation (6-104) can be easily generalized to include r rotations by writing $\mathsf{v} + (\mathsf{r}/2)$ for v, and the corresponding correction factor for $G_{\mathrm{vr}}(E)$ can be obtained by writing $\mathsf{r} + 2$ for r [cf. Eq. (6-50)]. By various manipulations Haarhoff truncated the series in (6-104) to obtain the more compact formula (in our notation)

$$\left.\begin{array}{ll}(k = 0) & \dfrac{N_{\mathrm{vr}}(E)\ (\mathrm{Morse})}{N_{\mathrm{vr}}(E)\ (\mathrm{harm})} \\ & \\ (k = 1) & \dfrac{G_{\mathrm{vr}}(E)\ (\mathrm{Morse})}{G_{\mathrm{vr}}(E)\ (\mathrm{harm})}\end{array}\right\} = C_k(E) \qquad (6\text{-}105)$$

$$C_k(E) = \left\{\left(1 + \frac{2}{\eta}\right)^{(\eta/2)[1+(\eta/2)]} \exp\left[-\frac{(\nu_{\mathrm{d}} - 1)}{3(1 + \eta)}\right]\right\}^{\mathsf{v}E_z/(\mathsf{n}+1)D}$$
$$\times \exp\left[M_2(1 + \eta)^2\left(\frac{E_z}{D}\right)^2 + M_3(1 + \eta)^3\left(\frac{E_z}{D}\right)^3\right] \qquad (6\text{-}106)$$

where D is the harmonic mean of the dissociation energies defined by $D^{-1} = \langle D_i^{-1}\rangle$;

$$M_2 = \mathsf{v}[4\mathsf{v} + \tfrac{5}{2}(\mathsf{r} + 2k)]/8(\mathsf{n} + 1)^2(\mathsf{n} + 2)$$
$$M_3 = \mathsf{v}[24\mathsf{v}^2 + \tfrac{59}{2}\mathsf{v}(\mathsf{r} + 2k) + \tfrac{37}{4}(\mathsf{r} + 2k)^2]/24(\mathsf{n} + 1)^3(\mathsf{n} + 2)(\mathsf{n} + 3)$$

and the other symbols have the same significance as in Eq. (6-81), namely

$$\mathsf{n} = \mathsf{v} + k - 1 + (\mathsf{r}/2), \qquad \eta = E/E_z, \qquad \nu_{\mathrm{d}} = \langle \nu^2\rangle/\langle \nu\rangle^2.$$

When tested (*50*) formula (6-106) has been found to perform very well in the case of a large molecule (cyclopropane, $\mathsf{v} = 21$) up to $7 \times 10^4\ \mathrm{cm}^{-1}$, but in the case of a small molecule ($\mathsf{v} = 5$) it has failed (*51*) above $\sim 3 \times 10^4\ \mathrm{cm}^{-1}$ (Fig. 6-2, curve b). It turns out that (6-106) is unreliable at energies which represent an appreciable fraction of E_{d}, the energy required to dissociate the entire molecule ($E_{\mathrm{d}} = \sum_{i=1}^{\mathsf{v}} D_i - E_z$). Therefore the usefulness of formula (6-106) is limited to roughly $E < 0.1E_{\mathrm{d}}$, which in a small molecule represents a rather narrow range of energies[19] since obviously a small molecule will have a low E_{d}. Another problem with the formula (6-106) is that D appears to be much too crude a parameter for describing a collection of

[19] Note that in the case of cyclopropane, $7 \times 10^4\ \mathrm{cm}^{-1} \approx 0.1E_{\mathrm{d}}$, while in the cited case of the small molecule (patterned on hydrogen peroxide) $3 \times 10^4\ \mathrm{cm}^{-1} \approx 0.25E_{\mathrm{d}}$.

Morse oscillators, since comparison with an essentially "exact" $C_k(E)$ obtained by steepest descents shows *(52)* that $C_k(E)$ is sensitive to individual D_i, i.e., $C_k(E)$ may be quite different for two different collections of Morse oscillators even if both have the same $D = \langle D_i^{-1} \rangle^{-1}$.

For a different, but less useful, approach to $C(E)$ see *(40,53,54)*.

Anharmonic $N_{vr}(E)$ and $G_{vr}(E)$ by the generalized method of steepest descents One might expect that a better result for the anharmonic correction factor $C(E)$ would be obtained by a suitable modification of an expression for density (or sum) of states which is based on the inversion of the quantum partition function. However, expression (6-80), excellent though it is, depends in a unique way on the expansion of a product of hyperbolic sines and cannot be usefully modified. The generalized method of steepest descents, on the other hand, offers a relatively easy way of calculating the anharmonic $N_{vr}(E)$ or $G_{vr}(E)$ directly. To this end, the partition function must be obtained first.

The vibrational levels of a Morse oscillator, referred to ground state, are given by Eq. (6-10); they do not extend to infinity but are finite in number. The quantum number of the highest level (n_{max}) is given by the solution of

$$(n + \tfrac{1}{2})h\nu - (n + \tfrac{1}{2})^2 x h\nu = D_e \tag{6-107}$$

which is

$$n_{max} = \tfrac{1}{2}(x^{-1} - 1) \tag{6-108}$$

so that the vibrational partition function for one Morse oscillator is

$$Q(z) = \sum_{n=0}^{n_{max}} z^{[n(1-x) - n^2 x]h\nu} \tag{6-109}$$

Note that because the summation is taken only to n_{max}, only levels below the dissociation limit are included, and (6-109) represents $Q(z)$ for bound vibrational states, as of course it should. For v such oscillators,

$$Q_v(z) = \prod_{i=1}^{v} \sum_{n_i=0}^{n_{i(max)}} z^{n_i'} \tag{6-110}$$

where $n_i' = [n_i(1 - x_i) - n_i^2 x_i]h\nu_i$; n_i is the vibrational quantum number of the ith oscillator and ν_i is its frequency. The logarithmic derivatives F, S, and T [Eq. (6-96)] then follow by performing the indicated operations on $Q_v(z)$ [see (*1*, Eq. (114))].

In all these formulas the polynomials in n_i must be summed term by term since there is no longer a simple formula for the sum as in the case of the harmonic oscillator, so that the computation is more complicated. The parameter θ is obtained by solving (6-97a); $G_v(E)$ and $N_v(E)$ then follow from (6-94) and (6-95), respectively. If separable rotations are also to be included, the partition function (6-110) can be easily modified by including the appropriate $Q_r(z)$ as a factor, which results in a trivial modification of the other equations. $G_v(E)$ calculated from (6-110) has been checked in the case of cyclopropane (*50*) and hydrogen peroxide (*51*) and found to give excellent results.

A somewhat different formulation of the Morse oscillator problem has been given by Hoare and Ruijgrok (*43*). They avoid the tedious polynomials in n_i of (6-110) and its derivatives by writing the single-oscillator partition function [Eq. (6-109)] in the form

$$Q(u) = \sum_n \exp\{-u[n(1-x) - n^2 x]\}, \quad u = h\nu/kT$$

$$= \sum_n [\exp(-un)\exp(nux + n^2 ux)] \tag{6-111}$$

If the exponential in x is expanded to first order,

$$\exp(nux + n^2ux) \approx 1 + nux + n^2ux \tag{6-112}$$

Eq. (6-111) becomes

$$Q(u) = \sum_n e^{-un}[1 + n(n+1)ux] \tag{6-113}$$

and even if the summation over n is now extended to infinity, the effect is to make (6-113) roughly equivalent to Eq. (6-111) truncated at $n_{\max}$. After some manipulation, (6-113) summed from $n = 0$ to $n = \infty$ becomes (our notation)

$$Q(z) = \frac{1}{1 - z^{h\nu}}\left[1 - \frac{2xz^{h\nu}\ln z^{h\nu}}{(1 - z^{h\nu})^2}\right] \tag{6-114}$$

which is in essence the harmonic partition function multiplied by a correction term [cf. Eq. (6-90)]. They further simplify the logarithm of Eq. (6-114) with the approximation $\ln(1 - y) \approx -y$, so that

$$\ln Q(z) \approx -\ln(1 - z^{h\nu}) - \frac{2xz^{h\nu}\ln z^{h\nu}}{(1 - z^{h\nu})^2} \tag{6-115}$$

For several oscillators, (6-115) is then summed over the individual oscillators to get the logarithm of the total partition function, and the logarithmic derivatives [cf. (6-96)] are then used in the usual way in Eqs. (6-97) to obtain θ and the derivatives of $\phi(\theta)$, from which $G(E)$ follows via Eq. (6-94).

The "exact" Morse $G_v(E)$, based on (6-110), and the "simplified" Morse $G_v(E)$, based on (6-115), are compared in Fig. 6-2, after dividing both by the harmonic $G_v(E)$; in other words, the comparison is in terms of the anharmonic

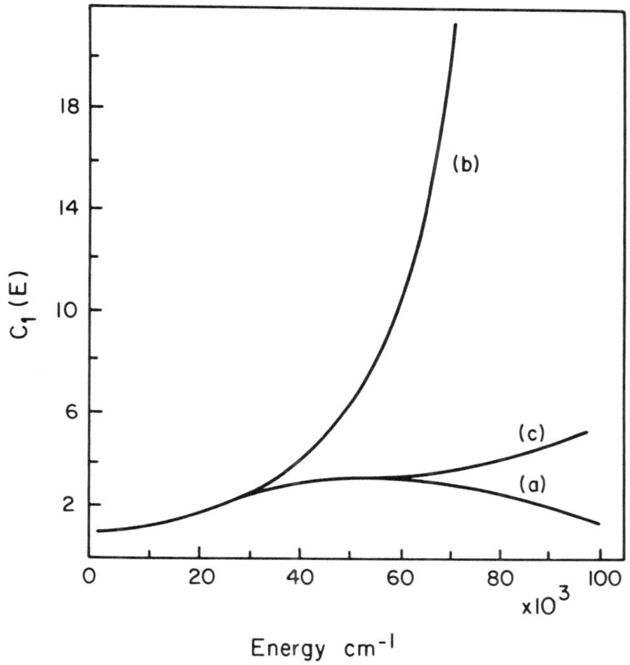

Fig. 6-2. *The anharmonic correction factor $C_1(E)$ as a function of energy.*

The factor shown is defined as $C_1(E) = G_v(E)(\text{Morse})/G_v(E)(\text{harmonic})$. Curve (a): "exact," as calculated by the method of steepest descents from (6-110) (Morse) and (6-90) (harmonic). Curve (b): calculated directly from formula (6-106) due to Haarhoff. Curve (c): calculated by the method of steepest descents from (6-115) (Morse) and (6-90) (harmonic). From Le-Khac Huy and W. Forst, unpublished results.

correction factor $C_1(E)$ [Eq. (6-105) for $k = 1$, r = 0]. It can be seen that Hoare and Ruijgrok's modification (curve c), which saves a good deal of machine time, is useful up to quite high energies ($\sim 0.5E_d$, several times the useful range of Haarhoff's correction factor), but at still higher energies the tedious sums in (6-110) cannot be avoided if good accuracy is desired.

The "exact" correction factor $C_1(E)$ (curve a in Fig. 6-2) shows graphically the two effects of anharmonicity discussed at the beginning of this section. At low energies, $C_1(E)$ increases because, relative to the harmonic case, the influence of the denser energy-level spacing in the Morse oscillator is predominant (point 1), whereas at high energies, $C_1(E)$ decreases again because now the major effect is the noninclusion of virtual states in the Morse count, which causes $G_v(E)$ (Morse) to reach a constant value at a sufficiently high E, while $G_v(E)$ (harmonic) keeps increasing (point 2). In the limit, therefore, as $E \to \infty$, $C_1(E) \to 0$.

Truncated harmonic oscillator Another model system, for which $N(E)$ and $G(E)$ can be easily obtained by the method of steepest descents, and which is useful in some applications, is the truncated harmonic oscillator. In this case, the spacing of energy levels is constant, as in the usual harmonic oscillator, but the number of levels is finite. If the quantum number of the highest level is again $n_{\max}$, the partition function for one truncated harmonic oscillator is

$$Q(z) = \sum_{n=0}^{n_{\max}} z^{nh\nu} = \frac{1 - z^{(n_{\max}+1)h\nu}}{1 - z^{h\nu}} \tag{6-116}$$

and the partition function for v such oscillators is

$$Q_{\text{v}}(z) = \prod_{i=1}^{\text{v}} \left(\sum_{n_i=0}^{n_{i(\max)}} z^{n_i h\nu_i} \right) = \prod_{i=1}^{\text{v}} \left(\frac{1 - z^{(n_{i(\max)}+1)h\nu_i}}{1 - z^{h\nu_i}} \right) \tag{6-117}$$

The rest of the calculations then follows in the usual way from Eqs. (6-94)–(6-97). For an earlier approach to the problem see (*68*).

Figure 6-3 shows the energy dependence of $G_{\text{vr}}(E)$ for a system of five oscillators and two rigid free rotors, assuming that the oscillators are harmonic [curve 1 calculated using $Q_{\text{v}}(z)$ and $Q_{\text{r}}(z)$ of (6-90)], or truncated-harmonic [curve 2 calculated using $Q_{\text{v}}(z)$ of (6-117), $Q_{\text{r}}(z)$ of (6-90)], or Morse [curve 3, calculated using $Q_{\text{v}}(z)$ of (6-110), $Q_{\text{r}}(z)$ of (6-90)]. In the last two cases a given oscillator, whether Morse or truncated-harmonic, is assumed to have the same dissociation energy. The curve shows the expected behavior: $G_{\text{vr}}(E)$ for the truncated-harmonic case is below the harmonic value at high energies, and the discrepancy increases with energy because the harmonic $G_{\text{vr}}(E)$ includes virtual states which are eliminated in the truncated-harmonic $G_{\text{vr}}(E)$ by cutting off each oscillator at its dissociation energy. The Morse $G_{\text{vr}}(E)$ is always higher than the other two because every energy level of the more numerous Morse levels combines with every one of the still more numerous levels of the rotor, which tends to exaggerate the denser spacing of levels in the Morse oscillator. In this respect, the Morse cases of Fig. 6-3 and Fig. 6-2 differ.

Anharmonic parameters in molecular systems The application of a Morse oscillator representation to an actual molecular system requires the assignment of x or D_{e} to each oscillator. Either parameter will do since they are related (*2*, p. 273 ff.) by

$$D_{\text{e}} = h\nu/4x \tag{6-118}$$

The parameter x is related to the spectroscopic normal frequency ω and the fundamental frequency ν by

$$x = (\omega - \nu)/2\omega \tag{6-119}$$

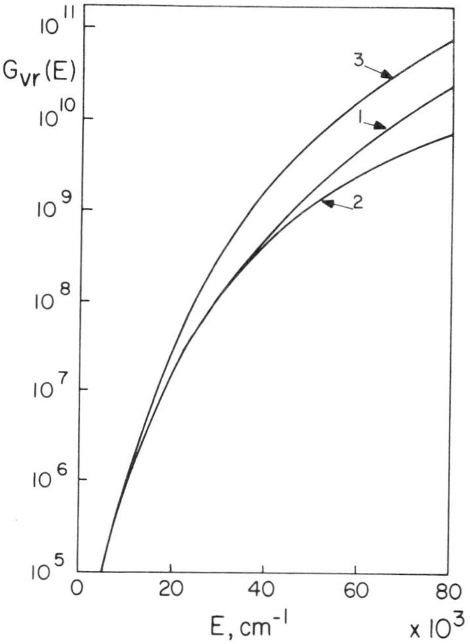

Fig. 6-3. *Energy Dependence of $G_{vr}(E)$.*

(1) Collection of harmonic oscillators; (2) collection of truncated harmonic oscillators; (3) collection of Morse oscillators. Coupled with one two-dimensional rigid free rotor in all three cases. The collection consists of oscillators with frequencies ν_i (in cm^{-1}) and anharmonicity coefficients x_i:

ν_i	x_i
880	0.0131
1296	0.0226
1440	0.0208
3774	0.0232
3788	0.0238

Each oscillator has the same dissociation energy (determined by x_i) in both the truncated-harmonic and Morse oscillator representation. The moment of inertia of the rotor is $I = 3.157 \times 10^{-39}$ g cm^2. Frequencies, anharmonicity coefficients, and moment of inertia are patterned on hydrogen peroxide. From Z. Prášil and W. Forst, unpublished results.

The difference $\omega - \nu$ is usually a small difference between two large numbers, and unless both ω and ν are known very accurately, (6-119) will yield a very unreliable x, and hence also a poor D_e. However, even if the ω's and ν's are very accurate, there is a problem (*55*). Spectroscopists usually calculate the ω's of a polyatomic molecule by the method of Dennison (*56*), which involves a more sophisticated relation between ω and x than Eq. (6-119), which is strictly valid only for a diatomic molecule. Using such ω's in (6-119) really amounts to adjusting each x to the somewhat artificial concept of an independent Morse oscillator. All this, of course, in addition to the assumption that the Morse function is an adequate representation for the vibration in question, which may not always be too evident.

If the vibration is a simple stretching vibration, D_e may also be calculated from the quadratic (f_2) and cubic (f_3) force constant, since in a Morse oscillator they are related by

$$D_e = (f_2)^3/2(f_3)^2 \tag{6-120}$$

The two force constants can be obtained from Badger's rule as modified by Herschbach and Laurie (*57*):

$$(-1)^j f_j = 10^{-(r_e - a_{xy})/b_{xy}}, \qquad j = 2 \text{ or } 3 \tag{6-121}$$

where r_e is the equilibrium bond length between atoms of rows x and y in the periodic table and a_{xy} and b_{xy} are tabulated semiempirical constants. If the dissociation energy of a stretching vibration is known from thermochemical data or other information, D_e may be obtained immediately. In such cases Eqs. (6-120) and (6-121) usually give much the same value for D_e, whereas it sometimes disagrees with D_e calculated from (6-118) and (6-119). In such a case D_e calculated from (6-120) is likely to be more reliable.

Since Eqs. (6-120) and (6-121) cannot be used for vibrations other than simple bond-stretching vibrations, Eq. (6-118) remains the only method for determining D_e of vibrational modes involving more complex motions, like skeleton bending vibrations and deformation vibrations. This is not very satisfactory because even if the difference $\omega - \nu$ [Eq. (6-119)] were known with sufficient accuracy for such complicated motions, the formula (6-118) is likely to apply only very approximately since it was derived only for the simple bond-stretching case. This is probably not too much of a handicap in thermal systems where the range of pertinent energies is relatively small, because when E is not too high, the anharmonic $N_{vr}(E)$ or $G_{vr}(E)$ is not overly sensitive to the D_e of individual oscillators (*55, 58–60*); in nonthermal systems, however, where the Boltzmann factor does not wipe out the contribution of the high-energy terms, the difficulty is more serious. Obviously, under such conditions we have reached the useful limit of the concept of a molecule as a collection of independent Morse oscillators. This perhaps artificial concept is not essential, for if we knew how to write the partition function for the molecule *as a whole*, and not merely as a product of single-oscillator partition functions, the inversion procedure of Section 8 would yield the appropriate anharmonic $N(E)$ and $G(E)$ immediately.

10. Conclusions

With the possible exception of the numerical inversion (*49*), no one approximation method gives sufficiently accurate results below about 3000 cm^{-1}. A closer comparison reveals that, in general, approximations given by Eqs. (6-80) and (6-89) get better at low energies as r increases, while the other approximations generally get worse. Therefore the lowest energy at which results are still within, say, 10% of the direct count depends on the approximation formula being used and on the complexity of the molecular system for which the number of states is being evaluated.

In any event, whatever the approximation formula used, and whatever the notion of an "acceptable" accuracy, $N_{vr}(E)$ and $G_{vr}(E)$ are preferably enumerated by direct count at low energies, which is why direct enumeration has been considered in some detail in Sections 2 and 6. This requirement is not a serious handicap and the two sections should help the interested investigator do the work easily by hand up to 3000 or 4000 cm^{-1}, at which point an approximation formula can usually take over. Computer programs for some of these formulas are given in the Appendix.

Computer calculations Insofar as the approximation formulas are concerned, the computational effort required of the computer is least for the semiclassical methods, in particular Eq. (6-63) with β_k given by Eq. (6-64), which gives best accuracy for least expenditure of computer time. Programming of the formula is also very easy. Equations (6-80) and (6-81) require somewhat more machine time, but while programming of Eq. (6-81) is easy, Eq. (6-80) is more difficult mainly because of the complicated coefficient $D_i^{(V)}$ [Eq. (6-72)]. However, since the machine need calculate a given set of $D_i^{(V)}$ only once, which are then usable at all energies, machine time for a given range of energies by formula (6-80) is nevertheless quite comparable to that for formula (6-81).

Equations (6-91) or Eqs. (6-94)–(6-95) are very easy to program but machine time is higher than for the previously mentioned formulas because a new θ must be calculated for every energy at which $N(E)$ and $G(E)$ is desired,[20] i.e., Eq. (6-93) or (6-97a) must be solved repeatedly for every E. Another factor that tends to increase machine time is that since θ is a number generally between 0.9999 and unity, it must be calculated to a precision of about 10^{-10} to obtain $G(E)$ and $N(E)$ with good accuracy. The Newton–Raphson method (*61*, p. 332) may be used to good advantage here; it takes about 15 iterations to find θ at the lowest (first) energy, and then using this θ as the input for iterations at the next higher E, etc., only four to seven iterations are usually sufficient at all other energies.

These remarks apply to the calculation of $N(E)$ or $G(E)$ for a collection of harmonic oscillators. With anharmonic oscillators, machine time for Eqs. (6-91) or (6-94)–(6-95) is higher again because of all the various summations that must be performed in (6-110), but not when Eq. (6-115) is used. The correction factor $C_K(E)$ [(6-106)], on the other hand, requires much less machine time.

Formula of Vestal We have not yet considered an approximation formula due to Vestal *et al.* (*62–65*) which was found (*16*) to give inferior results to all the other formulas at a much larger expenditure of computer time. However, it continues to be used by a number of investigators in the field of mass spectrometry, and therefore we include it here for completeness. It is designed to calculate $N_{vr}(E)$ and $G_{vr}(E)$ for a collection of *harmonic* oscillators and rigid free rotors; the point of departure is $G_v(E)$ of Eq. (6-9), which is then generalized to include free rotors and reduced to a manageable form with the help of the theory of symmetric functions. The approach therefore does not quite fit within the classification scheme of this chapter. The result is, in our notation,

$$\left.\begin{matrix} \underset{(k=0)}{N_{vr}(E)} \\ \\ \underset{(k=1)}{G_{vr}(E)} \end{matrix}\right\} = Q_r(h\langle\nu\rangle_g)^{\mathsf{n}-\mathsf{v}} \sum_{i=1}^{\mathsf{v}} \binom{\mathsf{v}}{i} \frac{\mathsf{U}_i^{\mathsf{v}-\mathsf{n}}}{\Gamma(i+\mathsf{n}-\mathsf{v})} \left(\frac{E}{h\langle\nu\rangle_g}\mathsf{U}_i - \frac{i-1}{2}\right)^{i+\mathsf{n}-\mathsf{v}} \tag{6-122}$$

where $\mathsf{n} = \mathsf{v} + k - 1 + (\mathsf{r}/2)$, $\langle\nu\rangle_g = \prod_{i=1}^{\mathsf{v}} (\nu_i)^{1/\mathsf{v}}$ is the geometric mean of

[20] Note that when Eqs. (6-94)–(6-95) are used $G(E)$ and $N(E)$ are both calculated with the same θ, whereas Eq. (6-91) requires a θ for $N(E)$ and another θ for $G(E)$, i.e., Eq. (6-93) must be solved twice at every E, once with $k = 0$ and then again with $k = 1$. In unimolecular rate calculations this is not much of a handicap because $N(E)$ is calculated for the molecule and $G(E)$ is calculated for the transition state at $E - E_0$, so that not only are the energies different, but the collection of oscillators and rotors for which the density and sum of states are calculated are also different, so that in any event two different θ's are required.

the v frequencies, and

$$\mathsf{U}_i = \left[\binom{\mathsf{v}}{i}\mathsf{K}_i\right]^{1/i}, \qquad \mathsf{R}_j = \sum_{i=1}^{\mathsf{v}} \left(\frac{\langle \nu \rangle_g}{\nu_i}\right)^j \tag{6-123}$$

$$\mathsf{K}_i = \sum \left\{ (-1)^{i + \sum_{j=1}^{i} m_j} \prod_{j=1}^{i} \left[\left(\frac{1}{m_j!}\right)\left(\frac{\mathsf{R}_j}{j}\right)^{m_j}\right]\right\} \tag{6-124}$$

where the summation in (6-124) is taken over all sets $(m_1, m_2, \ldots, m_k)$ of integers (zeros included) which satisfy $\sum_{j=1}^{i} jm_j = i$. The formula satisfies relations (6-51)–(6-52) and gives in general results [footnote 8 in (*16*)] that are too high relative to the exact count. In an improved version (*65*) the second term inside the square bracket of (6-122) is written as $i/2$ [instead of $(i - 1)/2$], which improves the approximation at low energies but makes it worse at high energies.

The various methods discussed in this chapter make it possible to calculate $G(E)$ and $N(E)$, and therefore also the microcanonical rate constant $k(E)$ [Eq. (4-34)], in a very general way for almost any kind of system. The rate constant $k(E)$ refers, however, only to vibrational potential, and we still have to generalize $k(E)$ further for the case of an effective potential, i.e., vibrational plus rotational. This is the subject of the next chapter where it turns out that insofar as $G^*(E)$ and $N(E)$ are concerned, the addition of the rotational potential affects only the energy at which $G^*(E)$ or $N(E)$ is calculated, but is otherwise without influence on the meaning, interpretation, and calculation of $G^*(E)$ and $N(E)$.

Exercises

1. Use Eqs. (6-15) and (6-16) to derive the well-known result for translational states of a particle of mass m in a one-dimensional box of length l:

$$N(E) = (l/h)(2m/E)^{1/2}, \qquad G(E) = (2l/h)(2mE)^{1/2}$$

Hint: Expand $[1 + (dE/E)]^{1/2} = 1 + \frac{1}{2}dE/E - \frac{1}{8}(dE/E)^2 + \cdots$ and drop terms second-order and higher in dE/E.

2. Show that Eqs. (6-62) can be obtained by the convolutions (6-32) and (6-35b) from (6-45) and (6-55). *Hint:* The classical convolution integral is related to the beta function $B(z, \omega)$ defined by

$$B(z, \omega) = \int_0^1 t^{z-1}(1 - t)^{\omega - 1}\, dt = \Gamma(z)\Gamma(\omega)/\Gamma(z + \omega)$$

See M. Abramowitz and I. A. Stegun (*66*, p. 258, Items 6.2.1 and 6.2.2).

The semiclassical formulas (6-63) (with $a_k = 1$) follow when the upper limit of the convolution integral is taken to $E + E_z$, but the density and integrated density so obtained is assigned to energy E (not $E + E_z$).

3. Show that the semiclassical formula [Eq. (6-63), with $a_k = 1$] is obtained by inverting Q_{vr}^{class} with the appropriate shift in the energy zero. *Hint:* It can be shown that if $\mathcal{L}^{-1}\{g(s)\} = f(E)$, then

$$\mathcal{L}^{-1}\{e^{bs}g(s)\} = 0 \qquad (E < -b)$$
$$= f(E + b) \qquad (E > -b)$$

Therefore evaluate by steepest descents $\mathcal{L}^{-1}\{e^{aE_z}Q_{vr}^{class}(s)/s^k\}$. Note that $(\mathrm{n})^{\mathrm{n}-(1/2)}(2\pi)^{1/2}e^{-\mathrm{n}}$ is Stirling's approximation to $\Gamma(\mathrm{n})$.

4. Show that the density of translational states of a particle of mass μ and energy E_t, confined in a three-dimensional volume V, is

$$(2^{5/2}\pi\mu^{3/2}V/h^3)E_t^{1/2}$$

See, for example, (*18*, p. 13, eq. 5.19).
Use this result in Eq. (ii) of Exercise 2 in Chapter 4 to obtain

$$G^*(E') = (8\pi\mu/h^2)\int_0^{E'} N^{\mathrm{f}}(E' - E_t)\,\sigma(E', E_t)\,E_t\,dE_t$$

In terms of the de Broglie wavelength $\lambda\!\!\!^{-2} = \hbar^2/2\mu E_t$ this can be written

$$\alpha G^*(E') = \int_0^{E'} N^{\mathrm{f}}(E' - E_t)\,\frac{\sigma(E', E_t)}{\pi\lambda\!\!\!^{-2}}\,dE_t \tag{a}$$

The cross-section $\sigma(E', E_t)$ refers to the association of fragments to form the *critical configuration* (in collision theory also referred to as *compound state*), not the actual product of the association which would be the unimolecular reactant [to whose transition state refers $G^*(E')$]. Following (*69*), define an average cross section by

$$\langle\sigma(E')\rangle = \int_0^{E'} N^{\mathrm{f}}(E' - E_t)\sigma(E', E_t)E_t\,dE_t \Big/ \int_0^{E'} N^{\mathrm{f}}(E' - E_t)E_t\,dE_t \tag{b}$$

Show that the integral in the denominator of (b) can be written as

$$\int_0^{E'} N^{\mathrm{f}}(E' - E_t)E_t\,dE_t = [G'^{\mathrm{f}}(E')]_{\mathrm{r}=2}$$

where $[G'^{\mathrm{f}}(E')]_{\mathrm{r}=2}$ represents *reduced* integrated density (cf. p. 102). If we now identify $8\pi\mu\langle\sigma(E')\rangle/h^2$ with $Q_{\mathrm{r}=2}'^{\mathrm{f}}$ [Eq. (6-41)], then, replacing E' with $E - E_0$, (a) becomes

$$G^*(E - E_0) = Q_{\mathrm{r}=2}'^{\mathrm{f}}\,[G'^{\mathrm{f}}(E - E_0)]_{\mathrm{r}=2} = [G^{\mathrm{f}}(E - E_0)]_{\mathrm{r}=2} \tag{c}$$

where $[G^{\mathrm{f}}(E - E_0)]_{\mathrm{r}=2}$ is the integrated density for the internal states of the *separated fragments plus two rotors* of moment of inertia $I^{\mathrm{f}} = \mu\langle\sigma(E - E_0)\rangle/\pi$. This is how a "loose" transition state can be interpreted in terms of additional rotors.

The form of the cross-section to be used in (a) can be obtained from the argument used in Exercise 1, Chapter 7 (p. 141). The classical cross-section corresponding to ℓ,

$b + db$ is $2\pi b\, db$; in terms of $\mathscr{L}$ this is $(\pi\hbar^2/2\mu E_t)(2\mathscr{L} + 1) = \pi\lambda\!\!\!^{-2}(2\mathscr{L} + 1)$. Since $\mathscr{L}$, $\mathscr{J}$ and $\mathscr{J}^f$ are related by $|\mathscr{L} - \mathscr{J}^f| \leqslant \mathscr{J} \leqslant \mathscr{L} + \mathscr{J}^f$, this mean that for a given $\mathscr{L}$, $\mathscr{J}^f$, the probability of obtaining a specified $\mathscr{J}$ is $P(\mathscr{J}) = (2\mathscr{J} + 1)/[(2\mathscr{L} + 1) \times (2\mathscr{J}^f + 1)]$; the cross-section is then

$$\frac{\sigma(E', E_t)}{\pi\lambda\!\!\!^{-2}} = \sum_{\mathscr{L},\mathscr{J}^f} (2\mathscr{L} + 1)P(\mathscr{J})$$

where the summation is taken over all allowed values of $\mathscr{L}$ and $\mathscr{J}^f$. See refs. (*25*) and (*26*) in Chapter 5 for actual calculations.

References

1. W. Forst, *Chem. Rev.* **71**, 339 (1971).
2. H. Eyring, J. Walter, and G. E. Kimball, "Quantum Chemistry." Wiley, New York, 1944.
3. J. E. Mayer and M. G. Mayer, "Statistical Mechanics." Wiley, New York, 1940.
4. R. E. Harrington, Ph.D. thesis, Appendix 1. Univ. of Washington (1960).
5. J. H. Current and B. S. Rabinovitch, *J. Chem. Phys.* **38**, 783 (1963).
6. G. Herzberg, "Molecular Spectra and Molecular Structure," Vol. II, Infrared and Raman Spectra of Polyatomic Molecules. Van Nostrand Reinhold, Princeton, New Jersey, 1952.
7. R. C. Tolman, "Principles of Statistical Mechanics." Oxford Univ. Press, London and New York, 1938.
8. O. K. Rice, "Statistical Mechanics, Thermodynamics, and Kinetics." Freeman, San Francisco, California, 1967.
9. S. H. Bauer, *J. Chem. Phys.* **6**, 403 (1938).
10. S. H. Bauer, *J. Chem. Phys.* **7**, 1097 (1939).
11. G. Doetsch, "Guide to the Application of Laplace Transforms." Van Nostrand Reinhold, Princeton, New Jersey, 1961.
12. G. E. Roberts and H. Kaufman, "Tables of Laplace Transforms." Saunders, Philadelphia, Pennsylvania, 1966.
13. R. Kubo, "Statistical Mechanics." North-Holland Publ., Amsterdam, 1965.
14. O. K. Rice and H. C. Ramsperger, *J. Amer. Chem. Soc.* **49**, 1617 (1927).
15. R. A. Marcus, *J. Chem. Phys.* **20**, 359 (1952).
16. W. Forst and Z. Prášil, *J. Chem. Phys.* **51**, 3006 (1969).
17. G. M. Wieder and R. A. Marcus, *J. Chem. Phys.* **37**, 1835 (1962).
18. R. A. Marcus and O. K. Rice, *J. Phys. Coll. Chem.* **55**, 894 (1951).
19. B. S. Rabinovitch and R. W. Diesen, *J. Chem. Phys.* **30**, 735 (1959).
20. B. S. Rabinovitch and J. H. Current, *J. Chem. Phys.* **35**, 2250 (1961).
21. C. Lifshitz and M. Wolfsberg, *J. Chem. Phys.* **41**, 1879 (1964).
22. G. Z. Whitten and B. S. Rabinovitch, *J. Chem. Phys.* **38**, 2466 (1963).
23. G. Z. Whitten and B. S. Rabinovitch, *J. Chem. Phys.* **41**, 1883 (1964).
24. J. L. Kinsey, *J. Chem. Phys.* **54**, 1206 (1971).
25. A. S. Dickinson and R. B. Bernstein, *Mol. Phys.* **18**, 305 (1970).
26. G. D. Mahan, *J. Chem. Phys.* **52**, 258 (1970).
27. A. I. Khinchin, "Mathematical Foundations of Statistical Mechanics." Dover, New York, 1949.
28. F. Lurçat and P. Mazur, *Nuovo Cimento* **31**, 140 (1964).

29. G. A. Korn and T. M. Korn, "Mathematical Handbook for Scientists and Engineers." McGraw–Hill, New York, 1961.
30. A. Erdelyi, W. Magnus, F. Oberhettinger, and F. G. Tricomi, "Higher Transcendental Functions." McGraw–Hill, New York, 1953.
31. N. E. Nörlund, *Acta Math.* **43**, 184 (1922).
32. N. E. Nörlund, *Acta Math.* **43**, 166 (1922).
33. N. E. Nörlund, *Acta Math.* **43**, 167 (1922).
34. N. E. Nörlund, "Vorlesungen über Differenzenrechnung." Chelsea, New York, 1954.
35. E. Thiele, *J. Chem. Phys.* **39**, 3258 (1963).
36. W. Forst, Z. Prášil, and P. St. Laurent, *J. Chem. Phys.* **46**, 3736 (1967).
37. W. Forst, Z. Prášil, and P. St. Laurent, *J. Chem. Phys.* **48**, 1431 (1968).
38. D. C. Tardy, B. S. Rabinovitch, and G. Z. Whitten, *J. Chem. Phys.* **48**, 1427 (1968).
39. P. C. Haarhoff, *Mol. Phys.* **7**, 101 (1963).
40. E. W. Schlag and R. A. Sandsmark, *J. Chem. Phys.* **37**, 168 (1962).
41. H. Jeffreys and B. S. Jeffreys, "Methods of Mathematical Physics." Cambridge Univ. Press, London and New York, 1956.
42. M. R. Hoare, *J. Chem. Phys.* **52**, 5695 (1970).
43. M. R. Hoare and T. W. Ruijgrok, *J. Chem. Phys.* **52**, 113 (1970).
44. S. H. Lin and H. Eyring, *J. Chem. Phys.* **39**, 1577 (1963).
45. S. H. Lin and H. Eyring, *J. Chem. Phys.* **43**, 2153 (1965).
46. J. C. Tou and S. H. Lin, *J. Chem. Phys.* **49**, 4187 (1968).
47. K. H. Lau and S. H. Lin, *J. Phys, Chem.* **75**, 981 (1971).
48. S. H. Lin and C. Y. Lin Ma, *Advan. Chem. Phys.* **21**, 143 (1971).
49. M. R. Hoare and P. Pal, *Mol. Phys.* **20**, 695 (1971).
50. W. Forst and Z. Prášil, *J. Chem. Phys.* **53**, 3065 (1970).
51. L.-K. Huy, W. Forst, and Z. Prášil, *Chem. Phys. Lett.* **9**, 476 (1971).
52. Z. Prášil, *Advan. Mass Spectrom.* **5**, 53 (1971).
53. E. W. Schlag, R. A. Sandsmark, and W. G. Valance, *J. Chem. Phys.* **40**, 1461 (1964).
54. K. A. Wilde, *J. Chem. Phys.* **41**, 448 (1964).
55. W. Forst and P. St. Laurent, *Can. J. Chem.* **45**, 3169 (1967).
56. D. M. Dennison, *Rev. Mod. Phys.* **12**, 175 (1940).
57. D. R. Herschbach and V. W. Laurie, *J. Chem. Phys.* **35**, 458 (1961).
58. F. W. Schneider and B. S. Rabinovitch, *J. Amer. Chem. Soc.* **84**, 4215 (1962).
59. W. Forst and P. St. Laurent, *Can. J. Chem.* **43**, 3052 (1965).
60. W. Forst and P. St. Laurent, *J. Chim. Phys.* **67**, 1018 (1970).
61. A. Ralston, "A First Course in Numerical Analysis." McGraw–Hill, New York, 1965.
62. M. Vestal, A. L. Wahrhaftig, and W. H. Johnston, U.S. Air Force Rep. ARL 62-426 (1962).
63. M. L. Vestal, *J. Chem. Phys.* **43**, 1356 (1965).
64. M. L. Vestal and G. Lerner, U. S. Air Force Rep. ARL 67-0114 (1967).
65. J. C. Tou and A. L. Wahrhaftig, *J. Phys. Chem.* **72**, 3034 (1968).
66. M. Abramowitz and I. A. Stegun, eds., "Handbook of Mathematical Functions." NBS Appl. Math. Ser. No. 55, Washington, D.C., 1965.
67. M. Šolc, *Coll. Czechoslov. Chem. Comm.* **36**, 2327 (1971).
68. M. L. Vestal and H. M. Rosenstock, *J. Chem. Phys.* **35**, 2008 (1961).
69. K. Morokuma, B. C. Eu, and M. Karplus, *J. Chem. Phys.* **51**, 5193 (1969).

CHAPTER 7

Unimolecular Rate with an Effective Potential

We now consider $k(E)$ for a system reacting under the influence of an effective potential, rather than merely a vibrational potential, as was assumed in Chapter 4. Recall that the effective potential (V_{eff}) was introduced in Chapter 3, Section 2 as an approximate way of handling rotational motions when the vibration–rotation couplings are neglected. It can be seen in Fig. 3-4 that the effective potential for $\mathscr{J} > 0$ has a form such that the distance from potential minimum to potential maximum, which we may call D_J, is less than D_e, the inequality $D_J < D_e$ increasing with increasing $\mathscr{J}$. Since the effective critical energy for dissociation is $D_J - \varepsilon_z$ (Fig. 7-1), we now have a critical energy which is decreasing with increasing $\mathscr{J}$. Therefore molecules which have initially a nonzero angular momentum ($\mathscr{J} \neq 0$) should decompose faster since they have a lower potential barrier to surmount, relative to molecules reacting under a vibrational potential ($\mathscr{J} = 0$).

An alternative way of looking at the same problem can be illustrated on the system $CH_4 \rightarrow CH_3 + H$. The reactant CH_4 is a spherical top; when one of the C—H bonds begins to increase in length the molecule becomes a prolate symmetric top $H_3C\cdots H$ whose moment of inertia I_b about an axis at right angle to the symmetry axis [cf. Eq. (5-3)] progressively increases as H moves away from CH_3, and ultimately becomes infinite when H is far enough. This means that in the process of dissociation, $E_r(\mathscr{J})$, the

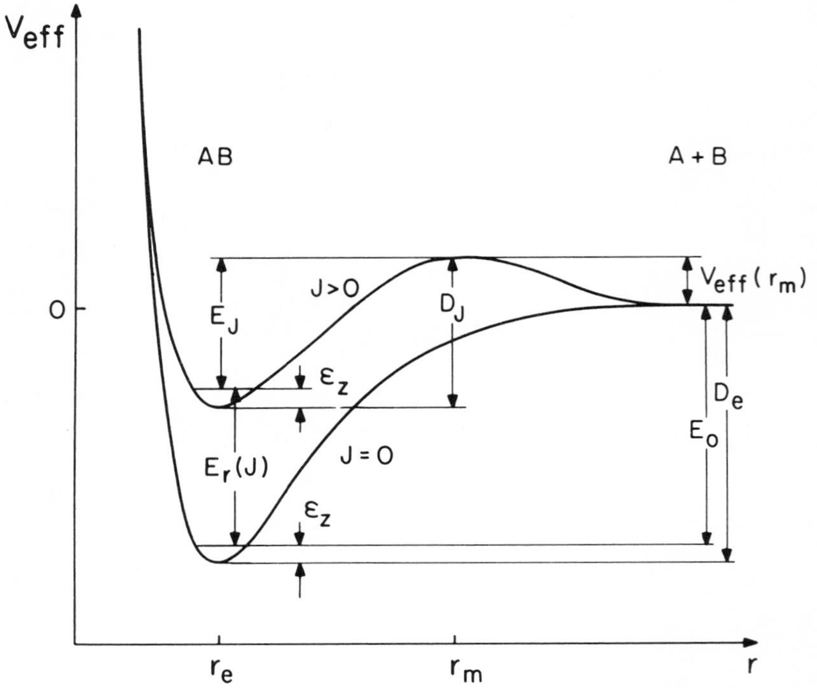

Fig. 7-1. *Effective potential V_{eff} for diatom $\mathcal{AB}$.*

This potential applies to type 1 reactions. The $\mathcal{A}$–$\mathcal{B}$ distance is r, the zero-point energy of $\mathcal{AB}$ is ε_z, and its rotational quantum number is $\mathcal{J}$. $E_r(\mathcal{J})$ is the rotational energy of $\mathcal{AB}$ and E_J is the effective critical energy for the dissociation $\mathcal{AB} \to \mathcal{A} + \mathcal{B}$. The zero of V_{eff} is at $r \to \infty$, while its maximum is at r_m. The curves are drawn so as to make their minima coincide at r_e, a slight misrepresentation (cf. Fig. 3-4). Here $E_0 \equiv D_0$ (cf. p. 137).

$\mathcal{J}$-dependent part of the rotational energy of the prolate symmetric top, "vanishes," i.e., does not appear in the products but is used up instead to surmount the potential energy barrier. This has been demonstrated to be at least partially the case in the photodissociation of (*1*)[1] $NO_2 \to NO + O$ and in the photoionization of methane[2] $CH_4^+ \to CH_3^+ + H$.

The $\mathcal{J}$-dependent rate constant To indicate that the critical energy is now a function of $\mathcal{J}$, we shall write E_J for E_0. The consequence is that through E_J the microcanonical rate constant $k(E)$ also becomes a function of $\mathcal{J}$,

[1] The reference list for this chapter can be found on p. 142.
[2] Chupka (*2*). Cf. also Chapter 10, Section 9.

which we shall indicate by writing $k(E, \mathscr{J})$ for $k(E)$. Thus Eq. (4-34) becomes

$$k(E, \mathscr{J}) = \alpha G^*(E - E_J)/hN(E) \tag{7-1}$$

We now have a threshold for reaction that can be in principle anywhere between E_0 (if $\mathscr{J} = 0$) and zero (for sufficiently large $\mathscr{J}$; cf. Fig. 7-2), so that if $\mathscr{J} \neq 0$, molecules with energy less than E_0 can decompose. This looks as if the barrier at E_0 were permeable to molecules with angular momentum, and therefore some authors (*3*) express the $\mathscr{J}$ dependence of the reaction threshold in terms of a "transmission coefficient." It must be emphasized that such "transmission coefficient" has nothing to do with the transmission coefficient $\kappa(E)$ defined in Chapter 4, Section 4.

The energy in (7-1) is referred to ground state of the reactant as zero. We shall find it more convenient to refer energy to the ground state of the transition state as zero; if we call this energy E', then $E' = E - E_J$ and

$$k(E', \mathscr{J}) = \alpha G^*(E')/hN(E' + E_J) \tag{7-2}$$

It is important to realize that the zero of E or E' is the ground vibrational ($\mathscr{n} = 0$) state of the $\mathscr{J}$th rotational state of the reactant or transition state, respectively. When $\mathscr{J} = 0$, E is at the same time also the total energy of the reactant, and heretofore was sometimes used in this sense; under an effective potential, however, it takes energy $E_r(\mathscr{J})$ [Eq. (5-6)] to prepare the reactant in state with angular momentum $\mathscr{J}$, so that, referred to the $\mathscr{n} = 0$, $\mathscr{J} = 0$ state, the reactant energy is $E + E_r(\mathscr{J})$, of which only E is available for randomization. Hence Eqs. (7-1) and (7-2) refer only to the randomizeable part of the total energy, as of course they must.

Effective potential The effective potential V_{eff} is given by the sum of the vibrational potential V_{vib} and the rotational potential V_r:

$$V_{eff} = V_{vib} + V_r \tag{7-3}$$

The essence of the problem of evaluating $k(E', \mathscr{J})$ by Eq. (7-2) is to obtain E_J in terms of E_0, $\mathscr{J}$, and molecular parameters of the reactant. To this end, it is necessary to have an expression for V_{eff} as a function of internuclear distances. In Section 1 a simplified model is discussed which is useful for obtaining an expression for V_r, valid for any type of vibrational potential. Approximations to a type 1 vibrational potential are considered in Section 2 and the corresponding effective potential is used in Section 3 to calculate the $\mathscr{J}$ dependence of E_J. Some supplementary remarks on the conservation of angular momentum are made here. The effective potential is then obtained for type 2 and type 3 reactions in Section 4, and E_J is calculated for these two cases.

1. The rotational potential

To simplify matters somewhat, we shall regard (*4*) a polyatomic molecule that upon dissociation yields fragments $\mathscr{A}$ and $\mathscr{B}$ as a quasi-diatomic molecule $\mathscr{A}\mathscr{B}$. The process of dissociation $\mathscr{A}\mathscr{B} \rightarrow \mathscr{A} + \mathscr{B}$ is then described by a single parameter r, the $\mathscr{A}$–$\mathscr{B}$ distance[3]; r should be interpreted, in principle, as the distance between the centers of gravity of $\mathscr{A}$ and of $\mathscr{B}$.

The justification for using r as the only parameter comes from data on dissociation of triatomic molecules ABC → AB + C: it is found (*5*) that the A–B distance in AB differs very little from its value in ABC, which means that the A–B distance does not change much as C moves away from B. This can also be seen by following the broken line on the potential energy surface of ABC shown in Fig. 3-1(a).

The rotational potential V_r is due to rotational energy associated with a constant $\mathscr{J}$ throughout the reaction. This is of course the same $\mathscr{J}$ that is fixed by angular momentum conservation, and involves therefore the two rotational degrees of freedom associated with two equal moments of inertia (Chapter 5, Section 1), so that the quasi-diatomic molecule approximation is quite appropriate. Since these two rotational degrees of freedom are not "pertinent," their effect is limited to influencing the critical energy. If μ_{AB} is the reduced mass of the quasidiatomic $\mathscr{A}\mathscr{B}$, its moment of inertia is $I = \mu_{AB}r^2$, and the corresponding rotational potential is, *as a function of r*,

$$V_r = \mathscr{J}(\mathscr{J} + 1)\hbar^2/2\mu_{AB}r^2 \tag{7-4}$$

If we let r_e be the distance between centers of gravity of $\mathscr{A}$ and $\mathscr{B}$ at equilibrium, we have, of necessity, $\mu_{AB}r_e^2 = I_b$, where $I_b = I_c$ are the two large moments of inertia of the prolate symmetric top, represented by the quasi-diatomic $\mathscr{A}\mathscr{B}$ for the purpose of this chapter.

2. Type 1 vibrational potential

The vibrational type 1 potential for the quasidiatomic $\mathscr{A}\mathscr{B}$ has the general shape exhibited by the lower curve in Fig. 7-1. There are a number of functions[4] that reproduce this shape but they all involve exponentials which

[3] This is a more stringent condition than if the motion $\mathscr{A} \leftrightarrow \mathscr{B}$ were associated with some sort of antisymmetric stretch normal mode vibration, since in principle *all* internuclear distances change during a normal mode motion, though not all to the same extent (cf. footnote 2 in Chapter 3).

[4] Varshni (*6*), Steele *et al.* (*7*), Varshni and Sukla (*8*), Levine (*9*), and Parr and Borkman (*10*); the latter consider the possibility of piecewise analyzing empirical potential energy curves over particular ranges of r to obtain effective "local" values of $\mathfrak{m}$ and $\mathfrak{n}$ in $V = (B/r^{\mathfrak{m}}) + (C/r^{\mathfrak{n}})$.

render the algebra of the present problem intractable. However, if we look at V_{eff} for a (true) diatomic AB (Fig. 3-4) and compare it with its V_{vib} (curve for $\mathscr{J} = 0$), we note that the hump in V_{eff} occurs at relatively large internuclear distances, especially when $\mathscr{J}$ is small. This means that the addition of V_{r} appreciably modifies only the long-range part of V_{vib}, i.e., the part where the shorter-range chemical forces (due to electronic overlap) do not come into play. This part of V_{vib} behaves therefore like a potential between two stable species $\mathscr{A}$ and $\mathscr{B}$ and, as a result, can be approximated by a van der Waals potential (V_w) which may be represented in a number of ways.

The most general approach (*11*, p. 27) considers V_w as consisting of three parts,

$$V_{\text{vib}} \sim V_w = V_{\text{es}} + V_{\text{ind}} + V_{\text{disp}} \tag{7-5}$$

where V_{es} is the electrostatic contribution arising from permanent charges and dipole moments on $\mathscr{A}$ and $\mathscr{B}$, V_{ind} is the inductive contribution due to the interaction of permanent charges and multipole moments on (say) $\mathscr{A}$ with the induced multipole moments on $\mathscr{B}$, and V_{disp} is the dispersion contribution arising from induced dipole moments on both molecules. These effects are orientation-dependent and when temperature-averaged over all orientations[5] one obtains, neglecting higher-order terms, a potential that decreases as $1/r^6$ for interaction between two neutral fragments and as $1/r^4$ for interaction between one charged fragment and one neutral fragment.[6] The formulas for the various contributions are summarized in Table 7-1. We thus arrive at the simple result

$$V_{\text{vib}} \sim -\mathscr{C}/r^{\mathfrak{n}}, \qquad r > r_{\text{e}} \tag{7-6}$$

where $\mathscr{C}$ is a constant having the dimension of (energy) $\times$ (distance)$^{\mathfrak{n}}$, with $\mathfrak{n} = 6$ for unimolecular reaction yielding two neutral fragments and $\mathfrak{n} = 4$ for a reaction producing one charged and one neutral fragment.[7]

Instead of calculating $\mathscr{C}$ from the formulas in Table 7-1, an alternative approach, useful for describing interaction between neutral fragments, consists in finding a semiempirical approximation to $\mathscr{C}$. This can be done by

[5] The term "van der Waals potential," as used in this chapter, is understood to apply to the potential obtained by averaging out the orientation dependence. The *instantaneous* potential depends on a lower power of $1/r$ than that given in connection with Eq. (7-6). Cf. (*11*, p. 26).

[6] The potential for interaction between two charged fragments is the usual Coulomb potential that decreases as $1/r$.

[7] If the two particles are an atom and its ion, the van der Waals potential is a series starting with $1/r^2$ if the neutral atom is in a low-lying excited state that can set up a perturbation by interacting with the ground state. This is a rather special case. See (*12*, p. 381).

TABLE 7-1

CONTRIBUTIONS TO VAN DER WAALS POTENTIAL[a,b] $V_w\ (= V_{es} + V_{ind} + V_{disp})$ BETWEEN $\mathscr{A}$ AND $\mathscr{B}$

	V_{es}		V_{ind}		V_{disp}	
$\mathscr{A}$ charged, $\mathscr{B}$ neutral	$-e^2\mu_b/3kTr^4$	(1)	$-\alpha_b e^2/2r^4$	(2)	0	
$\mathscr{A}$ neutral, $\mathscr{B}$ neutral	$-2\mu_a^2\mu_b^2/2kTr^6$	(3)	$-(\alpha_b\mu_a^2 + \alpha_a\mu_b^2)/r^6$	(4)	$-(3\alpha_a\alpha_b/2r^6)V'$	(5)

[a] In this table α_a, α_b are polarizabilities; μ_a, μ_b are dipole moments; e is the electronic charge ($= 4.8 \times 10^{-10}$ esu); a_0 is the radius of the first Bohr orbit ($= 0.5292 \times 10^{-8}$ cm); N_a, N_b are the numbers of electrons in the outer shell of $\mathscr{A}$, $\mathscr{B}$; and $\mathscr{I}_a$, $\mathscr{I}_b$ are ionization potentials. Subscripts a and b distinguish properties of $\mathscr{A}$ from those of $\mathscr{B}$.

[b] The constant V' in the last column is variously given as (i) $V' = e^2 a_0^{1/2}/[(\alpha_a/N_a)^{1/2} + (\alpha_b/N_b)^{1/2}]$ [J. C. Slater and J. G. Kirkwood, *Phys. Rev.* **37**, 682 (1931); K. S. Pitzer, *J. Amer. Chem. Soc.* **78**, 4565 (1956)] and (ii) $V' = \mathscr{I}_a\mathscr{I}_b/(\mathscr{I}_a + \mathscr{I}_b)$ [E. Gorin, *Acta Physicochim. URSS* **9**, 691 (1938)]. For other interpretations of V' see H. K. Kramer and D. R. Herschbach, *J. Chem. Phys.* **53**, 2792 (1970) and references therein.

assuming that the long-range interaction of two (neutral) fragments is given by the interaction of two similar stable molecules as determined from their transport properties. Thus the long-range potential between two methyl radicals, for instance, would be represented by second virial coefficient or viscosity data of methane (*13*, p. 518).

A variant of this approach is to take $-\mathscr{C}/r^6$ as the attractive part of a Lennard-Jones potential; then $\mathscr{C}$ is related to the depth of the potential well D_e by (*6*)

$$\mathscr{C} = D_e r_e^6\, \mathrm{m}/(\mathrm{m} - 6) \tag{7-7}$$

where r_e (cf. Fig. 7-1) is the equilibrium $\mathscr{A}$–$\mathscr{B}$ distance in $\mathscr{A}\mathscr{B}$, and m is the power of r in the $1/r^{\mathrm{m}}$ *repulsive* part of the potential.

It is undesirable to have $\mathscr{C}$ dependent on the form of the repulsive potential, and therefore a Sutherland potential (*14*, p. 148)

$$\begin{aligned} V_w &= -\mathscr{C}/r^{\mathrm{n}} \qquad & r \geqslant r_e \\ &= \infty & r \leqslant r_e \end{aligned} \tag{7-8}$$

is preferable (*15*), with

$$\mathscr{C} = D_e r_e^{\mathrm{n}} \tag{7-9}$$

The advantage of this approach is that $\mathscr{C}$ can be related directly to parameters of the dissociating $\mathscr{A}\mathscr{B}$, rather than merely to parameters of some more or less analogous system, and can be used for both neutral–neutral ($\mathfrak{n} = 6$) and charged–neutral ($\mathfrak{n} = 4$) interactions. The potential (7-8) with $\mathscr{C}$ given by (7-9) is not quite the potential for a van der Waals-type force, which would have a shallow minimum and infinitely steep repulsion at much larger[8] values of r.

Table 7-2 shows values of $\mathscr{C}$ calculated by various methods for several two-fragment systems. The agreement among the different methods is fairly good for radical pairs involving the methyl radical, which does not have a dipole moment, and is much worse for the CF_3–CF_3 interaction, where the radicals do have a dipole moment. Note, however, that the available values of $E_{a\infty}$ for CH_4 and of $\mathscr{I}_b$ (ionization potential) for CF_3 are rather uncertain.

3. Effective potential for type 1 reaction

From the discussion in Sections 1 and 2 it follows that the effective potential V_{eff} for a type 1 reaction, with energy zero referred to separated fragments $\mathscr{A}$ and $\mathscr{B}$, is approximately

$$V_{\text{eff}} \simeq -\frac{\mathscr{C}}{r^{\mathfrak{n}}} + \frac{\mathscr{J}(\mathscr{J}+1)\hbar^2}{2\mu_{\text{AB}}r^2} \tag{7-10}$$

The maximum[9] in the effective potential is located at r_{m}, which is given by the conditions $dV_{\text{eff}}/dr = 0$ and $d^2V_{\text{eff}}/dr^2 < 0$; this gives

$$r_m = \left(\frac{\mathfrak{n}\mu_{\text{AB}}\mathscr{C}}{\mathscr{J}(\mathscr{J}+1)\hbar^2}\right)^{1/(\mathfrak{n}-2)} \tag{7-11}$$

The potential at r_{m} is

$$V_{\text{eff}}(r_m) = \frac{\mathfrak{n}-2}{2}\left(\frac{\mathscr{J}(\mathscr{J}+1)\hbar^2}{\mathfrak{n}\mu_{\text{AB}}\mathscr{C}^{2/\mathfrak{n}}}\right)^{\mathfrak{n}/(\mathfrak{n}-2)} \tag{7-12}$$

and the new critical energy E_J is (cf. Fig. 7-1)

$$E_J = E_0 + V_{\text{eff}}(r_{\text{m}}) - E_{\text{r}}(\mathscr{J}) \tag{7-13}$$

[8] Actually the value of $\mathscr{C}$ would be about the same for both a van der Waals potential and the sort of potential considered in Eq. (7-8) since the deeper well in the latter (depth D_{e}) is to a large extent compensated by the smaller r_{e} (which appears to sixth power in $\mathscr{C}$) at which the well is located. For example, from second virial coefficient data of methane [(*11*, Table I-A, p. 1110)] we have $D_{\text{e}} \equiv \epsilon = 2.07 \times 10^{-14}$ erg, $d = 3.8 \times 10^{-8}$ cm, and then from Eq. (8-119), $\mathscr{C} = 4\epsilon d^2 = 24.9 \times 10^{-59}$ erg cm^6, which, if assumed to represent the CH_3–CH_3, interaction, is of the same order of magnitude as the data of Table 7-2.

[9] Equation (7-11) shows that there is a maximum only if $\mathfrak{n} > 2$. In practice, this means that V_{eff} has no maximum only for a special type of ion–atom reaction (footnote 7) or when $\mathscr{AB}$ involves ionic binding (footnote 6). In the discussion that follows and also in Chapter 11, Section 2 these two cases are implicitly excluded.

TABLE 7-2

VALUE OF CONSTANT $\mathscr{C}$ IN THE VAN DER WAALS POTENTIAL[a] BETWEEN FRAGMENTS $\mathscr{A}$ AND $\mathscr{B}$

Fragments	Molecular parameters of $\mathscr{B}$				$\mathscr{A}$, $\mathscr{B}$ neutral $\mathscr{C}$ (erg cm^6 × 10^{59})			$\mathscr{A}^+$, $\mathscr{B}$	Parameters of $\mathscr{AB}$	
$\mathscr{A}$, $\mathscr{B}$	μ_b (× 10^{18} esu)	α_b (× 10^{24} cm^3)	$\mathscr{I}_b$ (eV)	N_b	(i)	(ii)	(iii)	$\mathscr{C}$ (erg cm^4 × 10^{43})	r_e (× 10^8 cm)	$(E_{a\infty} + h\nu/2)$ (× 10^{12} erg)
CH_3, H	0	0.67	13.6	1	9.3	6.7	1.25	0.77	1.09 (C—H)	7.45
CH_3, CH_3	0	2.2	9.86	7[b]	10.8	5.7	8.1	2.53	1.54 (C—C)	6.06
				9[c]	12.3					
CF_3, CF_3	1.5	2.6	10.2	25[b]	35.6	17.6	8.9	4.32	1.54 (C—C)	6.6
				33[c]	39.6					

[a] Molecular parameters of $\mathscr{B}$ from refs. (*b*) and (*c*). Parameters of $\mathscr{A}$, $\mathscr{B}$ neutral: column (i), calculated from Eq. (5), Table 7-1, with V' given by expression (i); column (ii), calculated from Eq. (5), Table 7-1, with V' given by expression (ii); column (iii), calculated from Eq. (7-14) in the text using $E_{a\infty}$ (Table 11-2) for E_0. $\mathscr{A}^+$ means $\mathscr{A}$ positively charged. For this case, $\mathscr{C}$ calculated from Eqs. (1) (when applicable) and (2) in Table 7-1. Parameters of $\mathscr{AB}$ refer, top to bottom, to CH_4, C_2H_6, and C_2F_6, respectively.

[b] E. Tschuikow-Roux, *J. Phys. Chem.* **72**, 1009 (1968).

[c] J. C. Walton, *J. Phys. Chem.* **71**, 2763 (1967).

which we may find useful for some purposes to write also as

$$E_J = E_0 - \left\{1 - \frac{n-2}{2}\left(\frac{2r_e^2}{n\mathscr{C}^{2/n}}\right)^{n/(n-2)} [E_r(\mathscr{J})]^{2/(n-2)}\right\} E_r(\mathscr{J}) \quad (7\text{-}14)$$

where $E_r(\mathscr{J}) = \mathscr{J}(\mathscr{J} + 1)\hbar^2/2\mu_{AB}r_e^2$. The assumption involved here is that the potential minimum remains at r_e when $\mathscr{J} > 0$. This is consistent with the potential (7-8), but more accurate computations with a better potential (cf. Fig. 3-4) show that as $\mathscr{J}$ increases, the minimum moves slightly toward larger values of r; therefore Eq. (7-13) is a slight overestimate.

A word now about zero-point energy. The problem in this section has been reduced exclusively to one of a quasi-diatomic $\mathscr{AB}$ decomposing under the influence of an effective potential, and the zero-point energy involved is therefore just the zero-point energy of the stretching vibration $\mathscr{A}$—$\mathscr{B}$, or, approximately, of the antisymmetric stretch in the original polyatomic reactant. To indicate that the zero-point energy refers to just this one vibration, we denote it in Fig. 7-1 by ε_z. The transition state is assumed to be located at the maximum of V_{eff}, i.e., at r_m, so that we still have just one coordinate specifying the critical surface, as postulated in Chapter 4, Section 1, except that r_m and therefore also the location of the critical surface now depend on $\mathscr{J}$. This aspect is further examined in Chapter 11, Section 2. In the present case the transition state is represented as if consisting of the quasi-diatomic molecule $\mathscr{AB}$ about to break up into $\mathscr{A} + \mathscr{B}$, so that there is no zero-point energy involved in the transition state, the $\mathscr{A}$—$\mathscr{B}$ vibration being the reaction coordinate. In actual fact, of course, $\mathscr{A}$ and $\mathscr{B}$ are both (in principle) polyatomic, but the zero-point energy in the transition state of vibrations other than the $\mathscr{A}$—$\mathscr{B}$ vibration has been taken into account in E_0. The portion of Fig. 7-1 at large r should not be confused with Fig. 4-2, since the latter is meant to give only a *schematic* representation for a polyatomic reactant and transition state. For a different approach, see Ex. 1.

If the frequency (in energy units) of the $\mathscr{A}$—$\mathscr{B}$ stretching vibration is ν_{AB}, we have $\varepsilon_z = \frac{1}{2}\nu_{AB}$ and the constant $\mathscr{C}$ [Eq. (7-9)] becomes

$$\mathscr{C} = (E_0 + \tfrac{1}{2}\nu_{AB})r_e^6 \quad (7\text{-}15)$$

Fig. 7-2(a) shows E_J calculated (*16*) from Eqs. (7-12) and (7-13) for one particular case of dissociation into two neutral fragments ($n = 6$) using the potential (7-8) and $\mathscr{C}$ given by (7-15). When $\mathscr{J}$ is sufficiently high the effective critical energy E_J becomes eventually zero, and at this point $\mathscr{AB}$ is prepared in a rotationally dissociated state so that $k(E', \mathscr{J})$ becomes meaningless (i.e., infinite). Figure 7-3 shows a plot of $k(E', \mathscr{J})$ [Eq. (7-2)] versus E' and $\mathscr{J}$, calculated (*16*) for a type 1 system patterned on the reaction $HOOH \rightarrow 2OH$; this is the same system whose E_J is shown in Fig. 7-2(a).

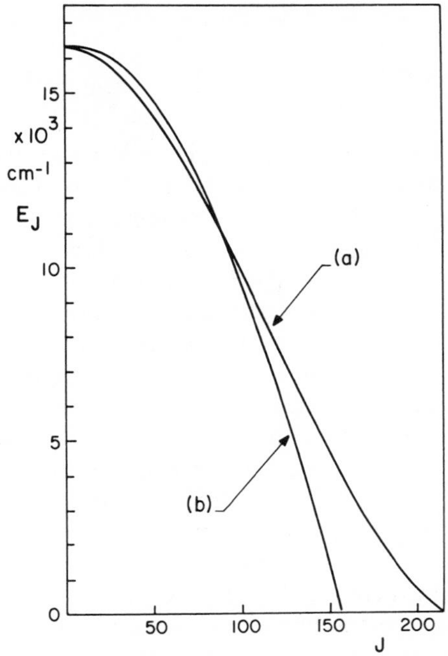

Fig. 7-2. *Effective critical energy E_J as a function of $\mathcal{J}$.*

(a) E_J as calculated from Eqs. (7-12)–(7-14), using $n = 6$, $\mu_{AB} = 1.422 \times 10^{-23}$ g, $r_e = 1.49 \times 10^{-8}$ cm, $\nu_{AB} = 880\ \text{cm}^{-1}$, $E_0 = 16349\ \text{cm}^{-1}$. These parameters apply to the process $H_2O_2 \rightarrow OH + OH$. (b) E_J as calculated from Eq. (7-21) assuming $r_e/r_0 = 0.5$; other parameters same as in case (a). W. Forst and Z. Prášil, *J. Chem. Phys.* **53**, 3065 (1970).

The assumption was made that angular momentum is strictly conserved, i.e., the two external rotations of the quasidiatom HO—OH are adiabatic and therefore contribute only to the decrease of E_0. The figure shows clearly how, at fixed excitation energy in the transition state, the rate constant increases with $\mathcal{J}$, owing to decreasing critical energy E_J.

4. Effective potential for type 2 and type 3 reactions

We shall find it convenient in this section to continue the representation of the reacting system as a dissociating quasidiatomic molecule $\mathcal{AB}$, except that in the case of isomerizations (type 3 reactions) we may think of $\mathcal{A}$—$\mathcal{B}$ as not actually separating to infinity but instead becoming bonded differently as $\mathcal{BA}$.

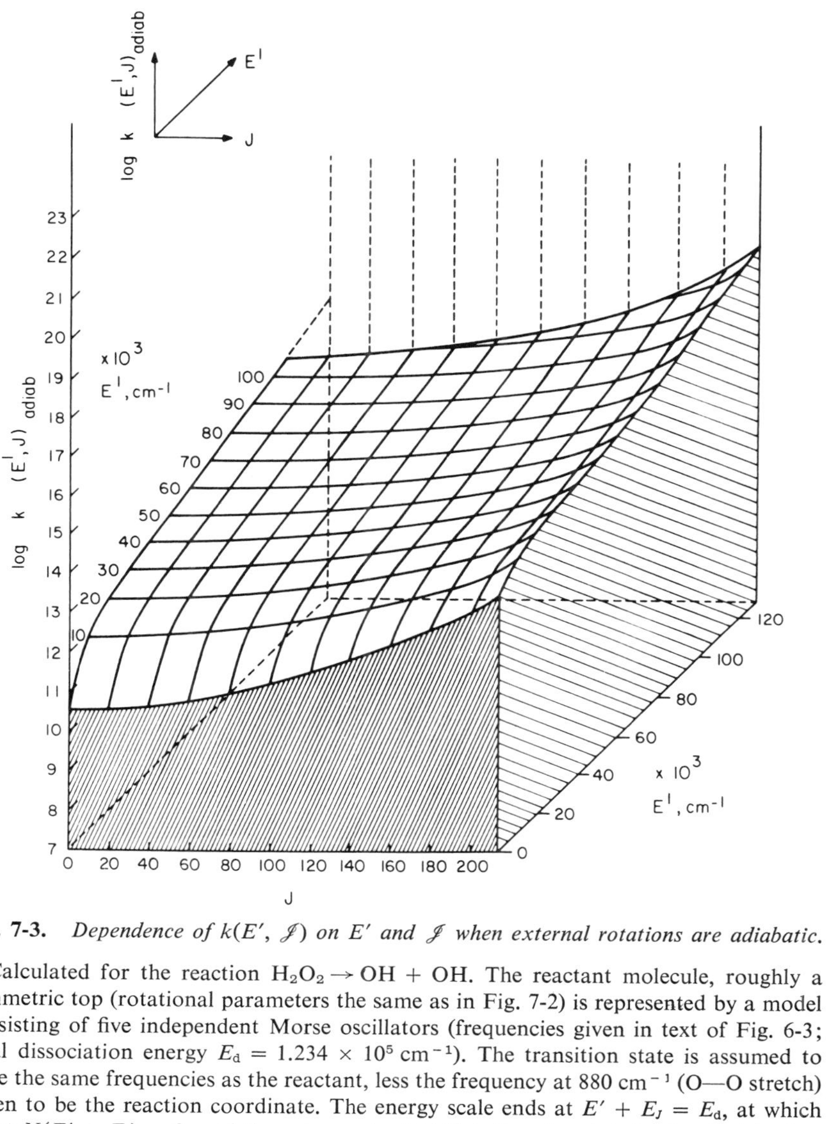

Fig. 7-3. *Dependence of $k(E', \mathscr{J})$ on E' and $\mathscr{J}$ when external rotations are adiabatic.*

Calculated for the reaction $H_2O_2 \rightarrow OH + OH$. The reactant molecule, roughly a symmetric top (rotational parameters the same as in Fig. 7-2) is represented by a model consisting of five independent Morse oscillators (frequencies given in text of Fig. 6-3; total dissociation energy $E_d = 1.234 \times 10^5$ cm^{-1}). The transition state is assumed to have the same frequencies as the reactant, less the frequency at 880 cm^{-1} (O—O stretch) taken to be the reaction coordinate. The energy scale ends at $E' + E_J = E_d$, at which point $N(E' + E_J) = 0$, and the rate constant (7-2) is therefore infinite (vertical broken lines in the figure). The $\mathscr{J}$ scale ends at $\mathscr{J} = 214$, since at this point $E_J = 0$ (Fig. 7-2a). Calculations by the method of steepest descents [Eqs. (6-110) and (6-94)–(6-97)]. W. Forst and Z. Prášil, *J. Chem. Phys.* **53**, 3065 (1970).

Type 2 and 3 reactions, as mentioned in Chapter 3, Section 4 involve an activation energy for the back reaction $\mathscr{A} + \mathscr{B} \rightarrow \mathscr{A}\mathscr{B}$ (or $\mathscr{B}\mathscr{A} \rightarrow \mathscr{A}\mathscr{B}$), and therefore the maximum in the potential for $\mathscr{A}\mathscr{B}$ exists even in the absence of angular momentum. For simplicity, we shall take this maximum, assumed to be located at $\mathscr{A}$–$\mathscr{B}$ distance r_0, to be the extremum of a parabolic potential (Fig. 7-4) which becomes infinitely steep at r_e ($r_e < r_0$). The vibrational potential, now for convenience referred to the maximum as zero, is

$$V_{\text{vib}} = -E_0(r - r_0)^2/(r_e - r_0)^2 \tag{7-16}$$

where E_0 is the critical energy for dissociation in the absence of angular momentum. The effective potential for rotational quantum number $\mathscr{J}$ is

$$V_{\text{eff}} = -\frac{E_0(r - r_0)^2}{(r_e - r_0)^2} + \frac{\mathscr{J}(\mathscr{J} + 1)\hbar^2}{2\mu_{\text{AB}} r^2} \tag{7-17}$$

The maximum in V_{eff} is located at r_m, which is the value of r satisfying $dV_{\text{eff}}/dr = 0$, i.e., is the solution of

$$r^3 r_0 - r^4 = \mathscr{J}(\mathscr{J} + 1)\hbar^2(r_e - r_0)^2/2\mu_{\text{AB}}E_0 \tag{7-18}$$

This equation is difficult to solve algebraically, but we may observe that numerically, for $E_0 > 10$ kcal mole^{-1}, $\hbar^2(r_e - r_0)^2/2\mu_{\text{AB}}E_0 < 10^{-35}$ cm^4, so that for all practical purposes the right-hand side of (7-18) is zero, and the solution is therefore $r_m = r_0$ for any reasonable value of $\mathscr{J}$ and for any E_0 of chemical interest.[10] It follows then that

$$V_{\text{eff}}(r_m) = \mathscr{J}(\mathscr{J} + 1)\hbar^2/2\mu_{\text{AB}}r_0^2 \tag{7-19}$$

and

$$\begin{aligned} E_J &= E_0 + [\mathscr{J}(\mathscr{J} + 1)\hbar^2/2\mu_{\text{AB}}r_0^2] - E_r(\mathscr{J}); \\ &= E_0 + \tfrac{1}{2}\mathscr{J}(\mathscr{J} + 1)\hbar^2[(1/I_0) - (1/I_b)] \end{aligned} \tag{7-20}$$

where I_0 is moment of inertia corresponding to $\mathscr{A}$–$\mathscr{B}$ distance of r_0 and I_b is moment of inertia corresponding to r_e. The result (7-19) or (7-20) is virtually independent of the assumption of a parabolic potential since the approximate solution $r_m = r_0$ indicates that it is only the potential in the immediate vicinity of r_0 that matters.

Since $r_0 > r_e$, we have also $I_0 > I_e$, and so the second term of (7-20) is negative and hence $E_J < E_0$, as in the case of type 1 reactions. On account of the moment of inertia I_0 being a fixed multiple of I_b, Eq. (7-20) can also be written as

$$E_J = E_0 - [1 - (r_e/r_0)^2]E_r(\mathscr{J}) \tag{7-21}$$

[10] If we had $r_m = 0.999r_0$, the right-hand side of (7-17) would still be quite large ($\sim 10^{-3}$) so that r_m must be a good deal closer to r_0 than this.

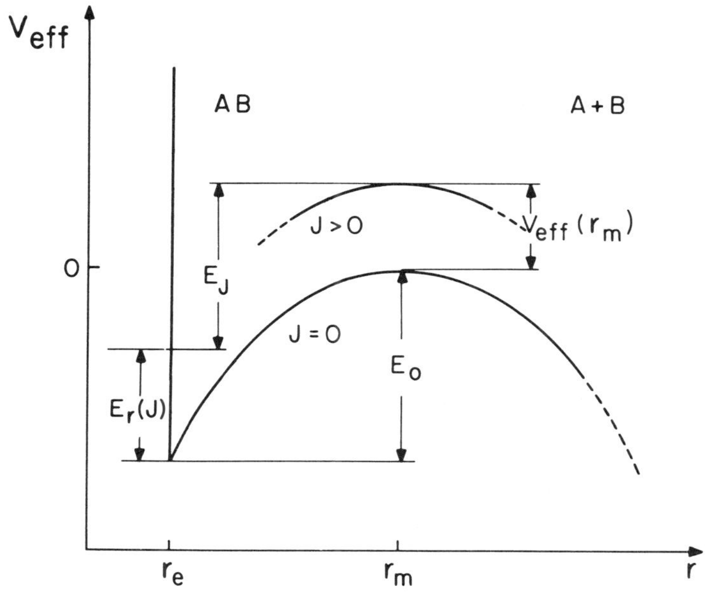

Fig. 7-4. *Parabolic effective potential for diatom $\mathscr{AB}$.*

This potential applies to type 2 and type 3 reactions. We have again $E_J < E_0$ when $\mathscr{J} > 0$. The zero of V_{eff} is at the maximum of the curve for $\mathscr{J} = 0$; the potential becomes infinitely steep at $r = r_e$ for all $\mathscr{J}$. Zero-point energy of $\mathscr{AB}$ not shown. Symbols have the same meaning as in Fig. 7-1.

A plot of E_J given by Eq. (7-21) versus $\mathscr{J}$ is shown in Fig. 7-2(b), assuming that $r_0 = 2r_e$. At low $\mathscr{J}$, the plot is quite similar to the one obtained for type 1 reactions, and therefore at low $\mathscr{J}$ we may expect $k(E', \mathscr{J})$ for type 2 and 3 reactions to behave in a manner similar to the one shown in Fig. 7-3.

Equation (7-1) represents the most general formulation of the microcanonical rate constant $k(E, \mathscr{J})$. Now that we have expressions (7-14) or (7-21) for E_J, the derivation of $k(E, \mathscr{J})$ is complete. However, the rate constant actually observable in an experiment is a more or less complicated function of $k(E, \mathscr{J})$. The relation between $k(E, \mathscr{J})$ of Eq. (7-1) and the observable rate constant is the subject of Chapters 8–10 in Part II of this book.

Exercise

1. Conditions regarding angular momentum in Type 1 dissociations can be considered in terms of the *reverse* association of fragments $\mathscr{A} + \mathscr{B} \rightarrow \mathscr{AB}$ (cf. Exercise 2, Chapter 4). In this view, V_{eff} [Eq. (7-10)] is then defined not by $\mathscr{J}$ but by $\mathscr{L}$, the orbital angular

momentum of separated fragments (*17*). If E_t is the relative translational energy of fragments and ℓ their impact parameter then $\mathscr{L}(\mathscr{L}+1)\hbar^2 = 2\mu_{AB}E_t\ell^2$.
(a) Show that since fragments associating with energy E_t must clear the centrifugal barrier at $V_{eff}(r_m)$, we must have $E_t \geqslant V_{eff}(r_m)$, which imposes the limits (for $n = 6$, i.e. neutral fragments)

$$\ell^2 \leqslant (3/2)^{2/3}(3\mathscr{C}/E_t)^{1/3} \quad \text{i.e. } \mathscr{L}(\mathscr{L}+1)\hbar^2 \leqslant 6\mu_{AB}\mathscr{C}^{1/3}(E_t/2)^{2/3} \tag{i}$$

(b) Let E' = fixed total energy of fragments, $E_v{}^f$, $E_r{}^f$ = their vibrational and external rotational energies, respectively; then $E' = E_t + E_v{}^f + E_r{}^f$. Consider a simple case where $\mathscr{A}$ = atom, $\mathscr{B}$ = diatomic molecule; thus $E_r{}^f = \mathscr{J}^f(\mathscr{J}^f + 1)\hbar^2/2I_B$. The requirement that E_t be positive imposes the restriction $E_r{}^f \leqslant E' - E_v{}^f$; hence $E_t \approx [(\mathscr{J}_m{}^f)^2 - (\mathscr{J}_m{}^f)^2]\hbar^2/2I_B$. Thus $\mathscr{J}^f$ and $\mathscr{L}$ are related via E_t. Show from (i) that

$$(\mathscr{L}/\mathscr{L}_m) = [1 - (\mathscr{J}^f/\mathscr{J}_m{}^f)^2]^{1/3} \tag{ii}$$

where $\mathscr{L}_m$ and $\mathscr{J}_m{}^f$ correspond to upper limits of $\mathscr{L}$ and $E_r{}^f$. The $\mathscr{L}$ and $\mathscr{J}^f$ are further related via $\boldsymbol{L} + \boldsymbol{M}_B = \boldsymbol{M}$, i.e.

$$|\mathscr{L} - \mathscr{J}^f| \leqslant \mathscr{J} \leqslant \mathscr{L} + \mathscr{J}^f \tag{iii}$$

where the total angular momentum $\boldsymbol{M}$ is conserved [Eq. (5-1)]. Insofar as the transition state for the *forward* process $\mathscr{AB} \to \mathscr{A} + \mathscr{B}$ is concerned, the energy of its bending modes correlates with $E_r{}^f$, which is subject to restrictions (ii) and (iii), so that proper consideration of angular momentum conservation imposes, in effect, a restriction on the allowable energy of certain vibrations. Cf. refs. (24-26) in Chapter 5.

References

1. J. N. Pitts, Jr., J. H. Sharp, and S. I. Chan, *J. Chem. Phys.* **40**, 3655 (1964).
2. W. A. Chupka, *J. Chem. Phys.* **48**, 2337 (1968).
3. P. F. Knewstubb, *Advan. Mass Spectrom.* **5**, 48 (1971).
4. O. K. Rice and H. Gershinowitz, *J. Chem. Phys.* **2**, 853 (1934).
5. K. Machida and J. Overend, *J. Chem. Phys.* **50**, 4437 (1969).
6. Y. P. Varshni, *Rev. Mod. Phys.* **29**, 664 (1967).
7. D. Steele, E. R. Lippincott, and J. T. Vanderslice, *Rev. Mod. Phys.* **34**, 239 (1962).
8. Y. P. Varshni and R. C. Sukla, *J. Chem. Phys.* **40**, 250 (1964).
9. I. N. Levine, *J. Chem. Phys.* **45**, 827 (1966).
10. R. G. Parr and R. F. Borkman, *J. Chem. Phys.* **46**, 3683 (1967).
11. J. O. Hirschfelder, C. F. Curtis, and R. B. Bird, "Molecular Theory of Gases and Liquids." Wiley, New York, 1964.
12. G. Herzberg, "Molecular Spectra and Molecular Structure," Vol. I, Spectra of Diatomic Moleculars. Van Nostrand Reinhold, Princeton, New Jersey, 1950.
13. O. K. Rice, "Statistical Mechanics, Thermodynamics, and Kinetics." Freeman, San Francisco, California, 1967.
14. E. A. Desloge, "Statistical Physics." Holt, New York, 1966.
15. W. Forst, *J. Chem. Phys.* **48**, 3665 (1968).
16. W. Forst and Z. Prášil, *J. Chem. Phys.* **53**, 3065 (1970).
17. P. Pechukas, J. C. Light, and C. Rankin, *J. Chem. Phys.* **44**, 794 (1966).

PART II

Introduction

The rate constant $k(E, \mathscr{J})$ which we have learned how to calculate in Part I does not represent a measurable property, for it is not feasible to observe the decay of an undisturbed isolated molecule of specified energy and angular momentum. In the real world of the laboratory two factors conspire against such a direct measurement: To begin with, it is not possible to prepare molecules within a narrow range of energies and angular momenta, and second, the decay of interest almost always occurs in competition with some other process.

One such kind of competition may involve the production and removal of molecules in reactive states. We can have removal by deexcitation to a nonreactive state (e.g., by collision), in which case there is no net loss of reactant, or to products by the decay process of interest, and then every molecule transformed to products is of course irretrievably lost. If neither of the two removal processes is very much faster than the other, a steady state is eventually set up in a time that is long compared to the time scale on which the individual production and removal processes operate. Which stage of the overall process is observed by experiment depends on the time resolution of the experimental arrangement. If the time resolution is "coarse," of the order of seconds or minutes, as in conventional kinetics, the system will be observed in its steady state; this is the case of the usual experiments in thermal and chemical activation systems, discussed in Chapters 8 and 9.

The competition may also be due to competing parallel and/or consecutive paths of decay, and if the time resolution is "fine," of the order of microseconds, measurements correspond to non-steady-state conditions; this is the case of systems discussed in Chapter 10.

Whatever the nature of the competing processes, the resulting kinetics is in general more or less complex. We can solve the kinetic scheme by treating molecules with each value of E and of $\mathcal{J}$ as a separate species. Suppose the species that decays in some complex fashion is $\mathsf{A}(E, \mathcal{J})$; we can define an *apparent* first-order rate constant $k(E, \mathcal{J})_{\text{app}}$ by

$$k(E, \mathcal{J})_{\text{app}} = -\frac{1}{[\mathsf{A}(E, \mathcal{J})]}\frac{[\mathsf{A}(E, \mathcal{J})]}{dt} \tag{II-1}$$

where $[\mathsf{A}(E, \mathcal{J})]$ is the instantaneous concentration of $\mathsf{A}(E, \mathcal{J})$. If the system is in steady state, the relation (II-1) defines the steady-state apparent rate constant $k(E, \mathcal{J})_{\text{app,ss}}$, which will be some function of the microcanonical rate constant $k(E, \mathcal{J})$ and of the rate constants $k_1, k_2, \ldots$ of the other steps in the kinetic scheme, *but not of time*:

$$k(E, \mathcal{J})_{\text{app,ss}} = f[k(E, \mathcal{J}), k_1, k_2, \ldots]_{\text{ss}} \tag{II-2}$$

A simple example is the strong-collision model of thermal unimolecular reactions, discussed in Chapter 8, Section 1, where $k_1 = \text{const}$ and $k_2 = \cdots = 0$ [cf. Eq. (8-11)]. If the system is not in steady state, $k(E, \mathcal{J})_{\text{app}}$ of Eq. (II-1) will be a function of $k(E, \mathcal{J}), k_1, k_2, \ldots$ and *of time*:

$$k(E, \mathcal{J})_{\text{app},t} = \mathfrak{f}[k(E, \mathcal{J}), k_1, k_2, \ldots, t] \tag{II-3}$$

A simple example of this kind of system is discussed in Chapter 10, Section 1, where $k_1 = k_2 = \cdots = 0$. Some, or all, of the other rate constants $k_1, k_2, \ldots$ may be also functions of E and $\mathcal{J}$. Equations (II-2) and (II-3) still represent a hypothetical case because $k(E, \mathcal{J})_{\text{app}}$, or its steady-state version, is not accessible to measurement. We shall next drop the subscript "app" as too cumbersome, with the understanding that $k(E, \mathcal{J})_{\text{ss}}$ and $k(E, \mathcal{J})_t$ refer to apparent rate constants.

The measurable property is only the apparent rate constant representing the *total* disappearance of A, i.e., the removal of $A(E, \mathcal{J})$ of all E's and $\mathcal{J}$'s produced by the excitation process. If $[\mathsf{A}] = \sum_{E,\mathcal{J}} [\mathsf{A}(E, \mathcal{J})]$, and if we let $\langle k(E, \mathcal{J})_{\text{ss}}\rangle$ represent the observable *apparent* steady-state rate constant, we have

$$\langle k(E, \mathcal{J})_{\text{ss}}\rangle = -\frac{1}{[\mathsf{A}]}\frac{d[\mathsf{A}]}{dt} \tag{II-4}$$

This is a prescription for obtaining $\langle k(E, \mathcal{J})_{\text{ss}}\rangle$ experimentally if measurements of $[\mathsf{A}]$ are done under steady-state conditions. To calculate the total removal of A, we sum Eq. (II-2) or (II-3) over all E's and $\mathscr{J}$'s, each E and $\mathscr{J}$ weighted by the probability that a given E and $\mathscr{J}$ shall be produced by the excitation process. Let $P(E, \mathscr{J})\, dE\, d\mathscr{J}$ be the probability that the reactant molecule will be formed initially with energy in the range E, $E + dE$ and with angular momentum in the range $\mathscr{J}$, $\mathscr{J} + d\mathscr{J}$; then

$$\langle k(E, \mathscr{J})_{ss}\rangle_P = \iint f[kE, \mathscr{J}), k_1, k_2, \ldots]_{ss} P(E, \mathscr{J})\, dE\, d\mathscr{J} \qquad \text{(II-5)}$$

This is a prescription for calculating $\langle k(E, \mathscr{J})_{ss}\rangle$. Thermal (Chapter 8) and chemical activation (Chapter 9) systems differ only in the forms of the functions $P(E, \mathscr{J})$ and f; in both cases $k_1 = \text{const}$ and $k_2 = \cdots = 0$.

In a similar fashion, if we let $\langle k(E, \mathscr{J})_t\rangle$ represent the observable apparent rate constant under non-steady-state conditions, the experimental prescription for $\langle k(E, \mathscr{J})_t\rangle$ is again [cf. Eq. (II-4)]

$$\langle k(E, \mathscr{J})_t\rangle = -\frac{1}{[\mathsf{A}]}\frac{d[\mathsf{A}]}{dt} \qquad \text{(II-6)}$$

if the instantaneous concentration [A] is measured under non-steady-state conditions. The prescription for calculating $\langle k(E, \mathscr{J})_t\rangle$ then follows from (II-3):

$$\langle k(E, \mathscr{J})_t\rangle_P = \iint \mathfrak{f}[k(E, \mathscr{J}), k_1, k_2, \ldots, t] P(E, \mathscr{J})\, dE\, d\mathscr{J} \qquad \text{(II-7)}$$

The functions $P(E, \mathscr{J})$ that appear in (II-5) and (II-7) are the same since for the purpose of the present discussion it was assumed that both equations refer to the same process, the only difference being the length of the time during which the system is under observation (steady-state versus non-steady-state conditions). If Eqs. (II-5) and (II-7) each refer to a different system (i.e., to different reactions, or to the same reaction but with different mechanisms of excitation), the functions $P(E, \mathscr{J})$ will be of course different.

Case when** $k(E, \mathscr{J})$ **factors out Suppose now that the functions f and $\mathfrak{f}$ are of a simple enough form such that $k(E, \mathscr{J})$ is a factor; thus

$$f[k(E, \mathscr{J}), k_1, k_2, \ldots]_{ss} = k(E, \mathscr{J}) \times f_a[k(E, \mathscr{J}), k_1, k_2, \ldots]_{ss} \qquad \text{(II-8)}$$

$$\mathfrak{f}[(E, \mathscr{J}), k_1, k_2, \ldots, t] = k(E, \mathscr{J}) \times \mathfrak{f}_b[k(E, \mathscr{J}), k_1, k_2 \ldots, t] \qquad \text{(II-9)}$$

where f_a and $\mathfrak{f}_b$ represent f and $\mathfrak{f}$, respectively, from which $k(E, \mathscr{J})$ has been factored out. We can then write $f_a \times P = P_{ss}$ and $\mathfrak{f}_b \times P = \mathfrak{P}$ and the two integrals in (II-5) and (II-7) become

$$\langle k(E, \mathscr{J})\rangle_{ss} = \iint k(E, \mathscr{J}) P[E, \mathscr{J}, k(E, \mathscr{J}), k_1, k_2, \ldots]_{ss}\, dE\, d\mathscr{J} \qquad \text{(II-10)}$$

$$\langle k(E, \mathscr{J})\rangle_t = \iint k(E, \mathscr{J})\mathfrak{P}[E, \mathscr{J}, k(E, \mathscr{J}), k_1, k_2, \ldots, t]\, dE\, d\mathscr{J} \tag{II-11}$$

The integrands in (II-5) and (II-10), or in (II-7) and (II-11), are identical, but the interpretations differ: Eqs. (II-10) and (II-11) can be interpreted as the averages of $k(E, \mathscr{J})$ over the steady-state distribution function P_{ss} or over the non-steady-state distribution function, $\mathfrak{P}$, respectively. Therefore $\langle k(E, \mathscr{J})_{ss}\rangle_P = \langle k(E, \mathscr{J})\rangle_{ss}$ and $\langle k(E, \mathscr{J})_t\rangle_P = \langle k(E, \mathscr{J})\rangle_t$.

It should be noted that any averaging inevitably obscures some of the detail in $k(E, \mathscr{J})$, the more so the wider the distribution function, and therefore tends to render the comparison between theory and experiment less significant. The worst case from this point of view is $P(E, \mathscr{J})$ for thermal systems (Chapter 8), which extends from zero to infinity for both E and $\mathscr{J}$. In chemical activation systems (Chapter 9) $P(E)$ is narrower, principally because of the low temperature at which these experiments can be performed; not much is known about the distribution of $\mathscr{J}$'s. Ionic systems (Chapter 10) may be prepared with a $P(E)$ that is essentially a delta function, which makes these systems particularly attractive; however, there are other problems.

The functions f and $\mathfrak{f}$ we will be dealing with in Part II mostly satisfy Eqs. (II-8) and (II-9), so that it is then immaterial whether we choose to regard the observable rate constant $\langle k(E, \mathscr{J})_{ss}\rangle$ as the average of f over P or as the average of $k(E, \mathscr{J})$ over P_{ss}. For convenience we use the latter interpretation in Chapter 8, Section 3. Similarly, only convenience dictates whether we wish to regard $\langle k(E, \mathscr{J})_t\rangle$ as the average of $\mathfrak{f}$ over P or as the average of $k(E, \mathscr{J})$ over $\mathfrak{P}$. The former interpretation is used in Chapter 10, Section 1.

Experimental information available for the majority of systems discussed in Chapter 10, however, is less detailed, in that the measured property is not some overall rate but merely the total fractional abundance of A at some time t after excitation. We can start with Eq. (II-1); after integration we get for the fractional abundance of $\mathsf{A}(E, \mathscr{J})$ after time t

$$\frac{[\mathsf{A}(E, \mathscr{J})]}{[\mathsf{A}(E, \mathscr{J})]^0} = \exp\left[-\int_0^t k(E, \mathscr{J})_{\text{app}}\, dt\right] \tag{II-12}$$

where $[\mathsf{A}(E, \mathscr{J})]^0$ is the concentration of $\mathsf{A}(E, \mathscr{J})$ at $t = 0$. We can then write, in view of (II-3),

$$\exp\left[-\int_0^t k(E, \mathscr{J})_{\text{app}}\, dt\right] = \mathfrak{F}[k(E, \mathscr{J}), k_1\, k_2, \ldots, t] \tag{II-13}$$

where $\mathfrak{F}$ is a function different from $\mathfrak{f}$. The total fractional abundance of A is a weighted sum of (II-13) over all E's and $\mathscr{J}$'s, i.e.,

$$[\mathsf{A}]/[\mathsf{A}]^0 = \iint \mathfrak{F}[k(E, \mathscr{J}), k_1, k_2, \ldots, t] P(E, \mathscr{J})\, dE\, d\mathscr{J} \qquad \text{(II-14)}$$

which is an equation very similar to (II-7), since $\mathfrak{F}$ is related[1] to $\mathfrak{f}$ through (II-3) and (II-13). The function $\mathfrak{F}$ is obtained directly by solving the differential equations of the kinetic scheme and it is not necessary to proceed from $\mathfrak{f}$ via Eq. (II-13). The form of the function $\mathfrak{F}$ for specific cases is discussed in Chapter 10, Section 2.

The purpose of these derivations is to impress on the reader that the relationships between time, concentration, and rate constant discussed in Chapter 1 are valid only for the elementary unimolecular decay events of the species $\mathsf{A}(E, \mathscr{J})$, and that in any actual kinetic system the relationship between $k(E, \mathscr{J})$ and an observable property is more complex.[2] Chapters 8–10 of Part II examine the various functions that appear in the integrals (II-5), (II-10), and (II-14) with the object of connecting $k(E, \mathscr{J})$ with observable properties. Chapter 11 then summarizes the experimental evidence insofar as it has a bearing on those aspects of the theory that the developments in Part I have left insufficiently defined, i.e., on the transition state.

[1] Since $\mathfrak{F} = \exp(-\int_0^t \mathfrak{f}\, dt)$, we have $\mathfrak{f} = -d \ln \mathfrak{F}/dt$.

[2] In this Introduction, and in Part II in general, $\mathscr{J}$ is often referred to as the angular momentum. This is true classically, and $\mathscr{J}$ is then a continuous variable. Quantum mechanically, however, $\mathscr{J}$ is the angular momentum *quantum number*, and the square of the actual quantum mechanical angular momentum is $\mathscr{J}(\mathscr{J} + 1)\hbar^2$; here $\mathscr{J}$ is a discrete variable. For the most part, we shall use a semiclassical approach, taking the quantum-mechanical form of the angular momentum, but treating $\mathscr{J}$ as a continuous variable.

CHAPTER 8

Thermal Reactions

A thermal reaction is one where the macroscopic state of the reacting system is described by its temperature; on the microscopic level, it represents an infinite collection of states with all possible values of E and $\mathcal{J}$ between zero and (in principle) infinity. This represents a very considerable collection of states indeed, with the result that the thermal distribution function is very broad, and the averaging therefore smears out much, if not all, of the detail in $k(E, \mathcal{J})$. In this respect, therefore, thermal systems provide a very insensitive test of the theory; however, despite this limitation, much of unimolecular theory (and reaction rate theory in general) has developed on the basis of thermal data, since until fairly recently thermal systems were the only kind of systems accessible experimentally.[1] Thus thermal systems will be considered first, mostly for historical reasons.

In thermal systems, molecules accumulate energy neccessary for reaction by collisions. Discussed in Section 1 are those features of the simple Lindemann model of collisional activation that are quite general and independent of assumptions concerning the details of the unimolecular process. In many ways, the discussion here is an extension of arguments presented in Chapter 1. Averaging of $k(E, \mathcal{J})$ over the equilibrium distribution of E and $\mathcal{J}$ is taken

[1] A sad commentary on the vicissitudes of chemical kinetics is the fact that most experimental systems on which theoretical models were tried eventually turned out not to be unimolecular at all.

up in Section 2, and the averaging over the steady-state distribution is discussed in Section 3. It is shown that it is sufficient to consider only averaging over E, since averaging over $\mathscr{J}$ involves multiplication by a correction factor and therefore can be done separately. The evaluation of this factor is considered in Section 4. The pressure dependence of the average rate constant $\langle k(E, \mathscr{J})\rangle$ is considered in Section 5, and a consistency theorem is derived linking high- and low-pressure data; the temperature dependence of the average rate constant is discussed next in Section 6. Both the pressure and temperature dependences of the average rate constant provide a connection with experimental observables, which is discussed in some detail in Section 7. Also discussed here are the inherent limitations on the sort of information that can be extracted from experimental data in thermal systems. Finally, Section 8 deals with the generalized Lindemann mechanism and the non-equilibrium distribution at the low-pressure limit.

1. Collisional activation: general features of Lindemann's simple mechanism

In the old days (*ca.* 1920), unimolecular reactions were thought to be immutably first order kinetically, which led inevitably to the question of how reactant molecules are raised to reactive states. Much confusion reigned until a clear distinction was made between order and molecularity. An interesting account of these early speculations may be found in the book by Daniels (*1*, p. 31 ff.).[2]

The now generally accepted mechanism was first proposed by Lindemann (*2*) and bears his name. In its simple form, it is a one-step collisional mechanism, but with a time lag introduced between the instant a molecule has received sufficient energy to react and the instant it actually reacts.

Activation takes place by binary (inelastic) collisions between molecules of reactant A and heat bath molecules M:

$$\mathsf{A} + \mathsf{M} \xrightarrow{k_1} \mathsf{A}(X) + \mathsf{M} \tag{8-1}$$

where $\mathsf{A}(X)$ represents molecules A in a reactive state specified by parameter(s) X. The identity of M is immaterial; it can be A itself, its product formed in (8-3), or some inert foreign gas. The molecule $\mathsf{A}(X)$ can be either deactivated by collision

$$\mathsf{A}(X) + \mathsf{M} \xrightarrow{k_2} \mathsf{A} + \mathsf{M} \tag{8-2}$$

or it can react

$$\mathsf{A}(X) \longrightarrow \text{product(s)} \tag{8-3}$$

[2] The reference list for this chapter can be found on p. 197.

The thermal reaction is thus viewed as proceeding by two consecutive processes: first, creation of a reactive state by collision (process 1, kinetically second order), followed by the chemical transformation (process 3, kinetically first order). The overall kinetics of the process will therefore depend on which of the two processes is slowest.

The kinetic scheme consisting of Eq. (8-1)–(8-3) can be solved exactly (*3*) for the general case if a rate constant is assigned to process (8-3), and yields a result of the form (square brackets indicate concentrations)

$$[\mathrm{A}] = C_1 \exp(-kt) + C_2 \exp(-k't) \tag{8-4}$$

where C_1, C_2, k, and k' are complicated coefficients which need not be given explicitly here; they involve [M] and the individual rate constants. Differentiating (8-4) with respect to time and using (8-4) to eliminate $C_1 e^{-kt}$ gives

$$-d[\mathrm{A}]/dt = C_2(k' - k) \exp(-k't) + k[\mathrm{A}] \tag{8-5}$$

The solution thus consists of a transient term and a steady-state term. It can be shown (*3,4*) that except for the very fastest reactions, k' is sufficiently large to ensure that the transient term dies out quickly, so that the steady-state solution $k = -(1/[\mathrm{A}])\, d[\mathrm{A}]/dt$ is a very good approximation under "coarse" time resolution (cf. the introduction to Part II), but not under "fine" time resolution. Shock-tube experiments represent a borderline case, in that the reaction is fast and the time resolution is almost "fine," so that the transient may not be always negligible. Neglect of the transient without special justification renders such unimolecular data from shock-tube experiments somewhat suspect.

It is important to realize that in Lindemann's mechanism $\mathrm{A}(X)$ is assumed to be formed in *one step* from ordinary A molecules, so that the rate of formation of $\mathrm{A}(X)$ is proportional to the concentration of A molecules. This can be generalized to a mechanism where $\mathrm{A}(X)$ is formed in n steps, but such a mechanism is deferred until Section 8. A consequence of the one-step formation of $\mathrm{A}(X)$ is that every collision $\mathrm{A}(X) + \mathrm{A}$ must lead to deactivation of $\mathrm{A}(X)$, i.e., must render $\mathrm{A}(X)$ incapable of reacting; if this were not true, then some molecules A in reactive states would be formed by collision from other A molecules in some less excited reactive states, and we would no longer have a one-step activation, contrary to assumption. Thus in Lindemann's simple mechanism, the rate $k_2[\mathrm{M}]$ for the process in Eq. (8-2) must be essentially the collision frequency $\omega = Z[\mathrm{M}]$, where Z is the kinetic theory collision number and [M] is the concentration or pressure of M (i.e., the total pressure in the system). The equation $k_2[\mathrm{M}] = \omega$ constitutes the so-called strong collision assumption.

A more general treatment It is instructive to derive (*5,6*) an expression for the steady-state rate constant k, incorporating the strong-collision assumption, but without making any assumption about the process (8-3), i.e., without specifying the distribution of lifetimes of molecules A(X) with respect to decomposition.

Suppose molecule A has just suffered a collision at time $t = 0$; let $P(t)$ be the probability of A not decomposing in the interval $(0, t)$ in the absence of collisions. Not much is known about the state of A produced at $t = 0$; if it is not in a reactive state, obviously $P(t) = 1$; hence $P(t) \neq 1$ only if the collision at $t = 0$ produced A(X), and the nontrivial form of $P(t)$ therefore necessarily measures the lifetime of A(X) with respect to decomposition [cf. Eq. (1-14)]. Since collisions are random, the probability of *no* collision in $(0, t)$ is[3] $e^{-\omega t}$, and the probability of A suffering a collision next in $(t, t + dt)$ is $\omega\, dt$. Thus the probability of A(X) not dissociating and not suffering a collision in $(0, t)$ is $P(t)e^{-\omega t}$, and therefore the probability that such an "intact" A(X) shall suffer a collision next in $(t, t + dt)$ is $P(\omega)\, dt = \omega P(t)e^{-\omega t}\, dt$. We are interested only in the steady-state behavior of the system, such as would appear to an observer able to perceive only events on the time scale of the order of minutes. For convenience, we take the time of observation to be infinite (cf. Chapter 1, Section 4), and we shall call $\int_0^\infty P(\omega)\, dt = P(\omega)_{\mathrm{abs}}$, the absolute probability of "intact" A(X) suffering a collision. Hence

$$P(\omega)_{\mathrm{abs}} = \int_0^\infty \omega P(t)e^{-\omega t}\, dt = 1 + \int_0^\infty e^{-\omega t}(dP/dt)\, dt \tag{8-6}$$

where dP/dt is written for the more cumbersome $dP(t)/dt$.

We can look on $P(\omega)_{\mathrm{abs}}$ as the absolute probability of A(X) being present intact at the time of the next collision, for Eq. (8-6) refers to the probability of collision of a species that has not changed its chemical identity from the time of its formation at $t = 0$. Therefore $1 - P(\omega)_{\mathrm{abs}}$ is the absolute probability of "intact" A(X) *not* being present at the time of the next collision. The species A(X) can be absent at the time of the next collision for two reasons: Either its internal state has changed because it has suffered a previous collision, but this is contrary to the hypothesis that $1 - P(\omega)_{\mathrm{abs}}$ refers to the *first* or next collision; or A(X) has changed its chemical identity, i.e., has dissociated. Therefore

$$1 - P(\omega)_{\mathrm{abs}} = -\int_0^\infty e^{-\omega t}(dP/dt)\, dt \tag{8-7}$$

is necessarily the absolute probability of A(X) forming products.

The rate of formation of A(X) in a thermal system, assuming strong collisions, is[4] ω[A(X)], and the rate of A(X) forming products at steady state

[3] See, for example, Morse (7, p. 157).

[4] We have from (8-1), for the rate of formation of A(X), d[A(X)]$/dt = k_1$[A][M]; at equilibrium, this is equal to its rate of deactivation, which from (8-2) is k_2[M][A(X)]. However, even if equilibrium does not obtain, the rate of formation of A(X) is still the equilibrium rate k_2[M][A(X)] $= \omega$[A(X)]. Cf. footnote 7 and text following Eq. (8-43).

is $\omega[\mathrm{A}(X)]\{1 - P(\omega)_{\mathrm{abs}}\}$; hence the steady-state rate constant $k(X)_{\mathrm{ss}}$ for the reaction of molecules A(X) is

$$k(X)_{\mathrm{ss}} = -\frac{1}{[\mathrm{A}(X)]}\frac{d[\mathrm{A}(X)]}{dt} = -\omega\int_0^\infty e^{-\omega t}(dP/dt)dt = -\mathfrak{L}\left\{\frac{dP}{dt}\right\} \tag{8-8}$$

where, as usual, $\mathfrak{L}\{\ \}$ indicates the Laplace transform. We have dropped for simplicity the designation of $k(X)_{\mathrm{ss}}$ as the "apparent" rate constant, used in the introduction to Part II to describe the similar equation (II-2).

Equation (8-8), with $-dP/dt$ replaced by what we have called $P_{\mathrm{norm}}(t)$ in Chapter 1 [Eq. (1-22)], has been originally derived by Slater (*9*, p. 188) and forms the basis of his "new approach to rate theory." More recently, this version of Eq. (8-8) has also been used by Bunker (*10,11*).

In principle, an explicit expression for $P(t)$ could be derived on the basis of density operator formalism (*6*). If we assume for simplicity, as in Chapter 1, Section 4, that the lifetime is random, i.e., that it has the exponential form

$$P(t) = e^{-k(X)t} \tag{8-9}$$

with a single rate constant $k(X)$ characterizing the process (8-3), then it follows from (8-8) and (8-9) by integration that

$$k(X)_{\mathrm{ss}} = \omega k(X)/\{k(X) + \omega\} \tag{8-10}$$

The rate constant $k(X)_{\mathrm{ss}}$ is what we have called noncommittally k in Eq. (8-5). If we further identify $k(X)$ with $k(E, \mathcal{J})$ of Chapter 7, we have

$$k(E, \mathcal{J})_{\mathrm{ss}} = \omega k(E, \mathcal{J})/\{k(E, \mathcal{J}) + \omega\} \tag{8-11}$$

which is the steady-state rate constant for the rate of disappearance of A molecules of energy E and angular momentum $\mathcal{J}$. The right-hand side of (8-11) gives the explicit form of the function $f[k(E, \mathcal{J}), k_1, k_2, \ldots]_{\mathrm{ss}}$ [Eq. (II-2)] for the case of a thermal reaction.

Two extreme cases can now be envisaged, depending on the relative magnitude of the two terms in the denominator of (8-11) or (8-10). When $\omega \gg k(E, \mathcal{J})$, we have $k(E, \mathcal{J})_{\mathrm{ss}} = k(E, \mathcal{J})$, so that the steady-state rate constant is merely the microcanonical rate constant itself. Here, the unimolecular process (8-3) is the slowest part of the sequence and therefore controls the overall (steady-state) rate. When, on the other hand, $k(E, \mathcal{J}) \gg \omega$, this corresponds to low-pressure conditions, and then $k(E, \mathcal{J})_{\mathrm{ss}} = \omega$, i.e., the rate of collisional energy transfer is the slowest, and therefore rate-controlling, part of the sequence.

The general rate[5] (8-11), or either of its extreme cases just considered,

[5] We shall often use the term "rate" as abbreviation for "rate constant"; strictly speaking, of course, rate = rate constant × (concentration)^n if the reaction is of order n.

accounts for the competition of the three processes (8-1)–(8-3) as they affect an A molecule of specified E and $\mathscr{J}$ but does not represent the observable rate because collisions produce A molecules with a whole distribution of values of E and $\mathscr{J}$. To obtain the observable rate, the distribution function for E, $\mathscr{J}$ must first be determined and then the rate constant $k(E, \mathscr{J})_{ss}$ of Eq. (8-11) must be averaged over this distribution. This averaging will now be considered [cf. Eq. (II-5)].

2. Averaging over equilibrium distribution

We shall start by assuming that the chemical transformation is the slowest process [i.e., that $k(E, \mathscr{J}) \ll \omega$ for *all* E and $\mathscr{J}$], which is the easiest case to consider. The assumption is equivalent to assuming that the loss of A(X) by reaction is insignificant, i.e., that collisions form molecules A(X) much more rapidly than they disappear by chemical transformation; therefore an equilibrium population is maintained, not only of A(X), but of molecules of all energies and angular momenta. In ensemble theory, such a state of affairs is described by the canonical ensemble of Gibbs [see, e.g., Tolman (*12*, p. 58)].

The distribution function over which $k(E, \mathscr{J})$ is to be averaged is then the equilibrium probability $P(E, \mathscr{J})_e$, which expresses the probability that the energy of A(X) will fall in a range of energies between E and $E + dE$ and its overall angular momentum in a range of $\mathscr{J}$ values between $\mathscr{J}$ and $\mathscr{J} + d\mathscr{J}$. Since the overall rotations are assumed to be completely decoupled from the degrees of freedom involved in E (see Chapter 5, Section 4), $P(E, \mathscr{J})_e$ is simply the product of two equilibrium distributions

$$P(E, \mathscr{J})_e = P(E)_e P(\mathscr{J})_e \tag{8-12}$$

where

$$P(E)_e = Q^{-1}N(E)e^{-E/kT} \tag{8-13}$$

$$P(\mathscr{J})_e = Q_r^{-1}(2\mathscr{J} + 1)e^{-E_r(\mathscr{J})/kT} \tag{8-14}$$

which are the familiar (normalized) Boltzmann distributions for E and $\mathscr{J}$ characterized by temperature T, the heat bath temperature. Here, Q is the partition function for degrees of freedom involved in $N(E)$ [cf. Eq. (6-3)] and Q_r is the partition function for the two-dimensional rotor of energy $E_r(\mathscr{J})$ [cf. Eqs. (5-6) and (6-40)], which, as shown in Chapter 7, Section 1, is used to approximate the rotational potential of the decomposing molecule. The symmetry number for the rotor appears in both the numerator and denominator of Eq. (8-14) and therefore cancels out. Hence

$$P(E, \mathscr{J})_e = (1/QQ_r)(2\mathscr{J} + 1)N(E)e^{-[E+E_r(\mathscr{J})]/kT} \tag{8-15}$$

which is likewise normalized.

The average of $k(E, \mathscr{J})$, as given by (7-1), is then

$$\langle k(E, \mathscr{J})\rangle_{\mathrm{e}} = \frac{\alpha}{QQ_{\mathrm{r}}} \iint (2\mathscr{J} + 1)N(E) \frac{G^*(E - E_J)}{hN(E)} \exp\left[-\frac{E + E_{\mathrm{r}}(\mathscr{J})}{kT}\right] dE\, d\mathscr{J} \quad (8\text{-}16)$$

Note that $N(E)$ drops out. The limits of integration are, in principle, zero to infinity with respect to E; there is an upper limit on $\mathscr{J}$ corresponding to the $\mathscr{J}$ value for which $E_J = 0$, but since this occurs generally at $\mathscr{J} \sim 10$ or 10^2 (cf. Fig. 7-2a), the limit can be extended to infinity as an approximation[6] since, through $E_{\mathrm{r}}(\mathscr{J})$, the negative exponential involves $\mathscr{J}^2$. At low E, the integrand drops to zero when $E < E_J$; thus the limit of integration with respect to E may be extended down to zero, it being understood that the integrand is zero between $E = 0$ and $E = E_J$.

To simplify, we change to the variable $E' = E - E_J$, so that (8-16) becomes

$$\langle k(E, \mathscr{J})\rangle_{\mathrm{e}} = \frac{\alpha}{hQQ_{\mathrm{r}}} \int_0^{\infty}\int_0^{\infty} (2\mathscr{J} + 1)G^*(E') \exp\left[-\frac{E' + E_J + E_{\mathrm{r}}(\mathscr{J})}{kT}\right] dE'\, d\mathscr{J} \quad (8\text{-}17)$$

The energy is now referred to the lowest vibrational level of the $\mathscr{J}$th rotational state of the transition state as zero (cf. Fig. 7-1) and the integrand is non-vanishing all the way down to $E' = 0$.

Integration with respect to $\mathscr{J}$ The term $E_J + E_{\mathrm{r}}(\mathscr{J})$ does not involve E', as shown below, and therefore integrations with respect to E' and $\mathscr{J}$ can be performed independently. The integration with respect to E' can be done by parts (*13*, p. 561), or, more simply, we may note that the integral is the Laplace transform of $G^*(E')$:

$$\int_0^{\infty} G^*(E') \exp(-E'/kT)\, dE' = \mathcal{L}\{G^*(E')\} \quad (8\text{-}18)$$

From Eq. (6-27), we have immediately the result

$$\mathcal{L}\{G^*(E')\} = kTQ^* \quad (8\text{-}19)$$

where Q^* is the partition function of the transition state.

The discussion of Chapter 7 shows that the explicit form of E_J depends on the type of reaction. However, Eqs. (7-13) and (7-20) indicate that we can write quite generally

$$E_J + E_{\mathrm{r}}(\mathscr{J}) = E_0 + \phi(\mathscr{J}) \quad (8\text{-}20)$$

where $\phi(\mathscr{J})$ is a function of $\mathscr{J}$ which is different for type 1 and type 2 or 3

[6] However, see Section 3, where this approximation turns out to be a bad one.

reactions, but which need not be given explicitly at this point. The integral with respect to $\mathscr{J}$ is therefore

$$\text{integral}(\mathscr{J}) = f_\infty\, e^{-E_0/kT} \tag{8-21}$$

where

$$f_\infty = Q_r^{-1} \int_0^\infty (2\mathscr{J} + 1)e^{-\phi(\mathscr{J})/kT}\, d\mathscr{J} \tag{8-22}$$

The reason for the subscript infinity will appear shortly.

Collecting terms from the various integrations, we thus obtain for $\langle k(E, \mathscr{J})\rangle_e$ of Eq. (8-17)

$$\langle k(E, \mathscr{J})\rangle_e = \alpha f_\infty(kT/h)(Q^*/Q)e^{-E_0/kT} \tag{8-23}$$

which, apart from the factor αf_∞, is the standard expression of transition-state theory (*14*, p. 286, Eq. 77).

Equation (8-23) is of limited interest and we shall not have much occasion to use it. Much more interesting is the fact that the integration with respect to $\mathscr{J}$ yields always the result $f_\infty e^{-E_0/kT}$ [Eq. (8-21)] and it is worth while to consider it more closely. The term $e^{-E_0/kT}$ represents merely the change in zero of energy from the lowest vibrational level of the reactant to the lowest vibrational level of the transition state *in the absence of centrifugal potential*; therefore the entire effect of centrifugal potential is accounted for by the factor f_∞. Hence, with the benefit of hindsight, Eq. (8-16) could have been written as much simpler integral over E only:

$$\langle k(E, \mathscr{J})\rangle_e = (f_\infty/Q) \int k(E)N(E)e^{-E/kT}\, dE \tag{8-24}$$

where $k(E)$ is given by

$$\begin{aligned} k(E) &= 0 && \text{if } E \leqslant E_0 \\ &= \alpha G^*(E - E_0)/hN(E) && \text{if } E \geqslant E_0 \end{aligned} \tag{8-25}$$

which is of course $k(E, \mathscr{J})$ for $\mathscr{J} = 0$ [Eq. (4-34)]. We may interpret Eq. (8-24) essentially as f_∞ times the average of $k(E)$ [Eq. (8-25)] over a Boltzmann distribution in E:

$$\langle k(E, \mathscr{J})\rangle_e = f_\infty \int k(E)P(E)_e\, dE = f_\infty \langle k(E, \mathscr{J} = 0)\rangle_e \tag{8-26}$$

where $P(E)_e$ is given by Eq. (8-13). Since $k(E)$ is zero below $E = E_0$, it does not matter if the lower limit of integration in (8-24) and (8-26) is taken as $E = 0$ or $E = E_0$; the upper limit may again be taken as infinity for convenience.

If Eq. (8-26) is written as $f_\infty = \langle k(E, \mathscr{J})\rangle_e / \langle k(E, \mathscr{J} = 0)\rangle_e$ the factor f_∞ may be interpreted as a correction for centrifugal effects.

3. Averaging over steady-state distribution

When the processes (8-1)–(8-3) compete on equal terms, an equilibrium population of A(X) cannot be maintained and the equilibrium distribution function (8-12) no longer applies. However, a steady state is established, and our problem is therefore to obtain an expression for the steady-state distribution function. The argument below follows fairly closely the derivation originally given by Slater (*15*). The point of view is that which led to Eq. (II-10).

Let ω be the collision frequency and [A] the total concentration of reactant molecules A of all energies and angular momenta. The number of molecules A raised by collisions into the energy range E, $E + dE$ and into the range of angular momentum quantum numbers $\mathcal{J}$, $\mathcal{J} + d\mathcal{J}$ is always given by[7] the equilibrium rate

$$[\mathrm{A}]\omega P(E, \mathcal{J})_{\mathrm{e}}\, dE\, d\mathcal{J} \quad \text{molecules cm}^{-3}\,\text{sec}^{-1} \tag{8-27}$$

whether equilibrium actually obtains or not; the reason is that the molecules cannot "know," while colliding, what their ultimate fate is. Relation (8-27) is valid for all E and $\mathcal{J}$. By contrast, the steady-state population of A(X), i.e., of molecules with energy in a range dE above $E = E_J$ (but not *below* $E = E_J$) will be smaller than the equilibrium population because some of these molecules are irretrievably lost by chemical transformation. Let this steady-state distribution be $P(E, \mathcal{J})_{\mathrm{ss}}$; then the number of molecules A(X) lost by chemical transformation from the range E, $E + dE$ ($E \geqslant E_J$) and $\mathcal{J}$, $\mathcal{J} + d\mathcal{J}$ is

$$[\mathrm{A}]k(E, \mathcal{J})P(E, \mathcal{J})_{\mathrm{ss}}\, dE\, d\mathcal{J} \quad \text{molecules cm}^{-3}\,\text{sec}^{-1} \tag{8-28}$$

and the number of molecules lost by deactivating collisions, using the strong-collision assumption, is

$$[\mathrm{A}]\omega P(E, \mathcal{J})_{\mathrm{ss}}\, dE\, d\mathcal{J} \quad \text{molecules cm}^{-3}\,\text{sec}^{-1} \tag{8-29}$$

By the principle of detailed balancing, the gain and loss of A(X) must be balanced for every range dE and $d\mathcal{J}$, so that

$$[\mathrm{A}]\omega P(E, \mathcal{J})_{\mathrm{e}} = [\mathrm{A}]k(E, \mathcal{J})P(E, \mathcal{J})_{\mathrm{ss}} + [\mathrm{A}]\omega P(E, \mathcal{J})_{\mathrm{ss}} \tag{8-30}$$

from which

$$P(E, \mathcal{J})_{\mathrm{ss}} = \frac{\omega}{k(E, \mathcal{J}) + \omega} P(E, \mathcal{J})_{\mathrm{e}} \tag{8-31}$$

[7] See footnote 4. The assumption, which goes all the way back to Rice and Ramsperger (*16*), is that any molecule in a reactive state, whenever it collides, loses enough energy so that its probability of reaction becomes negligible. In other words, (8-27) embodies the strong-collision assumption.

This defines $P[E, \mathscr{J}, k(E, \mathscr{J}), k_1, k_2, \ldots]_{ss}$ of Eq. (II-10) for the specific case of a thermal reaction. At high pressures, i.e., in the equilibrium region, we can write $k(E, \mathscr{J})_e \equiv k(E, \mathscr{J})$; then from Eq. (8-11), we have formally

$$k(E, \mathscr{J})_{ss} = \frac{\omega}{k(E, \mathscr{J}) + \omega} k(E, \mathscr{J})_e \tag{8-32}$$

i.e., the equilibrium rate is reduced in the steady-state region by the factor $\omega/[k(E, \mathscr{J}) + \omega]$. Similarly, we see from Eq. (8-31) that the steady-state distribution is the equilibrium distribution reduced by the same factor. The argument makes it clear that the steady-state distribution refers only to molecules in reactive states; in other words, molecules with energies *below* reaction threshold maintain their equilibrium population and are therefore distributed according to $P(E, \mathscr{J})_e$.

The average of $k(E, \mathscr{J})$ over the steady-state distribution is thus

$$\langle k(E, \mathscr{J})\rangle_{ss} =$$

$$\frac{1}{QQ_r} \iint \frac{\omega k(E, \mathscr{J})}{\omega + k(E, \mathscr{J})} (2\mathscr{J} + 1)N(E) \exp\left[-\frac{E + E_r(\mathscr{J})}{kT}\right] dE\, d\mathscr{J} \tag{8-33}$$

The same comments apply about the limits of integration as those following Eq. (8-16). Before considering Eq. (8-33) in more detail, it is useful first to consider two special cases.

The high-pressure limit When $\omega \gg k(E, \mathscr{J})$ for all E and $\mathscr{J}$, i.e., when $\omega \to \infty$, we have

$$\lim_{\omega \to \infty} \langle k(E, \mathscr{J})\rangle_{ss} = \langle k(E, \mathscr{J})\rangle_e \tag{8-34}$$

where $\langle k(E, \mathscr{J})\rangle_e$ is given by Eq. (8-24). It is customary to write

$$\lim_{\omega \to \infty} \langle k(E, \mathscr{J} = 0)\rangle_{ss} = \langle k(E, \mathscr{J} = 0)\rangle_e = k_\infty \tag{8-35}$$

so that Eqs. (8-26) and (8-34) can be written

$$\lim_{\omega \to \infty} \langle k(E, \mathscr{J})\rangle_{ss} = k_\infty f_\infty = k_\infty' \tag{8-36}$$

The subscript infinity on f introduced previously becomes self-explanatory. The rate constant k_∞ is generally referred to as the high- (or infinite-) pressure rate constant; the prime signifies that a correction for centrifugal effects has been made.

The condition $\omega \to \infty$ is a mathematical convenience that ignores the physics of the situation, in the sense that under such conditions, the Hamiltonian $\mathscr{H}$ of the system (cf. Chapter 4, Section 1) is no longer a single-molecule

Hamiltonian but should be some sort of effective single-molecule Hamiltonian (*17*) which takes into account the high-density environment[8]; the rate constant is then no longer a simple average of $k(E, \mathscr{J})$ over the equilibrium distribution. If the effect of high-density environment is partially taken into account by correcting (*19*) for nonideality at high pressures, it is found that the correction is small, and therefore some of the unusual rate constant pressure dependence observed at extremely high pressures (cf. Ref. 25 in Chapter 2) is no doubt due to more profound effects of the high-density environment. In most thermal studies the condition $\omega \to \infty$ is reached by extrapolating moderately high-pressure data (cf. Section 7) so that the high-density complications are more or less side-stepped.

The low-pressure limit The other special case is when $\omega \ll k(E, \mathscr{J})$ for all E and $\mathscr{J}$, i.e., when $\omega \to 0$. We then have from (8-33), for E_J as the lower limit for integration with respect to E,

$$\lim_{\omega \to 0} \langle k(E, \mathscr{J}) \rangle_{ss} = (\omega / Q Q_r) \times \text{integral}(E, \mathscr{J}) \tag{8-37}$$

where

$$\text{integral}(E, \mathscr{J}) = \int_0^\infty \left\{ \int_{E_J}^\infty N(E) e^{-E/kT} \, dE \right\} (2\mathscr{J} + 1) e^{-E_r(\mathscr{J})/kT} \, d\mathscr{J} \tag{8-38}$$

Integration with respect to the two variables E and $\mathscr{J}$ cannot be performed separately because the integral in E has a $\mathscr{J}$-dependent lower limit; however, the expression (8-38) can be reduced if integrated by parts (*20*). On remembering that $E_r(\mathscr{J})/kT = \mathscr{J}(\mathscr{J} + 1)/Q_r$ [cf. Eqs. (6-40) and (5-6)], and that E_J is a function of $\mathscr{J}$ [Eqs. (7-13) and (7-20)], integration by parts yields

$$- Q_r \left[e^{-E_r(\mathscr{J})/kT} \int_{E_J}^\infty N(E) e^{-E/kT} \, dE \right]_{\mathscr{J}=0}^{\mathscr{J}=\infty}$$

$$+ Q_r \int_0^\infty e^{-E_r(\mathscr{J})/kT} \frac{d}{d\mathscr{J}} \left(\int_{E_J}^\infty N(E) e^{-E/kT} \, dE \right) d\mathscr{J} \tag{8-39}$$

The exponential in the first term of (8-39) vanishes strongly at the upper limit, and at the lower limit we have $E_J(\mathscr{J}=0) = E_0$. The derivative in the second term can be evaluated from the general relation

$$d \int_{g(x)}^\infty F(y) \, dy = -F[g(x)] \, d[g(x)] \tag{8-40}$$

[8] This can be done rigorously by the application of the Green's function technique. See Paul and Coombe (*18*).

Thus

$$\text{integral}(E, \mathscr{J}) = Q_r\left[\int_{E_0}^{\infty} N(E)e^{-E/kT}\,dE - \int_{E_0}^{0} N(E_J)e^{-[E_J+E_r(\mathscr{J})]/kT}\,dE_J\right] \tag{8-41}$$

The second integral in (8-41) is in effect over $\mathscr{J}$, from $\mathscr{J} = 0$ to (approximately) $\mathscr{J} = \infty$; however, we prefer to write it as an integral with respect to E_J to underline the fact that the upper limit $\mathscr{J} = \infty$ is inapropriate in this case, since E_J would then be negative and this, of course, has no physical significance; moreover, $N(x)$ is not defined for negative x. Therefore the upper limit is $E_J = 0$, meaning a large but not infinite $\mathscr{J}$, and the lower limit is E_0, which is the value of E_J for $\mathscr{J} = 0$.

Equation (8-37) can now be written

$$\lim_{\omega\to 0} \langle k(E, \mathscr{J})\rangle_{ss} = k_0 - (\omega/Q)\int_{E_0}^{0} N(E_J)e^{-[E_J+E_r(\mathscr{J})]/kT}\,dE_J \tag{8-42}$$

where

$$k_0 = \lim_{\omega\to 0} \langle k(E, \mathscr{J} = 0)\rangle_{ss} = (\omega/Q)\int_{E_0}^{\infty} N(E)e^{-E/kT}\,dE \tag{8-43}$$

which is simply $\omega \times P(E)_e$ integrated between E_0 and infinity, so that k_0 may be interpreted as the number of molecules colliding per unit time, times the *equilibrium* fraction of molecules in reactive states. The fact that it is the equilibrium fraction is of course a consequence of the assumption, mentioned earlier in Section 1 [see also remarks in footnote 4], that all molecules in any one of the reactive states come originally from ordinary molecules in nonreactive states, so that if the concentration of the latter is not greatly perturbed by the reaction, the rate at which ordinary molecules are raised into reactive states is the same equilibrium rate whether or not net reaction takes place, i.e., whether the system is in equilibrium or not.[9] A more general case of activating mechanism which results in a nonequilibrium distribution below, as well as above, threshold is treated in Section 8.

The integral $\int_{E_0}^{\infty} N(E)e^{-E/kT}\,dE$ may be looked upon as the partition function of reacting molecules. It can be evaluated easily enough numerically by generating $N(E)$ by one of the better methods of Chapter 6. If E_0 is high and the molecule is not too large, the first one or two terms of Eq. (6-80) may be used for $N(E)$ with sufficient accuracy, and then the integral in (8-43) can be easily evaluated using tabulated values of the exponential integral (*22*). An approximation which is not helpful for numerical calculations but which we shall find useful for other purposes in Sections 6 and 8 is to express $N(E)$

[9] For further discussion on this point see Forst and St. Laurent (*8*) and Troe and Wagner (*21*).

in terms of the semiclassical formula (6-56); then using the hint of Exercise 3, Chapter 10, we find that

$$\int_{E_0}^{\infty} N(E)e^{-E/kT}\,dE \approx kTN(E_0)e^{-E_0/kT} \tag{8-43a}$$

Another version of (8-43) is the subject of Exercise 1.

We may define a low-pressure centrifugal correction factor by

$$f_0 = \frac{\lim_{\omega\to 0}\langle k(E, \mathscr{J})\rangle_{ss}}{\lim_{\omega\to 0} k(E, \mathscr{J} = 0)_{ss}} = 1 - \frac{\int_{E_0}^{0} N(E_J)e^{-[E_J + E_r(\mathscr{J})]/kT}\,dE_J}{\int_{E_0}^{\infty} N(E)e^{-E/kT}\,dE} \tag{8-44}$$

so that Eq. (8-37) can now be written

$$\lim_{\omega\to 0} \langle k(E, \mathscr{J})\rangle_{ss} = k_0 f_0 = k_0' \tag{8-45}$$

which is the low-pressure equivalent of Eq. (8-36). Thus, once again, the average rate can be written as the rate in absence of centrifugal potential multiplied by a centrifugal correction factor. Since the integral in the numerator of (8-44) is negative, the second term in (8-44) is positive and f_0 is larger than unity, as one would expect.

The general-pressure rate When ω is neither very small nor very large compared with $k(E, \mathscr{J})$, the integral (8-33) cannot be further simplified by integration by parts, because both the integrand and the lower limit of the integral with respect to E are now $\mathscr{J}$-dependent, so that we have to go back to Eq. (8-33) and calculation is therefore more laborious. On substituting for $k(E, \mathscr{J})$ in the numerator of the integrand and letting $E - E_J = E'$, Eq. (8-33) can be written as [cf. Eq. (8-17)]

$$\langle k(E, \mathscr{J})\rangle_{ss} =$$

$$\frac{\alpha}{hQQ_r}\int_0^{\infty}\int_0^{\infty} \frac{(2\mathscr{J} + 1)G^*(E')}{1 + [k(E', \mathscr{J})/\omega]} \exp\left[-\frac{E' + E_J + E_r(\mathscr{J})}{kT}\right] dE'\,d\mathscr{J} \tag{8-46}$$

This equation is exact and involves a rather tedious double integration with respect to E' and $\mathscr{J}$ that must be repeated at every value of ω. However, Waage and Rabinovitch (*23*) have found that to a good approximation the integrations with respect to E' and $\mathscr{J}$ can be performed separately if $k(E', \mathscr{J})$ is replaced by an average (of sorts) over $\mathscr{J}$ defined by

$$\langle k(E', \mathscr{J})\rangle_J = \frac{\alpha f_\infty}{f_0} \times \frac{G^*(E')}{hN(E' + E_0)} \tag{8-47}$$

Note that except for a different energy zero, $\langle k(E', \mathscr{J})\rangle_J$ of Eq. (8-47) is just $k(E)$ of Eq. (8-25) multiplied by f_∞/f_0. Replacing $k(E', \mathscr{J})$ in (8-46) by its average (8-47) and writing $\phi(\mathscr{J}) + E_0$ [Eq. (8-20)] for $E_J + E_r(\mathscr{J})$ yields

$$\langle k(E, \mathscr{J})\rangle_{ss} \approx \frac{\alpha \exp[-E_0/kT]}{hQQ_r} \int_0^\infty \int_0^\infty \frac{(2\mathscr{J} + 1)G^*(E')}{1 + [\langle k(E', \mathscr{J})\rangle_J/\omega]} \times \exp\left[-\frac{E' + \phi(\mathscr{J})}{kT}\right] dE'\, d\mathscr{J} \tag{8-48}$$

Integrations with respect to $\mathscr{J}$ and E' can now be performed separately; in fact, integration with respect to $\mathscr{J}$ yields, by Eq. (8-22), merely f_∞, so that $\langle k(E', \mathscr{J})\rangle_{ss} \approx$

$$\frac{\alpha f_\infty \exp(-E_0/kT)}{hQ} \int_0^\infty \frac{G^*(E')}{1 + [\langle k(E', \mathscr{J})\rangle_J/\omega]} \exp(-E'/kT)dE' = k_{\text{uni}} \tag{8-49}$$

This is really an interpolation formula, since it is easy to verify that (8-49) reduces to the appropriate expressions at the two limits: at the high-pressure limit, when $\langle k(E', \mathscr{J})\rangle_J/\omega \ll 1$, Eq. (8-49) reduces to (8-24); and at the low-pressure limit where

$$\frac{\langle k(E', \mathscr{J})\rangle_J}{\omega} = \frac{f_\infty \alpha G^*(E')}{f_0 \omega h N(E' + E_0)} \gg 1 \tag{8-50}$$

(8-49) reduces to (8-45). The dummy variable E' represents energy referred to the ground level of the transition state as zero, and is identical with the variable x used in Eq. (8-62). With the help of Eq. (8-49), calculations in the fall-off region permit inclusion of centrifugal effects without any undue mathematical labor once f_∞ and f_0 have been determined.

The rate constant k_{uni} of Eq. (8-49) can be rewritten as (see Exercise 2)

$$k_{\text{uni}} = \frac{f_0}{Q} \int \frac{(f_\infty/f_0)k(E)N(E)}{1 + [f_\infty k(E)/f_0\omega]} e^{-E/kT}\, dE \tag{8-51}$$

We thus have the result that the unimolecular rate constant, averaged over thermal distribution of energies and angular momenta, can be expressed over the entire range of pressures as k_{uni}, which is just f_0 times the average of $k'(E)$ over the thermal distribution of energies, where [cf. Eq. (8-47)]

$$k'(E) = (f_\infty/f_0)k(E) \tag{8-52}$$

The prime in Eq. (8-52), and also in Eqs. (8-36) and (8-45), signifies that the quantity involved has been corrected for centrifugal effects by multiplication by an appropriate factor. It is therefore sufficient in thermal reactions to focus attention only on averaging or integration with respect to energy, which will help to simplify the ensuing arguments. However, before proceeding further, we shall first obtain expressions for the factors f_0 and f_∞.

4. Evaluation of centrifugal correction factors

The centrifugal correction factors f_∞ and f_0 were left in the form of integrals with respect to $\mathscr{J}$ [cf. Eqs. (8-22) and (8-44)]. It now behoves us to complete the integrations. Since the rotational potential is different for different types of reactions, the reaction type must be specified. Recall that type 1 refers to unimolecular dissociations without formal activation energy for the reverse association, so that there is only a centrifugal barrier to be surmounted, the position and height of which is $\mathscr{J}$-dependent (Chapter 7, Section 3). Reaction types 2 and 3 refer to reactions with activation energy for the reverse of the unimolecular step; here, the rotational potential influences only the height of the barrier but not its position (Chapter 7, Section 4). We shall start by considering the high-pressure centrifugal correction factor first.

High-pressure correction From Eqs. (7-12) and (7-13), we have for a type 1 reaction

$$\phi(\mathscr{J}) = E_J + E_r(\mathscr{J}) - E_0 = \frac{\mathfrak{n}-2}{2}\left[\frac{\mathscr{J}(\mathscr{J}+1)\hbar^2}{\mathfrak{n}\mu_{AB}\mathscr{C}^{2/\mathfrak{n}}}\right]^{\mathfrak{n}/(\mathfrak{n}-2)} \tag{8-53}$$

and therefore f_∞ of Eq. (8-22) becomes

$$f_\infty = Q_r^{-1}\int_0^\infty (2\mathscr{J}+1)\exp\left\{-\frac{\mathfrak{n}-2}{2kT}\left[\frac{\mathscr{J}(\mathscr{J}+1)\hbar^2}{\mathfrak{n}\mu_{AB}\mathscr{C}^{2/\mathfrak{n}}}\right]^{\mathfrak{n}/(\mathfrak{n}-2)}\right\} d\mathscr{J} \tag{8-54}$$

The substitutions

$$\mathscr{J}(\mathscr{J}+1) = \omega, \qquad \frac{\hbar^2}{\mathfrak{n}\mu_{AB}\mathscr{C}^{2/\mathfrak{n}}}\left(\frac{\mathfrak{n}-2}{2kT}\right)^{(\mathfrak{n}-2)/\mathfrak{n}} = a, \qquad \frac{\mathfrak{n}-2}{\mathfrak{n}} = \mathfrak{n}'$$

brings this to a standard form,[10] and remembering that $Q_r = 2\mu_{AB}r_e^2kT/\hbar^2$, we get the result

$$f_\infty(\text{type 1}) = \frac{1}{r_e^2}\left[\left(\frac{\mathfrak{n}-2}{2kT}\right)\mathscr{C}\right]^{2/\mathfrak{n}}\Gamma\left(\frac{\mathfrak{n}-2}{\mathfrak{n}}\right) \tag{8-55}$$

which for $\mathfrak{n} = 6$ (neutral fragments) reduces to

$$f_\infty(\text{type 1}) = \frac{\Gamma(2/3)}{r_e^2}\left(\frac{2\mathscr{C}}{kT}\right)^{1/3} \tag{8-56}$$

For type 2 and 3 reactions, we have from Eq. (7-20),

$$\phi(\mathscr{J}) = \mathscr{J}(\mathscr{J}+1)\hbar^2/2\mu_{AB}r_0^2 \tag{8-57}$$

[10] Dwight (*24*), number 860.18:

$$\int_0^\infty \exp(-a\omega)^{\mathfrak{n}'}\,d\omega = \Gamma(1/\mathfrak{n}')/\mathfrak{n}'a$$

so that f_∞ of Eq. (8-22) becomes

$$f_\infty = Q_r^{-1} \int (2\mathscr{J} + 1) \exp\left[-\frac{\mathscr{J}(\mathscr{J}+1)\hbar^2}{2\mu_{AB} r_0^2 kT}\right] d\mathscr{J} \tag{8-58}$$

The integral here is analogous to that which appears in Eq. (6-40), so that we get immediately

$$f_\infty(\text{type 2 or 3}) = (r_0/r_e)^2 \tag{8-59}$$

Low-pressure correction In evaluating the integral (8-44) for the low-pressure centrifugal correction factor f_0, it is convenient to start with type 2 and type 3 reactions.

Let us make the substitution $E_J = -E_0 y$; then $dE_J = -E_0\,dy$, and from Eq. (7-20), we have for type 2 and type 3 reactions

$$E_J + E_r(\mathscr{J}) = \rho E_0 y + (\rho + 1)E_0 \tag{8-60}$$

where $\rho = r_e^2/(r_0^2 - r_e^2)$. Hence,

$$\int_{E_0}^{0} N(E_J) e^{-[E_J + E_r(\mathscr{J})]/kT}\, dE_J = -E_0 e^{-(\rho+1)E_0/kT} \int_{-1}^{0} N(-E_0 y) e^{-\rho E_0 y/kT}\, dy \tag{8-61}$$

Changing to a new variable $x = E - E_0$ in the denominator of (8-44), we get

$$f_0(\text{type 2 or 3}) = 1 + \frac{E_0 e^{-\rho E_0/kT} \int_{-1}^{0} N(-E_0 y) e^{-\rho E_0 y/kT}\, dy}{\int_0^\infty N(x + E_0) e^{-x/kT}\, dx} \tag{8-62}$$

This equation is identical with one [*23*, Eq. (33)] derived on the basis of a somewhat different argument. Up to this point, everything is exact. Further evaluation[11] of (8-62) requires the use of an approximation for $N(E)$, except in the case of a diatomic molecule, where $N(E) = \text{const}$; (8-62) then yields

$$f_0 = 1 - (e^{-\rho E_0/kT} - 1)/\rho$$

In all cases of physical interest, $e^{-\rho E_0/kT} \ll 1$, so that

$$f_0(\text{type 2 or 3, diatomic}) = 1 + (1/\rho) = (r_0^2/r_e^2) \tag{8-63}$$

which is the same as f_∞ of Eq. (8-59). Thus in the special case of a diatomic molecule, the centrifugal correction factor is pressure independent (*20*).

In the considerably more frequent case of a polyatomic molecule, $N(E)$ must be represented by the semiclassical formula (6-63) if the algebra is to

[11] For a review of the various approximations based on the semiclassical density formula, see Waage and Rabinovitch (*23*).

become tractable. In this way, Waage and Rabinovitch (*25,26*) reduced (8-62) to

$$f_0(\text{type 2 or 3}) \approx \frac{r_0^2}{r_e^2}\left[1 + \frac{(\mathsf{v} - 1 + \frac{1}{2}\mathsf{r})kT}{\rho \times (E_0 + a_0 E_z)}\right]^{-1} \tag{8-64}$$

where v and r are the numbers of vibrational and rotational degrees of freedom, respectively, among which energy is randomized in the reactant molecule, and a_0 is the energy-level density correction factor (a_k for $k = 0$) evaluated at $E = E_0$. By comparison with the "exact" equation (8-62), the two cited authors found[12] that (8-64) represents a very good approximation. We may note, however, that since in a unimolecular calculation, "good" values of $N(E)$ have to be generated anyway for calculating $k(E)$, exact evaluation of (8-62) does not normally increase mathematical labor very much.

For type 1 reactions, let us make the substitution

$$E_r(\mathcal{J}) = \mathcal{J}(\mathcal{J} + 1)\hbar/2\mu_{AB}r_e^2 = z \tag{8-65}$$

Then, from Eq. (8-53), we have

$$E_J = E_0 + bz^{\mathfrak{n}/(\mathfrak{n}-2)} - z \tag{8-66}$$

where

$$b = \frac{\mathfrak{n} - 2}{2}\left(\frac{2r_e^2}{\mathfrak{n}\mathscr{C}^{2/\mathfrak{n}}}\right)^{\mathfrak{n}/(\mathfrak{n}-2)} \tag{8-67}$$

Then Eq. (8-44) becomes

$$\begin{aligned} f_0(\text{type 1}) &= 1 - \left\{\int_0^{z_0} f(z)\, dz \Big/ \int_0^{\infty} f(x)\, dx\right\} \\ f(x) &= N(x + E_0)e^{-x/kT} \\ f(z) &= N(E_0 + bz^{\mathfrak{n}/(\mathfrak{n}-2)} - z)\left(\frac{\mathfrak{n}}{(\mathfrak{n} - 2)}\, bz^{2/\mathfrak{n}-2)} - 1\right)\exp\left(\frac{-bz^{\mathfrak{n}/(\mathfrak{n}-2)}}{kT}\right) \end{aligned} \tag{8-68}$$

where z_0 is the solution of

$$E_0 + bz^{\mathfrak{n}/(\mathfrak{n}-2)} - z = 0 \tag{8-69}$$

Once again, infinity in the upper limit of the integral in the numerator of (8-68) is inappropriate because for $z > z_0$ the density function $N(E)$ would be for a negative argument and therefore undefined. A new variable x has been introduced in the denominator, as in Eq. (8-62). Except for the integration limit, Eq. (8-68) is identical with that derived by Waage and Rabinovitch

[12] Their "exact" result was obtained using $N(E)$ of Eq. (6-63) with a fixed a_k; thus an approximate $N(E)$ was used, and the result is therefore "exact" only in the sense that, unlike in Eq. (8-64), the resulting algebra was not further simplified by means of approximations.

(*23*, Eq. (38)). As in the case of Eq. (8-62), Eq. (8-68) can be easily evaluated by numerical integration using the same set of "good" $N(E)$'s.

Insofar as numerical values are concerned, we have quite generally, for any type of reaction,

$$f_0 \leqslant f_\infty \tag{8-70}$$

where the equality sign applies to diatomic reactant and the inequality to polyatomic reactant. Usually f_∞ is less than ten and f_0 is less than five.

5. Pressure dependence

The steady-state distribution function $P(E, \mathscr{J})_{ss}$ [Eq. (8-31)], in addition to being a function of E and $\mathscr{J}$, is also pressure- (or concentration-) dependent through ω; recall that $\omega = Z[\mathrm{M}]$. Pressure is one of only two parameters that are under the control of the experimenter (the other is temperature, considered next in Section 6), and therefore consideration of the pressure dependence of the rate, besides its intrinsic interest, will be also useful to provide a connection between theory and experiment.

We shall make use of the simplification introduced at the end of Section 3, namely that only averaging with respect to energy need be considered, centrifugal effects being taken care of by applying the appropriate correction factor. It is therefore convenient to introduce a new (normalized) distribution function $R(E)$ which is a function of energy only and is defined by

$$R(E) = f_0 k'(E)P(E)_e/k_\infty' = k(E)P(E)_e/k_\infty \tag{8-71}$$

The factor $(k_\infty')^{-1}$ arises from normalization, since clearly, by (8-26)

$$f_0 \int_0^\infty k'(E)P(E)_e \, dE = k_\infty' \tag{8-72}$$

An example of $R(E)$ as a function of E is shown in Fig. 8-1, calculated for one specific case over a part of the energy range. Since $R(E)$ is in fact the integrand of the integral leading to k_∞', the plot of $R(E)$ versus E shows the contribution of each energy range to k_∞'. Note that the function has a high-energy tail that drops off only very slowly, i.e., $R(E)$ encompasses an appreciable range of energies. Increase in temperature would decrease the height of the maximum and further spread the function along the energy axis. We can now express k_{uni}/k_∞' of Eq. (8-51) as an average of $\omega/[\omega + k'(E)]$ over the function $R(E)$:

$$\frac{k_{\text{uni}}}{k_\infty'} = \int_0^\infty \frac{\omega}{\omega + k'(E)} R(E) \, dE = \left\langle \frac{\omega}{\omega + k'(E)} \right\rangle_R \tag{8-73}$$

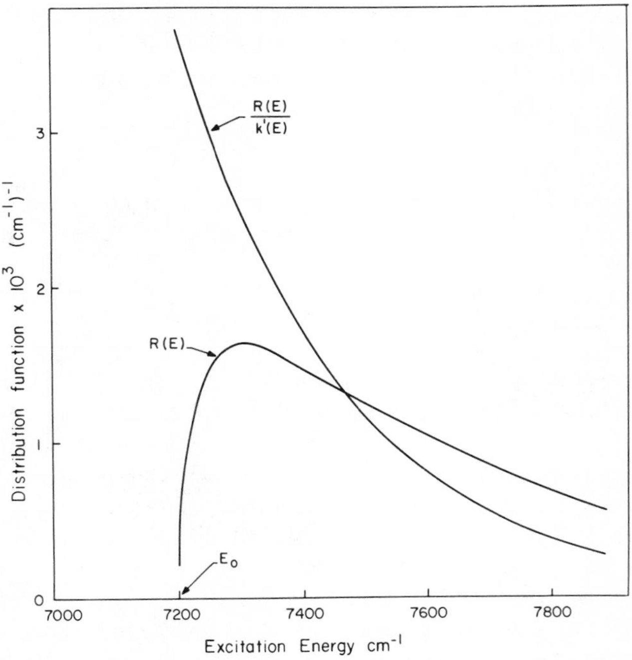

Fig. 8-1. *Distribution functions R(E) and R(E)/k′(E) as a function of excitation energy in reactant, calculated for the reaction* $C_2H_4Cl \rightarrow C_2H_4 + Cl$.

Critical energy E_0 is 7200 cm^{-1} in this case. Both functions are normalized to unity. Temperature is 310°K. After Le-KhacHuy, W. Forst, J. A. Franklin and G. Huybrechts, *Chem. Phys. Letters* **3**, 307 (1969), and Le-Khac Huy, M.Sc. Thesis, Univ. Laval 1968.

Division by k_∞' normalizes k_{uni} to unity at the high-pressure limit, which provides a convenient point of reference.

On differentiating (8-73) with respect to ω, we find

$$\frac{\partial}{\partial\omega}\left\langle\frac{\omega}{\omega + k'(E)}\right\rangle_R = \left\langle\frac{k'(E)}{[\omega + k'(E)]^2}\right\rangle_R \qquad \text{(slope)}$$
$$\frac{\partial^2}{\partial\omega^2}\left\langle\frac{\omega}{\omega + k'(E)}\right\rangle_R = -\left\langle\frac{2k(E)}{[\omega + k'(E)]^3}\right\rangle_R \qquad \text{(curvature)} \qquad (8\text{-}74)$$

Thus the plot of k_{uni}/k_∞' versus ω (i.e., versus pressure) has a positive slope and negative curvature everywhere. This conclusion is quite general and is independent of the nature of the function $R(E)$. At $\omega \rightarrow \infty$, the slope is zero, and hence $k_{uni} \leq k_\infty'$, the inequality increasing with decreasing ω; the rate constant k_{uni}/k_{∞}' therefore "falls off" as ω decreases. The curvature of the fall-off depends on the energy dependence of $k'(E)$ [note that $R(E)$ also depends on $k'(E)$], while the location of the fall-off region on the pressure axis depends essentially on the ratio $k'(E)/\omega$. Since the centrifugal correction factor f_0 is smaller than f_{∞}, centrifugal effects therefore contribute somewhat to the pressure fall-off.

Inasmuch as complex molecules generally have lower $k'(E)$ than small molecules, a given value of the ratio $k'(E)/\omega$, i.e., a given amount of fall-off, is realized at lower pressure in the case of the more complex molecule. An extreme case is the simplest molecule of all, the diatomic molecule, and it is of some interest to make a rough calculation here to locate the region of fall-off. We have seen in Chapter 4, Section 3 [Eq. (4-27)] that $k(E)$ for a harmonic diatomic molecule is constant and equal to oscillator frequency; thus $k(E) \sim 10^{13}$ sec^{-1}. We may arbitrarily consider that the region of fall-off is located in the neighborhood of a pressure such that $k(E)/\omega = 1$. For a diatomic molecule below 1000°K, the collision number Z is roughly 10^{-10} cm^3 molecule^{-1} sec^{-1}. Thus $k(E)/\omega$ will reach unity at a concentration of $\sim 10^{23}$ molecules cm^{-3} i.e. at a pressure of some 20,000 atm. The high-pressure limit of the unimolecular decomposition of a diatomic molecule is therefore unattainable at practical pressures, and at the more usual pressures of a few atmospheres, the decomposition will be essentially at its low-pressure limit. The fall-off region for molecules containing ten atoms or more is generally located below 100 Torr.

Operational definition of pressure limits Equation (8-73) can be used to obtain an operational definition of the high- and low-pressure limits. For large ω, the term $[1 + k(E)/\omega]^{-1}$ can be expanded in inverse powers of ω to yield

$$k_{uni} = k_{\infty}'\left\{1 - \left\langle\frac{k'(E)}{\omega}\right\rangle_R + \left\langle\left(\frac{k'(E)}{\omega}\right)^2\right\rangle_R - \cdots\right\}$$
$$= k_{\infty}' \int_{E_0}^{\infty} \left\{\sum_{n=0}^{\infty} (-1)^{n+2} \left[\frac{k'(E)}{\omega}\right]^n\right\} R(E)\, dE \tag{8-75}$$

Therefore, remembering that $\omega = Z[\text{M}]$, k_{uni} is a linear function of $1/[\text{M}]$ in the vicinity of the high-pressure limit, with intercept k_{∞}' and slope $-\text{M}$, where

$$\text{M} = (k_{\infty}'/Z) \int_{E_0}^{\infty} k'(E)R(E)\, dE \tag{8-76}$$

For small ω, the term $[1 + k'(E)/\omega]^{-1}$ can be expanded in powers of ω to yield, after division by [M],

$$\frac{k_{\text{uni}}}{[\text{M}]} = \frac{k_\infty'}{[\text{M}]}\left\{\left\langle\frac{\omega}{k'(E)}\right\rangle_R - \left\langle\left[\frac{\omega}{k'(E)}\right]^2\right\rangle_R + \cdots\right\}$$

$$= k_\infty' \int_{E_0}^{\infty}\left\{\sum_{n=1}^{\infty}(-1)^{n+1}[\text{M}]^{n-1}\left[\frac{Z}{k'(E)}\right]^n\right\} R(E)\, dE \qquad (8\text{-}77)$$

We may note that the first term of the expansion (8-77) is

$$k_\infty' Z \int_{E_0}^{\infty} \frac{R(E)}{k'(E)}\, dE = \frac{k_0'}{[\text{M}]} = k_{\text{bi}} \qquad (8\text{-}78)$$

where [cf. Eqs. (8-43) and (8-71)] k_{bi} is the limiting low-pressure *second-order* rate constant. Thus $k_{\text{uni}}/[\text{M}]$ is a linear function of [M] in the vicinity of the low-pressure limit, with intercept k_{bi} and slope $-\text{N}$, where

$$\text{N} = k_\infty' Z^2 \int_{E_0}^{\infty} \{R(E)/[k'(E)]^2\}\, dE \qquad (8\text{-}79)$$

The integrand of k_{bi} in Eq. (8-78) is

$$R(E)/k'(E) = f_0 P(E)_{\text{e}}$$

so that the energy-dependent part of the integrand is just the tail end of the equilibrium distribution function at energies above E_0. A plot of $f_0 P(E)_{\text{e}}$ as a function of energy, from E_0 onward, is shown in Fig. 8-1 for the same reaction for which the integrand of k_∞' is shown. The function $f_0 P(E)_{\text{e}}$ is here normalized, i.e., the area under the curve from $E = E_0$ to $E \to \infty$ is equal to unity. The part of $P(E)_{\text{e}}$ above E_0, normalized in the described fashion, will be denoted by $P(E_0, E)_{\text{e,norm}}$.

Consistency relations Following Johnston (*27*), let us now define two new functions $\mathscr{f}$ and $\mathscr{g}$, where

$$\mathscr{f} = [R(E)k'(E)/Z]^{1/2}, \qquad \mathscr{g} = [ZR(E)/k'(E)]^{1/2} \qquad (8\text{-}80)$$

From Schwarz's inequality (*28*, p. 135, Eq. (3–113)), it follows for any two real functions $\mathscr{f}$ and $\mathscr{g}$, which are functions of E, that

$$\left(\int \mathscr{f}^2\, dE\right) \times \left(\int \mathscr{g}^2\, dE\right) \geqslant \left(\int \mathscr{f}\mathscr{g}\, dE\right)^2 \qquad (8\text{-}81)$$

Expressing the integrals of $\mathscr{f}$ and $\mathscr{g}$ in terms of slope M and rate constant k_{bi}, the inequality (8-81) yields, after rearrangement,

$$k_{\text{bi}} \geqslant (k_\infty')^2/\text{M} \qquad (8\text{-}82)$$

If f and g are defined as

$$f = Z[R(E)]^{1/2}/k'(E), \qquad g = [R(E)]^{1/2} \tag{8-83}$$

then, expressing the integrals of f and g in terms of slope N and rate constant k_{bi}, the inequality (8-81) yields for the newly defined f and g

$$k_{\infty}' \geqslant k_{bi}^2/\mathrm{N} \tag{8-84}$$

The relations (8-82) and (8-64) are quite general and are independent of the nature of the functions f and g. As a result, the collision frequency, which is part of most of these functions, could just as well have been taken as a function of E without in any way affecting the inequalities (8-82) or (8-84); in other words, the two inequalities do not necessarily assume the validity of the strong-collision assumption. We shall find these inequalities useful in the treatment of experimental data.

6. Temperature dependence

Temperature, after concentration (or pressure), is the only other parameter variable at will in a thermal system, and therefore after having considered pressure dependence in the previous section, it is time now to consider the temperature dependence of rate.

Since the Arrhenius (or experimental) activation energy E_a is by definition kT^2 times the logarithmic derivative of rate constant with respect to temperature, it is convenient, for later comparison of experimental and calculated activation energies, to use this sort of derivative.

A quite general interpretation of the meaning of E_a in a thermal system under steady-state conditions can be obtained by going back to Eq. (8-33). From Eqs. (8-31) and (8-12)–(8-14), we have, if we neglect[13] the temperature dependence of ω with respect to $e^{-E/kT}$

$$\begin{aligned} E_a &= kT^2 \frac{\partial[\ln\langle k(E, \mathcal{J})\rangle_{ss}]}{\partial T} \\ &= \frac{\iint [E + E_r(\mathcal{J})]k(E, \mathcal{J})P(E, \mathcal{J})_{ss}\, dE\, d\mathcal{J}}{\iint k(E, \mathcal{J})P(E, \mathcal{J})_{ss}\, dE\, d\mathcal{J}} \\ &\quad - \frac{\iint [E + E_r(\mathcal{J})]P(E, \mathcal{J})_e\, dE\, d\mathcal{J}}{\iint P(E, \mathcal{J})_e\, dE\, d\mathcal{J}} \end{aligned} \tag{8-85}$$

[13] Since ω depends on $T^{\pm 1/2}$ [Eq. (8-100)], the neglect of the temperature dependence of ω does not lead to serious error.

The first term is $E + E_r(\mathcal{J})$ averaged over $k(E, \mathcal{J})P(E, \mathcal{J})_{ss}$, i.e., over the steady-state distribution of molecules dissociating per unit time, and the second term, arising from the logarithmic differentiation of the partition function product QQ_r in the denominator of $P(E, \mathcal{J})_e$, is $E + E_r(\mathcal{J})$ averaged over the equilibrium distribution. This equilibrium distribution, even though integrated over the full range of E and $\mathcal{J}$ encompassing all molecules, refers essentially to nonreacting molecules, since they represent by far the major part of the total. We may therefore write E_a as the difference between average internal and (external-adiabatic) rotational energy of reacting (per unit time) and nonreacting molecules:

$$E_a = \langle E + E_r(\mathcal{J})\rangle_{\text{reacting, sec}} - \langle E + E_r(\mathcal{J})\rangle_{\text{nonreacting}} \tag{8-86}$$

Equation (8-86), originally due to Tolman[14] in a slightly different form, is perfectly general and is independent of the form of $k(E, \mathcal{J})$ and $P(E, \mathcal{J})_{ss}$. Since, for any kind of energy, $\partial\langle E\rangle/\partial T = C_p$, the experimental activation energy E_a will be independent of temperature only if the heat capacities C_p (vibrational and rotational) of reacting and nonreacting molecules are the same. This is not the case in general, but the difference $C_{p(\text{reacting/sec})} - C_{p(\text{nonreacting})}$ is usually so small that over a small temperature range it is lost in experimental error. It would therefore take experimental rate data over a temperature range of several thousand degrees to make the temperature dependence of E_a noticeable.

In order to transcribe (8-86) for the specific form of $P(E, \mathcal{J})_{ss}$ that we have been using, it is necessary to return to k_{um} of Eq. (8-51). Aside from the temperature-dependent exponential term, other temperature-dependent factors are Q, f_0, f_∞, and ω, the last three of which appear inside the integrand, which complicates matters considerably. More useful expressions are obtained if we consider only the temperature dependence of k_{uni} at the two extremes of pressure.

The high-pressure case At $\omega \to \infty$, k_{uni} becomes k_∞' and the distribution function over which $\langle\cdots\rangle_{\text{reacting/sec}}$ is evaluated is clearly the function $R(E)$ shown in Fig. 8-1. For our purpose k_∞' is best given by Eq. (8-23); differentiating, we get for the high-pressure experimental activation energy $E_{a\infty}$

$$E_{a\infty} = kT^2\frac{\partial(\ln k_\infty')}{\partial T} = kT^2\frac{\partial(\ln f_\infty)}{\partial T} + kT + kT^2\frac{\partial(\ln Q^*)}{\partial T} + E_0 - kT^2\frac{\partial(\ln Q)}{\partial T} \tag{8-87}$$

In interpreting the various terms in this equation, it is helpful to recall that in a canonical ensemble there exists the general relation [see, e.g., (*12*)]

$$kT^2\,\partial(\ln Q)/\partial T = \langle E\rangle \tag{8-88}$$

[14] Tolman (*29*). The relation (8-86) requires only that $P(E, \mathcal{J})_{ss}$ have exponential temperature dependence.

where $\langle E \rangle$ is the average energy of the degrees of freedom involved in the partition function Q. Taking the first term in (8-87), the derivative of f_∞, it is easy to see with the help of (8-88), (8-55), and (8-59) that

$$\begin{aligned} kT^2\, \partial(\ln f_\infty)/\partial T &= \langle E_r(\mathscr{J}) \rangle_{\text{reacting/sec}} - \langle E_r(\mathscr{J}) \rangle_{\text{nonreacting}} \\ &= -(2/\mathfrak{n})kT \qquad \text{(type 1 reaction)} \\ &= 0 \qquad \text{(type 2 or 3 reaction)} \end{aligned} \tag{8-89}$$

In fact, if we return to the definition of f_∞ [Eq. (8-22)], we may write it as $f_\infty = Q_r^*/Q_r$, where the numerator is the partition function for the two external (adiabatic) rotations in the transition state, i.e., in a molecule about to react, and Q_r is the partition function for the same rotations in the reactant (i.e., nonreacting) molecule. Note that for type 1 reactions the temperature coefficient of f_∞ is negative, meaning that the average rotational energy in *non*reacting molecules is slightly larger than in reacting molecules. The last term in (8-87) is clearly $\langle E \rangle_{\text{nonreacting}}$, i.e., average energy in the degrees of freedom of the nonreacting molecule among which energy is randomized, and similarly $kT^2\, \partial(\ln Q^*)/\partial T = \langle E^* \rangle$, where $\langle E^* \rangle$ is average energy in degrees of freedom of the transition state among which energy is randomized. Hence

$$\langle E \rangle_{\text{reacting/sec}} = E_0 + kT + \langle E^* \rangle \tag{8-90}$$

so that average energy of molecules reacting per unit time is $kT + \langle E^* \rangle$ above threshold; here kT is the average translational energy in the reaction coordinate (Cf. Exercise 3 in Chapter 10).

The low-pressure case In the low-pressure region, when $\omega \to 0$, the interpretation of experimental activation energy as given by Eq. (8-86) is still valid, except for one small but important difference. We saw in Section 3 that we may look upon the low-pressure rate constant k_0' as essentially the number of molecules colliding, times the fraction of molecules in reactive states, i.e., times the fraction of molecules reacting. Note that we do not say "reacting per unit time," since the pressure is by definition so low at the low-pressure limit that *every* molecule in a reactive state has sufficient time to decompose before the next collision. Since ω involves the collision number Z, which is proportional to $T^{\pm 1/2}$, we can write for E_{a0}, the experimental activation energy at the low-pressure limit,

$$E_{a0} = \pm\tfrac{1}{2}kT + \langle E + E_r(\mathscr{J}) \rangle_{\text{reacting}} - \langle E + E_r(\mathscr{J}) \rangle_{\text{nonreacting}} \tag{8-91}$$

The positive sign in front of $kT/2$ applies if Z (or k_{bi}) is expressed in concentration units and the negative sign if Z (or k_{bi}) is expressed in pressure units [cf. Eq. (8-100)]. The distribution function over which $\langle \cdots \rangle_{\text{reacting}}$ is evaluated is of course the function $R(E)/k'(E)$ shown in Fig. 8-1.

At $\omega \to 0$, k_{uni} becomes k_0' of Eq. (8-45), or, after division by [M], the second-order rate constant k_{bi} [Eq. (8-78)]. On differentiating either with the help of Eq. (8-43), we get for the low-pressure activation energy

$$E_{a0} = \frac{\pm \ell T}{2} + \ell T^2 \frac{\partial(\ln f_0)}{\partial T} + E_0 + \frac{\int_0^\infty xN(E_0 + x)e^{-x/\ell T}\,dx}{\int_0^\infty N(E_0 + x)e^{-x/\ell T}\,dx} - \ell T^2 \frac{\partial(\ln Q)}{\partial T} \tag{8-92}$$

In a manner analogous to Eq. (8-89), we may identify

$$\ell T^2\, \partial(\ln f_0)/\partial T = \langle E_r(\mathcal{J})\rangle_{\text{reacting}} - \langle E_r(\mathcal{J})\rangle_{\text{nonreacting}} \tag{8-93}$$

The third and fourth terms in (8-92) can be written as

$$\ell T^2\,(\partial/\partial T)\left\{\ln\left(\int_{E_0}^\infty N(E)e^{-E/\ell T}\,dE\right)\right\} = \langle E\rangle_{\text{reacting}} \tag{8-94}$$

The integral in (8-94), when evaluated with a "good" $N(E)$, turns out[15] to be slightly larger (*8*) than $E_0 + \ell T$, which means that at the low-pressure limit the average energy of reacting molecules barely exceeds the minimum energy required for reaction. Finally, the last term in (8-92) is, as before, $\langle E\rangle_{\text{nonreacting}}$.

Differentiation of f_0 is easy only in the case of type 2 or 3 reactions if we use Eq. (8-64); thus

$$\ell T^2 \frac{\partial(\ln f_0)}{\partial T}\,(\text{type 2 or 3}) \approx -\frac{(\mathrm{v} - 1 + \frac{1}{2}\mathrm{r})\,\ell T}{(\mathrm{v} - 1 + \frac{1}{2}\mathrm{r}) + [\rho \times (E_0 + a_0 E_z)/\ell T]} \tag{8-95}$$

This expression yields about $-\ell T/2$ for $r_0^2/r_e^2 \sim 10$ and about $-\ell T/10$ for $r_0^2/r_e^2 \sim 2$, so that the effect is quite small. Since for type 1 reactions, a simple, tractable expression for f_0 is not available, it is more convenient to start with [cf. Eqs. (8-45), (8-43), and (8-68)]

$$k_0'(\text{type 1}) = \frac{\omega \exp(-E_0/\ell T)}{Q}$$
$$\times \left\{\int_0^\infty N(x + E_0)\, e^{-x/\ell T}\,dx\right.$$
$$\left. - \int_0^{z_0} N(E_0 + bz^{n/(n-2)} - z)\left(\frac{n}{n-2}\,bz^{2/(n-2)} - 1\right)\exp\left(\frac{-bz^{n/(n-2)}}{\ell T}\right)dz\right\} \tag{8-96}$$

[15] Equation (8-94) is really a special case of Eq. (8-88) if we look on the integral in the former as the partition function for reacting molecules. Replacing the integral in (8-94) by its approximate value (8-43a), we find that $\langle E\rangle_{\text{reacting}}$ equals $E_0 + \ell T$ exactly.

which yields on differentiation

$$E_{a0}(\text{type 1}) = \not{k}T^2 \frac{\partial[\ln k_0'(\text{type 1})]}{\partial T} = \pm\frac{\not{k}T}{2} + E_0 - \not{k}T^2 \frac{\partial(\ln Q)}{\partial T}$$

$$+\Bigg\{\Bigg[\int_0^\infty xN(x+E_0)\,e^{-x/\not{k}T}\,dx - \int_0^{z_0} bz^{n/(n-2)}$$

$$\times N(E_0 + bz^{n/(n-2)} - z)\left(\frac{n}{n-2}\,bz^{2/(n-2)} - 1\right)\exp\left(\frac{-bz^{n/(n-2)}}{\not{k}T}\right)dz\Bigg]$$

$$\times\Bigg[\int_0^\infty N(x+E_0)\,e^{-x/\not{k}T}\,dx$$

$$-\int_0^{z_0} N(E_0 + bz^{n/(n-2)} - z)\left(\frac{n}{n-2}\,bz^{2/(n-2)} - 1\right)\exp\left(\frac{-bz^{n/(n-2)}}{\not{k}T}\right)dz\Bigg]^{-1}\Bigg\} \tag{8-97}$$

Expression (8-97) looks sufficiently cumbersome to strike terror in the heart of the timid, although for a given E_0, it would be easy enough to evaluate on a computer; however, no such machine calculation is on record. The integrals with respect to z are actually negative, so that despite the negative sign, (8-97) involves the ratio of a *sum* of integrals. This ratio is very likely $\not{k}T$ or less, since the ratio of integrals with respect to x is generally slightly more than $\not{k}T$ [see comments following Eq. (8-94)], and the ratio of integrals with respect to z is related to f_0 which always has a small, negative temperature coefficient.

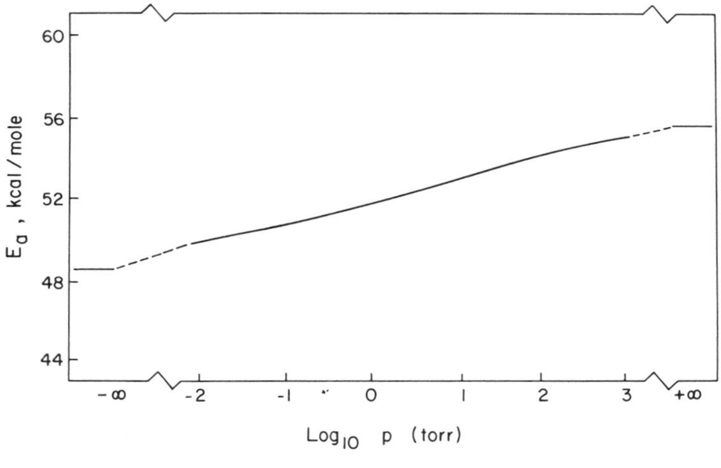

Fig. 8-2. *Calculated pressure dependence of experimental activation energy E_a for the decomposition of azomethane.*
W. Forst and P. St. Laurent, unpublished results.

Combining (8-87) and (8-91), we can write

$$E_{a\infty} - E_{a0} = \langle E + E_r(\mathscr{J})\rangle_{\text{reacting/sec}} - \langle E + E_r(\mathscr{J})\rangle_{\text{reacting}} \pm \tfrac{1}{2}kT \quad (8\text{-}98)$$

Since it takes a higher average energy for a molecule to react in unit time than for it just to react, whatever the time interval, we have $E_{a\infty} > E_{a0}$, i.e., the experimental activation energy falls off with pressure. It should be emphasized that $E_{a\infty}$ and E_{a0} are related to, but distinct from, the critical energy E_0, which is an intrinsic property of the reaction system on a microscopic level, while $E_{a\infty}$ and E_{a0} are macroscopic properties arising from the two different distributions of energies that exist at the two pressure limits. Figure 8-2 shows the calculated dependence of E_a on pressure for the thermal decomposition of azomethane (*30*); in this particular case, $E_{a\infty} - E_{a0} \sim 7$ kcal mole^{-1} ($\sim \frac{7}{2}kT$).

7. Theoretical parameters and experimental observables

If all the theoretical parameters were known, either by calculation from first principles or at least from nonkinetic information, k_{uni} could be calculated *ab initio* and then compared with the experimental k_{uni}. Such is not the case, unfortunately, since some parameters can be obtained only from experimental observables. It is therefore more than the intrinsic interest of the comparison that leads us to consider the connection between theory and experiment in a fairly detailed way.

We shall take the theoretical k_{uni} in the simplified form given by Eq. (8-51). The following theoretical quantities are necessary for the calculation of this k_{uni}:

1. The critical energy E_0, which is the lower limit of the integral in (8-51); it appears also in $k(E)$ [Eq. (8-25)], f_0 [Eq. (8-62) or (8-64)], and in f_∞ [Eq. (8-56)] through the constant $\mathscr{C}$ [Eq. (7-15)].

2. The reactant density of states $N(E)$, at energies from E_0 onward. $N(E)$ appears in the integral (8-51) and in the denominator of $k(E)$ [Eq. (8-25)] and in Q. The states to which $N(E)$ refers are of course those states of the reactant that are involved in internal randomization of energy, i.e., that belong to "pertinent" degrees of freedom; as mentioned in Chapter 4, especially Section 4, *all* internal degrees of freedom and some external rotations are assumed to be pertinent. The parameters required to calculate $N(E)$ by one of the methods of Chapter 6 in the independent harmonic oscillator, rigid free rotor approximation are the vibrational frequencies of

the individual oscillators[16] and moments of inertia of the rotors.[17] Q can be obtained directly from these parameters without the necessity of passing through $N(E)$ first. We can write $Q = Q_v Q_r$; the rotational partition function Q_r follows by substituting the moment(s) of inertia into the formulas (6-39) and (6-40); the vibrational partition function Q_v can be obtained from tables of thermodynamic functions of the harmonic oscillator,[18] since

$$\ln Q_v = -\sum_i (F_i^0 - E_{i0}^0)/RT \tag{8-99}$$

where $F_i^0 - E_{i0}^0$ is the free energy function of the ith oscillator.

3. Integrated density $G^*(E)$ for the transition state, at energies from zero onward. $G^*(E)$ appears in $k(E)$ [Eq. (8-25)]. The same comments apply about the calculation of $G^*(E)$ as about $N(E)$.

4. Reaction path degeneracy α, which appears in $k(E)$ [Eq. (8-25)].

5. The exponent $\mathfrak{n}$ in the long-range part of the potential for type 1 reactions, which appears in f_∞ [Eq. (8-55)] and f_0 [Eq. (8-68)].

6. Interatomic distance r_0 at potential maximum and equilibrium interatomic distance r_e for type 2 and 3 reactions; the ratio of the two appears in f_∞ [Eq. (8-59)] and f_0 [Eq. (8-64)].

7. Collision number Z, which enters the collision frequency ω. From kinetic theory, Z is given by[19]

$$\begin{aligned} Z &= \pi d^2(8kT/\pi\mu)^{1/2} \quad \text{cm}^3\ \text{molecule}^{-1}\ \text{sec}^{-1} \\ &= \pi d^2(8/\pi\mu kT)^{1/2} \quad \text{Torr}^{-1}\ \text{sec}^{-1} \end{aligned} \tag{8-100}$$

where μ is reduced mass (equal to $M/2$ for like collision partners, each of mass M) and πd^2 is collision cross section.

Parameters 2 and 5–7 can be obtained from nonkinetic information. Vibrational frequencies and moments of inertia of the reactant are generally obtainable from vibrational and rotational spectroscopy, and, in some instances, also from normal mode coordinate analysis; the exponent $\mathfrak{n}$ can be determined by arguments described in Chapter 7, Section 2; the interatomic distance r_e can be obtained from the known geometry of the reactant, and even a very rudimentary knowledge of the potential energy surface is sufficient to estimate r_0; in the absence of such information, r_0 can be taken, at least in isomerizations, as a distance intermediate between r_e and the

[16] For a useful compilation of vibrational frequencies, see Shimanouchi (*31*).

[17] Moments of inertia can be calculated from rotational constants compiled by Cord *et al.* (*32*) or from interatomic distances compiled in "Tables of Interatomic Distances and Configurations in Molecules and Ions" (*33,34*).

[18] See, for example, Hilsenrath and Ziegler (*35*). A shorter table may be found in Abramowitz and Stegun (*36*, p. 999).

[19] See, for example, Laidler (*37*), p. 65. The reduced mass μ refers here to A + M, the collision partners in process (8-2).

similarly defined distance in the product; finally, the collision cross section πd^2 is usually equated with cross sections determined from transport properties.[20]

Model-dependent parameters Parameters 1, 3, and 4 depend on the structure of the transition state. Since properties of the transition state are not accessible to direct measurement, it is necessary to construct a model for the transition state, and therefore we shall call parameters 1, 3, and 4 "model dependent." The model dependence of parameters 3 and 4 (cf. Chapter 4, Section 5) is fairly obvious, but the case of E_0 merits a separate comment. With reference to Fig. 4-2, we have $E_0 = D_0 + E_z^*$, where E_z^* is the zero-point energy of the transition state. In type 1 reactions, which involve simple bond breaking, D_0 is the bond dissociation energy,[21] which is in general obtainable from nonkinetic information, and in other types of reactions, D_0 is often also obtainable from nonkinetic information or, at least in principle, calculable; hence the model dependence of E_0 arises from E_z^*, a point that is not usually considered. The zero-point energy E_z^* is appreciable for large molecules and cannot be ignored with respect to D_0; on the other hand, E_z^* is small for small molecules and therefore it matters less if D_0 is large. Therefore in a small molecule, we can write $E_z^* \approx E_z - \varepsilon_z$, where E_z is the zero-point energy of the reactant (known) and ε_z is the zero-point energy of the reactant vibration which becomes the reaction coordinate in the transition state (also known). In such a special case, E_0 may be considered as practically model independent.

Experimental data are necessary to determine, or to check on, or otherwise obtain clues about, model-dependent parameters, and eventually to check on the applicability of the theory as a whole. Experimental data consist of the pressure and temperature dependences of the experimentally determined k_{uni}; unfortunately, these data are rarely complete. Because of practical difficulties, they never span the entire range of pressures[22] (zero to infinity) and temperatures,[23] so that the investigator must make do with experimental

[20] For a thorough discussion of transport properties see Hirschfelder *et al.* (*38*, Ch. 7).

[21] "Bond energies" derived from kinetic information [see, for example, Benson (*39*)] are unsuitable since they are generally equated with E_{a0} or, worse, with $E_{a\infty}$, which is much too crude an approximation for the present purpose.

[22] The difficulty is compounded by the fact that both the high- and low-pressure limits are approached only asymptotically, i.e., very slowly; moreover, the actual pressure at which the two limits are reached are often so high and so low, respectively, as to be unrealizable experimentally.

[23] Usual thermal data are limited to temperatures at which the rate is not too fast and which are below the softening point of reactor material, both generally well below 1000°K (strain point of Pyrex at ~800°K). Shock tube experiments generally furnish very high-temperature data (1000–2000°K) near the low-pressure limit, but there is no

data that in one manner or another are always deficient. Since with respect to determination of model-dependent parameters the deficiency, or information content, of different sorts of experimental data is different, we shall discuss two typical cases to see which parameter can and which cannot be determined. The assumption will be made that information concerning reactant is available from other sources, i.e., that model-independent parameters 2 and 5–7 are known.

Experimental data available at and near the low-pressure limit only Ideally such data would consist of k_{uni} over a range of temperatures and pressures such that k_{uni} has a pressure-independent experimental activation energy, and $k_{uni}/[M]$ as a function of [M] is linear over a range of pressures sufficient to obtain the slope N [Eq. (8-79)] and, by extrapolation, k_{bi}. In practice, there are usually not enough experimental points at extreme low pressures to allow a determination of the slope N; instead, the low-pressure limit is considered (*41*) to have been reached when k_{uni} is a linear function of [M] over a range of pressures within experimental error, and k_{bi} is then the slope of this line.[24]

Thus in the majority of cases, k_{bi} and E_{a0} are the two experimental observables at the low-pressure limit. The theoretical expression for k_{bi} is given by Eq. (8-78), where k_0' is given by Eqs. (8-43) and (8-45); it includes E_0 as the only model-dependent parameter. Thus experimental data at the low-pressure limit permit the unambiguous determination of E_0, but otherwise offer no information whatever about other model-dependent parameters. Alternatively, in the special case that E_0 can be considered as model independent and therefore known, comparison of calculated k_0' with the experimental value may be used as a test of the assignment of randomizable degrees of freedom in the reactant, and of the strong-collision assumption.

The energy E_0 is calculable from the experimental activation energy E_{a0} by means of Eq. (8-92). This is not very easy since E_0 occurs inside the integral with respect to x, and also inside the integral with respect to z in the case of type 1 reactions [Eq. (8-97)], where, through z_0, it also appears in the upper limit of the same integral. A solution is possible by successive approximations,

---

convenient technique to cover the intermediate temperature range between thermal and shock tube ranges. An unusual technique is "very low-pressure pyrolysis" (VLPP) [Benson and Spokes (*40*)], where activation occurs primarily by molecule collisions with the wall, rather than by the usual collisions among the molecules themselves. If the wall is sufficiently inert, homogeneous unimolecular rates in the fall-off region presumably can be measured in this way at quite high temperatures.

[24] The slope of the logarithmic plot is usually less than 45°, which means that reaction order is slightly less than two. The above discussion assumes that heterogeneous effects, which plague low-pressure measurements, have been eliminated.

although, depending on the accuracy of the experimental E_{a0} (often quite poor), cruder approximations might be adequate.

The vibrational part of the term $kT^2\,\partial(\ln Q)/\partial T$, where $Q = Q_v Q_r$, may be evaluated directly from vibrational frequencies by the use of tables of thermodynamic functions of the harmonic oscillator, since

$$kT^2\,\partial(\ln Q_v)/\partial T = kT \sum_i (H_i^0 - E_{0i}^0)/RT \tag{8-101}$$

where $H_i^0 - E_{0i}^0$ is the enthalpy function of the ith oscillator. For the rotational part, we have

$$kT^2\,\partial(\ln Q_r)/\partial T = \mathfrak{r}kT/2 \tag{8-102}$$

for $\mathfrak{r}$ classical free rotors.

If low-pressure data are sufficiently extensive to permit the determination of the slope N, we see by Eq. (8-79) that N involves $k'(E)$, which in turn involves model-dependent parameters 3 and 4. It is therefore possible in such a case, by comparing experimental and calculated slope N, to obtain a check on parameters 3 and 4. Such a check, however, assumes the validity of the strong-collision assumption which is incorporated in Eq. (8-79). It is therefore generally preferable to check model-dependent parameters on high-pressure data, which are more closely related to the unimolecular process itself than low-pressure data, which reflect primarily the rate of accumulation of randomizable energy.

Experimental data available at and near the high-pressure limit only Ideally, such data would consist of k_{uni} over a range of temperatures and pressures such that k_{uni} has a pressure-independent experimental activation energy, and k_{uni} is a linear function of 1/[M] over a range of pressures sufficient to obtain the slope M [Eq. (8-76)] and, by extrapolation, k_∞'. In practice, pressures at which the cited linearity would be observed are often extremely high, so that this extrapolation procedure for k_∞' cannot be usefully applied. It is therefore necessary to obtain k_∞' by linearizing existing data by some empirical procedure; Oref and Rabinovitch (*42*) suggest plotting k_{uni}^{-1} versus $[M]^{-\zeta}$, where ζ is chosen empirically to make the plot linear, so that k_∞' can be obtained by linear extrapolation.

Deconvolution If k_∞' is determined experimentally over a range of temperatures, it is known in the form

$$k_\infty' = A_\infty e^{-E_{a\infty}/kT} \tag{8-103}$$

where A_∞ is the temperature-independent preexponential factor.[25] The appropriate theoretical expression for $k_\infty{}'$ is given by Eq. (8-26):

$$k_\infty{}' = (f_\infty/Q)\int_0^\infty k(E)N(E)e^{-E/kT}\,dE \tag{8-104}$$

where the integral is written deliberately with limits $(0, \infty)$ if the critical energy E_0 is not known, which is the usual case. We may observe that the integral in (8-104) is in fact the Laplace transform of the function $k(E)N(E)$:

$$k_\infty{}' = (f_\infty/Q)\,\mathfrak{L}\{k(E)N(E)\} \tag{8-105}$$

It should therefore be possible to recover (*43*) the the product function $k(E)N(E)$ by inversion, and since $N(E)$ refers to reactant and hence is known, $k(E)$ follows. Combining the experimental [Eq. (8-103)] and theoretical [Eq. (8-105)] forms of $k_\infty{}'$, we have, with $s = 1/kT$ as the transform parameter,

$$\mathfrak{L}\{k(E)N(E)\} = (A_\infty/f_\infty) \times [Q(s)e^{-sE_{a\infty}}] \tag{8-106}$$

from which, by inversion,

$$k(E)N(E) = (A_\infty/f_\infty)\,\mathfrak{L}^{-1}\{Q(s)e^{-sE_{a\infty}}\} \tag{8-107}$$

where we write $Q(s)$ for Q to indicate that the reactant partition function also depends on s.

We know from Eq. (6-26) that $\mathfrak{L}^{-1}\{Q(s)\} = N(E)$; the function being inverted in (8-107) is merely $Q(s)$ with zero of energy shifted to E_a (cf. Exercise 3 of Chapter 6), so that the result of the inversion is

$$k(E)N(E) = (A_\infty/f_\infty)N(E - E_{a\infty})\mathfrak{H}(E - E_{a\infty}) \tag{8-108}$$

where $\mathfrak{H}(x)$ is the Heaviside step function

$$\begin{aligned} \mathfrak{H}(x) &= 0 \qquad (x < 0) \\ &= 1 \qquad (x > 0) \end{aligned} \tag{8-109}$$

It follows therefore that

$$\begin{aligned} k(E) &= \frac{A_\infty}{f_\infty}\,\frac{N(E - E_{a\infty})}{N(E)} \qquad (E > E_{a\infty}) \\ &= 0 \qquad\qquad\qquad\qquad\quad (E < E_{a\infty}) \end{aligned} \tag{8-110}$$

The function $k(E)$ is thus recovered from its average, a process sometimes referred to as deconvolution.

The result of the deconvolution may be stated as follows: if available information is A_∞ and $E_{a\infty}$, (1) the best estimate of critical energy is $E_0 \equiv E_{a\infty}$; (2) the energy dependence of $k(E)$ is given by Eq. (8-110).

[25] Note that because $E_{a\infty}$ is in the exponential, A_∞ is quite sensitive to any experimental error in $E_{a\infty}$.

Involved in the result (8-110) are several assumptions which are discussed at some length in Forst (*43*). The principal of these is that (8-103) is assumed to be exact and valid over all temperatures. This is true only approximately, and therefore the result (8-110), while mathematically correct, is also true only approximately because it makes use of imperfect experimental information. The result (8-110) also neglects the temperature coefficient of f_{∞}, which is either zero or $-(2/\mathfrak{n})kT$ [Eq. (8-89)], and this is much less than the likely error in $E_{a\infty}$.

The main point is that $k(E)$ of Eq. (8-110) contains all the information that can be extracted from high-pressure experimental data, and in the absence of additional information, or of additional assumptions, it is therefore impossible to obtain a better[26] estimate of E_0 or of $k(E)$. The test of the theory is complete if $k(E)$ of Eq. (8-110), when substituted into k_{uni} of Eq. (8-51), reproduces the experimental fall-off of the rate constant over most of the pressure range. However, the fall-off cannot be successfully calculated from (8-110) all the way to low pressures because the correct low-pressure rate constant k_0' requires the accurate knowledge of E_0 [cf. Eq. (8-43)], of which $E_0 \equiv E_{a\infty}$ is much too crude an estimate. Thus high-pressure experimental data alone are insufficient to characterize the unimolecular thermal reaction over the whole range of pressures. To do that, additional information would be necessary (such as the experimental determination of k_0').

In practical terms, the interest of (8-110) is that it involves only the density of states of the *reactant*. Hence (8-110) offers a test of unimolecular rate theory on the basis of high-pressure thermal data and the properties of the reactant molecule, *without* the necessity of explicit reference to the properties of the transition state. Inasmuch as the properties of the transition state are always subject to uncertainty, if not outright adjustment, while those of the reactant are known, Eq. (8-110) offers a more significant, if limited, test of unimolecular theory than the customary curve fitting of fall-off curves (*93*).

It is still useful, however, to interpret the high-pressure thermal results in terms of properties of the transition state, if for no other reason than to discover general rules for the formulation of transition states that can be used in reactions with other modes of activation; such rules will be considered in more detail in Chapter 11. In the present context, it must be emphasized that formulation of a transition-state structure does not add any new knowl-

[26] For example, $k(E)$ of Eq. (8-110) has a somewhat pathological behavior at threshold if A is much larger than $\sim 10^{13}$ sec^{-1}. This merely illustrates the fact that recovering a function from its average yields a function having the correct average behavior but not necessarily the correct behavior at some particular energy. In other words, there is a limited amount of information to be extracted from thermal data; specifically, there is no information in such data on threshold behavior.

edge beyond items 1 and 2 deduced from deconvolution, and is therefore essentially an exercise in curve-fitting.

Basically, one tries to guess a structure for the transition state such that the corresponding calculated $k(E)$ has an energy dependence that corresponds roughly to that given by Eq. (8-110). If we change to a new variable E' defined, as before, as energy in excess of critical energy, we have from (8-110) for the rate constant from deconvolution

$$k(E') = A_\infty N(E')/f_\infty N(E' + E_{a\infty}) \tag{8-111}$$

while from (8-25) we get for the theoretical rate constant

$$k(E') = \alpha G^*(E')/hN(E' + E_0) \tag{8-112}$$

Equating the E'-dependent parts of (8-111) and (8-112), we have

$$(\alpha/h)G^*(E') = (A_\infty/f_\infty)N(E') \tag{8-113}$$

so that the structure of the transition state should be such that its integrated density is related to the density of states of reactant by Eq. (8-113). This approach to the transition-state structure is computationally difficult, although it offers some interesting insights (*44*, p. 75).

Thermodynamic interpretation A simpler, and in the end probably more useful, approach is to exploit the thermodynamic interpretation of reaction rates. In a unimolecular reaction, A_∞ is related to entropy of activation ΔS^* by[27]

$$A_\infty = (\alpha f_\infty e\ell T/h) \exp(\Delta S^*/\ell) \tag{8-114}$$

[27] The general derivation may be found in Glasstone *et al.* (*14*, p. 195 ff.). It is perhaps worthwhile to sketch the derivation for the slightly more general case considered here If we return to Eq. (8-23), we can write it as $(Q^*/Q)e^{-E_0/\ell T} = K^*$, where K^* can be looked upon as the thermodynamic equilibrium constant for the formation of the transition state. We have, by the application of standard thermodynamic relations, $\ln K^* = -\Delta F^*/\ell T = -(\Delta H^*/\ell T) + (\Delta S^*/\ell)$, where ΔF^*, ΔH^*, and ΔS^* are the standard free energy, enthalpy, and entropy of activation, respectively. Equation (8-23) thus becomes

$$k_\infty' = \alpha f_\infty(\ell T/h) \exp(-\Delta H^*/\ell T) \exp(\Delta S^*/\ell) \tag{8-23^1}$$

Since $kT^2\, \partial(\ln K^*)/\partial T = \Delta H^*$, we can write Eq. (8-87), using (8-89), as

$$E_{a\infty} = \ell T[1 - (2/\mathrm{n})] + \Delta H^*$$

and substituting for ΔH^* in (8-23[1]), we have

$$k_\infty' = \alpha f_\infty(\ell T/h) \exp(-E_{a\infty}/\ell T) \exp[1 - (2/\mathrm{n})] \exp(\Delta S^*/\ell)$$

Comparing this last expression for k_∞' with the experimental form (8-103), we see that the factor multiplying $e^{-E_{a\infty}/\ell T}$ is

$$A_\infty = \alpha f_\infty(\ell T/h) \exp[1 - (2/\mathrm{n})] \exp(\Delta S^*/\ell)$$

which yields Eq. (8-114) if we neglect the temperature dependence of f_∞ in type 1 reactions, which contributes the factor $e^{-2/\mathrm{n}}$.

where $\Delta S^* = S^* - S$, S^* being the entropy of the transition state and S the entropy of the reactant. The entropies refer, of course, to pertinent or randomizable degrees of freedom.

If only (separable) vibrational degrees of freedom are involved, their entropy can be obtained from tables of thermodynamic functions of the harmonic oscillator; this means starting with the known S_v for reactant, adjusting frequencies in the transition state until $S_v^* = S_v + \Delta S^*$. Once the frequencies of the transition state are known, $kT^2\, \partial(\ln Q_v^*)/\partial T$ can be obtained from the same tables [Eq. (8-101)], and E_0 can then be calculated from Eq. (8-87). The critical energy E_0 is thus a derived quantity, adjusted so that the combined temperature dependence of f_∞, Q^*, Q, and kT yields the experimental value. This completes the determination of model-dependent parameters.

Rotational degrees of freedom If pertinent degrees of freedom are assumed to be both vibrational and rotational, matters become more complicated. We have, quite generally, for entropy in a canonical ensemble characterized by temperature T:

$$TS = kT^2[\partial(\ln Q)/\partial T] + kT \ln Q \tag{8-115}$$

where in the present instance, $Q = Q_v Q_r$; an identical equation involving S^* and Q^* can be written for the transition state. Note that the first term of the right-hand side of (8-115) is the average energy [cf. Eq. (8-88)]. It follows from (8-115) that

$$\begin{aligned} T\Delta S^* &= T(S^* - S) \\ &= kT^2\left[\frac{\partial(\ln Q_v^* Q_r^*)}{\partial T} - \frac{\partial(\ln Q_v Q_r)}{\partial T}\right] + kT \ln \frac{Q_v^*}{Q_v} + kT \ln \frac{Q_r^*}{Q_r} \end{aligned} \tag{8-116}$$

If there are r (pertinent) rotational degrees of freedom in the reactant and r* in the transition state, the average energy is $kT/2$ per rotation and the total average energy in the reactant is $\langle E_v \rangle + \frac{1}{2}\mathrm{r}kT$, and similarly in the transition state it is $\langle E_v^* \rangle + \frac{1}{2}\mathrm{r}^*kT$. If we now use Eqs. (6-41) and (6-42) for the rotational partition function, we have $\ln Q_r = (\mathrm{r}/2)\ln(\text{const} \times TI)$, where I is the moment of inertia product for the reactant and T is temperature. Similarly, we have for the transition state $\ln Q_r^* = (\mathrm{r}^*/2)\ln(\text{const} \times TI^*)$. Equation (8-116) thus becomes (*30*)

$$T\Delta S^* = \frac{\mathrm{r}^* - \mathrm{r}}{2} kT\left(1 + \ln \frac{I^*}{I}\right) + \langle E_v^* \rangle - \langle E_v \rangle + kT \ln \frac{Q_v^*}{Q_v} \tag{8-117}$$

We now suppose, as always, that all parameters involving the reactant (i.e., r, I, $\langle E_v \rangle$, Q_v) are known; ΔS^* is also known from experiment via Eq. (8-114). It now requires an assumption to decide on the frequencies in the transition state, which will fix the value of $\langle E_v^* \rangle$ and of Q_v^*, but Eq. (8-117) still cannot be solved for r* unless an assumption is made about I^*. Conversely, I^*/I cannot be determined from (8-117) unless an assumption is made about r* − r, i.e., how many rotations there are in the transition state. Thus the assignment of rotations in the transition state is on shakier grounds[28] than the assignment of vibrations. Since, in final analysis, it is only the density of states that matters, and since a low-frequency vibration can be made to yield roughly the same density of states as a rotation, it seems preferable to take the pertinent degrees of freedom as being only vibrational in nature. The question will be taken up again in Chapter 11, Section 2; we will mention at this point only that while replacing one or more vibrations by rotations, and vice versa, changes the energy dependence of $k(E)$, the effect is negligible on the usual fall-off plot of log k_{uni} versus log[M], which is quite insensitive to all but the crudest manipulations of parameters.

The calculation of the slope M from Eq. (8-76) does not offer any advantage over the conventional comparison of calculated and experimental fall-off curves, particularly if extensive high-pressure data are lacking, and k_∞' must be obtained by empirical extrapolation.

If experimental data comprise both the low- and high-pressure limits, consistency relations (8-82) and (8-84) are useful to check whether data at the two limits are mutually consistent, assuming the data are extensive enough to permit the determination of the slopes M and N. Insofar as model-dependent parameters are concerned, low-pressure data yield E_0, and with a known E_0, high-pressure data then permit $\langle E_v^* \rangle$ to be calculated from Eq. (8-87) if all pertinent degrees of freedom are taken as vibrational, in accordance with the recommendation made above. With $\langle E_v^* \rangle$ known, and with r = r* = 0, Eq. (8-117) then yields Q_v^*. Thus the vibrational partition function of the transition state and its logarithmic derivative with respect to temperature are known, which puts some constraints on the permissible vibrational frequency assignment in the transition state.

8. Generalized Lindemann's mechanism and the nonequilibrium distribution

The theory of thermal reactions as outlined in Sections 1–7 accounts reasonably well for all experimental observations, except in the low-pressure

[28] The argument assumes that const is the same in both reactant and transition state, i.e., that both contain the same number of one-dimensional rotors; otherwise, still another assumption would be required to decide how many of the r* rotations are one-dimensional and how many belong to two-dimensional rotors. Symmetry numbers for I and I^* are assumed to have been included in reaction path degeneracy α (cf. Chapter 6, end of Section 2) and therefore are not considered here.

limit, where k_{bi} is found experimentally to depend on the nature of M much more than can be accounted for by the dependence of Z [Eq. (8-100)] on the nature of collision partners.[29] Clearly, then, our theoretical expression for k_0 [Eq. (8-43)] is deficient in some important way, since Z is the only M-dependent factor. This section is devoted to considerations that might remedy this discrepancy, which, as it turns out, are intimately connected with the strong collision assumption.

Collisions under more realistic potential The formulas (8-100) for Z give the collision number for two particles colliding under a hard-sphere potential, i.e., a potential that is zero everywhere except at interparticle distance d (mean collision diameter), where it is infinite. Alternatively, we might look at d as the critical impact parameter b such that there is unit probability of collision if $b < d$ and zero probability if $b > d$. The collision diameter, as usually determined from viscosity measurements or second virial coefficient data, is about 3–4 Å, so that Z for most collision partners is roughly the same,[30] which accounts for the insensitivity of our k_0 to the nature of M.

We might try to improve the formulation of Z by acknowledging that molecules are not hard spheres but interact under the influence of some more realistic potential. If we choose the effective potential (7–10) with $n = 6$, appropriate for two neutral species colliding, it can be shown (*46*, p. 144) that the collision number Z_{L-J} for molecules colliding under such potential is

$$Z_{L-J} = Z\left(\frac{2\mathscr{C}}{kT}\right)^{1/3} \frac{\Gamma(2/3)}{d^2} \tag{8-118}$$

The factor which multiplies Z arises from averaging over impact parameters, and since these are related to angular momenta, we get a factor analogous to f_∞ [Eq. (8-56)]. The constant $\mathscr{C}$ is customarily taken as the attractive part of a Lennard-Jones (6–12) potential (thus explaining the subscript L-J) rather than of a Sutherland potential as in Chapter 7; we then have to distinguish between the interparticle distance r_e at which is located the potential minimum (depth D_e), and the interparticle distance d at which the potential is zero while rising steeply; this is the analog of the hard-sphere diameter. For such a potential, we have (*47*, p. 322 and Fig. 15-2) $r_e = 2^{1/6}d$, so that we get from Eq. (7-7) for $m = 12$

$$\mathscr{C} = 4\epsilon d^6 \tag{8-119}$$

[29] For a review of extensive low-pressure experimental data in the unimolecular isomerization of CH_3NC, see Chan *et al.* (*45*).

[30] An exception is hydrogen because of its low mass.

where ϵ is written for D_e in deference to the customary notation used in the literature dealing with transport phenomena. After evaluation of numerical factors, (8-118) and (8-119) yield

$$Z_{L-J} = 2.708Z(\epsilon/kT)^{1/3} \tag{8-120}$$

so that the hard-sphere collision rate Z is increased by about a factor of two since usually $\epsilon < kT$.

Alternatively, using a somewhat more sophisticated nonequilibrium transport theory (*38*, Chapter 7), it is found that d^2 is increased by the factor $\Omega^{(2,2)}$, which is the collision integral that appears in the calculation of transport coefficients. It is tabulated as a function of the reduced temperature $T^\star = kT/\epsilon$ for the Lennard-Jones potential (*38*, Appendix I, Table M), applicable for interactions between nonpolar molecules, and also for a Stockmayer-type 6–12 potential (*38*, Appendix IX), applicable to interactions between polar molecules. Thus

$$Z_{L-J} = Z\Omega^{(2,2)} \tag{8-121}$$

In general, Z_{L-J} calculated by (8-121) is somewhat smaller than that given by Eq. (8-120), and either modifies the temperature dependence of Z. Since the literature data are mostly available only for pure gases, parameters for mixtures of two gases a and b are calculated from empirical combining rules (*38*, p. 68)

$$d_{ab} = \tfrac{1}{2}(d_a + d_b); \qquad \epsilon_{ab} = (\epsilon_a\epsilon_b)^{1/2} \tag{8-122}$$

The collision integral of a mixture is then taken to be $\Omega^{(2,2)}$ corresponding to $T^*_{ab} = kT/\epsilon_{ab}$.

We may now define an experimental, or pressure-for-pressure, efficiency β_p of gas M relative to reactant A as

$$\beta_p = (k_{bi})_M/(k_{bi})_A \tag{8-123}$$

where $(k_{bi})_M$ is the slope of k_0' versus pressure of M at constant reactant pressure, and similarly $(k_{bi})_A$ is the slope of k_0' versus reactant pressure. We may similarly define a relative collision number Z_{rel} by

$$Z_{rel} = (Z_{L-J})_M/(Z_{L-J})_A \tag{8-124}$$

where $(Z_{L-J})_M$ is calculated from (8-120) or (8-121) for collisions A–M, and $(Z_{L-J})_A$ for collisions A–A. If transport properties were the main factor determining the efficiency of M relative to A in transferring energy by collision, there should be some parallelism between β_p and Z_{rel}. Table 8-1 shows that there is not, whether Z_{rel} is calculated by (8-120) or (8-121), even if one takes into account uncertainties in the values of the transport parameters. With the benefit of hindsight, such a lack of correlation is not too surprising because

Z_{L-J} (or Z) involves the cross section for transfer of momentum, whereas we want the collision rate involving the cross section for interconversion of translational and vibrational–rotational energy, and the two kinds of cross section are likely to be different.

A quantity that is more closely related to the desired cross section for transfer of energy is obtained by dividing out the momentum dependence in β_p and defining a collision-per-collision efficiency β_c:

$$\beta_c = \beta_p / Z_{rel} \tag{8-125}$$

The efficiency β_c (or β_p) is obtained from $(k_{bi})_M$, which is the slope of k_0' versus [M] *at constant* [A] [cf. (8-78)], while k_0' is a function of the total pressure of [A] + [M]. The result is that β_c is a function of [A]/[M], i.e., of dilution (*48*), so that to get a dilution-independent quality, we have to extrapolate β_c as [A]/[M] $\rightarrow 0$, i.e., to infinite dilution, which is experimentally realized (see footnote 29) at [M]/[A] $\sim$ 50. The β_c values of Table 8-1 refer to infinite dilution.

A β_c that is less than one—and a glance at Table 8-1 shows that β_c for small molecules is always less than one, the underlying assumption being that A–A collisions have β_c of unity—suggests that the simple one-step Lindemann mechanism no longer applies; instead, various M apparently

TABLE 8-1

RELATIVE EFFICIENCIES OF M IN ISOMERIZATION OF CH_3NC AT 553°K[a]

M	β_p	Z_{rel}, Eq. (8-120)	Z_{rel}, Eq. (8-121)	β_c	Boiling point (°K)	$\langle \Delta E \rangle / kT$, Eq. (8-148)	$E_{a0} - E_{a0}^{\ddagger}$ (kcal mole^{-1}), Eq. (8-149)
CH_3NC	(1.00)	(1.00)	(1.00)	(1.00)	332	—	—
He	0.171	0.71	0.71	0.242	3.3	0.7	1.27
Xe	0.119	0.68	0.51	0.231	166	0.5	1.43
H_2	0.28	1.50	1.21	0.23	20.5	1.1	1.03
CO_2	0.32	0.77	0.59	0.54	195	1.3	0.95
CH_4	0.44	0.96	0.72	0.61	112	2.0	0.73
C_2F_6	0.43	0.60	0.67	0.65	194	1.9	0.74
C_3F_8	0.58	0.59	0.57	1.02	234	3.2	0.52
n-C_4F_{10}	0.62	0.57	0.62	1.00	272	3.7	0.46
n-C_5F_{12}	0.69	0.57	0.69	1.00	303	4.9	0.37
n-C_6F_{14}	0.75	0.56	0.75	1.01	330	6.7	0.28
n-$C_{10}H_{22}$	1.28	0.76	1.25	1.02	447	—	—

[a] Experimental values of β_p from Ref. 45. β_c based on Z_{rel} of Eq. (8-121); $\langle \Delta E \rangle$ is the average energy transferred per deactivating collision; $E_{a0} - E_{a0}^{\ddagger}$ is the decrease in experimental low-pressure activation energy relative to reactant.

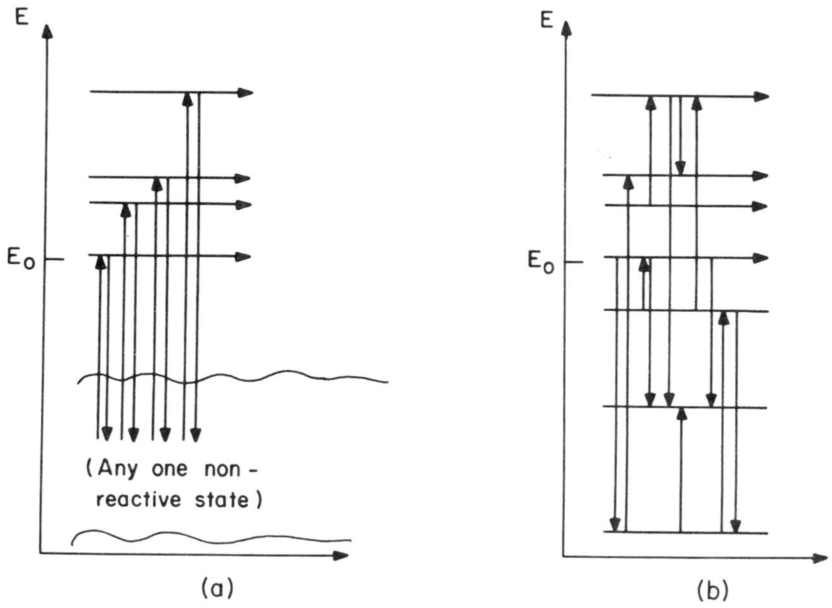

Fig. 8-3. *Schematic comparison of collisional one-step and multistep activation mechanisms.*
(a) Simple one-step Lindemann mechanism; (b) general multistep Lindemann mechanism. All states with $E > E_0$ are reactive states and horizontal arrows indicate reaction. Vertical arrows indicate transitions between levels of the system. Only some of all the possible transitions are shown for case (b).

remove varying amounts of energy, generally insufficient for deactivation, such that on the average, $1/\beta_c$ collisions are necessary to bring a molecule from a reactive state to a nonreactive one. Therefore we have to consider a more general Lindemann mechanism with a less coarse energy scale, one in which all the allowed energy levels are taken into consideration. The simple one-step Lindemann mechanism and the generalized multistep Lindemann mechanism are schematically compared in Fig. 8-3.

Stepwise activation The kinetic scheme for the multistep mechanism may be written as

$$\mathrm{A}_i + \mathrm{M} \underset{\omega\varphi_{ji}}{\overset{\omega\varphi_{ij}}{\rightleftharpoons}} \mathrm{A}_j + \mathrm{M}, \qquad \mathrm{A}_j \xrightarrow{k_j} \text{products} \qquad (8\text{-}126)$$

Here A_i and A_j represent reactant species A in states i and j, and φ_{ij} and φ_{ji} are probabilities of transition *per collision* between states i and j, or j and i, respectively. We may think of ω as $\omega = Z[\mathrm{M}]$, or as the more elaborate

$\omega = Z_{L-J}[M]$. The essential difference between the schemes (8-1)–(8-3) and (8-126) is that A_j does *not* necessarily represent a reactive state, and that A_i *may* be a reactive state; however, $k_j \neq 0$ only for reactive states, as before, and $k_j = 0$ otherwise. The mechanism (8-126) leads to the master equation

$$d[A_j]/dt = \omega \sum_i \varphi_{ij}[A_i] - \omega[A_j] \sum_i \varphi_{ji} - k_j[A_j] \qquad (8\text{-}127)$$

Essentially, this is a system of rate equations involving all input and output processes for each state j. A formal approximate solution (*49,50*) shows that after an induction period, analogous to the transient of Eq. (8-5), the process is described by a single, time-independent, steady-state rate constant which is the lowest eigenvalue of a relaxation matrix that describes all the microscopic transitions among all the levels of the system undergoing reaction. A more detailed solution of (8-127) is very difficult (*51,52*) because, first, a priori calculation of the matrix coefficients φ_{ij} and φ_{ji} is involved even in simple cases (*51–53*); second, the matrix is very large. However, it is possible to obtain a fairly straightforward solution under two simplifying assumptions:

1. If colliding partner M is present in sufficient excess so that it has a fixed internal state, unchangeable by collision, we have a heat reservoir whose properties remain constant in time, and as a result, the transition probabilities φ_{ij} and φ_{ji} are also time independent.[31] It is further necessary to assume that the diluent M is itself so dilute that only binary collisions occur, and that the time between collisions is long enough compared to the period of molecular rotations and vibrations. These conditions are satisfied by a unimolecular reaction in the second-order region at infinite dilution.

2. It is assumed that the reaction proceeds essentially in a steady state, i.e., that the transient dies out quickly. With reference to Eq. (8-127), this means that $d[A_j]/dt = 0$.

For reactive states defined by energy, as introduced in Chapter 2 and used since, states i and j are energy states of energies E_i and E_j, respectively; in particular, k_j becomes $k(E_j)$, which is zero unless $E_j \geqslant E_0$. We consider only distribution of energies, inasmuch as distribution of angular momenta can always be taken into account ex post facto by multiplying the end result by the appropriate centrifugal correction factor, as done in Section 3 of this chapter. Since in polyatomic molecules, energy states are densely spaced, we may replace summations over i by integrations with respect to E_i.

The problem of stepwise activation followed by dissociation as described by the master equation (8-127) may be considered as a multistate relaxation of a system containing a negative source (or sink) of particles, and can be

[31] An exact solution for this case is given by Cooper and Hoare (*54*) using a transition probability derived on the assumption that in each collision, the internal energy is randomly redistributed between the collision partners. Cf. (*55*).

described as a random walk through the range of bound states involving an absorbing barrier.[32] The solution may be sought either by deriving an expression for the mean first passage time, i.e., the time a random walker takes to reach the absorbing barrier for the first time, i.e., or by finding a nonequilibrium steady-state distribution function of the multistage process [called S(E) later] to replace $P(E)_{ss}$ [Eq. (8-31)] of the simple one-stage process. The latter approach is more convenient for the present purpose (*21, 56,57,58*, p. 30 ff., 118ff.; *59–61*); for matrix formulation see (*94*).

The nonequilibrium distribution Returning to Eq. (8-127), we now define a distribution function for molecules of energy E_k:

$$S(E_k) = [A(E_k)] \Big/ \sum_i [A(E_i)], \qquad k = i, j \qquad (8\text{-}128)$$

where $[A(E_k)]$ is the instantaneous concentration of molecules A with energy E_k. Under steady-state conditions and with time-independent coefficients q_{ij} and q_{ji}, Eq. (8-127) becomes, taking $\{\sum_i [A(E_i)]\}^{-1} d[A(E_i)]/dt \approx 0$

$$\omega \int_0^\infty q(E_i, E_j)S(E_i)\, dE_i - \omega S(E_j) \int_0^\infty q(E_j, E_i)\, dE_i - k(E_j)S(E_j) = 0 \quad (8\text{-}129)$$

from which

$$S(E_j) = \frac{\omega \int_0^\infty q(E_i, E_j)S(E_i)\, dE_i}{\omega \int_0^\infty q(E_j, E_i)\, dE_i + k(E_j)} \qquad (8\text{-}130)$$

By the principle of detailed balancing, the probabilities for activation and deactivation must be related by

$$q(E_i, E_j)P(E_i)_e = q(E_j, E_i)P(E_j)_e, \qquad E_j > E_i \qquad (8\text{-}131)$$

where $P(E_k)_e$ $(k = i, j)$ is the equilibrium distribution function (8-13). If we now define a new distribution function $H(E_k)$,

$$H(E_k) = S(E_k)/P(E_k)_e, \qquad k = i, j \qquad (8\text{-}132)$$

which measures the fractional deviation S from the equilibrium value, then after eliminating $q(E_i, E_j)$ by means of (8-131), (8-130) yields

$$H(E_j) = \frac{\omega \int_0^\infty q(E_j, E_i)H(E_i)\, dE_i}{\omega \int_0^\infty q(E_j, E_i)\, dE_i + k(E_j)} \qquad (8\text{-}133)$$

---

[32] For a recent review, see Oppenheim *et al.* (*56*).

where we now have the transition probability $\mathscr{P}(E_j, E_i)$ in both numerator and denominator.

At high pressure, $k(E_j) \ll \omega \int_0^\infty \mathscr{P}(E_j, E_i)\, dE_i$ for all $E_j \geqslant E_0$; the solution of (8-132) is then the trivial one $\mathsf{H}(E_j) = 1$, i.e., $\mathsf{S}(E_j) = P(E_j)_e$ or, in other words, the distribution is the equilibrium distribution, as we of course already know. More interesting is the form that $\mathsf{H}(E_j)$ takes at the other pressure extreme when $k(E_j) \gg \omega \int_0^\infty \mathscr{P}(E_j, E_i)\, dE_i$ for all $E_j \geqslant E_0$. It is now convenient to proceed in two steps, first taking into account nonreactive states [$E_j \leqslant E_0$, for which $k(E_j) = 0$], and then the reactive states. For j states which have energies below reaction threshold, (8-133) becomes

$$\mathsf{H}(E_j) = \frac{\int_0^{E_0} \mathscr{P}(E_j, E_i)\mathsf{H}(E_i)\, dE_i}{\int_0^\infty \mathscr{P}(E_j, E_i)\, dE_i}, \qquad E_j \leqslant E_0 \tag{8-134}$$

while above reaction threshold, we have

$$\mathsf{H}(E_j) = [\omega/k(E_j)] \int_0^{E_0} \mathscr{P}(E_j, E_i)\mathsf{H}(E_i)\, dE_i \qquad E_j \geqslant E_0 \tag{8-135}$$

The integration in the numerators of (8-134) and (8-135) is taken only to threshold since in the low-pressure limit, every molecule with energy of at least E_0 is assured of dissociation, and therefore the population of states above E_0 is small.

The transition probability The distribution function $\mathsf{H}(E_j)$ can be obtained from the integral equation (8-134) if the transition probability $\mathscr{P}(E_j, E_i)$ is known. At low levels of excitation, involving the first few vibrational levels, as in ultrasonic dispersion, theoretical expressions for the transition probability are available (*62–64*) which are in reasonable agreement with experiment (*65–67*).[33] In a unimolecular reaction, however, the level of excitation is high, involving as it does vibrational levels near the dissociation limit, and here the low-level results are inapplicable. Stochastic analysis of experimental results on the deactivation efficiency β_c yields $\mathscr{P}(E_j, E_i)$ if $\langle \Delta E \rangle$, the average amount of energy transferred per deactivating collision, is postulated, or, inversely, $\langle \Delta E \rangle$ can be obtained for an assumed form of $\mathscr{P}(E_j, E_i)$, but it is not possible to deduce both $\langle \Delta E \rangle$ and $\mathscr{P}(E_j, E_i)$ from experiment (*48,69*). We shall therefore have to *assume* a form for $\mathscr{P}(E_j, E_i)$, and following Troe and Wagner (*21,60,61*), we shall take an exponential (*72*) form[34] such that the transition probability depends only on the amount of energy transferred

[33] For a review, see Cottrell and McCoubrey (*68*).

[34] There are indications that for inefficient deactivators, like He, H_2, CO, and other small molecules, the collisional transition probability is indeed exponential, while for

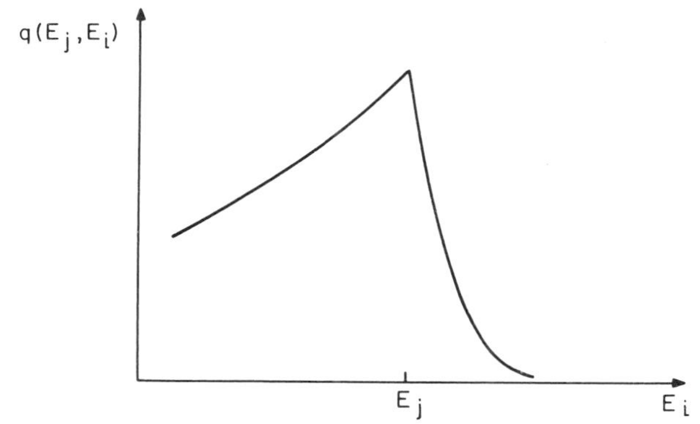

Fig. 8-4. *Schematic drawing of the transition probability $q(E_j, E_i)$ of Eq. (8-136).* Above E_j: transition probability for activation ($E_i > E_j$). Below E_j: transition probability for deactivation ($E_i < E_j$). Because of the additional factor $e^{-\Delta E/kT}$, the transition probability for activation drops off faster as a function of ΔE than the probability for deactivation; this must be so to satisfy detailed balance. The area under the curve above E_j is $kT/(2kT + \gamma)$, and the area under the curve below E_j is $(kT + \gamma)/(2kT + \gamma)$, so that the combined area is unity. "Below E_j" is to be understood as $-\infty \leqslant E_i \leqslant E_j$; "above" as $E_j \leqslant E_i \leqslant +\infty$ (cf. footnote 35). Note that the probability has a maximum at $\Delta E = E_j - E_i = 0$, i.e. for an *elastic* collision.

and is otherwise independent of the energy of the initial state:

$$q(E_j, E_i) = \frac{1 + (\gamma/kT)}{\gamma[2 + (\gamma/kT)]} \exp\left(-\frac{|E_j - E_i|}{\gamma}\right) \qquad \text{(deactivation, } E_j > E_i\text{)}$$

$$\text{(8-136)}$$

$$q(E_j, E_i) = \frac{1 + (\gamma/kT)}{\gamma[2 + (\gamma/kT)]} \exp\left[-|E_j - E_i|\left(\frac{1}{\gamma} + \frac{1}{kT}\right)\right] \text{(activation, } E_i > E_j\text{)}$$

This transition probability is shown graphically in Fig. 8-4. It drops off more rapidly for activation, but in the limit $\gamma \ll kT$, becomes symmetric. The probability satisfies completeness (*51,52*) and is therefore normalized to unity, i.e., the area under the curve of Fig. 8-4 is equal to unity.[35] The

---

more efficient, larger deactivators, it is perhaps Gaussian. See Georgakakos *et al.* (*70*); Rynbrandt and Rabinovitch (*71*). The term "Gaussian" is actually somewhat misleading in that the essential difference between the two transition probabilities is that in the exponential case the probability has a maximum at $\Delta E = 0$ (see legend of Fig. 8-4), while in the Gaussian case the probability is symmetric about a maximum located at $\Delta E \neq 0$, i.e. unlike the exponential case, there is now maximum probability for an *inelastic* collision.

[35] The transition probability (8-136) satisfies the completeness requirement in the form

$$\int_{-\infty}^{E_j} q(E_j, E_i)\, dE_i + \int_{E_j}^{\infty} q(E_j, E_i)\, dE_i = 1$$

where the lower integration limit of minus infinity on the first integral is a mathematical necessity, although a physically meaningful lower limit is obviously $E_i = 0$.

average energy transferred per deactivating collision is, with $\Delta E = E_j - E_i$,

$$\langle \Delta E \rangle = \left[\int_0^\infty (\Delta E) e^{-\Delta E/\gamma}\, d(\Delta E)\right] \Big/ \int_0^\infty e^{-\Delta E/\gamma}\, d(\Delta E) = \gamma \tag{8-137}$$

and the average energy transferred per activating collision is

$$\langle \Delta E \rangle_a = \frac{\int_0^\infty \Delta E \exp(-\gamma' \Delta E)\, d(\Delta E)}{\int_0^\infty \exp(-\gamma' \Delta E)\, d(\Delta E)} = \frac{\gamma k T}{\gamma + kT}, \qquad \gamma' = \frac{1}{\gamma} + \frac{1}{kT} \tag{8-138}$$

The transition probabilities (8-136) also satisfy detailed balance in the form

$$\mathcal{P}(E_i, E_j) e^{-E_i/kT} = \mathcal{P}(E_j, E_i) e^{-E_j/kT} \tag{8-139}$$

provided that

$$N(E_i) \approx N(E_j) \tag{8-140}$$

If we take $E_j = E_0$, $E_j - E_i = \langle \Delta E \rangle_a$, we find that

$$\frac{N(E_i)}{N(E_0)} = 1 - \frac{\langle \Delta E \rangle_a}{N(E_0)} \left[\frac{dN(E)}{dE}\right]_{E=E_0} \tag{8-141}$$

and therefore (8-139) and (8-140) is satisfied as long as

$$\frac{\langle \Delta E \rangle_a}{N(E_0)} \left[\frac{dN(E)}{dE}\right]_{E=E_0} \ll 1 \tag{8-142}$$

which is generally true for a molecule that is not too large.[36]

The form (8-136) of the transition probability is admittedly crude, and has been chosen mainly to permit a relatively straightforward solution of the integral equation (8-134). However it does embody the principal conclusion of calculations involving diatomic molecules in lower vibrational levels (see footnote 38), viz., that $\mathcal{P}(E_j, E_i)$ should be some sort of exponentially decreasing function of $|E_j - E_i|$.

The solution Solution[37] of the integral equation (8-134) yields for the distribution function $\mathsf{H}(E)$, after dropping the subscript j on E_j (Exercise 3),

$$\mathsf{H}(E) = 1 - \frac{e^{-(E_0 - E)/kT}}{1 + (\gamma/kT)} \qquad (E \leqslant E_0) \tag{8-143}$$

[36] Using the semiclassical expression (6-56) for $N(E)$, we have

$$[\langle \Delta E \rangle_a / N(E_0)][dN(E)/dE]_{E=E_0} \approx (\mathsf{v} - 1)\langle \Delta E \rangle_a / (E_0 + aE_z)$$

which will be small if v (i.e., the molecule) is not too large.

[37] The author is indebted to Prof. J. Troe for correspondence on this matter.

and

$$\mathsf{H}(E) = \frac{\omega e^{-(E-E_0)/\gamma}}{k(E)[1 + (\not{k}T/\gamma)]} \qquad (E \geqslant E_0) \qquad \text{(8-144)}$$

The form (8-143) of $\mathsf{H}(E)$ satisfies the boundary condition $\mathsf{H}(0) = 1$, meaning that the system in ground state has the equilibrium distribution, and both forms of $\mathsf{H}(E)$ satisfy $\{\partial[\ln \mathsf{H}(E)]/\partial E\}_{E=E_0} = -1/\gamma$, which assures the continuity of the logarithmic derivative at $E = E_0$, so that the two forms of $\mathsf{H}(E)$ join smoothly at threshold. It is interesting to note that Levitt (*73*) has found that the deviation from equilibrium in the dissociation of NO_2 behind a shock front is roughly of the form (8-143). If $\gamma \gg \not{k}T$, the amount of energy transferred per collision is large, and $\mathsf{H}(E)$ of Eq. (8-143) is unity, i.e., the distribution function below threshold is the equilibrium distribution, as of course we would expect since we are then back to the simple one-step Lindemann mechanism of activation.

Using $\mathsf{H}(E)$ of Eq. (8-144), the low-pressure rate constant becomes

$$k_0{}^{\dagger} = \int_{E_0}^{\infty} k(E)\mathsf{S}(E)\,dE = \int_{E_0}^{\infty} k(E)P(E)_{\mathrm{e}}\mathsf{H}(E)\,dE$$

$$= \frac{\omega \exp(E_0/\gamma)}{Q[1 + (\not{k}T/\gamma)]} \int_{E_0}^{\infty} N(E) \exp(-\gamma' E)\,dE, \qquad \gamma' = \frac{1}{\gamma} + \frac{1}{\not{k}T} \qquad \text{(8-145)}$$

where the dagger superscript will be used henceforth to distinguish nonequilibrium quantities from equilibrium ones. The integral in (8-145) can be easily evaluated numerically using a "good" $N(E)$, but it is illustrative to evaluate it approximately by the same method used to derive Eq. (8-43a); the result is (cf. Exercise 4)

$$k_0{}^{\dagger} \approx \frac{\omega \not{k}T N(E_0) e^{-E_0/\not{k}T}}{Q[1 + (\not{k}T/\gamma)]^2} \qquad \text{(8-146)}$$

which can be written as

$$\frac{k_0{}^{\dagger}}{k_0} \approx \frac{1}{[1 + (\not{k}T/\gamma)]^2} \qquad \text{(8-147)}$$

In other words, the nonequilibrium rate is smaller than k_0, and the two become equal in the limit as $\gamma \gg \not{k}T$.

We can now establish a connection with β_c of Eq. (8-125). Assuming that centrifugal effects approximately cancel out, i.e., that $f_0{}^{\dagger} \sim f_0$, we have

$$\beta_c = k_0{}^{\dagger}/Z_{\mathrm{rel}}k_0 \approx \{Z_{\mathrm{rel}}[1 + (\not{k}T/\gamma)]^2\}^{-1} \qquad \text{(8-148)}$$

since by definition the reactant itself behaves as a strong collider. Using the experimental value of β_c, we can calculate γ from (8-148), subject of course to the validity of the transition probability model and all the other approximations. One thus finds from the data in Table 8-1, using Z_{rel} based on Eq. (8-121), that $\gamma \equiv \langle \Delta E \rangle$ varies from 270 cm^{-1} (= 0.7$\mathscr{k}T$) for a weak collider like He, to 1400 cm^{-1} (= 3.7$\mathscr{k}T$) for a strong collider like n-C_4F_{10}. These numbers are roughly what has been found (*48*) for the exponential transition probability model in the CH_3NC system by a more complicated analysis of the data.

If we put $\mathscr{k}T^2\, \partial(\ln k_0{}^{\dagger})/\partial T = E^{\dagger}_{a0}$, we find from (8-148) by differentiating with respect to T,

$$\mathscr{k}T^2\, \partial(\ln \beta_c)/\partial T = E^{\dagger}_{a0} - E_{a0} \approx -2(\mathscr{k}T)^2/(\gamma + \mathscr{k}T) \tag{8-149}$$

if we assume that the temperature coefficient of Z_{rel} is small [it is zero for Z_{rel} expressed by means of Z_{L-J} of Eq. (8-120)]. Thus $E^{\dagger}_{a0}$, the low-pressure experimental activation energy for activation with a weak collider, should be smaller than the corresponding activation energy for a strong collider like the reactant itself. For the cited case of He ($\gamma = 0.7\mathscr{k}T$), as heat bath molecule in the CH_3NC system, Eq. (8-149) gives $E^{\dagger}_{a0} - E_{a0} = -1.27$ kcal mole^{-1}, compared with the experimental value (*74*) ~ -1.5 kcal mole^{-1}. Experimental evidence from the few other systems where activation efficiency has been determined as a function of temperature is less clear: In the low-pressure unimolecular decomposition of NOCl, there was no discernible effect (*75–78*), within experimental error, on the experimental activation energy with M = NOCl, N_2, NO, or H_2; in the case of N_2O, experimental activation energy was found (*79,80*) to increase slightly with M = CF_4, CO_2, and SO_2, and in the case of hydrogen peroxide (*81*), although the experimental activation energies are all roughly equal within the rather large experimental error, it seems nevertheless that they decrease in the order M = $H_2O_2 > H_2O > N_2 > Ar \sim He$, which is the order of collisional activation efficiencies. The common difficulty is that the difference $E^{\dagger}_{a0} - E_{a0}$ is quite small, often less than experimental error in the activation energies, so that it takes an exceptionally well-behaved system and very careful work to keep the experimental error within a few hundred calories [cf. *95*, *96*].

The arguments presented thus far permit us to analyze experimental data on the basis of the assumed transition probability model but do not throw much light on the more fundamental question of what it is on the molecular level that governs the form of the transition probability. It is obviously of more than passing interest to know if small amounts of energy may be more likely to be transferred than large ones, as on the exponential transition probability model, which appears to fit small and inefficient colliders, and why, inversely, large amounts of energy seem to be more likely to be transferred than small ones in the case of large and efficient colliders which appear to follow the Gaussian probability (see footnote 34). The problem is essentially one of collision dynamics as it affects the conversion of translational-to-vibrational energy for monoatomic heat bath gases, and of vibrational-to-vibrational energy for polyatomic heat bath gases.[38] The difficulty here is very much the same as with unimolecular rate theory itself: For physically meaningful systems, the collision dynamics in highly excited reactive systems is too complicated,[39] and therefore it is necessary to forego much of the detail.

If one seeks a correlation between β_c and some molecular property of the heat bath gas, it is found (*45*) that the only property that correlates reasonably well is the boiling point (cf. Table 8-1). This suggests that some sort of weak binding is involved between substrate and heat bath molecules, perhaps in the form of a collision complex which involves statistical redistribution of internal energy. This idea has been around for some time (*84–86*), and Hoare (*55*) has used it to make some calculations of the transition probability for a few simple cases. More recently, Rabinovitch *et al.* (*87,88*) have treated the CH_3NC data on this model.

In essence, the model focuses attention on the new vibrational modes that the collision complex (A–M) possesses relative to A itself. These modes [called transitional modes (*87,88*)] correlate with relative translational and rotational degrees of freedom of the separating partners A + M; conservation of angular momentum limits the allowable energy in these modes, and thus emerges as an important factor governing the overall efficiency of M. If M is polyatomic, internal modes of M may also become involved in the internal redistribution, resulting in much increased efficiency of M. Calculations of the transition probabilities on the basis of this model (excluding internal modes of M) indicate that for inefficient (monoatomic) M, small down-transitions are more probable than large ones (exponential probability), while for more efficient (polyatomic) M, the distribution is more nearly Gaussian; this is roughly in accord with experimental observations (see footnote 34).

Exercises

1. Show that if the classical expression (6-55) is used for $N(E)$, Eq. (8-43) yields the Hinshelwood formula

$$k_0 \approx \frac{\omega}{\Gamma(v)}\left(\frac{E_0}{kT}\right)^{v-1} e^{-E_0/kT}$$

[38] For a review, see Rapp and Kassal (*82*).

[39] For a qualitative discussion of some of the factors involved, see Pritchard (*83*, p. 368, Chapter 5).

Use Hint of Exercise 2 in Chapter 1. See also Fowler and Guggenhein (*89*, p. 495 ff.). For numerical data, see Sivertz and Goldsack (*90*).

2. Show that for a collection of v classical oscillators, the so-called Kassel integral is

$$\frac{k_{\text{uni}}}{k_{\infty}} = \frac{1}{\Gamma(\text{v})} \int_0^{\infty} \frac{x^{\text{v}-1} e^{-x}\, dx}{1 + (A_{\infty}/\omega)[x/(x+b)]^{\text{v}-1}}$$

where $A_{\infty} = \prod^{v} \nu_i / \prod^{v-1} \nu_i^*$ and $b = E_0/kT$. Use Eq. (8-24) for k_{∞} and Eq. (8-51) for k_{uni}, assuming $f_{\infty} = f_0 = 1$, in both and taking the result of Exercise 1 in Chapter 4 for $k(E)$. Numerical values of the Kassel integral have been tabulated by Emanuel (*91*).

3. Solve Eq. (8-134) for the transition probability (8-136). We shall give an outline of the solution. Since $q(E_j, E_i)$ of Eq. (8-136) is normalized (between $-\infty$ and $+\infty$), we need not worry about the denominator of (8-134). Let us write (8-134) in the form

$$f(x) = C \int_{-\infty}^{x} f(y) \exp\,[-(x-y)/\gamma]\, dy$$
$$+ C \int_{x}^{E_0} f(y) \exp\,[-(y-x)\gamma']\, dy$$

where $x = E_j$, $y = E_i$. Differentiation with respect to x brings this to the form

$$f''/f' = (kT)^{-1}$$

where the prime and double prime indicate first and second derivatives, respectively. The solution of this differential equation is $f(x) = C_1 e^{x/kT} + C_2$, where C_1 and C_2 are constants to be determined from the boundary conditions. These are (see text): (i) $f(0) = 1$, from which $C_2 = 1 - C_1$; and (ii) $(df/dx)_{x=E_0} = -f(E_0)/\gamma$; which yields

$$C_1 = \{1 - [1 + (\gamma/kT)]e^{E_0/kT}\}^{-1}$$

and Eq. (8-143) in the text follows with $x \equiv E$. Substituting (8-143) into (8-135), (8-144) is obtained if the integration limits in (8-135) are taken $(-\infty, E_0)$.

4. (a) Starting with Eq. (8-130), show that $k_0^{\dagger}$ of Eq. (8-145) can also be written as the "upward flow"

$$k_0^{\dagger} = \omega \int_{E_0}^{\infty} \left\{ \int_0^{E_0} q(E_i, E_j) \mathrm{S}(E_i)\, dE_i \right\} dE_j$$

For a slightly different derivation of this result see Bak and Lebowitz (*92*).

(b) Using (8-136) for $\mathscr{q}(E_i, E_j)$ and (8-143) for $\mathsf{H}(E_i)$ $[=\mathsf{S}(E_i)/P(E_i)_e]$, show that the above expression for $k_0{}^{\dagger}$ leads to Eq. (8-146) if the lower limit of the first integration is extended to $-\infty$. Use $P(E_i)_e \approx N(E_0)e^{-E_i/kT}/Q$. Note that now $k_0{}^{\dagger}$ is represented as the "downward flow."

References

1. F. Daniels, "Chemical Kinetics." Cornell Univ. Press, Ithaca, New York, 1938.
2. F. A. Lindemann, *Trans. Faraday Soc.* **17**, 598 (1922).
3. C. J. Jachimowski and M. E. Russell, *Z. Phys. Chem.* (*Frankfurt*) **48**, 102 (1966).
4. H. K. Shin and J. C. Giddings, *J. Phys. Chem.* **65**, 1164 (1961).
5. E. Thiele, *J. Chem. Phys.* **45**, 491 (1966).
6. R. D. Levine, *J. Chem. Phys.* **48**, 4556 (1968).
7. P. M. Morse, "Thermal Physics." Benjamin, New York, 1965.
8. W. Forst and P. St. Laurent, *Can. J. Chem.* **45**, 3169 (1967).
9. N. B. Slater, "Theory of Unimolecular Reactions." Cornell Univ. Press, Ithaca, New York, 1959.
10. D. L. Bunker, *J. Chem. Phys.* **37**, 393 (1962).
11. D. L. Bunker, *J. Chem. Phys.* **40**, 1946 (1964).
12. R. C. Tolman, "The Principles of Statistical Mechanics." Oxford Univ. Press, London and New York, 1938.
13. O. K. Rice, "Statistical Mechanics, Thermodynamics, and Kinetics." Freeman, San Francisco, California, 1967.
14. J. Glasstone, K. J. Laidler, and H. Eyring, "The Theory of Rate Processes." McGraw-Hill, New York, 1941.
15. N. B. Slater, *Phil. Trans. Roy. Soc.* (*London*) **A246**, 57 (1953).
16. O. K. Rice and H. C. Ramsperger, *J. Amer. Chem. Soc.* **49**, 1617 (1927).
17. F. P. Buff and D. J. Wilson, *J. Chem. Phys.* **45**, 1447 (1966).
18. R. Paul and D. A. Coombe, *J. Chem. Phys.* **54**, 416 (1971).
19. F. W. Schneider and B. S. Rabinovitch, *J. Amer. Chem. Soc.* **84**, 4215 (1962).
20. W. Forst, *J. Chem. Phys.* **48**, 3665 (1968).
21. J. Troe and H. Gg. Wagner, *Ber. Bunsenges.* **71**, 937 (1967).
22. W. Forst, *Chem. Phys. Lett.* **1**, 687 (1968).
23. E. V. Waage and B. S. Rabinovitch, *Chem. Rev.* **70**, 377 (1970).
24. H. W. Dwight, "Tables of Integrals and Other Mathematical Data." Macmillan, New York, 1965.
25. E. V. Waage and B. S. Rabinovitch, *J. Chem. Phys.* **52**, 5581 (1970).
26. E. V. Waage and B. S. Rabinovitch, *J. Chem. Phys.* **53**, 3389 (1970).
27. H. S. Johnston, *J. Chem. Phys.* **20**, 1103 (1952).
28. H. Margenau and G. M. Murphy, "The Mathematics of Physics and Chemistry," 2nd ed. Van Nostrand Reinhold, Princeton, New Jersey, 1956.
29. R. C. Tolman, *J. Amer. Chem. Soc.* **42**, 2506 (1920).
30. W. Forst, *J. Chem. Phys.* **44**, 2349 (1966).
31. T. Shimanouchi, "Tables of Molecular Vibrational Frequencies," Parts I, II, and III. NSRDS-NBS, Nos. 6, 11, and 17, respectively. Washington, D.C., 1967–68.

32. M. S. Cord, J. D. Petersen, M. S. Lojko, and R. H. Haas, "Microwave Spectral Tables," Vols. I, III, and IV. NBS Monographs 70. Washington, D.C., 1968–69.
33. "Tables of Interatomic Distances and Configurations in Molecules and Ions." Chem. Soc. Special Publ. 11, 1958.
34. "Tables of Interatomic Distances and Configurations in Molecules and Ions." Chem. Soc. Special Publ. 18, 1956.
35. J. Hilsenrath and G. G. Ziegler, "Tables of Einstein Functions. Vibrational Contributions to the Thermodynamic Functions." NBS Monograph 49, Washington, D.C., 1962.
36. M. Abramowitz and I. A. Stegun, eds., "Handbook of Mathematical Functions." NBS Appl. Math. Ser. No. 55, Washington, D.C., 1964.
37. K. J. Laidler, "Chemical Kinetics." McGraw-Hill, New York, 1965.
38. J. O. Hirschfelder, C. F. Curtiss, and R. B. Bird, "Molecular Theory of Gases and Liquids." Wiley, New York, 1964.
39. S. W. Benson, "Thermochemical Kinetics." Wiley, New York, 1968.
40. S. W. Benson and G. N. Spokes, *J. Amer. Chem. Soc.* **89**, 2525 (1967).
41. F. J. Fletcher, B. S. Rabinovitch, K. W. Watkins, and D. J. Locker, *J. Phys. Chem.* **70**, 2823 (1966).
42. I. Oref and B. S. Rabinovitch, *J. Phys. Chem.* **72**, 2, 4488 (1968).
43. W. Forst, *J. Phys. Chem.* **76**, 342 (1972).
44. W. Forst, *in* "Reaction Transition States" (J. E. Dubois, ed.). Gordon and Breach, New York, 1972.
45. S. C. Chan, B. S. Rabinovitch, J. T. Bryant, L. D. Spicer, T. Fujimoto, Y. N. Lin, and S. P. Pavlou, *J. Phys. Chem.* **74**, 3160 (1970).
46. H. S. Johnston, "Gas Phase Reaction Rate Theory." Ronald Press, New York, 1966.
47. N. Davidson, "Statistical Mechanics." McGraw-Hill, New York, 1962.
48. D. C. Tardy and B. S. Rabinovitch, *J. Chem. Phys.* **48**, 1282 (1968).
49. W. G. Valance and E. W. Schlag, *J. Chem. Phys.* **45**, 216 (1966).
50. W. G. Valance and E. W. Schlag, *J. Chem. Phys.* **45**, 4280 (1966).
51. E. W. Montroll and K. E. Shuler, *Advan. Chem. Phys.* **1**, 361 (1958).
52. B. Widom, *Advan. Chem. Phys.* **5**, 353 (1963).
53. E. E. Nikitin and G. H. Kohlmaier, *Ber. Bunsenges.* **72**, 1021 (1968).
54. R. D. Cooper and M. R. Hoare, *Chem. Phys. Lett.* **12**, 123 (1971).
55. M. Hoare, *Mol. Phys.* **4**, 465 (1961).
56. I. Oppenheim, K. E. Shuler, and G. H. Weiss, *Advan. Mol. Relaxation Processes* **1**, 13 (1967).
57. A. I. Osipov and E. V. Stupochenko, *Sov. Phys.-Usp.* **6**, 47 (1963). E. E. Nikitin, *Teoret. Eksperim. Khim.* **2**, 19 (1966).
58. E. E. Nikitin, "Theory of Thermally Induced Gas Phase Reactions." Indiana Univ. Press, Bloomington, Indiana, 1966.
59. J. Troe, *Ber. Bunsenges.* **72**, 908 (1968).
60. J. Troe and H. Gg. Wagner, *in* "Recent Advances in Aerothermochemistry" (I. Glassman, ed.), Vol. 1, p. 21. AGARD Conf. Proc. No. 12, 1967.
61. H. A. Olschewski, J. Troe, and H. Gg. Wagner, *Ber. Bunsenges.* **70**, 450 (1966).
62. L. Landau and E. Teller, *Phys. Z. Sowjetunion* **10**, 34 (1936).
63. Z. Slawsky, R. N. Schwartz, and K. F. Herzfeld, *J. Chem. Phys.* **20**, 1591 (1952).
64. R. N. Schwartz and K. F. Herzfeld, *J. Chem. Phys.* **22**, 767 (1954).
65. B. Stevens, "Collisional Activation in Gases." Pergamon, Oxford, 1967.
66. G. H. Kohlmaier, *in* "Chemische Elementarprozesse" (H. Hartmann, ed.), p. 117. Springer-Verlag, Berlin, 1968.

67. A. B. Callear, *Appl. Opt. Suppl.* **2**, 145 (1965).
68. T. L. Cottrell and J. C. McCoubrey, "Molecular Energy Transfer in Gases." Butterworths, London and Washington, D.C., 1961.
69. D. C. Tardy and B. S. Rabinovitch, *J. Chem. Phys.* **45**, 3720 (1966).
70. J. H. Georgakakos, B. S. Rabinovitch, and E. J. McAlduff, *J. Chem. Phys.* **52**, 2143 (1970).
71. J. D. Rynbrandt and B. S. Rabinovitch, *J. Phys. Chem.* **74**, 1679 (1970).
72. F. P. Buff and D. J. Wilson, *J. Chem. Phys.* **32**, 677 (1960).
73. B. P. Levitt, *Trans. Faraday Soc.* **59**, 59 (1963).
74. S. C. Chan, J. T. Bryant, and B. S. Rabinovitch, *J. Phys. Chem.* **74**, 2055 (1970).
75. P. G. Ashmore and M. S. Spencer, *Trans. Faraday Soc.* **55**, 1868 (1959).
76. P. G. Ashmore and M. G. Burnett, *Trans. Faraday Soc.* **57**, 1315 (1961).
77. P. G. Ashmore and M. G. Burnett, *Trans. Faraday Soc.* **58**, 1801 (1962).
78. W. Forst and P. St. Laurent, *J. Chim. Phys.* **67**, 1018 (1970).
79. T. N. Bell, P. L. Robinson, and A. B. Trenwith, *J. Chem. Soc.* 1440 (1955).
80. T. N. Bell, P. L. Robinson, and A. B. Trenwith, *J. Chem. Soc.* 1474 (1957).
81. A. Tessier and W. Forst, *Advan. Mass Spectrom.* **6**, 281 (1971).
82. D. Rapp and T. Kassal, *Chem. Rev.* **69**, 61 (1969).
83. H. O. Pritchard, *in* "Transfer and Storage of Energy by Molecules" (G. M. Burnett and A. M. North, eds.). Wiley, New York, 1969.
84. B. Stevens and M. Boudart, *Ann. N. Y. Acad Sci..* **67**, 570 (1957).
85. R. E. Harrington, B. S. Rabinovitch, and M. R. Hoare, *J. Chem. Phys.* **33**, 744 (1960).
86. B. Stevens, *Mol. Phys.* **3**, 589 (1960).
87. Y. N. Lin and B. S. Rabinovitch, *J. Phys. Chem.* **74**, 3151 (1970).
88. F. M. Wang, T. Fujimoto, and B. S. Rabinovitch, *J. Phys. Chem.* **76**, 1935 (1972).
89. R. Fowler and E. A. Guggenheim, "Statistical Thermodynamics." Cambridge Univ. Press, London and New York, 1960.
90. C. Sivertz and D. Goldsack, *J. Chem. Phys.* **36**, 569 (1962).
91. G. Emanuel, *Int. J. Chem. Kinet.* **4**, 591 (1972).
92. T. A. Bak and J. L. Lebowitz, *Phys. Rev.* **131**, 1138 (1963).
93. H. M. Frey, R. G. Hopkins, and I. C. Vinall, *J.C.S. Faraday I* **68**, 1874 (1972).
94. R. G. Gilbert and I. G. Ross, *J. Chem. Phys.* **57**, 2299 (1972).
95. R. G. Gilbert, *Chem. Phys. Lett.* **11**, 146 (1971).
96. D. C. Tardy, *Chem. Phys. Lett.* **17**, 431 (1972).

CHAPTER 9

Chemical Activation Systems

In thermal systems, the deposition of energy in reactant molecules is left to the chance of molecular encounters. By contrast, chemical activation is a technique whereby the energy necessary for reaction is deposited in the reactant molecules in the course of their production by chemical reaction. For example, vibrationally excited *sec*-butyl radicals can be prepared by the reaction (*1*)[1]

$$\text{H} + \text{butene} \longrightarrow (\textit{sec}\text{-butyl radical})^{\text{v}} \qquad (9\text{-}1)$$

where the superscript v represents vibrational excitation. This is the same reaction system mentioned previously in Chapter 2, Section 4. The reaction is exothermic and the entire heat of reaction, which is about 40 kcal mole^{-1} (the precise value depends on which of the three butenes is the reactant), appears in the product radical; since this is about 7 kcal mole^{-1} more than necessary for breaking a C—C bond in the butyl radical, process (9-1) is followed by the unimolecular reaction

$$(\textit{sec}\text{-butyl radical})^{\text{v}} \longrightarrow CH_3 + C_3H_6 \qquad (9\text{-}2)$$

In other words, the chemical activation process can be used to "prepare"

[1] The reference list for this chapter can be found on p. 254.

butyl radicals with at least 7 kcal $mole^{-1}$ internal energy (plus some thermal energy) in excess of threshold for reaction into methyl radical and propylene.

There is quite a variety of processes which produce a reactant chemically activated for unimolecular reaction. Most of the systems mentioned in Chapter 2, Section 4 belong to this class, and many similar ones leading to vibrationally or electronically excited neutral reactants are known.[2] A number of ion–molecule reactions yield a "persistent complex"[3] that has all the characteristics of a chemically activated reactant, for example (*4*) $O_2^+ + D_2 \rightarrow D_2O_2^+$, which then decomposes further unimolecularly. A "persistent complex" is also formed in several crossed-molecular-beam reactions involving cesium,[4] as in $Cs + SF_6 \rightarrow CsSF_6$, which then breaks up into CsF and the SF_5 radical. By stretching the point a little, a unimolecular reaction following photon absorption can also be considered as a case of chemical activation, for example (*8*,*9*), $CF_3COCF_3 + h\nu \rightarrow (CF_3COCF_3)^{e,v}$ where the superscripts e,v signify electronic and vibrational excitation, followed by $(CF_3COCF_3)^{e,v} \rightarrow CF_3CO + CF_3$. A very sophisticated version of this approach uses crossed molecular and laser beams, a technique called "photofragment spectroscopy".[5] Decomposition of charged species following excitation by electron impact or charge exchange may be also regarded as an example of chemical activation.

The various chemical activation systems are useful for unimolecular rate theory calculations only insofar as the distribution of energy deposited in the reactant is known or is calculable, since otherwise it is not possible to relate calculated and experimental rates.[6] Due to lack of the necessary thermochemical data, reaction systems involving a "persistent complex" are not yet useful for a test of unimolecular rate theory, but the method has obvious potential for ultimately yielding the most detailed information yet about the unimolecular process. "Photofragment spectroscopy" shows the same potential, with the added advantage that the known energy of the photon beam simplifies the energy balance in the system.

Plan of chapter At present, this leaves as suitable for consideration only systems other than molecular beam systems. A rather arbitrary division is

[2] For a review, see Rabinovitch and Flowers (*2*).

[3] Reviewed by Wolfgang (*3*).

[4] Freund *et al.* (*5*); Bennewitz *et al.* (*6*). A general theoretical analysis of condition leading to statistical complexes in inelastic collisions has been given, in terms of the *S*-matrix theory, by George and Ross (*7*). Persistent complex formation has now been also shown in the reaction of K, Rb, and Cs with $SnCl_4$ and SF_6 (*114*) and in the reaction of F + ethylene (*115*); these results have been interpreted in terms of RRKM theory.

[5] For an outline of the experiments, see *Physics Today* (*10*). Cf. also footnote 9 in Chapter 1.

[6] In molecular beam studies, the experimental observables are energy and angular momentum disposal among the products.

made for convenience; this chapter deals with moderately excited systems involving neutral species (including photoactivated species), and the next chapter deals mostly with highly excited charged species. The main features of chemically activated neutral systems are surveyed in Section 1. The distribution of energy deposited in the reactant by the activating process is dealt with in Section 2. It is shown there that as a consequence of a distribution of energies, in contrast with a monoenergetic system, the observable rate is both pressure and temperature dependent; these dependences are discussed in Sections 3 and 4, respectively. The strong-collision assumption is abandoned in Section 5 and the effect of multistep deactivation on the low-pressure behavior of the rate constant is examined using the simple stepladder cascade model. Section 6 considers chemical activation by photon absorption and briefly outlines some of the mechanistic problems in photochemistry. Finally, Section 7 ties in theoretical and experimental parameters and Section 8 sketches an outline for consideration of centrifugal effects.

1. Nature of chemical activation

There are only a few selected reactions for which all the ancillary thermochemical information is accessible, and which therefore are potentially useful examples of a chemically activated unimolecular process. Basically, these reactions belong to one of two classes[7]: (1) the activating reaction *is not* the reverse of the unimolecular reaction of interest, examples of which are reactions (9-1) and (9-2). Figure 9-1(a) shows schematically a section through the potential energy surface for this process. (2) The activating reaction *is* the reverse of the unimolecular process of interest; an example here is the decomposition of various chlorinated ethyl radicals[8] in the sequence

$$C_2H_{5-n}Cl_{n-1} + Cl \longrightarrow (C_2H_{5-n}Cl_n)^v \longrightarrow C_2H_{5-n}Cl_{n-1} + Cl$$

It may be noted from Fig. 9-1(b) that in this case it does not matter whether the reaction producing the vibrationally excited species is endothermic or exothermic since the species, once formed, is automatically activated for the reverse process of decomposition. This chapter will be concerned mostly with class 1 processes, which are the better known; since all photoactivated reactions belong of necessity to class 1, all general comments relating to

[7] The concept of reaction *class* introduced here should not be confused with reaction *type* (cf. Chapter 3) which refers to the properties of the potential energy surface for decomposition.

[8] For $n = 1$, see Le-Khac Huy *et al.* (*11*). For $n = 3$, see Knox (*12*); Tardy and Rabinovitch (*13*). This case is somewhat more complicated than the others because 1,2-dichloroethylene exists in cis and trans isomers. For calculations on the entire series, see Beadle *et al.* (*14*).

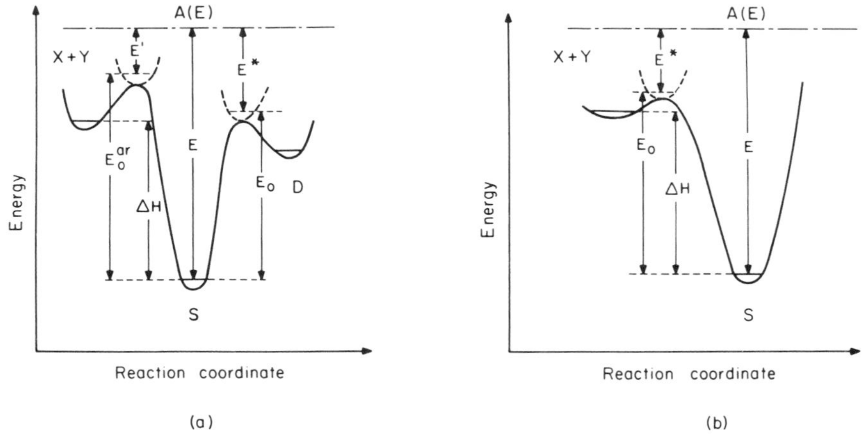

Fig. 9-1 *Energy relationships in chemical activation systems (schematic).* (a) Class 1: activating reaction is *not* the reverse of the unimolecular process of interest. (b) Class 2: activating reaction *is* the reverse of the unimolecular process of interest. The dot-dash line represents the excitation energy present in the system. The dashed lines represent part of the potential energy curves of the various transition states. All energies referred to ground state of stabilized A (= **S**) as zero. ΔH is the enthalpy of the activating reaction at 0°K; E is the internal (= randomizable) energy of A, the unimolecular reactant; E' is the internal energy of the transition state for the activating reaction; and E^* is the internal energy of the transition state for unimolecular fragmentation of A. If X and Y are radicals, the maximum separating X + Y from **S** in (a) and (b) may be vanishingly small.

class 1 reactions apply equally well to photoactivated systems. Specific differences are left for Section 6.

Experimentally, the chemical activation technique is somewhat more complicated than for similar thermal reactions, since one must first prepare the reactants, one or both of which are radicals, and then allow them to react together. What does one get in return for the experimental complication?

First of all, the reaction producing the chemically activated reactant proceeds generally at low temperatures (room temperature and below), which helps to prevent undesirable secondary reactions, insofar as they require an activation energy. The second important point is that, as we shall see, energy is deposited in the reactant with a distribution which has a "width" (i.e., variance) very similar to that of the Boltzmann distribution at the reaction temperature. This means that while the observable rate in a chemical activation system still involves an average over a distribution of energies, the average is now over a distribution that is much narrower than in the typical thermal reaction because of the much lower reaction temperature in the chemical activation system. Therefore the measured chemical activation rate should, in principle, preserve a little more of the detail originally contained in the detailed (microcanonical) rate constant $k(E)$ than in a thermal system. A third point is that unlike in the case of thermal reactions where the average energy of reacting species can be manipulated only by changing the temperature of the heat bath, the average energy deposited in the reactant by means of the chemical activation technique can be manipulated, in addition, by modifying the nature of the reactants. For example, substituting D atoms for H atoms in reaction (9-1) changes the average excitation energy of the butyl radical by 1.8 kcal mole^{-1}; a further change can be accomplished by taking for the other reactant butene-1, *cis*-butene-2, or *trans*-butene-2, giving a total of six variants of reaction (9-1). In this fashion, *sec*-butyl radicals have been prepared[9] with average energies ranging from 7.2 kcal mole^{-1} (from H + *trans*-butene-2 at −78°C) to 12.9 kcal mole^{-1} (from D + butene-1 at 25°C).

In practice, several circumstances conspire to prevent the chemical-activation technique from realizing its full potential. One are side reactions, notably in the butyl radical and cyclopropane systems, despite the low temperatures at which these experiments are run, and the other, perhaps more important, is the generally insufficient accuracy of thermochemical data necessary for characterizing the various energy parameters of the system. This problem is considered more fully in Section 7. Nevertheless, the technique provides useful insights into the unimolecular decay of highly excited species. A review of both theoretical and practical aspects of the chemical activation technique has been given by Setser (*16*).

Decomposition versus stabilization One aspect of the chemical activation technique which it has in common with thermal reactions is that the unimolecular process is experimentally detectable only in competition with a collisional stabilization process. In other words, process (9-2), for example, which is the decomposition process (**D** for short), is always accompanied by the stabilization process (**S** for short)

$$(sec\text{-butyl radical})^{v} + \mathsf{M} \longrightarrow \text{stabilized radical} + \mathsf{M} \qquad (9\text{-}3)$$

which is the exact analog of the thermal deactivation process (8-2). Here, M represents, as usual, any molecule that acts as an energy sink. The stabilized radical, having by definition insufficient energy for fragmentation, undergoes

[9] For a summary, see Rabinovitch and Setser (*15*). See also Table 9-1 of this chapter. It is worthwhile noting that in these experiments, the hydrogen or deuterium atoms are essentially thermalized. The reaction of nonthermalized atoms (recoil tritium) with butene-1 and *cis*-butene-2 has also been studied (*116*).

secondary reactions (recombination, disproportionation, and the like) ultimately leading to stable products that are different from final stable products resulting from the decomposition process (9-1), so that the ratio **D**/**S** can be obtained directly by analysis of the products.

The importance of the decomposition process **D** relative to the stabilization process **S** is a function of total pressure in the reacting system, and is therefore variable at will, within limits. Hence at high pressures, the chemical activation system can be used to yield information about the unimolecular process **D** [Eq. (9-2)], and at low pressures, about the energy transfer process **S** [Eq. (9-3)]; this is discussed in more detail in Section 5.

To keep the argument general, we shall henceforth write the chemical activation process considered in this chapter quite generally as

$$\mathsf{X} + \mathsf{Y} \xrightarrow{k} \mathsf{A} \tag{9-4}$$

$$\mathsf{A} \xrightarrow{k_1} \mathbf{D} \tag{9-5}$$

$$\mathsf{A} + \mathsf{M} \xrightarrow{k_2} \mathsf{A}\,(\text{stabilized} = \mathbf{S}) + \mathsf{M} \tag{9-6}$$

where X and Y are the reactants for the activating reaction and A is the reactant for the unimolecular reaction. For class 2 chemical activation reactions, we have **D** = X + Y. As written, this reaction scheme refers to total (macroscopic) rates, i.e., represents the overall rate of formation, decomposition, and stabilization of reactant A of all energies. On the microscopic level, we have to consider the rate at which reacts A of specified energy E. Such A shall be designated by A(E); then the rate constant of process (9-5) is $k(E)$, the microcanonical rate constant for the unimolecular reaction of interest, and the processes (9-4) and (9-6) represent the formation and stabilization, respectively, of A(E).

The symbol A will be used to represent the chemically activated reactant of all energies; occasionally, when the distinction is important, **S** will be used to denote *stabilized* A, i.e., A having energies below threshold for reaction.

2. Determination of the energy distribution function

In any chemical activation system in which there is competition between the decomposition (**D**) and stabilization (**S**) processes, the distribution of energies of the reacting species A can be formally derived by a procedure similar to that used in Section 3 of Chapter 8.

The steady-state distribution $F(E)_{ss}$ The distribution is viewed as a steady-state distribution, designated below by $F(E)_{ss}$, characterizing the steady-state population of A(E)'s, set up as a result of the competition between the **D** and **S** processes in a time that is assumed to be short compared with the duration of observation on **D** and **S**; in short, the rate of disappearance of A(E) is supposed to have a transient term that dies out quickly. The steady-state distribution function [cf. Eq. (II-10)] then follows by equating the gain and loss of A(E) for every range dE.

Let $(d[\mathrm{A}]/dt)F(E)\,dE$ (molecules $\mathrm{cm}^{-3}\,\mathrm{sec}^{-1}$) be the number of molecules A(E) formed per second and per unit volume in the energy range E, $E + dE$ by the activating reaction (9-4). Here, $d[\mathrm{A}]\,dt$ is the total rate of formation of molecules A of *all* energies and $F(E)$ is the distribution function for the *formed* molecules of energy E, i.e., their fractional amount $[\mathrm{A}(E)]/[\mathrm{A}]$ per unit energy range. The distribution $F(E)$ is not maintained in the course of the reaction because there is a net loss of A(E)'s. If $k(E)$ is the rate constant for the decomposition of A(E) in the process (9-5), the number of molecules A(E) decomposing per second and per unit volume in (9-5) at steady state is

$$(d[\mathrm{A}]/dt)k(E)F(E)_{\mathrm{ss}}\,dE \quad \text{molecules cm}^{-3}\,\text{sec}^{-1} \tag{9-7}$$

If k_2 is the rate constant for the stabilization process (9-6), the number of molecules A(E) stabilized per second and per unit volume in (9-6) at steady state is

$$(d[\mathrm{A}]/dt)k_2[\mathrm{M}]F(E)_{\mathrm{ss}}\,dE \quad \text{molecules cm}^{-3}\,\text{sec}^{-1} \tag{9-8}$$

where [M] is the concentration of heat bath molecules M. Equating the gain and loss of molecules A(E) for every range dE, we have

$$F(E) = \{k(E) + k_2[\mathrm{M}]\}F(E)_{\mathrm{ss}} \tag{9-9}$$

so that[10]

$$F(E)_{\mathrm{ss}} = F(E)/\{k(E) + k_2[\mathrm{M}]\} \tag{9-10}$$

Equation (9-10) can be easily generalized to include more than one mode of decomposition of A(E). In such a case, $k(E)$ in Eqs. (9-7)–(9-10) is to be interpreted as the sum of rate constants over the parallel decomposition processes, i.e., $k(E)$ becomes $\sum_i k_i(E)$, where $k_i(E)$ is the rate constant of the ith decomposition process. There is one mode of decomposition that should, in principle, always be considered in addition to (9-5), namely the reverse reaction A(E) → X + Y, the rate constant of which is later referred to as $k^{\mathrm{ar}}(E)$. Consequently, even if there is no decomposition process parallel with (9-5), $k(E) + k^{\mathrm{ar}}(E)$ should always be written for $k(E)$. It turns out,

[10] The dimension of $F(E)$ is energy^{-1} and that of $F(E)_{\mathrm{ss}}$ is second × energy^{-1}; thus both are to be understood as distribution functions "per unit energy range at E." To convert these to distribution functions "at E" involves the trivial operation of multiplying by dE.

however, that the potential energy surface is mostly of the kind shown in Fig. 9-1(a), where $E_0^{\mathrm{ar}} > E_0$, and the higher activation energy for the reverse of the activating reaction then ensures (*17*) that $k^{\mathrm{ar}}(E) \ll k(E)$, so that Eq. (9-10), as written, is a good approximation.

We shall now simplify things somewhat by making the equivalent of the strong-collision assumption in thermal systems (cf. Chapter 8, Section 1); we shall put $k_2[\mathrm{M}] = \omega$, meaning that every gas-kinetic collision removes from $\mathrm{A}(E)$ sufficient energy to prevent it from decomposing. This assumption, which will be relaxed later in Section 5, can be justified at this stage by noting that it is true in thermal systems for heat bath molecules M that are not too small (cf. Table 8-1, Chapter 8). Thus we have

$$F(E)_{\mathrm{ss}} = F(E)/[k(E) + \omega] \tag{9-11}$$

Now that the rate constant of the collisional stabilization process has been assigned the known value $k_2 = Z$, where Z is given by (8-100), we can use the stabilization reaction as an "internal clock" against which the rate of the unimolecular reaction is measured. If we write k_{a} for the observable, or overall, rate of the decomposition reaction, then, following Rabinovitch and co-workers (*15*), it is convenient to define it as[11]

$$k_{\mathrm{a}}/\omega = \mathbf{D}/\mathbf{S} \tag{9-12}$$

where **D** and **S** stand for the total amount of products resulting from the decomposition and stabilization processes, respectively. The ratio **D**/**S** is given by the ratio of Eqs. (9-7) and (9-8), each integrated over reaction time τ and over all energies:

$$\frac{\mathbf{D}}{\mathbf{S}} = \frac{\int_0^\tau (d[\mathrm{A}]/dt)\, dt \int_0^\infty k(E)F(E)_{\mathrm{ss}}\, dE}{\omega \int_0^\tau (d[\mathrm{A}]/dt)\, dt \int_0^\infty F(E)_{\mathrm{ss}}\, dE} \tag{9-13}$$

Note that integrations over time and over energy can be performed separately since a steady-state distribution is by definition time independent. Thus although the total mounts of **D** and **S** depend, individually, on reaction time τ, i.e., on how long the reaction is allowed to run, their ratio does not, since integration over time leads to a common factor that cancels out in **D**/**S**, as shown in Eq. (9-13). If we had not integrated in (9-13) over reaction time, i.e., if we had interpreted **D** and **S** as the *rates* of formation of **D** and **S**, respectively, the result for **D**/**S** would still be the same, so that insofar as the ratio **D**/**S** is concerned, it is immaterial whether **D**/**S** is interpreted as the ratio of rates of formation or as the ratio of actual amounts formed, although

[11] If several decomposition modes compete, we have for the *i*th mode $k_{\mathrm{a}i} = \omega \mathbf{D}_i/\mathbf{S}$. For an actual example, see, for instance, Tardy *et al.* (*18*).

of course it is the latter interpretation that is the closest to an actual experimental observable. A probabilistic interpretation of **D** and **S** is given in Section 5. From (9-11)–(9-13), we have

$$k_a = \omega \frac{\int_0^\infty \{k(E)F(E)/[k(E) + \omega]\}\, dE}{\int_0^\infty \{\omega F(E)/[k(E) + \omega]\}\, dE} = \langle k(E)\rangle_{ss} \tag{9-14}$$

where, as usual, we omit for brevity the designation of $\langle k(E)\rangle$ as the *apparent* steady-state rate constant (cf. the introduction to Part II). In other words, the observable rate constant k_a turns out to be $k(E)$ averaged over the *normalized* distribution $F(E)_{ss}$. The derivation of $\langle k(E)\rangle_{ss}$ is now only formally complete because $F(E)$ is not specified.

The distribution $F(E)$ ***of formed molecules*** If the A(E)'s were monoenergetic, i.e., if all were formed with the same value of E, say $E = E_{mono}$, $F(E)$ would be a delta function, i.e., $F(E_{mono}) = 1$, and $F(E) = 0$ for all other values of E. In that event, (9-14) reduces simply to $\langle k(E)\rangle_{ss} = k(E_{mono})$, so that the observable rate constant, as defined by (9-12) would be pressure independent. Unfortunately, the A(E)'s are never quite monoenergetic, so that $F(E)$ has a more complicated form than a delta function, with the result that $\langle k(E)\rangle_{ss}$ becomes pressure dependent. Thus the pressure dependence of $\langle k(E)\rangle_{ss}$ is due entirely to the spread of excitation energies in the reactant produced by the activating reaction. The immediate concern now is to find the form of $F(E)$.

Rabinovitch and Diesen (*19*; *20*, p. 58 ff.) have derived a general form of $F(E)$ by first considering the rate of the reverse of the activating reaction and then applying the principle of detailed balancing to get the rate of the forward process. The reason for this procedure is that the activating reaction is a thermal bimolecular reaction, while its reverse is simply the unimolecular decomposition of A(E), which is somewhat easier to deal with.

It is useful to recall first that the distribution function $F(E)$ is the distribution that would exist if A(E) were not irretrievably lost by reaction. Therefore, since it is $F(E)$ that we are after, let us consider the activating reaction as starting with the reactants X + Y and ending with the collision-stabilized (but not decomposed) A, and let us assume that this hypothetical reaction has come to equilibrium. Under these conditions, we have the following scheme (cf. footnote 12):

$$\mathrm{X} + \mathrm{Y} \underset{k^{ar}(E)}{\overset{k^{af}(E)}{\rightleftharpoons}} \mathrm{A}(E) \underset{}{\overset{k_2[\mathrm{M}]}{\rightleftharpoons}} \mathrm{A}\ (\text{stabilized} = \mathbf{S}) \tag{9-15}$$

The second-order rate constant $k^{af}(E)$ measures the rate of the activating reaction in the forward sense (hence the superscript "af") leading to the formation of A(E) in the range E, $E + dE$, and the first-order rate constant $k^{ar}(E)$ measures the rate of activating reaction in the reverse direction (hence the superscript "ar"). To keep the discussion quite general, the treatment assumes that we are dealing with class 1 reactions (where the activating reaction is *not* the reverse of the unimolecular reaction of interest), since formulas applicable to class 2 reactions can be easily deduced later as a special case by a trivial modification. Since we have equilibrium,

$$[\mathrm{A}(E)]/[\mathrm{A}] = P(E)_{\mathrm{e}} \tag{9-16}$$

where $P(E)_{\mathrm{e}}$ is the equilibrium distribution (8-13) and square brackets indicate concentrations, as usual. We have also

$$k^{af}(E)[\mathrm{X}][\mathrm{Y}] = k^{ar}(E)[\mathrm{A}(E)] = k^{ar}(E)P(E)_{\mathrm{e}}[\mathrm{A}] \tag{9-17}$$

if [A(E)] is eliminated by means of Eq. (9-16). If we write $[\mathrm{A}]/[\mathrm{X}][\mathrm{Y}] = K$, where K is the equilibrium constant for the overall reaction $\mathrm{X} + \mathrm{Y} \rightleftharpoons \mathrm{A}$, (9-17) becomes

$$k^{af}(E) = Kk^{ar}(E)P(E)_{\mathrm{e}} \tag{9-18}$$

Under nonequilibrium conditions, the processes leading from A to X + Y may be neglected[12] and the rate of formation of A(E) is

$$d[\mathrm{A}(E)]/dt = k^{af}(E)[\mathrm{X}][\mathrm{Y}] \tag{9-19}$$

While *rates* are equal only at equilibrium [Eq. (9-17)], the relations between *rate constants* [Eq. (9-18)], although derived at equilibium, are valid whether equilibrium actually obtains or not, so that it is permissible to substitute Eq. (9-18) for $k^{af}(E)$ in Eq. (9-19). Furthermore, we may observe that

$$[\mathrm{X}][\mathrm{Y}]K = [\mathrm{A}]_{\mathrm{e}} \tag{9-20}$$

where $[\mathrm{A}]_{\mathrm{e}}$ is the numerical value of the equilibrium concentration of A if the concentrations of X and Y are [X] and [Y], respectively. Thus (9-19) becomes

$$d[\mathrm{A}(E)]/dt = k^{ar}(E)P(E)_{\mathrm{e}}[\mathrm{A}]_{\mathrm{e}} \tag{9-21}$$

i.e., the rate of formation of A(E) under nonequilibrium conditions is numerically equivalent to its rate of disappearance at overall equilibrium. It should be noted that since [X] and [Y] vary with time, so does $[\mathrm{A}]_{\mathrm{e}}$, in a way such that Eq. (9-20) is satisfied.

[12] See the discussion following Eq. (9-10). It is perhaps useful to point out that throughout this treatment, the concentration of stabilized A is taken as essentially equal to the total concentration of A, it being implicitly assumed that A's in reactive states represent only a very small part of the total.

Since Eq. (9-21) gives the rate of formation of molecules A in the specified energy range, the (normalized) distribution function $F(E)$ is clearly

$$C \times F(E) = \frac{1}{[\mathrm{A}]_e} \frac{d[\mathrm{A}(E)]}{dt} = k^{ar}(E)P(E)_e \qquad (9\text{-}22)$$

where C is the normalization factor

$$C = \int_0^\infty k^{ar}(E)P(E)_e \, dE \qquad (9\text{-}23)$$

which ensures that $F(E)$ have the proper dimension of energy^{-1}. The unimolecular rate constant for the reverse of the activating reaction can be formulated in the manner of Eq. (8-25) as

$$k^{ar}(E) = \alpha^{ar}\mathscr{G}^*(E - E_0^{ar})/hN(E) \qquad (E \geqslant E_0^{ar}) \qquad (9\text{-}24)$$

where α^{ar} and E_0^{ar} (cf. Fig. 9-1) are the reaction path and critical energy, respectively, for the decomposition of A into X + Y, and $\mathscr{G}^*(E - E_0^{ar})$ is the integrated density for the transition state of this reaction at excitation energy $E - E_0^{ar}$. The form of the equilibrium distribution is given by Eq. (8-13) as

$$P(E)_e = N(E)e^{-E/kT}/Q \qquad (9\text{-}25)$$

In the present context, $N(E)$ in both (9-24) and (9-25) is the density of states of A at excitation energy E (cf. Fig. 9-1) and Q is the partition function of A. Thus from Eqs. (9-22)–(9-25),

$$F(E) = \frac{\mathscr{G}^*(E - E_0^{ar})e^{-E/kT}}{\int_0^\infty \mathscr{G}^*(E - E_0^{ar})e^{-E/kT} \, dE} \qquad (E \geqslant E_0^{ar}) \qquad (9\text{-}26)$$

The distribution $F(E)$ is thus temperature dependent, as of course one would expect since the activating reaction and its reverse are thermal reactions. It is not necessary to evaluate the denominator of (9-26) if $F(E)$ is to be used in Eq. (9-14), since the denominator then drops out. However, after a change of variable to $y = E - E_0^{ar}$, it is easy to see immediately, using Eqs. (8-18) and (8-19), that

$$\int_0^\infty \mathscr{G}^*(E - E_0^{ar}) \exp(-E/kT) \, dE = kT\mathscr{Q}^* \exp(-E_0^{ar}/kT) \qquad (9\text{-}27)$$

where $\mathscr{Q}^*$ is the partition function for the transition state of the activating reaction.

$F(E)$ depends, besides temperature, on the magnitude of $\mathscr{G}^*(E - E_0^{ar})$, i.e., on how closely spaced the energy levels of the transition state are. The

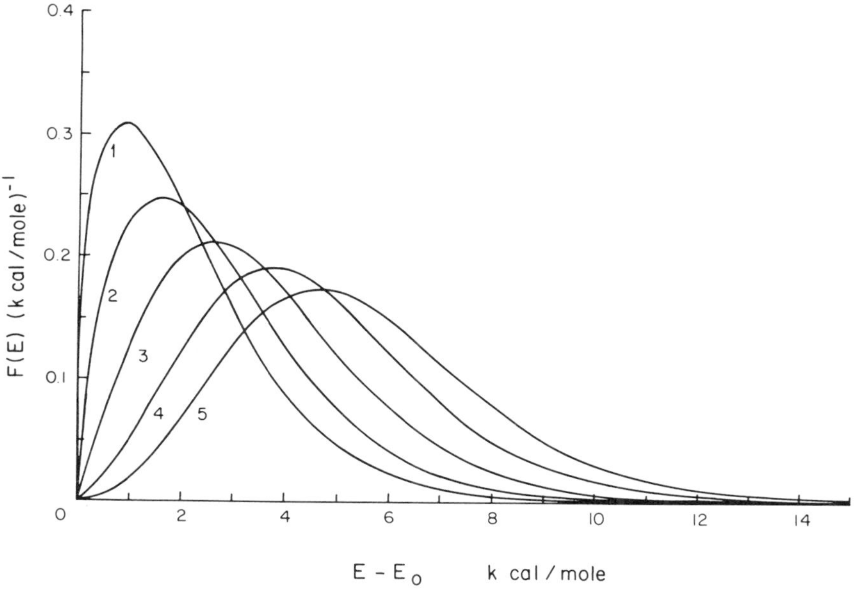

Fig. 9-2 *The distribution function F(E) for various chemically activated chloroethyl radicals (class 2 systems).*

(1) C_2H_4Cl; (2) $C_2H_3Cl_2$; (3) $C_2H_2Cl_3$; (4) C_2HCl_4; (5) C_2Cl_5. On replacement of each H in the ethyl radical by Cl, a C—H vibration in the transition state becomes a C—Cl vibration, the frequency of which is four or five times lower; thus $\mathscr{G}^*(E)$ increases from monochloroethyl to pentachloroethyl. There is also at the same time an increase in the moment of inertia and a change in the critical energy E_0, neither of which has an effect on the shape of $F(E)$. From P. C. Beadle, J. H. Knox, F. Placido and K. C. Waugh, *Trans. Faraday Soc.* **65**, 1571 (1969), Fig. 2. Note that in class 2 systems $E_0^{\text{ar}} = E_0$.

effect of closer spacing of energy levels in the transition state resulting from the lowering of several vibrational frequencies is shown in Fig. 9-2, from which it may be seen that closer spacing of vibrational levels has the effect of broadening the distribution $F(E)$. In principle, $F(E)$ is also a function of the overall angular momentum, and so are $k^{\text{ar}}(E)$ and $k(E)$. Consideration of such centrifugal effects is postponed until Section 8 for reasons that will appear later.

Although $F(E)$ of Eq. (9-26) depends on the model of the transition state for the activating reaction, this model dependence is not too serious inasmuch as (9-26) contains only the *ratio* of $\mathscr{G}^*(E)$'s. We can also observe that (9-26) expresses the plausible notion that $F(E)$ depends on the effective total number of states of the combined system (X + Y) at $E - E_0^{\text{ar}}$ [this is, in essence, the meaning of $\mathscr{G}^*(E - E_0^{\text{ar}})$], weighted by the thermal factor $e^{-E/kT}$. Since the nature of the effective degrees of freedom of the separate systems X and Y can be determined, or guessed at, the total number of states of the combined system can be obtained by performing the convolution indicated in Eqs. (6-35) and (6-36). Note that the convolution involves the *density* for one of the constituent systems and the *integrated density* for the other constituent system.

The concept of $F(E)$ representing the total number of states available to the combined system (X + Y) is behind the reasoning used by Knox (*12*)[13] to derive the form of $F(E)$ in the chemical activation system Cl + dichloroethylene → trichloroethyl radical → **D** or **S**. If we assume that only one translational degree of freedom of the chlorine atom intervenes (i.e., the one along the line of centers) in the activating reaction, then the integrated density at energy x for this one translation is proportional[14] to $x^{1/2}$. If the density of the v vibrational states of dichloroethylene at $E - x$ is $N_v(E - x)$, then the integrated density for the combined system (Cl + dichloroethylene) at E is given by the convolution integral

$$G(E) = \text{const} \times \int_0^E x^{1/2} N_v(E - x)\, dx \tag{9-28}$$

This is of the same form as the function $G(E)$ for a system of v oscillators and one free rotor [cf. Eq. (6-48)], which is a plausible model for the transition state in this reaction (*13*). The constant in front of the integral (9-28) is of course different if x refers to a translation or a rotation, but this is immaterial because when this above $G(E)$ is substituted[15] for $\mathcal{G}^*(E)$ in Eq. (9-26), the constant drops out. This shows that (fortunately) $F(E)$ is not overly sensitive to the details of the function $\mathcal{G}^*(E)$.

Comparison between $F(E)$ ***and the equilibrium distribution*** It is now interesting to compare $P(E)_e$ of Eq. (9-25) and $F(E)$ of Eq. (9-26). First of all, we may note that since $\mathcal{G}^*(E - E_0^{ar})$ is defined only for $E \geqslant E_0^{ar}$, $F(E)$ is zero everywhere from $E = 0$ to $E = E_0^{ar}$, and is finite only above E_0^{ar}. By contrast, $P(E)_e$ is finite over the entire range of (positive) energies. To continue the comparison, we next locate the maximum of each function on the energy axis and then determine their respective variance σ^2, which is a measure of the "width" of each function.

To find the energy E_{max} at which is located the maximum of each function, we let the derivative of each function vanish. Thus in the case of $F(E)$, we have

$$\frac{dF(E)}{dE} = 0 = \frac{e^{-E/kT}}{\text{DEN}}\left[\mathcal{N}^*(E - E_0^{ar}) - \frac{\mathcal{G}^*(E - E_0^{ar})}{kT}\right] \tag{9-29}$$

[13] The author is grateful to Prof. Knox for correspondence on this matter.
[14] See Chapter 6, Exercise 1.
[15] Read $E - E_0^{ar}$ for E when substituting in (9-23) or (9-26).

Here, DEN is the denominator of Eq. (9-26) as given by Eq. (9-27), and $\mathcal{N}^*(E - E_0^{\rm ar})$ is $d\mathcal{G}^*(E - E_0^{\rm ar})/dE$, i.e., the density of states for the transition state[16] of the activating reaction. Classically,[15] $\mathcal{G}^*(E) \approx \text{const} \times E^{\mathsf{n}^*}$, where n^* is the number of vibrational degrees of freedom plus one-half the number of pertinent rotational degrees of freedom involved in the transition state of the activating reaction. Consequently,[15] $\mathcal{N}^*(E) \approx \text{const} \times \mathsf{n}^* E^{\mathsf{n}^*-1}$, and the solution of Eq. (9-29) is approximately

$$E_{\max,\mathrm{F}} \approx \mathsf{n}^* kT + E_0^{\rm ar} \tag{9-30}$$

Similarly, we have in the case of $P(E)_{\rm e}$

$$\frac{dP(E)_{\rm e}}{dE} = 0 = \frac{e^{-E/kT}}{Q}\left[\frac{dN(E)}{dE} - \frac{N(E)}{kT}\right] \tag{9-31}$$

Now, classically, $N(E) = \text{const}' \times E^{\mathsf{n}-1}$, where const′ is a constant different from that which appears in $\mathcal{N}^*(E)$, and n is the number of vibrational degrees of freedom plus one-half of the pertinent rotational degrees of freedom in A. Hence $dN(E)/dE = \text{const}' \times (\mathsf{n} - 1)E^{\mathsf{n}-2}$, and the solution of Eq. (9-31) is therefore[17]

$$E_{\max,P} \approx (\mathsf{n} - 1)kT \tag{9-32}$$

Since the transition state of A has one degree of freedom less than A itself, we have $\mathsf{n}^* = \mathsf{n} - 1$, and Eqs. (9-30) and (9-32) therefore show that both functions $F(E)$ and $P(E)_{\rm e}$ have their maximum at about $(\mathsf{n} - 1)kT$ above their respective thresholds. It can also be fairly easily shown (Exercise 1) that the heights of the two functions at their respective maxima are roughly the same, i.e., that

$$F(E_{\max,F}) \approx P(E_{\max,P})_{\rm e} \tag{9-33}$$

The "width," or variance σ^2, of $F(E)$ is defined as

$$\sigma^2 = \langle E^2 \rangle - \langle E \rangle^2 = \int_0^\infty E^2 F(E)\, dE - \left[\int_0^\infty EF(E)\, dE\right]^2 \tag{9-34}$$

and the variance of $P(E)_{\rm e}$ is defined in a similar way. The integrations involved in the calculations of σ^2 can be readily evaluated if the classical form

[16] At the risk of repeating the obvious, note that the activating reaction and its reverse necessarily involve the same transition state by virtue of microscopic reversibility.

[17] See also Benson (*21*, p. 230 and Fig. XI.2). The results (9-30), (9-32), and (9-33) can be easily shown to remain valid when the semiclassical approximation (6-63) (with $a_K = 1$) for $G^*(E)$ and $N(E)$ is used, provided the respective zero-point energies E_z^* and E_z are not too different.

of the density functions is used, as in the previous paragraph. It is left to the reader as Exercise 2 to show that the variance of $F(E)$ is $\sigma_F{}^2 \approx (\mathsf{n}^* + 1)(kT)^2$ and that the variance of $P(E)_e$ is $\sigma_P{}^2 \approx \mathsf{n}(kT)$. Since $\mathsf{n}^* = \mathsf{n} - 1$, we have therefore $\sigma_F{}^2 \approx \sigma_P{}^2$, showing that $F(E)$ and $P(E)_e$ have substantially the same "width."

The result of the comparison between $F(E)$ and $P(E)_e$ is therefore, in essence, that $F(E)$ is not too dissimilar from the function $P(E)_e$ shifted along the energy axis by E_0^{ar}. Allowing for the fact that this has been deduced by means of a crude approximation (see, however, footnote 17), the similarity between $F(E)$ and $P(E)_e$ must not be taken too literally, but it does suggest that in the absence of any knowledge of the transition state for the activating reaction, we can take, as a first approximation,

$$F(E) \approx P(E - E_0^{ar})_e = \frac{N(E - E_0^{ar})\exp[-(E - E_0^{ar})/kT]}{Q} \qquad (E \geqslant E_0^{ar}) \tag{9-35}$$

or, as a slightly better approximation,

$$F(E) \approx \frac{G(E - E_0^{ar})\exp[-(E - E_0^{ar})/kT]}{kTQ} \qquad (E \geqslant E_0^{ar}) \tag{9-36}$$

where $N(E)$ and $G(E)$ refer to density and integrated density, respectively, of the reactant A. Figure 9-3 shows a comparison between $F(E)$ calculated from (9-26) and $F(E)$ calculated from (9-35) and (9-36) for reaction (9-1) forming the *sec*-butyl radical from H + *cis*-butene-2. The difference among the three forms of $F(E)$ is not large, mainly at low energy, where it matters least; such difference as there is between the form (9-26) of $F(E)$ and the other two arises from the assumption of a fairly "loose" transition state in reaction (9-1), i.e., one with several frequencies presumed to be lower than in the radical. Consequently $\mathscr{G}^*(E - E_0^{ar}) \gg G(E - E_0^{ar}) > N(E - E_0^{ar})$, and as a result, the form (9-26) of $F(E)$ is squatter than the other two and with a maximum shifted to slightly higher energy, in accord with the trend shown in Fig. 9-2.

The comparison, crude as it might be, also offers an insight into the essential difference between thermal and chemical activation systems. Owing largely to the exothermicity of the activating reaction, the formation of A proceeds at relatively low temperature with a distribution of energies that not only begins, but also reaches its maximum, near the threshold energy for reaction. If the maximum of the thermal distribution $P(E)_e$ were to be shifted to the vicinity of E_0^{ar}, i.e., if a distribution of energies somewhat similar to $F(E)$ were to be obtained by purely thermal means, the required increase in the temperature of the system would be substantial indeed. Suppose that $E_0^{ar}/kT \approx 40$, a typical figure. To shift the maximum of $P(E)_e$ from $(\mathsf{n} - 1)kT$ to $(\mathsf{n} - 1)kT + E_0^{ar}$ would require the temperature of the system to be

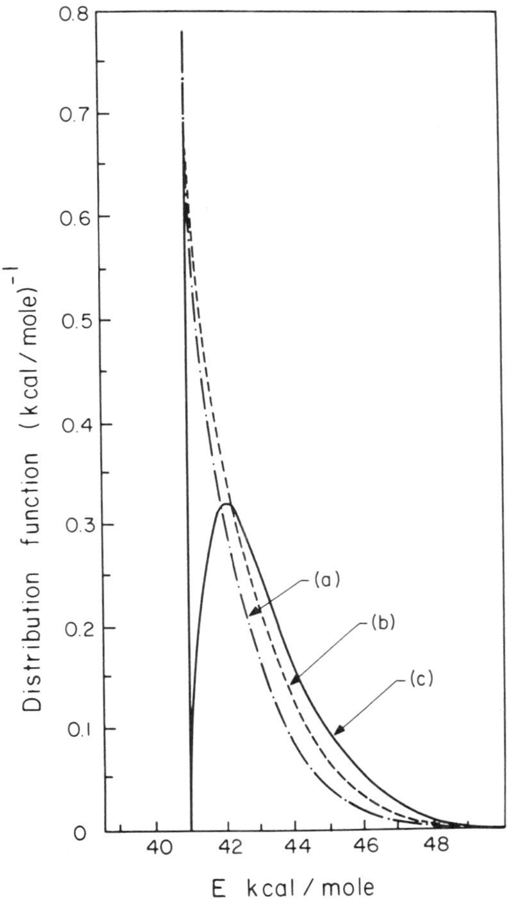

Fig. 9-3 *Distribution function F(E) for butyl radicals in the* H + *cis-butene-2 system at 298*°K.

(a) Calculated from Eq. (9-35), using frequencies of the butyl radical. This curve has the highest maximum. (b) Calculated from Eq. (9-36) using frequencies of the butyl radical. (c) Calculated from Eq. (9-26) for an assumed transition state. From R. E. Harrington, Ph.D. Thesis, University of Washington, 1960, Fig. 15. [Frequencies of butyl radical and transition state as given by B. S. Rabinovitch, R. F. Kubin and R. E. Harrington, *J. Chem. Phys.* **38**, 405 (1963).]

increased from T to $\{1 + [40/(n - 1)]\}T$, an increase of a size not readily attainable. However, even if such a temperature increase were possible to realize experimentally, $P(E)_e$ would become exceedingly flat, or wide, since the variance of the distribution function depends roughly on the *square* of the temperature [Eq. (9-34)]. In practice, thermal systems are investigated at

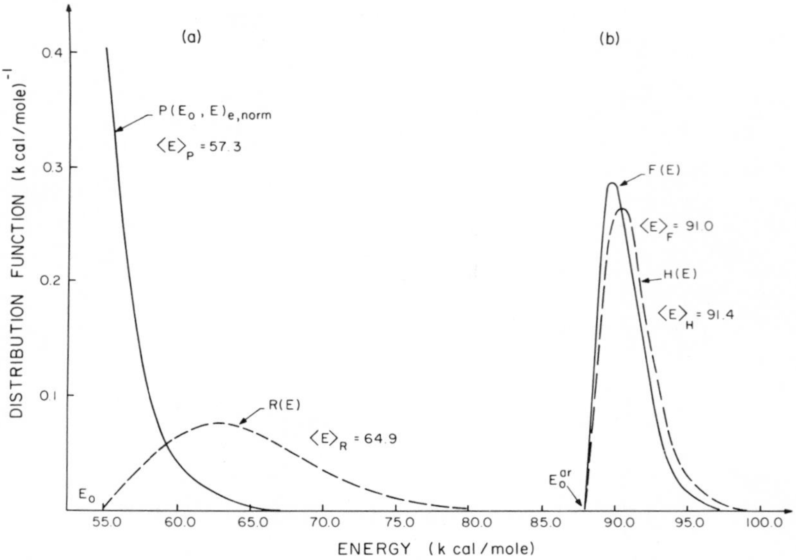

Fig. 9-4 *Comparison of thermally and chemically activated chloroethane systems.* (a) Thermal system $C_2H_5Cl + M \rightarrow (C_2H_5Cl)^v + M$ at 800°K; $E_0 = 55$ kcal mole^{-1}. (b) Chemical activation system $CH_3 + CH_2Cl \rightarrow (C_2H_5Cl)^v$ at 300°K; $E_0^{ar} = 88.3$ kcal mole^{-1}. The distribution function $P(E_0, E)_{e,norm}$ and $F(E)$ refer to the distribution of energies of $(C_2H_5Cl)^v$ *formed* by the thermal and chemical activation processes, respectively. The distribution functions $R(E)$ and $H(E)$ refer to the distribution of energies of $(C_2H_5Cl)^v$ *reacting* per unit time. From D. W. Setser, paper presented at Second Winter Course in Gas Kinetics, Lake Arrowhead, California, 1970.

temperatures where the reaction of interest proceeds at a convenient rate, mostly below 900°K, the softening point of Pyrex. At most, this is only about three times the temperature of a chemical activation system, so that in typical thermal experiments, the maximum of $P(E)_e$ is always at energies well below the threshold energy for reaction. Thus thermal unimolecular reactions amount to averaging $k(E)$ over the tail[18] of $P(E)_e$, while the averaging in the chemical activation systems involves the entire energy range of $F(E)$, not just its tail. These aspects of the comparison of $P(E)_e$ and $F(E)$, touched upon in Section 1, are shown graphically in Fig. 9-4.

The rate constant Returning now to Eq. (9-14), the rate constant $k(E)$ for the chemically activated unimolecular reaction of interest can be formulated in the manner of Eq. (8-25) as

$$k(E) = \alpha G^*(E - E_0)/hN(E) \qquad (E \geqslant E_0) \tag{9-37}$$

[18] The tail-end part of $P(E)_e$, appropriately normalized, is denoted in Chapter 8, Section 5 as $P(E_0, E)_{e,norm}$; this notation is also used in Fig. 9-4.

where α and E_0 are the reaction path degeneracy and critical energy, respectively, for the decomposition A $\rightarrow$ **D**, and $G^*(E - E_0)$ is the integrated density for the transition state of this reaction at excitation energy $E - E_0$. Note that E_0 and $G^*(E - E_0)$ in (9-37) are different from E_0^{ar} and $\mathscr{G}^*(E - E_0^{ar})$ in (9-26).

Since $k(E)$ is defined only for $E \geqslant E_0$ and $F(E)$ is defined only for $E \geqslant E_0^{ar}$, the integrand in (9-14) is zero over part of the energy range and the zero lower limit of integration has only formal significance. Since $E_0 \neq E_0^{ar}$, some care is necessary in determining the actual lower limit of integration. If we look at Fig. 9-1(a), which represents the case $E_0^{ar} > E_0$, it becomes immediately obvious that, referred to its ground state, the minimum energy of the decomposing A(E) is not E_0 but E_0^{ar}, since any system that gets over the barrier at E_0^{ar} has more than the minimum energy necessary to get across the barrier at E_0 by the amount of the difference $E_0^{ar} - E_0$. Similarly, if $E_0^{ar} < E_0$,[19] the minimum energy of the decomposing A(E) will be E_0. In short, then, the lower limit of integration in (9-14) is determined by whichever of the two quantities, E_0 and E_0^{ar}, is the *largest*.

The treatment given sofar is for class 1 reactions. In class 2 reactions, where the activating reaction is merely the reverse of the unimolecular reaction of interest, there is no need to distinguish between the two reactions and therefore the superscript "ar" becomes redundant. The result is a considerable simplification, since then $E_0^{ar} = E_0$, $G^*(E - E_0) = \mathscr{G}^*(E - E_0^{ar})$, so that Eqs. (9-24) and (9-37) become identical, $F(E)$ involves the same transition state as $k(E)$, and the lower limit of the integral in (9-14) is, of course, E_0.

Some, if not most, class 2 chemical activation systems present a practical problem in that the amount of the decomposition product **D** is not experimentally observable since the process of its formation (9-5) cannot be normally distinguished from the activation process (9-4). Hence the average rate constant k_a [Eq. (9-12)] cannot be measured experimentally and all that is accessible to measurement is the amount of the stabilized product **S**, or its rate of accumulation. This will be considered in more detail in Section 3.

A special group of class 2 systems where k_a can be measured is the competitive photochlorination of ethylene, or of its chlorinated derivatives, in the presence of an alkane (*22–24*); in these cases, the ratio **D**/**S** for the chemically activated chloroethyl radical can be extracted from the overall kinetics of the process. Another group of class 2 reactions where the rate constant for the decomposition of the chemically activated reactant can be extracted from the overall kinetics are the systems (*25*) H (or D) + acetylene $\rightleftharpoons (C_2H_3)^v$ and (*26*) H + ethylene $\rightleftharpoons (C_2H_5)^v$ and similar systems (*27*).

[19] In this case, however, $k^{ar}(E)$ cannot be neglected with respect to $k(E)$ in the denominator of $F(E)_{ss}$. See the paragraph following Eq. (9-10).

3. Pressure dependence of rate

It has been pointed out in Section 2 that the pressure dependence of $\langle k(E)\rangle_{ss}$ arises from the spread of energies of the chemically activated reactant A. Therefore by considering the pressure dependence of $\langle k(E)\rangle_{ss}$ in more detail, we can expect to obtain some information about the energetics of the unimolecular process of interest, particularly since pressure (and temperature) is a parameter that can be conveniently manipulated by the experimenter.

The general pressure rate expression (9-14) is too cumbersome, and, as in the case of thermal reactions, it is more useful to consider the two pressure limits. In the limit of high pressure (9-14) gives

$$\lim_{\omega\to\infty}\langle k(E)\rangle_{ss} = k_{a\infty} = \int_0^\infty k(E)F(E)\,dE = \langle k(E)\rangle_F \tag{9-38}$$

The symbol $k_{a\infty}$ is chosen to distinguish the *chemical activation* high-pressure rate constant from the *thermal* high-pressure rate constant k_∞, and the subscript F indicates that the averaging is over $F(E)$. Thus Eq. (9-38) shows that the observable rate constant in the limit of high pressure is simply $k(E)$ averaged over the distribution $F(E)$. Similarly, in the limit of low pressure, (9-14) yields

$$\lim_{\omega\to 0}\langle k(E)\rangle_{ss} = k_{a0} = \frac{1}{\displaystyle\int_0^\infty [F(E)/k(E)]\,dE} = \frac{1}{\langle 1/k(E)\rangle_F} \tag{9-39}$$

where k_{a0} represents the chemical activation low-pressure rate constant.

If $F(E)$ were a delta function, i.e., if the reactant molecules A(E) were formed with just one particular value of E, we would have $k_{a\infty}/k_{a0} = 1$. For a distribution $F(E)$ of finite width, the relation between $k_{a\infty}$ and k_{a0} can be found from Schwarz's inequality introduced previously in Chapter 8, Section 5, Eq. (8-81). If we let

$$f = [k(E)F(E)]^{1/2}, \qquad g = [F(E)/k(E)]^{1/2} \tag{9-40}$$

then

$$\int f^2\,dE = k_{a\infty}; \qquad \int g^2\,dE = 1/k_{a0} \tag{9-41}$$

and from Eqs. (8-81), (9-40), and (9-41), it follows that

$$k_{a\infty}/k_{a0} \geqslant 1 \tag{9-42}$$

Thus we can expect that in general k_{a0} will be smaller than $k_{a\infty}$ since the

TABLE 9-1

RATE CONSTANTS AND AVERAGE ENERGY OF *sec*-BUTYL RADICALS IN VARIOUS SYSTEMS AT 298°K[a]

Activating reaction	k_{a0} ($sec^{-1} \times 10^7$)	$k_{a\infty}$ ($sec^{-1} \times 10^7$)	$\langle E \rangle_F$ (kcal $mole^{-1}$)	$\langle E \rangle_H$ (kcal $mole^{-1}$)
H + *trans*-butene-2	—	0.82	8.57	10.3
H + *cis*-butene-2	0.90	1.48	9.88	11.41
H + butene-1	1.3	2.4	10.98	12.40
D + *trans*-butene-2	0.98	1.72	10.48	12.04
D + *cis*-butene-2	1.35	2.20	11.78	13.19
D + butene-1	3.1	4.2	12.88	14.19

[a] k_{a0} and $k_{a\infty}$ are experimental, $\langle E \rangle_F$ and $\langle E \rangle_H$ are calculated. E refers to energy in excess of E_0. Experimental data and calculations of B. S. Rabinovitch, R. F. Kubin and R. E. Harrington, *J. Chem. Phys.* **38**, 405 (1963).

reactant A is always formed with some spread of energies[20]; this is borne out by experiment, as shown in Table 9-1.

It is interesting to make a comparison between the average energies of molecules A decomposing per unit time at the two extremes of pressure. The energy distribution function for molecules decomposing per unit time and per unit energy range, at any pressure, is the integrand of $\langle k(E) \rangle_{ss}$, i.e. $k(E)F(E)/\int_0^\infty k(E)F(E)\,dE$. The average energy of molecules reacting per unit time at the low-pressure limit is therefore

$$\lim_{\omega \to \infty} \langle E \rangle = \left[\int_0^\infty EF(E)\,dE\right] \Big/ \int_0^\infty F(E)\,dE = \langle E \rangle_F \tag{9-43}$$

since under these conditions $F(E)_{ss}$ reduces to $F(E)/k(E)$. Thus the average energy of molecules reacting per second at low pressures is determined by the distribution of energies the molecules have as they are formed. This is understandable because at low pressure, there are not enough deactivating collisions to substantially modify the distribution of energies of the formed molecules before they react. At the high-pressure limit, $F(E)_{ss}$ reduces to $F(E)/\omega$, and therefore

$$\lim_{\omega \to \infty} \langle E \rangle = \left[\int_0^\infty Ek(E)F(E)\,dE\right] \Big/ \int_0^\infty k(E)F(E)\,dE = \langle E \rangle_H \tag{9-44}$$

where the subscript H indicates averaging over $H(E)$ given by

$$H(E) = k(E)F(E) \Big/ \int_0^\infty k(E)F(E)\,dE = k(E)F(E)/k_{a\infty} \tag{9-45}$$

[20] Note, however, that the result (9-42) is derived from Eq. (9-14), which involves the assumption that $k_2[M] = \omega$, i.e., that there is unit efficiency for deactivation. In a multilevel system (Section 5), the inequality is reversed, as may be seen from Fig. 9-6.

Since $H(E)$ is simply the function $F(E)$ weighted by the factor $k(E)$, which increases with E and therefore has the effect of giving more weight to higher energies, we can expect that $\langle E \rangle_H > \langle E \rangle_F$, the inequality increasing with the width of $F(E)$. Calculated values of $\langle E \rangle_H$ and $\langle E \rangle_F$ in the butyl radical system are shown in Table 9-1. The small difference between $\langle E \rangle_H$ and $\langle E \rangle_F$ attests to the small width of $F(E)$. The functions are also shown graphically in Fig. 9-4 for the chloroethane system, where the difference $\langle E \rangle_H - \langle E \rangle_F$ is even smaller.

Chemical activation versus thermal systems To provide a connection between thermal and chemical activation results on the same system,[21] we may note that $F(E)$ [Eq. (9-22)] has the same form as the function $R(E)$ of Eq. (8-71) introduced in connection with thermal reactions. That is, we can write

$$F(E) = k^{\mathrm{ar}}(E)P(E)_{\mathrm{e}} \Big/ \int_0^{\infty} k^{\mathrm{ar}}(E)P(E)_{\mathrm{e}}\, dE = R^{\mathrm{ar}}(E) \qquad (9\text{-}46)$$

so that, formally, the only difference between $R^{\mathrm{ar}}(E)$ and $R(E)$ is that the former refers to a reaction that is *not* the same reaction as the unimolecular reaction of interest. Such is the case for class 1 reactions, but for class 2 reactions, all difference between $R^{\mathrm{ar}}(E)$ and $R(E)$ disappears since $k^{\mathrm{ar}}(E)$ in (9-46) then does in fact refer to the unimolecular reaction of interest.[22] The function $R(E)$, which is shown in Fig. 8-1, has actually been calculated for a class 2 chemical activation system.[23]

It is therefore possible to establish a direct correspondence, at some common temperature T, between parameters of the thermal and chemical activation versions of the same class 2 reaction. The amount of stabilized product **S** (for any class of reaction) is the denominator of (9-13), where $F(E)_{\mathrm{ss}}$ is given by (9-11). In a class 2 reaction, $F(E) \equiv R(E)$, and therefore we can write

$$\mathbf{S} = \langle \omega/[k(E) + \omega] \rangle_R \times \int_0^{\tau} (d[\mathrm{A}]/dt)\, dt \qquad (9\text{-}47)$$

[21] The denominator in (9-46) is nothing else than the *thermal* high-pressure rate constant for the reverse of the activating reaction (case of class 1 reaction), or for the actual unimolecular reaction of interest (case of class 2 reaction).

[22] The implicit assumption here is that the thermal and chemical activation systems are compared at the same temperature, although in practice thermal reactions may require higher temperatures than chemical activation systems to proceed at a measurable rate. See Fig. 9-4 for the chloroethane system as an example.

[23] It must be carefully pointed out that despite their identity, the functions $R(E)$ serve different purposes in the two kinds of systems. In thermal systems, $R(E)$ is the actual integrand in the rate constant expression, i.e., represents the contribution of each energy range to the overall rate, while $R(E)$ in the chemical activation system appears only as a distribution function that still needs to be multiplied by $k(E)$ to get the actual integrand of the rate constant expression.

the subscript indicating averaging over $R(E)$. If centrifugal effects are neglected, this average is equal, by Eq. (8-73), to k_{uni}/k_{∞}. Now, $d[A]/dt$ measures the total rate of disappearance of A, so that

$$\int_0^{\tau} (d[A]/dt)\, dt = \mathbf{S} + \mathbf{D} \tag{9-48}$$

Hence (9-47) can be written

$$k_{uni}/k_{\infty} = \mathbf{S}/(\mathbf{S} + \mathbf{D}) = [1 + (\mathbf{D}/\mathbf{S})]^{-1} \tag{9-49}$$

Note that at a pressure such that $\mathbf{D}/\mathbf{S} = 1$, we have $k_{uni}/k_{\infty} = 0.5$. If $\mathbf{D}/\mathbf{S}$ is known as a function of pressure, the thermal fall-off curve giving the dependence of k_{uni}/k_{∞} on pressure can be directly constructed using Eq. (9-49), and vice versa. Equation (9-49) has been used to correlate thermal experimental data with calculated $\mathbf{S}/\mathbf{D}$ values in the dissociation of hydrazine.[24]

The connection between thermal and chemical activation results on the same class 2 reaction in which $\langle k(E)\rangle_{ss}$ (i.e., $\mathbf{D}/\mathbf{S}$) is experimentally obtainable can be pursued further. We have designated in Chapter 8, Section 5 the slope of a plot of the (thermal) k_{uni} versus $1/[M]$ by $\textsc{m}$, where $\textsc{m}$ is given by Eq. (8-76), which can be written, disregarding centrifugal effects, as

$$\textsc{m} = (k_{\infty}/Z) \int_0^{\infty} k(E)R(E)\, dE \tag{9-50}$$

Since for a class 2 reaction $F(E) \equiv R(E)$, (9-38) can be written

$$k_{a\infty} = \int_0^{\infty} k(E)R(E)\, dE \tag{9-51}$$

Hence the chemical activation parameter $k_{a\infty}$ is related to the thermal parameters $\textsc{m}$ and k_{∞} by

$$k_{a\infty} = Z\textsc{m}/k_{\infty} \tag{9-52}$$

Since the slope $\textsc{m}$ is in general difficult to obtain, relation (9-52) is not likely to be useful in practice. Of greater potential import are relations connecting the low-pressure thermal and chemical activation rate constants. For class 2 reactions, $k^{ar}(E) \equiv k(E)$, and using (9-46), Eq. (9-39) becomes

$$k_{a0} = \left[\int_{E_0}^{\infty} k(E)P(E)_e\, dE\right] \bigg/ \int_{E_0}^{\infty} P(E)_e\, dE \tag{9-53}$$

[24] Setser and Richardson (*28*). More recent data for the $H—NH_2$ bond dissociation energy suggest that the threshold energy used in these calculations may have been too low.

while the thermal second-order rate constant k_{bi}, disregarding centrifugal effects, is [cf. Eq. (8-43) or (8-78)]

$$k_{bi} = \bar{Z} \int_{E_0}^{\infty} P(E)_e \, dE \tag{9-54}$$

Hence

$$k_{a0} = Zk_{\infty}/k_{bi} \tag{9-55}$$

Relations (9-52) and (9-55) could be used to check the mutual consistency of the thermal and chemical activation versions of a class 2 reaction.

If the amount of the decomposition product **D** is not experimentally observable in a class 2 chemical activation system, only the rate of accumulation of the stabilized product $d[\mathrm{S}]/dt$ is accessible to measurement. From (9-47), we have

$$\frac{d[\mathrm{S}]}{dt} = \left\langle \frac{\omega}{k(E) + \omega} \right\rangle_R \times \frac{d[\mathrm{A}]}{dt} \tag{9-56}$$

The total rate of formation of A is

$$d[\mathrm{A}]/dt = k[\mathrm{X}][\mathrm{Y}] \tag{9-57}$$

so that

$$\frac{d[\mathrm{S}]}{dt} = k[\mathrm{X}][\mathrm{Y}] \times \left\langle \frac{\omega}{k(E) + \omega} \right\rangle_R \tag{9-58}$$

Equation (9-58) states simply that $d[\mathrm{S}]/dt$ is equal to the total rate of formation of A (of all energies), times the fraction of those at energies above E_0 collisionally deactivated to S at pressure $\omega = zp$ (or $Z[\mathrm{M}]$). Thus the rate of formation of **S** is in general of third order, unless $\omega \gg k(E)$, i.e., unless the pressure is sufficiently high.

A class 2 system for which there exists extensive information, and which would therefore be useful for testing the various equations relating thermal and chemical activation data, is the system $CH_3 + CH_3 \rightleftharpoons C_2H_6$. The recombination of methyl radicals (the forward process) is a chemical activation system, and the decomposition of collisionally activated C_2H_6 (the reverse process) is the thermal system. A recent review (*29*; cf. *30*, *117*) of all available data unfortunately reveals a number of discrepancies between the forward and reverse processes, which seems to arise from the unreliability of experimental data rather than from the failure of the theory. (See also the end of Section 2 in Chapter 11.)

A less direct, but sometimes useful, comparison of thermal and chemical activation data on the same class 2 reaction can be done by means of Eq.

(9-27). For a class 2 chemical activation system, the transition-state partition function $\mathcal{Q}^*$, calculable from Eq. (9-27), refers to the same transition state as Q^* in the thermal version of the same system; i.e., $\mathcal{Q}^* = Q^*$. Hence Q^* and Q, the partition function for the reactant, can be used to obtain $(kT/h)Q^*/Q$, which, by Eqs. (8-23) and (8-103), is roughly equal to the thermal high-pressure preexponential factor A_∞. Thus if A_∞ is known from experiment, one can check if the transition state used to interpret chemical activation results is compatible with thermal data.

Identifying A_∞ directly with $(kT/h)Q^*/Q$ ignores the difference between $E_{a\infty}$ and E_0, but if the accuracy of the experimental data justifies it, this could be easily taken care of by calculating $E_{a\infty}$ from Eq. (8-87) first and then calculating A_∞ from Eq. (8-103); however, we still ignore α and f_∞.

Relating Q^*/Q to A_∞ is obviously not limited to class 2 reactions; in fact, such a comparison has been used to relate thermal and chemical activation data on several chemically activated C_4 and C_5 hydrocarbons (*31–34*), which are class 1 systems. However, such a comparison in a class 1 system offers no check on $\mathcal{Q}^*$, i.e., on $\mathcal{G}^*(E)$ and, by extension, on $F(E)$. More is given on this subject in Section 7.

4. Temperature dependence of rate

The experimental activation energy E_a in a chemical activation system is given by the usual logarithmic derivative of the observable rate constant, i.e., by

$$E_a = kT^2\, \partial[\ln\langle k(E)\rangle_{ss}/\partial T] \qquad (9\text{-}59)$$

We can write, using (9-12) and (9-13),

$$\langle k(E)\rangle_{ss} = \left[\int k(E)F(E)_{ss}\, dE\right]\Big/\int F(E)_{ss}\, dE \qquad (9\text{-}60)$$

From (9-11) and (9-26), we see that the only temperature-dependent factors in $F(E)_{ss}$ are ω and $e^{-E/kT}$. Proceeding as in Chapter 8, Section 6, we neglect the temperature dependence of ω relative to the exponential, so that logarithmic differentiation of (9-60) gives, at any pressure,

$$E_a \simeq \frac{\int Ek(E)F(E)_{ss}\, dE}{\int k(E)F(E)_{ss}\, dE} - \frac{\int EF(E)_{ss}\, dE}{\int F(E)_{ss}\, dE} \qquad (9\text{-}61)$$

The first term in (9-61) is E averaged over $k(E)F(E)_{ss}$, i.e., over the general pressure steady-state distribution of molecules reacting per unit time, and the second term is E averaged over $F(E)_{ss}$, i.e., over the energy distribution of all chemically activated molecules present and reacting at steady state.

Hence (9-61) can be rewritten in the form

$$E_a = \langle E \rangle_{\text{reacting/sec}} - \langle E \rangle_{\text{reacting}} \tag{9-62}$$

which is the formal analog of the thermal expression (8-86); in fact, the second term on the right-hand side of both (8-85) and (9-61) refers to the average energy of all reactant molecules A. Since in the thermal case, the energy of reactive states is located well past the maximum of the thermal distribution toward its tail, averaging over the whole of the thermal distribution amounts essentially to averaging over nonreactive states. This is quite unlike what we have in the second term on the right-hand side of Eq. (9-61) where averaging is exclusively over reactive states, with none of the nonreactive ones being included, since $F(E)_{ss}$ is located entirely on the high-energy side of the threshold energy for reaction.

We can therefore expect that $\langle E \rangle_{\text{reacting}}$ will be smaller than $\langle E \rangle_{\text{reacting/sec}}$ but not much smaller: it will certainly be much larger than $\langle E \rangle_{\text{nonreacting}}$ in the thermal case. In other words, the experimental activation energy E_a in a chemical activation system will be much less than E_a in a comparable thermal system. This is borne out by experiment; for instance, in the trichloroethyl radical system, Knox and Riddick (*22*) have measured $E_a \approx 1.3$ kcal mole^{-1}, a very small value compared with the expected ~ 20 kcal mole^{-1} in a similar thermal system.

At the high-pressure limit, Eq. (9-62) can be easily expressed in terms of averages over $F(E)$ and over the function $H(E)$ introduced previously. When $\omega \gg k(E)$, $F(E)_{ss}$ reduces to $F(E)/\omega$, and the high-pressure activation energy $E_{a\infty}$ is then, from Eq. (9-61), simply[25]

$$E_{a\infty} = \langle E \rangle_H - \langle E \rangle_F \tag{9-63}$$

For the butene system, $E_{a\infty}$, as given by (9-63), can be obtained directly from the calculated data of Table 9-1, showing again that it is quite small, about 1.5 kcal mole^{-1}.

It is easy to see that the magnitude of E_a is just another reflection of the energy spread in $F(E)_{ss}$, for if $F(E)_{ss}$ were a delta function, exponentially dependent on temperature, E_a [Eq. (9-61)] would be zero, i.e., the observable rate constant in a monoenergetic system would be temperature independent.

Thus "activation energy" in a chemical activation system measures only a small difference in average energies between two differently defined classes of reacting molecules. Therefore it does not have the physically simple and useful meaning it has in thermal systems, where it is a measure of the average

[25] Since ω drops out, the result (9-63), unlike (9-61), is exact.

excess energy of reacting molecules relative to nonreactive ones, and where for this reason it can be directly related to the critical energy, an important parameter. The smallness of the temperature coefficient of rate in a chemical activation system makes it likely that other factors, not directly related to the energetics of the process, may contribute; for example, the temperature dependence of ω which we have neglected and which reflects the temperature dependence of the collision cross section (35) [cf. Eqs. (8-120) and (8-121)], and, in multilevel systems (see Section 5), the temperature dependence of the average energy transferred per collision $\langle \Delta E \rangle$. On the experimental side, a small temperature coefficient of rate is difficult to measure experimentally, and an increase in temperature invariably introduces complications due to enhanced side reactions. For all these reasons, it is unprofitable, at least for the time being, to pursue further the temperature dependence of the chemical activation rate constant, on both theoretical and practical grounds.

5. Collisional deactivation in a multilevel system

The assumption was made in Section 2 that every gas-kinetic collision removes from $A(E)$ sufficient energy to stabilize it. This is a simple one-step deactivation mechanism reminiscent of the one-step Lindemann mechanism of thermal reactions. We now wish to generalize this simple deactivation mechanism by including all possible collisional transitions, both up and down.

The general scheme to be considered is

$$X + Y \xrightarrow{\Re F_j} A_j$$

$$A_i + M \underset{\omega \mathcal{q}_{ji}}{\overset{\omega \mathcal{q}_{ij}}{\rightleftharpoons}} A_j + M \qquad (9\text{-}64)$$

$$A_j \xrightarrow{k_j} D$$

Here, the subscripts i and j are short-hand notations for states of energies E_i and E_j, respectively, $\mathcal{q}_{ij}$ and $\mathcal{q}_{ji}$ are probabilities of transition per collision, k_j is the rate constant for decomposition of species A having energy E_j, $\Re$ is the total rate of formation of A by process (9-4), and F_j is the fraction of A's having[26] energy E_j. The rate constant $k_j \neq 0$ only for reactive states and $k_j = 0$ otherwise; molecules A_j for which $k_j = 0$ are classified as stabilized reactant S. This mechanism is quite similar to the multistep Lindemann mechanism (8-126), except that the activating reaction (9-4) now provides an additional source of A_j.

[26] Thus $\Re F_j$ is the term $(d[A]/dt)F(E_j)\,dE_j$ of Section 2 in more compact notation, with $\Re = d[A]/dt$ given by Eq. (9-57).

This scheme leads to the master equation

$$d[A_j]/dt = \omega \sum_i q_{ij}[A_i] - \omega[A_j] \sum_i q_{ji} - k_j[A_j] + \Re F_j \qquad (9\text{-}65)$$

which is again quite similar to the master equation (8-127) except for the term $\Re F_j$. In the present formulation, the total rate of formation of the decomposition product **D** is

$$\mathbf{D} = \sum_j k_j[A_j] \qquad (9\text{-}66)$$

and the total rate of formation of the stabilized reactant **S** is

$$\mathbf{S} = \Re - \sum_j k_j[A_j] \qquad (9\text{-}67)$$

so that the observable chemical activation rate constant can be written

$$\langle k(E) \rangle_{\text{ss}} = \frac{\omega \mathbf{D}}{\mathbf{S}} = \left(\omega \sum_j k_j[A_j] \right) \Big/ \left(\Re - \sum_j k_j[A_j] \right) \qquad (9\text{-}68)$$

in which use is made of the interpretation of **D/S** as the ratio of the respective (total) rates of formation.

Before embarking on the actual solution of Eq. (9-65), we make the same simplifying assumptions as in Chapter 8, Section 8, namely that q_{ij} and q_{ji} are time independent and that a steady state is set up shortly after the beginning of the reaction, i.e., that $d[A_j]/dt = 0$ is satisfied during all the time history of the system that matters. The solution of (9-65) could now proceed in a manner quite similar to that employed in Chapter 8, by finding the nonequilibrium steady-state distribution. However, given the nature of chemical activation systems and the sort of information that one can obtain from them, it is more convenient to proceed in a different way. The reason is the presence of an external source of A_j's which forms them very much in excess of the rate by which they would be formed by collisions, so that, in contrast with a thermal system, collisions in a chemical-activation system serve not so much to populate the active region above E_0 as to *depopulate* it.

The step-ladder model Basically, the information we want is to see how the equations that we have derived so far in this chapter have to be modified if **S** is formed in more than one step. The amount of the stabilized reactant **S** depends on the steady-state populations $[A_j]$ obtained by the solution of (9-65), and the solution depends in turn on the model for the transition probabilities q_{ij} and q_{ji}. Since experimental data in general are not accurate enough to distinguish among several possible models, we shall use the simplest model of them all, used extensively by Rabinovitch, Setser and co-workers (*36–39*), one which involves a simple stepladder cascade through the region of reactive states. To this end, several simplifying assumptions are necessary.

1. We shall assume that there are n discrete energy levels in the active region above E_0, and that the activating reaction produces monoenergetic A's in the topmost nth level of energy E_n; hence $F_j = 1$ for $j = n$ and $F_j = 0$ for $j \neq n$. As a further simplification, we shall take these levels to be evenly spaced, i.e., $E_i - E_{i-1} = \Delta E =$ const independent of i.

2. We shall neglect all "up" transitions, i.e., we shall assume that $\varphi_{ij} = 0$ for $j > i$. This is justified by the high level of excitation in chemical activation systems, which is very much in excess of thermal (i.e., collision-produced) energy, and therefore makes it unlikely that many energy levels above E_0 are populated by collisions. Although, strictly speaking, the neglect of "up" transitions violates detailed balance, this is not too serious if the step size is sufficiently large (i.e., the collider is sufficiently efficient) so that $\exp(-\Delta E/kT) \sim 0$.

3. For the "down" transitions, we shall take the simple probability[27]

$$\begin{aligned} \varphi_{ji} &= \xi \qquad \text{if} \quad j = i + \mathrm{s} \\ &= 0 \qquad \text{if} \quad j \neq i + \mathrm{s} \end{aligned} \tag{9-69}$$

Here ξ ($\xi < 1$), assumed to be independent of i and j, is the probability of inelastic collision and therefore represents essentially an efficiency factor; if φ_{ii} is the probability of elastic collision, then $\xi + \varphi_{ii} = 1$, since the probabilities must satisfy completeness. With $\xi < 1$, the probability is bimodal, meaning it has two maxima, one at $j = i + \mathrm{s}$ and the other at $j = i$. Physically, $\xi < 1$ is tantamount to a reduced collision cross section. Obviously $\xi = 1$ if no elastic collisions are allowed. The transition probability (9-69) permits only transitions of the specified step size to occur, determined by the parameter s, and all other step sizes are ruled out. In other words, deactivation from the active region takes place in a succession of constant energy decrements, hence the term "stepladder cascade." The cascade ends on reaching the zeroth level just below E_0, at which point molecules are classified as stabilized. In this sense, therefore, the zeroth level acts as an absorbing barrier.

Since $k(E)$ declines very strongly with energy (cf. Fig. 7-3), molecules are deactivated (at a given ω) as soon as $k(E)$ becomes smaller than ω in the course of a cascade. Hence molecules are effectively deactivated before they have actually reached the zeroth level.

A graphical representation of a system having the simplifying features 1–3 is shown in Fig. 9-5. As a preliminary attempt at solution, we shall first

[27] In the original literature (Refs. *36–39*), the notation $\varphi_{ji} = \beta$ is used. However, this β is different from the β introduced in Chapter 8 [Eqs. (8-123) and (8-125)], and to avoid confusion, the symbol ξ is used here.

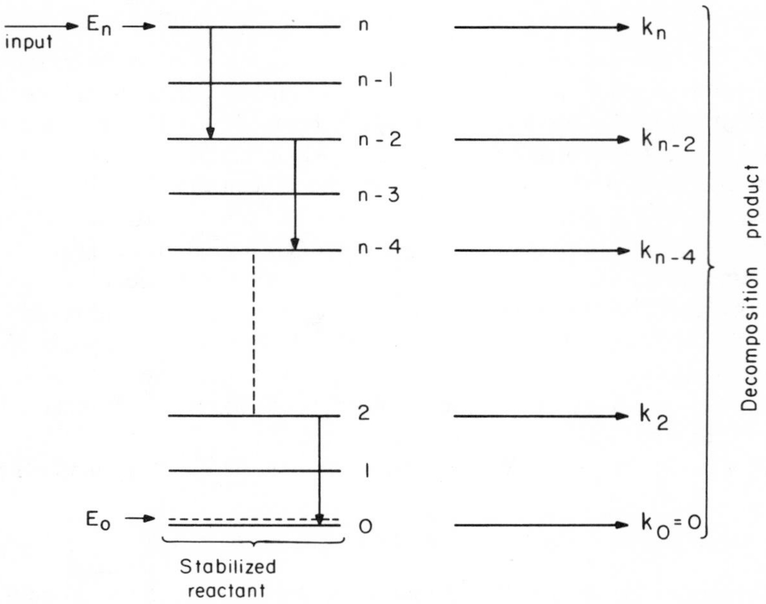

Fig. 9-5 *Schematic representation of stepladder cascade deactivation.* Input process is chemical activation producing vibrationally excited monoenergetic reactant of energy E_n, taken to be the highest (nth) level of the system. The first three and the last five levels of the systems are shown; all levels are equidistant. The zeroth level is just below the critical energy E_0. Vertical arrows represent transitions produced by collisional deactivation for an assumed step size of $\mathrm{s} = 2$. Horizontal arrows represent decomposition with rate constants k_n, k_{n-2}, etc. The rate constant for the zeroth level vanishes.

consider the simplest case, when stabilization takes place in just one step; this is the conventional strong-collision assumption. The step size in (9-69) is then $\mathrm{s} \geqslant n$, and Eq. (9-65) reduces at steady state to

$$d[\mathrm{A}_n]/dt = (k_n + \xi\omega)[\mathrm{A}_n] - \mathfrak{R} = 0 \tag{9-70}$$

from which it follows that $[\mathrm{A}_n] = \mathfrak{R}/(k_n + \xi\omega)$; hence we get for **S** and **D** from (9-66) and (9-67)

$$\mathbf{D} = \mathfrak{R}k_n/(k_n + \xi\omega); \qquad \mathbf{S} = \mathfrak{R}\xi\omega/(k_n + \xi\omega) \tag{9-71}$$

For the observable rate constant, we then get

$$k_a = \langle k(E)\rangle_{ss} = \xi\omega\mathbf{D}/\mathbf{S} = k_n \tag{9-72}$$

which is the same result that (9-14) yields when $F(E)$ is a delta function, as of course it should, since (9-14) incorporates the strong collision assumption.

To extend the argument to the case of a many-step, or "many-shot," deactivation, a term coined by Serauskas and Schlag,[28] we make use of the postulate of random incidence of collision and dissociation, alluded to previously in Chapters 1 and 8. A convenient point of departure is Eq. (8-6), which, applied to the present problem, gives a general relation for $P(\omega)_{\text{abs}}$, the absolute probability that A_n is present intact at the time of the first collision, using only the assumption that collisions are random but not assuming anything about the distribution of lifetimes of A_n with respect to dissociation. If, now, the lifetime of A_n with respect to dissociation (i.e., in absence of collisions) is also random, it is given by Eq. (8-9), which in the present instance, reads

$$P(t) = e^{-k_n t} \tag{9-73}$$

Therefore the absolute probability of A_n present intact at the first collision, assuming random incidence of collisions and of dissociation, is

$$P(\omega)_{\text{abs}} = \xi\omega \int_0^\infty e^{-(k_n + \xi\omega)t}\, dt = \xi\omega/(k_n + \xi\omega) \tag{9-74}$$

where the slightly more general $\xi\omega$ is written for ω of Eq. (8-6). We can thus write (9-71) as

$$\mathbf{S}_1 = \mathfrak{R}P(\omega)_{\text{abs}}; \qquad \mathbf{D}_1 = \mathfrak{R}[1 - P(\omega)_{\text{abs}}] \tag{9-75}$$

By hypothesis, A_n becomes stabilized in the one-shot mechanism when it is present intact at the first collision, so that $P(\omega)_{\text{abs}}$ is in fact the absolute probability of stabilization and $1 - P(\omega)_{\text{abs}}$ the absolute probability of dissociation. The subscripts one in (9-75) indicate that the relations apply to the one-step process.

Since in forming the ratio **D**/**S** the factor $\mathfrak{R}$ drops out, we have yet another way of formulating **D**/**S** as the ratio of absolute probabilities of decomposition and stabilization, respectively.

Many-step deactivation Suppose now that the molecule A_n has to make N transitions[29] in the active region before it becomes stabilized, so that the step size s in Eq. (9-69) is now $\text{s} = n/N$. In other words, it now takes N collisions before the molecule can become stabilized. The absolute probability that the molecule A_n will be present, and therefore deactivated, at the first collision, is, by (9-74), $\xi\omega/(k_n + \xi\omega)$. Having lost energy equivalent to the step size, A_n becomes at this point $A_{n-\text{s}}$ which has the probability of decomposition per second (= rate constant) equal to $k_{n-\text{s}}$; then the absolute

[28] Serauskas and Schlag (*40,41*). These authors give a particularly clear statement of the problem.

[29] Rabinovitch and others (Refs. *36–39*) use the symbol T for the number of transitions in the active region. To avoid confusion with T for temperature, N is used here instead.

probability of A_{n-s} becoming in turn deactivated by a collision is $\xi\omega/(k_{n-s} + \xi\omega)$. The absolute probability of A_n suffering two successive deactivating collisions is $(\xi\omega)^2/(k_n + \xi\omega)(k_{n-s} + \xi\omega)$, a simple product, the probabilities of each successive collision being independent since they refer to random events. By induction, the absolute probability that A_n will suffer N collisions, i.e., will become stabilized, is

$$\mathbf{S}_N = \prod_{j=1}^{N} \frac{\xi\omega}{k_j + \xi\omega}; \qquad \mathbf{D}_N = 1 - \mathbf{S}_N \tag{9-76}$$

where k_j is the unimolecular rate constant for decomposition from level jn/N. The observable rate constant k_a is therefore

$$k_a = \frac{\omega \mathbf{D}_N}{\mathbf{S}_N} = \omega \frac{1 - \prod_{j=1}^{N} [\xi\omega/(k_j + \xi\omega)]}{\prod_{j=1}^{N} [\xi\omega/(k_j + \xi\omega)]} = \omega\left[\prod_{j=1}^{N}\left(1 + \frac{k_j}{\xi\omega}\right) - 1\right] \tag{9-77}$$

It should be noted that for small values of j such that $k_j \ll \omega$, the contribution of such terms to the product in (9-76) is essentially unity. Therefore the strong-collision rate constant k_a calculated by taking $N = 1$ is in practice indistinguishable from k_a calculated with a larger N. The reason is that k_j decreases very steeply with decreasing j (i.e., with decreasing energy), and therefore to stabilize a molecule of A, it suffices to knock it to an energy level having a small enough rate constant, not necessarily zero, which is tantamount to $N > 1$. The precise value of $N (> 1)$ that still reproduces the strong-collision value of k_a depends on pressure.[30] In the dichloroethane system (*39*), $k_{a\infty}$ calculated with $N \leqslant 3$ ($s \geqslant 12$ kcal mole^{-1}) is indistinguishable from the strong-collision result.

As usual, two special cases may be recognized at the two extremes of pressure. In the high-pressure limit, $\omega \gg k_j$ for every j, which, when substituted into (9-77), yields $k_{a\infty} = 0$. This merely expresses the fact that at a sufficiently high pressure, all reactant molecules are stabilized before they have a chance to react. To get a more interesting result, we have to use a less extreme high-pressure condition. If we first develop the product

$$\prod_{j=1}^{N}\left(1 + \frac{k_j}{\xi\omega}\right) = 1 + \sum_j \frac{k_j}{\xi\omega} + \sum_{i \neq j} \frac{k_i k_j}{(\xi\omega)^2} + \sum_{i \neq j \neq l} \frac{k_i k_j k_l}{(\xi\omega)^3} + \cdots \tag{9-78}$$

and then keep only the term in $(\xi\omega)^{-1}$, Eq. (9-77) yields the more useful result

$$k_{a\infty} = (1/\xi) \sum_{j=1}^{N} k_j \tag{9-79}$$

[30] The author is indebted to Prof. D. W. Setser for correspondence on this point.

i.e., $k_{a\infty}$ is defined as the slope of the plot of k_a versus ω^{-1} near $\omega^{-1} \to 0$. Experimentally, $k_{a\infty}$ is therefore obtained by plotting **D/S** versus pressure^{-1} and extrapolating to pressure$^{-1} \to 0$. The plot should go through the origin and **D/S** should approach zero along a straight line[31] the slope of which is $k_{a\infty}/k_2$ (recall that $\omega = k_2 \times$ pressure, cf. Section 2). By assigning a value to the collision number k_2 [for example, $k_2 = Z$ or Z_{L-J}, Eq. (8-120) or (8-121)], $k_{a\infty}$ follows.

Equation (9-79) is useful only for a limited comparison of energy transfer quantities, in a system containing an inert, with those of some standard system for which N (say, N_{std}) is known (from low-pressure data discussed later); often, a system containing only reactants (no inert) is used as a standard. We then have

$$k_{a\infty}/(k_{a\infty})_{std} = \left[(\xi)_{std} \sum_{j=1}^{N} k_j\right] \Big/ \left(\xi \sum_{j=1}^{N_{std}} k_j\right) \tag{9-80}$$

Assuming that $(\xi)_{std}/\xi = 1$, which is equivalent to assuming that the respective collision cross sections are exactly known, N can be calculated if N_{std} is known.

More interesting deductions can be made at low pressures. Here, $\omega \ll k_j$, and (9-77) yields directly

$$k_{a0} = \frac{1}{\omega^{N-1}} \prod_{j=1}^{N} \frac{k_j}{\xi} \tag{9-81}$$

We see from this equation that the low-pressure behavior of k_a depends on $1/\omega^{N-1}$, and therefore when $N \gg 1$, k_{a0} rises rapidly as pressure decreases. Such a low-pressure "turn-up" of k_a is an unmistakable sign of a cascade; inversely, the absence of a turn-up at low-pressure is a sign that N is small. Note that the conclusions that can be drawn from the low-pressure turn-up are independent of any assumption regarding ξ or the collision cross section πd^2, since any change in either ξ or d^2 merely changes the effective pressure. By matching experimental low-pressure data to the calculated curve giving the pressure dependence of k_a, it is possible to determine N.

In actual practice, the active region is usually divided into levels 1 kcal mole^{-1} or less, apart, and it is then more informative to indicate step size in kilocalories per mole rather than the number of transitions N. It is also more convenient to plot $k_a/k_{a\infty}$ versus **S/D**, which is obtained directly from experiment, rather than plot just k_a versus pressure. Figure 9-6 shows an example of matching experimental and calculated low-pressure data in the case of collisional deactivation of chemically activated 1,2-dichloroethane (*31*) by CH_2Cl_2, CH_3Cl, and CF_4.

[31] For an example of such a plot in the dichloroethane system, see Hassler, *et al.* (*42*, Fig. 2). For the difluoroethane system, see Ref. 35, Figs. 1 and 2.

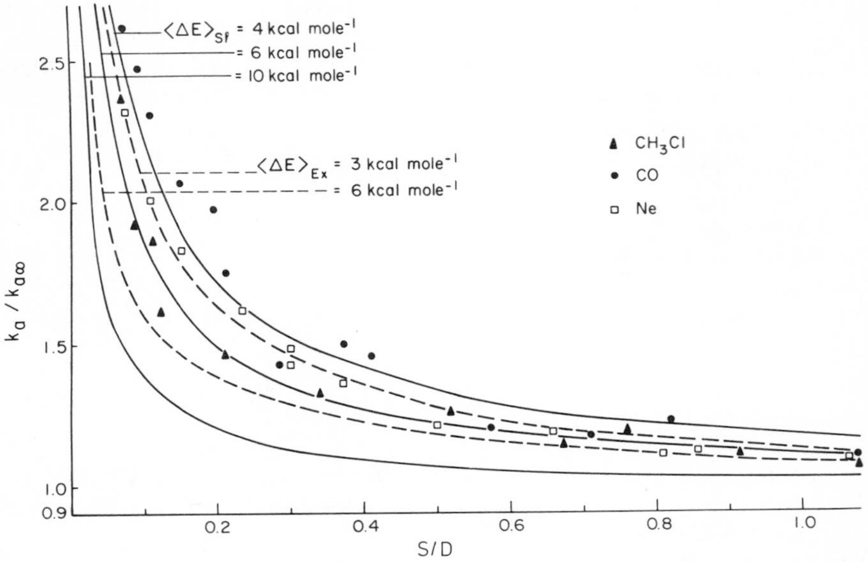

Fig. 9-6 *Matching of calculated and experimental data in the system* $(ClCH_2CH_2Cl)^v + M \rightarrow ClCH_2CH_2Cl$ (*stabilized*) $+ M$.

The figure contrasts results for weak colliders [CO(●), Ne(□)] with those for a strong collider [CH_3Cl)(▲)]. The points are experimental data, the curves are calculated results, full lines for the stepladder model, broken lines for the exponential model of the transition probability. The high-pressure rate constants $k_{a\infty}$ for the data shown are (in sec^{-1}) 2.9×10^8 (CH_3Cl), 4.0×10^8 (CO), and 8.2×10^8 (Ne). Good fit for the $k_a/k_{a\infty}$ plot and for the experimental $k_{a\infty}$ are obtained for the following models: CH_3Cl: $\langle \Delta E \rangle_{Sl} = 6$ kcal mole^{-1}, $k_{a\infty} = 3.1 \times 10^8$ sec^{-1}; CO: $\langle \Delta E \rangle_{Sl} = 4$ kcal mole^{-1}, $k_{a\infty} = 4.0 \times 10^8$ sec^{-1}; Ne: $\langle \Delta E \rangle_{Ex} = 3$ kcal mole^{-1}, $k_{a\infty} = 7.7 \times 10^8$ sec^{-1}. The chemical reaction is $2CH_2Cl \rightarrow (ClCH_2CH_2Cl)^v \rightarrow HCl + C_2H_3Cl$. E_0 for HCl elimination was taken to be 60 kcal mole^{-1}; $(ClCH_2CH_2Cl)^v$ is formed with an average excitation energy of 89 kcal mole^{-1}, so that the active region spans 34 kcal mole^{-1} from E_0 upward. E. E. Siefert and D. W. Setser, unpublished results.

In chemical activation work, the parameter ξ is always (see footnote 30) taken equal to unity, meaning that the observed rates at both pressure extremes can be consistently accounted for, on the stepladder model, by using a straight gas-kinetic collision cross section. Note that N, as determined from low-pressure turn-up, also has an effect on the high-pressure rate $k_{a\infty}$, Eq. (9-79). Since N does not depend on ξ but $k_{a\infty}$ does, a calculated $k_{a\infty}$ that is too small would suggest $\xi < 1$; however, there does not exist such a case on record.[32]

[32] Jakubowski *et al.* (*43*) find that they cannot fit the curvature of *low*-pressure data on the collisional deactivation of chemically activated methyl cyclopropane by helium unless they take $\xi \approx 0.03$. The reason for the discrepancy here seems to be that as

The results obtained from low-pressure experimental data on several chemical activation systems are summarized in Table 9-2. The column headed by $\langle E \rangle_F$ gives the average energy of the formed excited reactant molecules, referred to reaction threshold as zero. The step size s may be considered as representing $\langle \Delta E \rangle$, the average energy transferred per collision, and is indicated as such in the next to last column of Table 9-2. For the purpose of comparison, the first entry in the table gives two typical results for a thermal system, which shows small $\langle \Delta E \rangle$ for a small $\langle E \rangle_F$, while all the chemical activation systems show much larger $\langle \Delta E \rangle$ for much larger $\langle E \rangle_F$. It is clear that in the more highly excited systems, fairly large amounts of energy are transferred per collision by the more efficient colliders. The table also shows the temperature at which each system was studied, since when comparing various systems, account must be taken of the fact that in general an increase in temperature can be expected to decrease (*35*) $\langle \Delta E \rangle$ somewhat.

Inasmuch as the step size s, i.e., $\langle \Delta E \rangle$, is fixed by the curvature of the $k_a/k_{a\infty}$ versus **S/D** plot, a change in threshold energy E_0 (e.g., as a result of better thermochemical data for the given reaction becoming available), $\langle \Delta E \rangle$ will also change. Thus some of $\langle \Delta E \rangle$ values of Table 9-2 may be subject to small upward or downward revision in the future.

The stepladder cascade model for the transition probability can be refined (*37, 38*) by including up-transitions and by taking account of the distribution of energies of the chemically activated reactant. Other transition probability models have also been used (*38*): the Gaussian model, with a symmetric distribution of step sizes about a mean, which amounts to a generalized stepladder model, and the exponential model used in Chapter 8, Section 8. The essential difference between the exponential and the other two models is that the exponential model assigns a higher probability to small step sizes than to large ones, unlike the stepladder or Gaussian model.

Since, as we have seen, energy transfer information comes from experiments at low pressures where good accuracy is difficult to achieve, experimental data have in general not been accurate enough to make a clear distinction between the simple stepladder cascade model described here and

---

helium is an inefficient deactivator, the exponential, rather than the stepladder model, would have been more appropriate (cf. footnote 34 in Chapter 8 and remarks at the end of this section); this is also suggested by the data on neon shown in Fig. 9-6. Using phase-lag fluorimetry, Schlag *et al.* (*44,45*) measured the absolute collision frequency for loss of internal energy in excited β-naphthylamine with propane and propylene as heat bath gases and found $\xi \approx 0.1$ relative to $Z_{\text{L-J}}$. Halpern and Ware (*46*) studied the fluorescence decay of hexafluoroacetone. Assuming a two-step cascade model, they obtain for the second-order rate constant for vibrational relaxation about one-third of the hard-sphere (Z) value, suggesting $\xi \approx 0.3$. Inasmuch as the average excitation in these systems was some 20 kcal mole^{-1} or less, which is smaller than the average excitation energy in most chemical-activation systems (Table 9-2), the photochemical and chemical activation results may not be directly comparable.

TABLE 9-2

ENERGY DECREMENTS PER COLLISION IN VARIOUS SYSTEMS BASED ON STEPLADDER CASCADE DEACTIVATION[a]

System	$\langle E \rangle_F$ (kcal mole^{-1})	Deactivator M	$\langle \Delta E \rangle$ (kcal mole^{-1})	Ref.
$CH_3NC + M \rightarrow (CH_3NC)^v$ at 546°K	1	He	0.8	*b*
		CF_4	1.6	
H + *cis*-butene-2 → (*sec*-butyl)^v at 300°K	10	CD_3F	9	*c*
		CH_3Cl	9	
		SF_6	9	
H + 3,3-dimethylhexene-1 → (dimethylhexyl)^v at 300°K	14	H_2	1.4	*d*
		CF_4	5.6	
$2CH_2Cl$ → (1,2-dichloroethane)^v at 300°K	34	C_4F_8	13	*e*
		CH_2Cl_2	12	
		CH_3Cl	10	
		CF_4	8	
		N_2	6	
		Ar	6	
		CO	6	*f*
$CF_3 + CH_3 \rightarrow (CF_3CH_3)^v$ at 300°K	34	C_2F_6	7	*g*
		$CF_3N_2CH_3$	7	
		cyclo-C_4F_8	7	
		n-C_6F_{14}	10	
		n-C_8F_{16}	10	
$2CH_2F$ → (1,2-difluoroethane)^v at 300°K	35	CH_2FCl	11	*h*
$CH_2 + C_2H_4$ → (cyclopropane)^v at 300°K	47	C_2H_4	10–28	*i*
		N_2	6	
		Ar	6	
		He	6	
H + *cis*-butene-2 → (dimethylcyclopropane)^v at 573°K	47	*cis*-butene-2	12	*i*

[a] $\langle E \rangle_F$ is average energy, above threshold for reaction, of activated molecules. $\langle \Delta E \rangle$ is average energy transferred per collision, i.e., the step size in the stepladder cascade. All energy values approximate.

[b] D. C. Tardy and B. S. Rabinovitch, *J. Chem. Phys.* **48**, 1282 (1968). $\langle \Delta E \rangle$ on the stepladder model given here is not much different from $\langle \Delta E \rangle$ on the exponential model shown in Table 8-1.

[c] G. H. Kohlmaier and B. S. Rabinovitch, *J. Chem. Phys.* **38**, 1709 (1963).

[d] C. W. Larson and B. S. Rabinovitch, *J. Chem. Phys.* **51**, 2293 (1969). This is a competitive system with several parallel modes of decomposition having different thresholds for reaction.

its more refined variant, or the other transition probability models. A recent reinvestigation of the dimethylcyclopropane system by Rynbrandt and Rabinovitch (*47*), using the most recent refinements of experimental and computational techniques, reveals that the onset and steepness of the k_a versus **S/D** curve at low pressure with *cis*-butene-2 as deactivator cannot be fitted by the exponential model, but does fit quite well the stepladder or Gaussian model, on the basis of which ~11.4 kcal mole^{-1} are removed on every collision, in agreement with the earlier results. However, in the case of CO, a less efficient deactivator, the exponential model seems to fit the data better, with about ~4.6 kcal mole^{-1} removed per collision on this model.[33] The conclusion seems to be that in general the stepladder cascade describes best systems involving efficient deactivators.

6. Photoactivation

It is convenient at this point to mention briefly a somewhat special variant of the chemical activation method, one which involves the formation of the activated reactant A by direct photoexcitation. The emphasis is on the word "direct" because some of the systems mentioned previously, e.g., $CH_2 + C_2H_4 \rightarrow (\text{cyclopropane})^v \rightarrow$ propylene, do make use of photoexcitation, but only indirectly, that is, only to generate CH_2 radicals; the primary excitation of the reactant cyclopropane still comes from the exothermicity of the chemical reaction producing it.

Direct excitation by photon absorption generally involves the formation of one or more excited electronic states, some or all of which may be vibrationally excited (vibronic states). These vibronic states then decay by a complex process that may involve emission of radiation (fluorescence), or a number of radiationless transitions such as internal conversion (excited singlet to ground-state singlet), intersystem crossing (singlet to triplet), dissociation or isomerization (from either singlet or triplet states, or from ground state), or collisional deactivation of vibronic states to their respective ground vibrational states, or maybe all the way down to the ground electronic

[33] Setser *et al.* (*48*) have given conclusive evidence for exponential transition probabilities for inefficient deactivators (H_2, D_2, and the monatomic gases) in chemically activated CH_3CF_3 and 1,2-dichloroethane systems. See, however, (*119*).

[e] D. W. Setser and J. C. Hassler, *J. Phys. Chem.* **71**, 1364 (1967). Newer results in (*48*).
[f] W. G. Clark, D. W. Setser and E. E. Siefert, *J. Phys. Chem.* **74**, 1670 (1970).
[g] H. W. Chang, N. L. Craig and D. W. Setser, *J. Phys. Chem.* **76**, 954 (1972).
[h] H. W. Chang and D. W. Setser, *J. Amer. Chem. Soc.* **91**, 7648 (1969).
[i] Data as summarized by J. W. Simons, B. S. Rabinovitch, and D. W. Setser, *J. Chem. Phys.* **41**, 800 (1964). The result for C_2H_4 depends on the source of CH_2.

state. This list of possible transitions is not exhaustive and is meant only to be indicative of the complexity of the problem. It is the province of photochemistry to assess the exact mechanism of decay[34] in any one particular case, and as such is therefore outside the scope of the present treatment.

It is possible, however, to find systems where, under appropriate conditions, the principal reactions reduce to the following:

$$A + h\nu \longrightarrow (A)^{e,v} \tag{9-82}$$

$$(A)^{e,v} \xrightarrow{k_1} \text{decomposition} \tag{9-83}$$

$$(A)^{e,v} + M \xrightarrow{k_2} (A)^{e,v'} + M \tag{9-84}$$

where[35] $(A)^{e,v}$ is the reactant A in vibrational level v of electronic state e, and $(A)^{e,v'}$ is the reactant A in the vibrational level v' ($v' < v$) of the same electronic state e, such that level v' is just below E_0, the critical energy for decomposition. Since the electronic state involved in the sequence (9-82)–(9-84) is the same state throughout, we have here the exact analog of the chemical activation scheme (9-4)–(9-6), i.e., a competition between decomposition and stabilization of a vibrationally excited species. The difference between photoactivated and chemically activated systems is due mainly to different experimental observables.

The Stern–Volmer plot In photochemical systems, the rate of formation of a product **D** is usually expressed in terms of its quantum yield Φ_D, defined as

$$\Phi_D = (d[\mathbf{D}]/dt)/\mathfrak{I}_a \tag{9-85}$$

where $\mathfrak{I}_a$ is the total number of quanta absorbed per cubic centimeter per second; $\mathfrak{I}_a$ is also the actual rate (not the rate constant) of process (9-82). The quantum yield is, in essence, the fraction of excited molecules going to products, and is therefore equivalent to the ratio $\mathbf{D}/(\mathbf{S} + \mathbf{D})$ in terms of the notation of the previous sections {or to $(d\mathbf{D}/dt)/[(d\mathbf{S}/dt) + (d\mathbf{D}/dt)]$, cf. Eq. (9-47)}. At steady state, the scheme (9-82)–(9-84) gives for the quantum yield Φ_D of the decomposition product formed in (9-83)

$$1/\Phi_D = 1 + (\omega/k_1) \tag{9-86}$$

where $\omega = k_2[M]$. Hence $1/\Phi_D$ should be a linear function of [M], i.e., of the concentration of the heat bath species M acting as an energy sink. This is the so-called Stern–Volmer (*50*) plot; it is indeed linear, at least within the

[34] See, for example, Calvert and Pitts (*49*, Chapter 5).

[35] The notation used in photochemistry is S_1 for first excited singlet, S_0 for ground state, T_1 for first excited triplet electronic state, etc.

common range of pressures, for a number of systems [for example, the azoalkanes (*51*)].

In general, however, plots of Eq. (9-85) turn out to be nonlinear when the pressure dependence of $1/\Phi_D$ is examined more carefully. The reasons for the nonlinearity of Stern–Volmer plots at high and at low pressures are quite different: it turns out that at high pressures, it is due to $(A)^{e,v}$ being produced in process (9-82) with a spread of internal energies, while at low pressures, it is due to multistep deactivation in process (9-84). These are precisely the same complications that account for the pressure dependence of the experimental observable **D/S** in chemical activation systems.

The nonlinearity of a Stern–Volmer plot due to a spread of energies in the excited reactant will be considered first. If the rate constant k_1 in the process (9-83) is energy dependent, Eq. (9-86) holds only for a small energy range δE, and then Φ_D represents the quantum yield of product arising from $[A(E)]^{e,v}$ in the range E, $E + \delta E$. In practice, of course, only the overall quantum yield is measurable, due to the decomposition of $A(E)^{e,v}$ having all energies of its energy spectrum, and therefore (9-86) no longer holds as such; instead, we have to sum the contributions to Φ_D over all energy ranges. We shall assume that the exciting light incident on the system [process (9-82)] is essentially monochromatic, and therefore is not the cause of the energy spread. Under such conditions, the only source of the spread of energies in the excited reactant is the thermal distribution of energies existing in ground-state A prior to the absorption of radiation, superimposed on which is the effect of the photochemical excitation process itself, which takes place with a probability given by the so-called Franck–Condon factor[36], which in general varies from transition to transition, depending on the vibrational levels of initial and final states. These effects cause the excited reactant $(A)^{e,v}$ to be formed with a distribution of energies (essentially internal vibrational energies) that is basically thermal, but may be appreciably distorted or widened.

The distribution of energies Let λ be the wavelength of the exciting radiation, assumed to be monochromatic, and let $E_\lambda = hc/\lambda$ be the equivalent energy. If, in the first approximation, we neglect all factors other than the thermal distribution of energies of the ground state, the distribution of energies in $(A)^{e,v}$ becomes merely the thermal distribution $P(E)_e$ of Eq. (8-13) shifted by E_λ along the energy axis; let us call this distribution $P(E - E_\lambda)_e$, with the understanding that it is defined for positive argument only. The situation is thus quite similar to the case of chemical activation,

[36] The Franck–Condon factor is the square of the vibrational overlap integral between the initial and final electronic states. See Chapter 10, Section 3, Eq. (10-49).

where it was shown in Section 2 that the distribution function $F(E)$ of activated molecules is essentially $P(E)_e$ shifted by E_0^{ar} along the energy axis [cf. Eq. (9-35)]. Note that both the photoactivation and chemical activation generally take place at room temperature, so that $P(E - E_\lambda)_e$ refers to room temperature and consequently has a relatively small width, similar to that of $F(E)$.

Under these conditions, the observable quantum yield Φ_D therefore refers to the yield of product averaged over $P(E - E_\lambda)_e$, for which the scheme (9-82)–(9-84) yields (*52*)

$$\Phi_D = \int_{E_0 \text{ or } E_\lambda}^{\infty} \frac{k(E)P(E - E_\lambda)_e \, dE}{k(E) + \omega} \tag{9-87}$$

where $k(E)$ is written for k_1 to show its dependence on energy explicitly. The lower limit of integration is E_λ or E_0 (critical energy for decomposition), whichever is larger. Equation (9-87) assumes that E, E_0, and E_λ refer to the same energy zero. Figure 9-7 shows the relation of the various energy quantities.

If the assumption is made that all the energy of the excited state in excess of E_0 or E_λ [whichever appears as the lower limit of the integral in (9-87)]

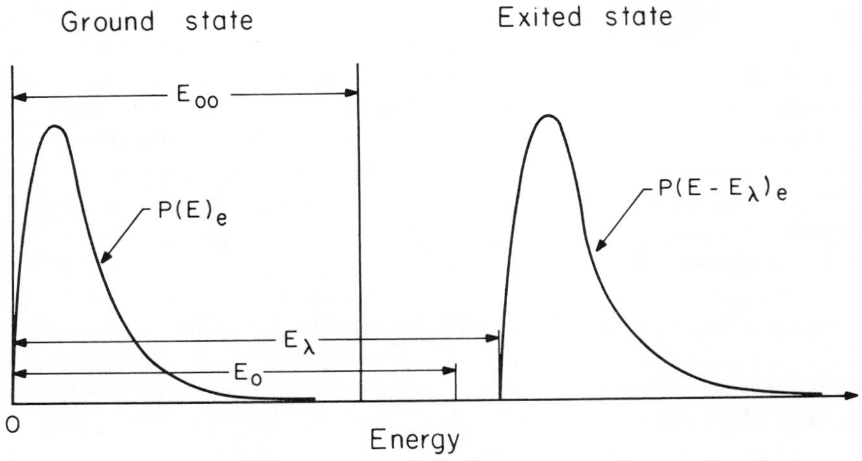

Fig. 9-7 *Energy diagram for photoactivation (schematic).*
Upon absorption of radiation of energy E_λ, a ground-state molecule, having a thermal distribution energies $P(E)_e$, yields an excited state whose distribution of energies in the first approximation is simply $P(E)_e$ shifted by E_λ along the energy axis. The figure represents the case when $E_\lambda > E_0$ (critical energy for dissociation). E_{00} is the energy difference between the ground vibrational levels of the excited and ground electronic states. The form of $P(E)_e$ as drawn is for clarity only. At room temperature, $P(E)_e$ would actually have a form more nearly like that of curve (a) or (b) in Fig. 9-3.

is randomized among the pertinent degrees of freedom of the excited state, $k(E)$ is calculable by the general method of Part I. Then Φ_D of Eq. (9-87) may also be calculated and compared with experiment, taking $k_2 = Z$, i.e., in the strong-collision approximation. The curvature of the calculated plot of $1/\Phi_D$ versus [M] turns out to be slightly negative, roughly what one gets from the experiment,[37] showing that there is only a small spread of energies, and that it appears to be fairly well accounted for by the thermal energy of the ground state. Note that since $P(E - E_\lambda)_e$ is temperature dependent, the quantum yield Φ_D, or its inverse, becomes also temperature dependent.

The high-pressure form of (9-87) is of some interest. When $k(E) \ll \omega$, Eq. (9-87) reduces to

$$\Phi_D = (1/\omega) \int_{E_0 \text{ or } E_\lambda}^{\infty} k(E)P(E - E_\lambda)_e \, dE = \langle k(E) \rangle_P / \omega \qquad (9\text{-}88)$$

In other words, the plot of $1/\Phi_D$ versus ω is linear, and the slope is $\langle k(E) \rangle_P^{-1}$. It is therefore possible to obtain the average rate constant of the unimolecular process from experiment and compare it with the value calculated from Eq. (9-88). Such calculations have been done in the photodissociation of ketene (*54*) and hexafluoroacetone (*55*). If the high-pressure Φ_D is known as a function of temperature, it is possible to determine the activation energy E_a corresponding to $\langle k(E) \rangle_P$. The energy may be interpreted in the same way as done in Section 4, so that by reason of the small width of the function $P(E - E_\lambda)_e$, the activation energy E_a can be expected to be small, about of the same order of magnitude as in chemical activation systems. In the photodissociation of azoethane, it was found (*56*) that $E_a \approx 2$ kcal mole^{-1}.

In most practical applications, it is probably quite sufficient to forego Eq. (9-87) or (9-88) and instead consider the system as essentially monoenergetic of energy E, with $E = E_\lambda + \langle E \rangle$, $\langle E \rangle$ being the average thermal energy.

Photoactivated systems Photoactivated systems have a potential for offering valuable information about unimolecular processes, because in principle the internal energy of the excited species is easily varied at will by changing the wavelength of the exciting radiation, and low experimental temperature ensures a small spread of internal energies and the absence of secondary reactions. Unfortunately, rate calculations suffer from the scanty knowledge of the parameters (including the critical energy E_0) of the excited electronic states involved, and therefore only order-of-magnitude calculated results may be hoped for. Furthermore, there may be mechanistic complications, in the sense that the scheme (9-82)–(9-84) may be only one of several

[37] See, for example, Bowers (*53*), who calculated Stern–Volmer plots in the photodissociation of azoethane.

parallel modes of decomposition, involving more than one excited state of the same species.[38]

There are, however, a few photochemical systems where decomposition following excitation occurs from the *ground* electronic state; that is, the initial excitation process (9-82) is followed by the internal conversion

$$(\mathsf{A})^{e,v} \longrightarrow (\mathsf{A})^{0,v} \qquad (9\text{-}89)$$

where $(\mathsf{A})^{0,v}$ represents vibrationally excited A in its ground electronic state. This is then followed by the exact analog of (9-83) and (9-84), i.e., $(\mathsf{A})^{0,v} \rightarrow$ decomposition, and $(\mathsf{A})^{0,v} + \mathsf{M} \rightarrow$ stabilization. The obvious advantage of such a system is of course that it involves the ground electronic state, the parameters of which are usually well known. One such case is the cyclobutanone system, where it was revealed on reinvestigation (*58*) that the process (9-91), mentioned later, is due to the decomposition of triplet cyclobutanone; however, there is a parallel process, due to an internal conversion, producing excited cyclobutanone in the ground electronic state which decays according to

$$\underset{\text{(singlet)}}{(\text{cyclobutanone})^{0,v}} \longrightarrow C_2H_4 + CH_2CO \qquad (9\text{-}90)$$

yielding products that are different from those produced in the decomposition of the triplet. Therefore the rate of the unimolecular decomposition of ground-state cyclobutanone is measurable, and also calculable, relative to the rate of stabilization. A similar system is the photochemical isomerization of 1,3,5-cycloheptatriene into toluene, mentioned toward the end of this section. In all such systems, assuming monochromatic irradiation, the ground-state molecules produced by the internal conversion are monoenergetic, except for the thermal contribution. These systems have not yet been exploited for detailed rate calculations.

Energy partitioning among fragments A rather special and interesting kind of photoexcitation, leading to a strongly nonthermal distribution of internal energies in the unimolecular reactant, arises when the species decomposing unimolecularly is one of two (or more) fragments formed in the photodissociation of the parent, for example (*59, 60*),

$$\text{cyclobutanone} + h\nu \longrightarrow \underset{\text{(triplet)}}{(\text{cyclobutanone})^{e,v}}$$

$$\longrightarrow (\text{cyclopropane})^{v} + (\text{CO})^{v} \qquad (9\text{-}91)$$

[38] Campbell *et al.* (*57*) show that such complication may be present even if the Stern–Volmer plot is linear over a large range of pressures.

What is interesting here is that the product cyclopropane contains no electronic, but only vibrational excitation,[39] and undergoes the usual isomerization to propylene, well known from thermal and chemical activation studies, in competition with collisional stabilization; hence the rate of this isomerization is amenable to calculation using the known parameters of ground-state cyclopropane. The internal excitation energy of $(\text{cyclopropane})^{v}$ represents only part of the total excess energy available for excitation of products upon the fragmentation of the excited cyclobutanone, since this energy is partitioned among the two products, cyclopropane and CO. As a result of such partitioning, and despite monochromatic excitation, $(\text{cyclopropane})^{v}$ is formed with a spread of internal energies that is considerably wider than that due to the thermal energy of ground-state reactant prior to photoexcitation, and this is reflected in the curvature of the Stern–Volmer plot for propylene (or of the propylene/cyclopropane plots versus ω^{-1}). The distribution function for the internal energy of the excited fragment is therefore no longer the thermal distribution simply shifted along the energy axis, but a function that reflects the partitioning of energy among the fragments.

At present, not much is known about the factors that govern the partitioning of energy among the fragments.[40] Since the unimolecular rate constant for the decay of the larger fragment can be calculated, then, by making an assumption about the mode of energy partitioning, the overall yield of the decay product (propylene) can be calculated and compared with experiment. Such systems therefore offer information about the partitioning of energy on fragmentation rather than about the unimolecular decay itself.

The simplest assumption to make is that the partitioning of energy is statistical. Such partitioning will be considered in more detail in Chapter 10, Section 2, but for the present purpose it is sufficient to outline the basic idea, which is this: Cyclobutanone has 27 vibrational degrees of freedom, cyclopropane has 21; assuming that all pertinent degrees of freedom (in the sense of unimolecular rate theory) are vibrational, statistical partitioning would correspond to cyclopropane carrying away 21/27 of the total available excess energy. Of the remaining six degrees of freedom, one is accounted for by the vibration in CO and five become degrees of freedom of relative motion. The actual energy distribution function corresponding to this partitioning can be calculated, and from the average of $k(E)$ over this distribution, the actual yields of cyclopropane and propylene can be computed and compared with

[39] The superscript zero on cyclopropane to indicate ground electronic state is dropped as redundant and fragments shall be presumed always to be formed in the ground electronic state unless specifically indicated otherwise.

[40] For calculations using simplified models, see Mitchell and Simons (*61*); Holdy *et al.* (*62*); Busch and Wilson (*63*). These calculations involve models for energy partitioning in the fragmentation of triatomic molecules NO_2, ICN, and NOCl. Cf. also (*118*).

experiment. It turns out that the partitioning of energy in the cyclobutanone photodissociation is quite nonstatistical, the cyclopropane carrying away *less* energy than it should have on the statistical model. Similar results have been obtained in the photodissociation of 2,3-diazobicyclo[2,2,1]hept-2-ene (*64*), which fissions into $(\text{bicyclo}[2,1,0]\text{pentane})^{v} + N_2$, and in the photodissociation of 4-methyl-1-pyrazoline,[41] which fragments into $(\text{methyl cyclopropane}^{v}) + N_2$. In the last two studies, an attempt was made to calculate the energy distribution function of the unimolecular reactant by assuming that the distribution was symmetric about a mean (i.e., Gaussian), and adjusting its maximum and its width to fit the experimentally observed distribution of the products. In the pyrazoline study, Dorer (*65*) thus found the square root of the variance of the energy distribution function for the excited methyl cyclopropane to be $\sigma \approx 14$–17 kcal mole^{-1}, whereas a purely thermal distribution would have $\sigma = \mathsf{n}^{1/2}kT$ [Eq. (9-36)], which works out to $\sim$3.3 kcal mole^{-1} for methyl cyclopropane ($\mathsf{n} = 30$) at room temperature; in addition, its maximum was located at 82 kcal mole^{-1}, which is much lower than if the partitioning had been statistical.[42]

Many-step deactivation Returning now to our original scheme (9-82)–(9-84), we shall examine its behavior at low pressures. We shall neglect the small spread of energies in $(\mathsf{A})^{e,v}$ due to the thermal energy of the ground state, and therefore will treat the process (9-83)–(9-84) as involving essentially a monoenergetic species of energy[43] E. Equation (9-86) can then be written as

$$\Phi_D = k(E)/[k(E) + \omega] = 1 - \{\omega/[k(E) + \omega]\} \qquad (9\text{-}92)$$

From (9-74) and (9-75), we see that $\omega/[k(E) + \omega]$ is the absolute probability of stabilization, if stabilization from level E takes place in one step, i.e., upon a single collision. If stabilization requires N steps, (9-92) becomes, using (9-76) and $\xi = 1$,

$$\Phi_D = 1 - \prod_{j=1}^{N} [\omega/(k_j + \omega)] \qquad (9\text{-}93)$$

[41] Dorer (*65*). Similar results on energy partitioning were obtained in the photolysis of pyrazoline itself [Cadman *et al.* (*66*)], of 3-vinyl-1-pyrazoline [Dorer *et al.* (*67*)] and of di-*tert*-butyl peroxide [Dorer and Johnson (*68*)]. On the other hand, studies of infrared luminescence from HX (X = F, Cl) formed by chemical activation in the process $(CHX_3)^{v} \rightarrow CX_2 + (HX)^{v'}$ show that the energy release in the products is nearly statistical [Chang *et al.* (*69*)]. Nonstatistical results on (9-91) are now in dispute (*120*).

[42] There are 36 vibrational degrees of freedom in pyrazoline and 30 in methyl cyclopropane, and the available excess energy is 132 kcal mole^{-1}. On the statistical model, methyl cyclopropane should carry away $132 \times 30/36 = 110$ kcal mole^{-1}.

[43] This energy is designated as $\langle E \rangle_F$ in Table 9-3.

where $k_j = k(E_0 + js)$, $s = \langle \Delta E \rangle$ being the step size such that $E_0 + N\langle \Delta E \rangle = E$, if E is referred to ground vibrational state as zero. It is actually in the above form that the stepladder cascade deactivation was first formulated by Porter and Connelly (*70*). When $1/\Phi_D$ of Eq. (9-93) is plotted against ω, it turns out that if $N \neq 1$, the plot shows positive curvature at low pressure, i.e., a curvature opposite to that due to a spread of energies. The curvature of the experimental plot can be fitted by computing theoretical curves for various step sizes, as discussed in connection with chemical activation studies in Section 5, and the best-fit theoretical curve then gives $\langle \Delta E \rangle$, the average energy removed per collision.[44] If, as is usually the case, the full kinetic scheme is more complicated than the scheme (9-82)–(9-84), a somewhat better test for the curvature in the Stern–Volmer plot at low pressures is provided by plotting against pressure a slightly more complicated function of the quantum yield (*72*).

Equation (9-93) was used by Atkinson and Thrush (*73, 74*) to determine $\langle \Delta E \rangle$ in the collisional stabilization of vibrationally excited ground-state 1,3,5-cycloheptatriene (CHT) and CHT-d_8 produced by internal conversion from an excited state following irradiation. They also made a parallel study of the thermal isomerization of CHT to obtain Arrhenius parameters from which information could be deduced regarding the transition state of the isomerization into toluene, which are necessary to calculate the rate constant k_j. Exciting radiation of several wavelengths was used, and four different deactivating gases were employed. The results for the extremes of the wavelength range are summarized in Table 9-3, and when they are compared with those of Table 9-2, one is immediately struck by the much smaller energy removed per collision in the photochemical system.

The chemical activation system closest to CHT, both in the structure of the reactant and in the average excitation energy, is the isomerization of excited cyclopropane, where (Table 9-3) helium was found to remove per collision almost 50 times the energy that it removes from excited CHT; for more complex deactivators, like ethylene or toluene, the difference is considerably less, but is still a factor of five to ten. In an older study on the deactivation by collision of photoexcited β-naphthylamine (Table 9-3), similar small amounts of energy transfer per collision were found.

The likely reasons for the discrepancy have been discussed recently (*75*). First of all, the curvature of the $1/\Phi_D$ versus pressure plot is very small, much smaller than in a chemical activation system at low pressure (Fig. 9-6), and therefore the procedure used in Ref. (*74*) is equivalent to extracting $\langle \Delta E \rangle$ from *high-pressure* chemical activation data, which makes the $\langle \Delta E \rangle$ so

[44] For a generalization of the cascade deactivation model to chopped excitation, see Schlag *et al.* (*71*).

TABLE 9-3

ENERGY DECREMENTS PER COLLISION IN TWO PHOTOACTIVATED SYSTEMS[a]

System	$\langle E\rangle_F$ (kcal mole^{-1})	Deactivator M	$\langle\Delta E\rangle$ (kcal mole^{-1})
1,3,5-cycloheptatriene (CHT) + $h\nu \rightarrow$ (CHT)^v at 300°K[b]	42.2	He	0.16
		CO_2	0.8
		SF_6	1.2
		toluene	1.9
		CHT	6.6
	75.8	He	0.14
		CO_2	0.9
		SF_6	1.3
		toluene	3.8
		CHT	4.0
β-naphthylamine + $h\nu \rightarrow$ (β-naphthylamine)^{e,v} at 420–460°K[c]		He	0.09
		CO_2	1.2
		SF_6	1.6

[a] $\langle E\rangle_F$ is the average energy, above threshold for reaction, of the activated molecules, taken equal to the energy to the energy of the exciting radiation plus the average thermal energy. $\langle\Delta E\rangle$ is the average energy transferred per collision.

[b] System treated as a stepladder cascade deactivation [Eq. (9-93)]. Data and calculations of R. Atkinson and B. S. Thrush, *Chem. Phys. Lett.* **3**, 684 (1969).

[c] System treated as a temperature relaxation, where collisions change vibrational temperature. Data and calculations of B. S. Neporent. *Zh. Fiz. Khim.* **21**, 111 (1947); **24**, 1219 (1950); M. Boudart and J. T. Dubois, *J. Chem. Phys.* **23**, 223 (1955). The data are summarized in B. Stevens, "Collisional Activation in Gases," p. 178, Table 6.6. Pergamon, Oxford, 1967.

obtained unreliable. The second point concerns the use of Eq. (9-93); it neglects up-transitions, which are important in calculating $\langle\Delta E\rangle$ for the case of weak deactivators, so that when a more sophisticated technique of computation is used (*47*) that takes up-transitions into account, the discrepancy between the photochemical and chemical activation $\langle\Delta E\rangle$ is much reduced.

In the study of fluorescence quenching of NO_2 using a somewhat cruder stepladder model and the simple Kassel formulation of $k(E)$, Schwartz and Johnston (*76*) have found that at least 4 kcal mole^{-1}, and probably more, was transferred per gas-kinetic collision, which is a good deal more than the figures of Table 9-3. These results were to a large extent confirmed in a later study (*77*). It is possible, however, that the fluorescence could be due also to collision-indiced predissociation (*78*), in which case $\langle\Delta E\rangle$ would really measure the cross section for intersystem crossing.

Thus the interpretation of results obtained in photoactivated systems is often beclouded by a wealth of complications.

7. Theoretical parameters and experimental observables

The basic equation relating observables to theory is Eq. (9-12) in a chemically activated system, and the very similar Eq. (9-87) in a photoactivated system. Either involves averaging of the rate constant $k(E)$ over a distribution function, and since the parameters required for the rate constant and the distribution function are not always the same, we shall discuss each separately.

First of all, it is necessary to know the following.

1. E_0, the critical energy for the unimolecular reaction of interest.
2. E_0^{ar}, the critical energy for the reverse of the activating reaction (not required in photoactivated system). In the discussion that follows, it will be assumed that $E_0^{ar} > E_0$, which is the more usual case; for greater clarity, refer to the energy diagram of Fig. 9-1(a). In photoactivated systems, it will be assumed that $E_\lambda > E_0$ (cf. Fig. 9-7); here, E_λ is given by the wavelength of the exciting radiation and is always known.

The rate constant $k(E)$ for the unimolecular reaction of interest, as given by Eq. (9-37), requires the knowledge of the following.

3. The density of states $N(E)$ of the reactant molecule, from $E = E_0^{ar}$ (or $E = E_\lambda$) onward. Note that if in the photoactivated system decomposition occurs from an electronically excited state, $N(E)$ refers to the density of levels in that excited state, with E_λ referred to the ground vibrational level of that electronically excited state as zero (which is at E_{00} above the lowest level of the ground electronic state; cf. Fig. 9-7). The reason is that E in $N(E)$ refers only to the randomizable part of the total energy, and therefore excludes E_{00}.
4. The function $G^*(E)$ for the transition state of the unimolecular reaction of interest, from $E_0^{ar} - E_0$ (or $E_\lambda - E_0$) onward.
5. Reaction path degeneracy α.

Next it is necessary to know the following.

6. The collision frequency $\omega = Z[\mathrm{M}]$ or $Z\not{p}$ ($\not{p}$ = pressure). We can use for Z the expression (8-100), or the somewhat more sophisticated $Z_{\text{L-J}}$ given by (8-120) or (8-121).

The chemical activation distribution function $F(E)$ [Eq. (9-26)] requires the knowledge of the following.

7. The function $\mathscr{G}^*(E)$ for the transition state of the activating reaction, from $E = 0$ onward. Observe that the reaction path degeneracy α^{ar} drops out and is not required. If $F(E)$ is approximated by Eq. (9-35) or (9-36) we need $N(E)$ [parameter 3] but from $E = 0$ onward. The corresponding $G(E)$ involves only a trivial modification of the calculations for $N(E)$ and therefore does not call for any new information.

The photoactivation distribution function $P(E - E_\lambda)_e$, when written out explicitly, is [cf. Eq. (9-35)]

$$P(E - E_\lambda)_e = N(E - E_\lambda)_e e^{-(E-E_\lambda)/kT}/Q \tag{9-94}$$

where $N(E)$ is the density of states of the reactant molecule in its ground electronic state, and is therefore identical with parameter 3 only if reaction also occurs from this state; otherwise we have as parameter

8. The density of states $N(E)$ of the ground electronic state (from $E = 0$ onward) if the unimolecular decomposition of interest occurs from an excited state. Since $N(E)$ is generally obtained from Q by inversion, the knowledge of Q is taken for granted. If instead of averaging over (9-94), the system is taken to be approximately monoenergetic of energy $E_\lambda + \langle E \rangle$, the average thermal energy $\langle E \rangle$ is given by the logarithmic derivative of Q with respect to temperature [Eqs. (8-88) and (8-101)], and therefore is based on the same information as $N(E)$ or Q.

Parameters 3, 6, and 8 depend on the properties of the reactant molecule A, which are generally known from a variety of sources. Parameters 4, 5, and 7, and to an extent also 1 and 2, are model dependent since they are a function of the structure of one or the other of the transition states involved. All eight parameters are quite analogous to parameters discussed in connection with thermal reactions in Chapter 8, Section 7, and the same general comments apply. Photoactivated systems where decomposition occurs from an electronically excited state may present a difficulty in that molecular dimensions and vibrational frequencies of such an excited state may not be known and have to be guessed at. As a first guess, it may be assumed that they are not too different from those of the ground electronic state.

Compared with thermal systems, the main difference is that in a chemically activated (or photoactivated) system, there does not exist a simple equation relating a model-dependent parameter to experimental data. There is of course the ultimate test of a comparison of calculated and experimental rates, but this measures the overall success of the entire enterprise and yields only indirect clues about the model of the transition state and parameters dependent thereon. In particular, there is no activation energy in the usual sense (Section 4), and therefore no direct estimate of E_0 or E_0^{ar}, two very important parameters.

The critical energies The parameter E_0^{ar} (cf. Fig. 9-1) is most easily determined as the difference $\Delta H - E_0^{af}$, where E_0^{af} is the critical energy of the activating reaction in the forward sense. Thermochemical data are necessary to obtain the enthalpy ΔH of reaction (9-4), which is given by

$$\Delta H = \Delta H_f^0(\mathsf{A}) - \Delta H_f^0(\mathsf{X}) - \Delta H_f^0(\mathsf{Y}) \qquad (9\text{-}95)$$

where $\Delta H_f^0(Z)$ represents the heat of formation of species Z in its standard state at 0°K. Heats of formation of stable compounds are generally obtainable (*79, 80*) but as Table 9-2 shows, in most chemical activation systems, either A or X and/or Y are radicals, the heats of formation of which are much less well known. Radical heats of formation depend on bond energies which may be obtained from a variety of sources (*81–83*), but ultimately data on polyatomic radicals come mostly from kinetic data in thermal systems (*84, 85*). There is then the additional problem that thermochemical data are normally referred to 300°K and must be reduced to 0°K with the help of known or estimated heat capacity data.

If X and Y are radicals, the critical energy E_0^{af} is zero or very small, and then in a class 2 chemical activation system, $E_0^{ar} \equiv E_0$ is essentially the bond strength X– Y. In this way, E_0 has been determined in the chloroethyl radical systems (see footnote 8). In the class 1 system $CH_3 + CH_2Cl \rightarrow (C_2H_5Cl)^v$ and variants thereof, Setser and collaborators (*86, 87*) have used $E_0^{af} = 1$ kcal mole^{-1}. An estimate of E_0^{af} in the H + *cis*-butene-2 system (*19,20*) has been obtained from the collision yield, which suggests $E_0^{af} \approx 3$ kcal mole^{-1}, admittedly a poor estimate, but one that is probably better than the heat of formation of the butyl radical, for which only a crude estimate is available. In the CH_2 + olefin systems, the uncertainty about the energy content of the CH_2 radical makes the thermochemistry so uncertain that E_0^{ar} has been adjusted to obtain agreement between calculated and observed rate constants (*88*). Given the accuracy of most thermochemical data and thermal activation energies, an error of ± 2 kcal mole^{-1} in E_0^{ar} is probably the best one may hope for in the most favorable case.

Since the final result for E_0^{ar} is based on data from many disparate sources that in most instances go back to activation energies of reactions involving various combinations of the atoms that constitute the chemical activation reactant A, the value of E_0^{ar} so determined refers to an unspecified transition state. In other words, this procedure leaves the model-dependent part of E_0^{ar}, i.e., the zero-point energy of the transition state, unknown and unknowable. Therefore E_0^{ar} is always treated as model independent, an approximation that can probably be justified by the modest accuracy with which E_0^{ar} is usually known. However, it should be recognized as an approximation, possibly not a very good one when very different transition states are tried with very different zero-point energies.

Transition state parameters Parameters relating to the transition state of the activating reaction, which are necessary for calculating $\mathscr{G}^*(E)$, can only be estimated by analogy with known parameters of a similar reaction, or, somewhat more crudely, $\mathscr{G}^*(E)$ may be replaced by $N(E)$ or $G(E)$ of the unimolecular reactant A itself, as discussed in connection with Eqs. (9-35) and (9-36). In principle, of course, E_0^{ar} and $\mathscr{G}^*(E)$ could be determined in a separate experiment from the activation energy and preexponential factor, respectively, of the thermal reaction A $\rightarrow$ X + Y, but this is of little practical help because the activating reaction normally cannot be made to proceed in the indicated sense and therefore direct experimental measurements are not possible.

If the unimolecular reaction of interest (9-4) has been studied in several differently activated systems, its parameters are transferable from system to system. In practice, because thermal systems are the most accessible, this means transferring the thermal value of E_0, and possibly also thermal information about the transition state, to the chemical activation system; an added advantage is that the model-dependent part of the thermal E_0 (i.e., the zero-point energy of the transition state) can be readily estimated. Such has been the case of E_0 in the chemically activated C_2H_5X systems (*86, 87*) (X = Cl, Br), and in most of the CH_2 + olefin systems (*89*), which produce various chemically activated cyclopropanes whose thermal decomposition has been extensively studied [cyclopropane: (*90–93*); dimethyl cyclopropane: (*94–96*)]. The compatibility of the chemical activation transition state with thermal data may be checked by calculating

$$\int_0^\infty G^*(E)e^{-E/kT}\,dE = kTQ^* \tag{9-96}$$

where Q^* is related to the thermal preexponential factor. This has been discussed in Section 3 in connection with class 2 chemical activation systems where the connection between thermal and chemical activation data is particularly direct.

Other sources of data If no parallel thermal data are available, E_0 of a chemical activation system may have to be adjusted to fit the data, which amounts of using RRKM theory to evaluate thermochemical data (*97, 98*). Since no simple equation exists that relates E_0 to experimental data, calculations of k_a have to be repeated in their entirety for several assumed values of E_0 and the E_0 giving the best overall agreement between calculated and observed rates is then picked as the "best" E_0. Fairly extensive data using this procedure exist for the $2CH_2F \rightarrow$ (1,2-difluoroethane)^{v} system, which has been studied with CH_2F radicals produced through the photolysis of 1,3-

difluoroacetone (*99–101*), or fluoroacetone (*102*, *103*), or by mercury photosensitization of chlorofluoromethane (*35*); the "best" E_0 was found by adjustment in each case, and the values fall between $E_0 = 61$ and $E_0 = 65$ kcal mole^{-1}, with the more recent results (*35*) in the 62–63 kcal mole^{-1} region. Such consistent results remove some of the arbitrariness of the adjustment procedure. In the C_2H_5F system, Setser and collaborators (*104*) found by adjustment $E_0 = 57$ kcal mole^{-1}, which was later confirmed in a thermal study (*105*).

It is sometimes possible to put to profit data on the reverse of the unimolecular reaction of interest.[45] In the *sec*-butyl radical system (*17*), E_0 for decomposition into CH_3 + propylene is not known, but the activation energy for the reaction CH_3 + propylene has been estimated at ~6 kcal mole^{-1}, which, combined with the ΔH of the reaction, yields an estimate for E_0. In the chloroethane system (*86*), the reaction is $(C_2H_5Cl)^v \rightarrow HCl + C_2H_4$. Benson and Haugen (*107*) have developed a semiempirical method for estimating the activation energy for the reverse reaction HX + olefin. It turns out, however, that the activation energy so obtained is based on a transition state that appears to be incompatible with the chemical activation results.

Insofar as the thermochemistry of electronically excited states of polyatomic molecules is concerned, the picture is quite bleak since with the possible exception of E_{00}, almost no data are available[46] and have to be borrowed from those of the ground electronic state, a rough approximation at best. Given the transient nature of these excited states, data can come only from spectroscopic observation, which in principle could provide the very desirable information about the dissociation energy D_0 (Fig. 4-1), i.e., data not dependent on any assumed transition state.

The enumeration of the various ways in which parameter values can be obtained is limited to the more frequent cases and is not meant to be exhaustive, so that in any one particular case, some ingenuity is required. It is clear, however, that evaluation of chemical activation parameters involves data from a great variety of systems, which is all for the better, since the data are more likely to be self-consistent than if they were evaluated from the very same data on which the calculated rates are then tested, as in thermal systems. The penalty, unfortunately, is that most of the required thermochemical and other data is only of modest accuracy since they of necessity refer to radicals and other species for which good data are hard to come by.

[45] For a useful compilation of radical plus molecule reactions, see Trotman-Dickenson and Milne (*106*).

[46] Most of the scarce data refer to π-systems, which are of little interest in the present context. A summary of all the data is given in Herzberg (*108*, Appendix VI). For a more recent discussion of molecular parameters of the excited states of ethane, see Lombos *et al.* (*109*); Lassettre *et al.* (*110*).

Hence the promise of chemical activation systems to provide a more meaningful and stringest test of unimolecular rate theory is to a considerable extent frustrated by the lack of ancillary information of sufficiently high accuracy. On the other hand, the technique makes possible the preparation of highly excited species and therefore, relative to thermal systems, widens considerably the domain of energies over which unimolecular reactions may be studied.

8. Centrifugal effects

Nothing has been said so far about a chemically activated reactant decomposing under the influence of an effective potential (cf. Chapter 7), because, in fact, there is very little information on the subject. This is not because the general principles are unknown, but because no actual calculations have been carried out.

There is little doubt that in a chemically activated system, as in any other system, overall angular momentum must be conserved, and therefore the same rotational quantum number $\mathcal{J}$ can be assigned to A throughout its history of formation and decomposition. Let us make clear that the rotations (and their quantum number $\mathcal{J}$) that we are referring to here are those that are excluded by conservation requirements from participating in the randomization of energy, as discussed in Chapter 5, Section 1. The activating reaction, however, being a thermal reaction, will not produce molecules A all of the same $\mathcal{J}$, but will produce a collection of molecules having all possible values of $\mathcal{J}$. Therefore the equilibrium distribution function introduced in Eq. (9-16) becomes $P(E, \mathcal{J})_e$, as given by (8-12), and the critical energy for the reverse of the activating reaction becomes $\mathcal{J}$ dependent, which we shall indicate by writing it E_J^{ar}. Therefore $F(E)$ of Eq. (9-26) becomes

$$F(E, \mathcal{J}) = \frac{(2\mathcal{J} + 1)\mathcal{G}^*(E - E_J^{ar}) \exp\{-[E + E_r^{a}(\mathcal{J})]/kT\}}{\iint (2\mathcal{J} + 1)\mathcal{G}^*(E - E_J^{ar}) \exp\{-[E + E_r^{a}(\mathcal{J})]/kT\} dE\, d\mathcal{J}} \tag{9-97}$$

where $E_r^{a}(\mathcal{J})$ is the $\mathcal{J}$-dependent part of the rotational energy of the activating reaction (hence the superscript "a").

The critical energy for the unimolecular reaction of interest is now likewise $\mathcal{J}$ dependent, and becomes E_J; the rate constant is then [cf. Eq. (7-1)]

$$k(E, \mathcal{J}) = \alpha G^*(E - E_J)/hN(E) \tag{9-98}$$

and the observable rate constant is then

$$k_a' = \langle k(E, \mathcal{J})\rangle_{ss} \tag{9-99}$$

where the average is now over [cf. (9-14)]

$$F(E, \mathscr{J})_{ss} = \frac{F(E, \mathscr{J})}{k(E, \mathscr{J}) + \omega} \Bigg/ \iint \frac{F(E, \mathscr{J})}{k(E, \mathscr{J}) + \omega}\, dE\, d\mathscr{J} \tag{9-100}$$

The prime signifies, as usual, that centrifugal effects are included. To avoid the somewhat messy algebra involved in the calculation of k_a' of Eq. (9-99), we shall concentrate, with only a slight loss of generality, on $k'_{a\infty}$, given by

$$k'_{a\infty} = \langle k(E, \mathscr{J}) \rangle_F \tag{9-101}$$

where [cf. Eq. (9-38)] the averaging is over $F(E, \mathscr{J})$ of Eq. (9-97).

Even so, calculation of $k'_{a\infty}$ is complicated enough to make it worthwhile to proceed in two steps. We shall first consider the easier of the two, which is the calculation of the denominator in (9-101), i.e., the denominator of $F(E, \mathscr{J})$ as given by Eq. (9-97). Changing to a new variable $E' = E - E_J^{ar}$ [cf. Fig. 9-1(a)], we have

$$E + E_r^a(\mathscr{J}) = E' + E_J^{ar} + E_r^a(\mathscr{J}) \tag{9-102}$$

Now, by Eq. (8-20),

$$E_J^{ar} + E_r^a(\mathscr{J}) = E_0^{ar} + \phi^a(\mathscr{J}) \tag{9-103}$$

where the superscript "a" reminds us that everything refers to the *activating* reaction; $\phi^a(\mathscr{J})$ is a function of $\mathscr{J}$ whose form depends on whether the activating reaction takes place over a type 1 or type 2 potential energy surface. With the new variable E' integrations over E and over $\mathscr{J}$ can be done separately; thus from (8-18) and (8-19),

$$\int_0^\infty \mathscr{G}^*(E')\, e^{-E'/kT}\, dE' = kT\mathscr{Q}^* \tag{9-104}$$

where $\mathscr{Q}^*$ is the partition function for the active degrees of freedom in the transition state of the activating reaction, and from (8-22), we have

$$\int_0^\infty (2\mathscr{J} + 1) \exp[-\phi_a(\mathscr{J})/kT]\, d\mathscr{J} = f_\infty{}^a Q_r{}^a \tag{9-105}$$

where $f_\infty{}^a$ is the thermal high-pressure centrifugal correction factor for the activating reaction and $Q_r{}^a$ is the partition function for the rotor of energy $E_r{}^a(\mathscr{J})$. Thus, finally

$$\text{denominator of } \langle k(E, \mathscr{J}) \rangle_F = f_\infty{}^a kT\mathscr{Q}^* Q_r{}^a \exp(-E_0^{ar}/kT) \tag{9-106}$$

The numerator of (9-101), when written out explicitly, is, assuming that $E_J^{\text{ar}} > E_J$,

numerator of $\langle k(E'\mathscr{J})\rangle_F$

$$= \frac{\alpha}{h}\int_0^{\infty}\left\{\int_{E_J^{\text{ar}}}^{\infty} \frac{G^*(E - E_J)\mathscr{G}^*(E - E_J^{\text{ar}})}{N(E)} \exp\left(\frac{-E}{kT}\right) dE\right\}(2\mathscr{J}+1)\exp\left(\frac{-E_r^{\text{a}}(\mathscr{J})}{kT}\right) d\mathscr{J} \tag{9-107}$$

This integral is reminiscent of Eq. (8-38), but it cannot be evaluated by integration by parts as was done with (8-38) because both the integrand and the integration limit here depend on $\mathscr{J}$ through E_J and E_J^{ar}. It is therefore necessary to proceed with a rather tedious numerical double integration.

The $\mathscr{J}$ dependence of the critical energy for the unimolecular reaction of interest is given by an equation similar to (9-103):

$$E_J = E_0 + \phi(\mathscr{J}) - E_r(\mathscr{J}) \tag{9-108}$$

Equations (9-103) and (9-108) differ not only in that the former refers to the activating reaction and the latter to the unimolecular reaction, but also in a more subtle way. Recall that the essence of the treatment of rotations in Chapter 7 is to consider a unimolecular reactant A that dissociates into fragments $\mathscr{A}$ and $\mathscr{B}$, as a quasidiatomic molecule $\mathscr{A}\mathscr{B}$. In the present context, A is considered as diatom $\mathscr{X}\mathscr{Y}$ in the activating reaction, and as diatom $\mathscr{A}\mathscr{B}$ in the unimolecular reaction leading to the decomposition product(s) **D**. Thus $E_r^{\text{a}}(\mathscr{J})$ is the rotational energy of the diatom $\mathscr{X}\mathscr{Y}$ and $E_r(\mathscr{J})$ that of the diatom $\mathscr{A}\mathscr{B}$; although both diatoms are just A in disguise, $E_r^{\text{a}}(\mathscr{J})$ and $E_r(\mathscr{J})$ are not the same[47] because, in principle, the diatoms $\mathscr{X}\mathscr{Y}$ and $\mathscr{A}\mathscr{B}$ do not both have the same reduced mass and the same equilibrium "internuclear" distance r_e, which is really the distance between the centers of gravity of the two parts of the diatom. For example, in the chemically activated C_2H_5Cl system (*86*), the chloroethane would be considered as the diatom (CH_3)—(CH_2Cl) for the purpose of the activating reaction and as the diatom (C_2H_4)—(HCl) for the purpose of the unimolecular dissociation.

A more fundamental difference attaches to the two functions $\phi^{\text{a}}(\mathscr{J})$ and $\phi(\mathscr{J})$. They are different not only on account of the two corresponding diatoms being different, but also because the form of $\phi(\mathscr{J})$ depends on what type of potential energy surface is involved, and there is no reason to expect that both the activating reaction and the unimolecular reaction potential energy surfaces belong to the same type. In the cited example of C_2H_5Cl, the activating reaction is type 1 and the unimolecular decomposition is type 2. The function $\phi(\mathscr{J})$ is called $V_{\text{eff}}(r_m)$ in Chapter 7 and for a type 1 potential energy surface, it is given by Eq. (7-12),and for type 2, by Eq. (7-19).

[47] The rotational quantum number $\mathscr{J}$ is of course the same for both $E_r^{\text{a}}(\mathscr{J})$ and $E_r(\mathscr{J})$.

No calculations using Eq. (9-107) have been done so far on a chemical activation system, and there is some uncertainty as to whether the external rotations actually have an appreciable effect on the calculated rate constant. The fact that in (9-107) the integrand is $\mathscr{J}$ dependent, whereas in a similar equation which pertains to thermal systems it is not, would tend to make centrifugal effects more important in a chemical activation system; on the other hand, lower temperature in chemical activation means lower average rotational quantum number $\mathscr{J}$, which would tend to make rotational effects less important relative to thermal systems, where temperature, and hence average $\mathscr{J}$, is higher.

We can use the latter approach to obtain a very rough idea of the lower limit of the decrease in the critical energy E_0 (or E_0^{ar}) due to angular momentum. The average critical energy E_J is

$$\langle E_J \rangle = \int_0^\infty E_J P(\mathscr{J})_{\text{e}} \, d\mathscr{J} \tag{9-109}$$

where $P(\mathscr{J})_{\text{e}}$ is the equilibrium rotational distribution as given by Eq. (8-14). A similar equation applies to $\langle E_J^{\text{ar}} \rangle$. Let us assume for simplicity that we are dealing with a type 2 potential energy surface; then E_J is given by Eq. (7-21). Aside from constants, the only $\mathscr{J}$-dependent part of (7-21) is $E_{\text{r}}(\mathscr{J})$, so that the integral in (9-109) reduces to

$$\int_0^\infty E_{\text{r}}(\mathscr{J}) P(\mathscr{J})_{\text{e}} \, d\mathscr{J} = kT \tag{9-110}$$

a well-known result for the average energy of a two-dimensional rotor. Thus

$$\langle E_J \rangle = E_0 - [1 - (r_{\text{e}}/r_0)^2] kT \tag{9-111}$$

where r_0 is the "internuclear" distance in the diatom $\mathscr{AB}$ corresponding to the top of the barrier separating $\mathscr{AB}$ from $\mathscr{A} + \mathscr{B}$. At 300°K, $kT \approx 0.6$ kcal mole^{-1}; assuming that $r_0 \sim 2r_{\text{e}}$, probably an overestimate, we see that $\langle E_J \rangle$ is smaller than E_0 by about 0.5 kcal mole^{-1}, which is much less than the usual uncertainty in E_0. The same argument, with the same result, applies to $\langle E_J^{\text{ar}} \rangle$ relative to E_0^{ar}. It thus appears that centrifugal effects are likely to be lost in the uncertainty due to the energy parameters, and undoubtedly for this reason, there has been little inducement to carry through the more complicated rotational calculations sketched out at the beginning of this section.[48]

[48] A rough calculation of centrifugal effects in the isomerization of CH_2TNC prepared by hot-atom substitution has been reported by Bunker (*111*). This system is different from those considered in this chapter in that the activating reaction is not thermal. Using a somewhat different approach from that espoused in Chapter 7, Bunker finds that $\Delta E_{\text{r}}(\mathscr{J})$ [Eq. (5-8)] is large enough to drive the reaction, so that most of the energy required for the isomerization of CH_2TNC comes from (adiabatic) rotations and the reaction is therefore centrifugally controlled. Here, therefore, $\langle E_J \rangle \ll E_0$.

The situation is quite different in a photoactivated system. Intersystem crossing or internal conversion may result in a highly nonthermal distribution of rotational energies, so that either the preceding arguments do not apply or the rotational distribution may be more or less thermal but with a temperature very different from that of the heat bath. Whenever dissociation occurs from an excited state that has a different geometry from the ground state, rotational complications can be expected due to torques exerted in the excited state and due to constraints imposed by angular momentum conservation.[49]

Exercises

1. On the basis of Eqs. (9-26), (9-27), and (9-32), verify that, classically, $F(E_{\max,F}) \approx P(E_{\max,P})_e$. *Hint*: Clasically, const′ differs from const by $h\nu^*$, the energy of the vibration which becomes the reaction coordinate. Then $kT/h\nu^*$ is its classical partition function, and to the extent that $(kT/h\nu^*)\mathcal{Q}^* \approx Q$, Eq. (9-33) is satisfied, all the other factors cancelling out.

2. Using the classical approximation to $N(E)$ and $\mathcal{G}^*(E)$, show that the variance of $F(E)$ and $P(E)_e$ is given by $\mathsf{n}(kT)^2$. Note that in their classical form, both $F(E)$ and $P(E)_e$ are, within a constant factor, nothing but the so-called gamma distribution, the variance of which may be found in tables. See Abramowitz and Stegun (*113*, p. 930, item 26.1.32).

References

1. B. S. Rabinovitch, R. F. Kubin, and R. E. Harrington, *J. Chem. Phys.* **38**, 405 (1963).
2. B. S. Rabinovitch and M. C. Flowers, *Quart. Rev.* **18**, 122 (1964).
3. R. Wolfgang, *Accounts Chem. Res.* **3**, 48 (1970).
4. M. H. Chiang, E. A. Gislason, B. H. Mahan, C. W. Tsao, and A. S. Werner, *J. Phys. Chem.* **75**, 1426 (1971).
5. S. M. Freund, G. A. Fisk, D. R. Herschbach, and W. Klemperer, *J. Chem. Phys.* **54**, 2510 (1971).
6. H. G. Bennewitz, R. Haerten, and G. Müller, *Chem. Phys. Lett.* **12**, 335 (1971).
7. T. F. George and J. Ross, *J. Chem. Phys.* **56**, 5786 (1972).
8. P. G. Bowers, *Can. J. Chem.* **46**, 307 (1968).
9. D. A. Whytock and K. O. Kutschke, *Proc. Roy. Soc.* **A306**, 503 (1968) and subsequent papers.
10. Physics Today, p. 19 (April 1971).
11. Le-Khac Huy, W. Forst, J. A. Franklin, and G. Huybrechts, *Chem. Phys. Lett.* **3**, 307 (1969).

[49] Cf. footnote 65 in Chapter 10 and Evans and Rice (*112*).

12. J. H. Knox, *Trans. Faraday Soc.* **62**, 1206 (1966).
13. D. C. Tardy and B. S. Rabinovitch, *Trans. Faraday Soc.* **64**, 1844 (1968).
14. P. C. Beadle, J. H. Knox, F. Placido, and K. C. Waugh, *Trans. Faraday Soc.* **65**, 1571 (1969).
15. B. S. Rabinovitch and D. W. Setser, *Advan. Photochem.* **3**, 1 (1964), Table XXI.
16. D. W. Setser, *in* "Chemical Kinetics" (J. C. Polyani, ed.), p. 1. Butterworths, London, 1972.
17. B. S. Rabinovitch, R. F. Kubin, and R. E. Harrington, *J. Chem. Phys.* **38**, 405 (1963), Appendix II.
18. D. C. Tardy, B. S. Rabinovitch, and C. W. Larson, *J. Chem. Phys.* **45**, 1163 (1966).
19. B. S. Rabinovitch and R. W. Diesen, *J. Chem. Phys.* **30**, 735 (1959).
20. R. W. Diesen, Ph.D. Thesis, Univ. of Washington (1958).
21. S. W. Benson, "The Foundation of Chemical Kinetics." McGraw-Hill, New York, 1960.
22. J. H. Knox and J. Riddick, *Trans. Faraday Soc.* **62**, 1190 (1966).
23. J. A. Franklin, P. Goldfinger, and G. Huybrechts, *Ber Bunsenges.* **72** 173 (1968).
24. J. H. Knox and K. C. Waugh, *Trans. Faraday Soc.* **65**, 1585 (1969).
25. K. Hoyermann, H. Gg. Wagner, and J. Wolfrum, *Ber. Bunsenges.* **72**, 1004 (1968).
26. J. A. Cowfer, D. G. Keil, J. V. Michael, and C. Yeh, *J. Phys. Chem.* **75**, 1584 (1971).
27. H. Gg. Wagner and R. Zellner, *Ber. Bunsenges.* **76**, 440, 518, 667 (1972).
28. D. W. Setser and W. C. Richardson, *Can. J. Chem.* **47**, 2593 (1969).
29. E. V. Waage and B. S. Rabinovitch, *Int. J. Chem. Kinet.* **3**, 105 (1971).
30. P. C. Kobrinsky, G. O. Pritchard, and S. Toby, *J. Phys. Chem.* **75**, 2225 (1971).
31. R. L. Johnson, W. L. Hase, and J. W. Simons, *J. Chem. Phys.* **52**, 3911 (1970).
32. W. L. Hase and J. W. Simons, *J. Chem. Phys.* **54**, 1277 (1971).
33. G. W. Taylor and J. W. Simons, *Int. J. Chem. Kinet.* **3**, 24 (1971).
34. G. W. Taylor and J. W. Simons, *Int. J. Chem. Kinet.* **3**, 453 (1971).
35. H. W. Chang and D. W. Setser, *J. Amer. Chem. Soc.* **91**, 7648 (1969).
36. R. E. Harrington, B. S. Rabinovitch, and M. R. Hoare, *J. Chem. Phys.* **33**, 744 (1960).
37. G. H. Kohlmaier and B. S. Rabinovitch, *J. Chem. Phys.* **38**, 1962 (1963).
38. D. W. Setser, B. S. Rabinovitch, and J. W. Simons, *J. Chem. Phys.* **40**, 1751 (1964).
39. D. W. Setser and J. C. Hassler, *J. Phys. Chem.* **71**, 1364 (1967).
40. R. V. Serauskas and E. W. Schlag, *J. Chem. Phys.* **42**, 3009 (1965).
41. R. V. Serauskas and E. W. Schlag, *J. Chem. Phys.* **43**, 898 (1965).
42. J. C. Hassler, D. W. Setser and R. L. Johnson, *J. Chem. Phys.* **45**, 3231 (1966).
43. E. Jakubowski, H. S. Sandhu, and O. P. Strausz, *J. Amer. Chem. Soc.* **93**, 2710 (1971).
44. E. W. Schlag, H. von Weyssenhoff, and M. C. Starzak, *Ber. Bunsenges.* **72**, 153 (1968).
45. E. W. Schlag, H. von Weyssenhoff, and M. C. Starzak, *J. Chem. Phys.* **47**, 1860 (1967).
46. A. M. Halpern and W. R. Ware, *J. Chem. Phys.* **53**, 1969 (1970).
47. J. D. Rynbrandt and B. S. Rabinovitch, *J. Phys. Chem.* **74**, 1679 (1970).
48. D. W. Setser and E. E. Siefert, *J. Chem. Phys.* **57**, 3623 (1972).
49. J. G. Calvert and J. N. Pitts, "Photochemistry." Wiley, New York, 1966.
50. O. Stern and M. Volmer, *Phys. Z.* **20**, 183 (1919).
51. G. O. Pritchard, W. A. Mattinen, and J. R. Dacey, *Int. J. Chem. Kinet.* **2**, 191 (1970) and references therein.

52. P. G. Bowers and G. B. Porter, *J. Phys. Chem.* **70**, 1622 (1966).
53. P. G. Bowers, *J. Phys. Chem.* **74**, 952 (1970).
54. P. G. Bowers, *J. Chem. Soc.* **A**466 (1967).
55. P. G. Bowers, *Can. J. Chem.* **46**, 307 (1968).
56. H. Cerfontain and K. O. Kutschke, *Can. J. Chem.* **36**, 344 (1958).
57. R. J. Campbell, E. W. Schlag, and B. W. Ristow, *J. Amer. Chem. Soc.* **89**, 5098 (1967).
58. N. E. Lee and E. K. C. Lee, *J. Chem. Phys.* **50**, 2094 (1969).
59. R. J. Campbell and E. W. Schlag, *J. Amer. Chem. Soc.* **89**, 5103 (1967).
60. R. F. Klemm, D. N. Morrison, P. Gilderson, and A. T. Blades, *Can. J. Chem.* **43**, 1934 (1965).
61. R. C. Mitchell and J. P. Simons, *Discuss. Faraday Soc.* **44**, 208 (1967).
62. K. E. Holdy, L. C. Klotz, and K. R. Wilson, *J. Chem. Phys.* **52**, 4588 (1970).
63. G. E. Busch and K. R. Wilson, *J. Chem. Phys.* **56**, 3655, 3626 (1972).
64. T. F. Thomas, C. I. Sutin, and C. Steel, *J. Amer. Chem. Soc.* **89**, 5107 (1967).
65. F. H. Dorer, *J. Phys. Chem.* **73**, 3109 (1969).
66. P. Cadman, H. M. Meunier, and A. F. Trotman-Dickenson, *J. Amer. Chem. Soc.* **91**, 7640 (1969).
67. F. H. Dorer, E. Brown, J. Do, and R. Rees, *J. Phys. Chem.* **75**, 1640 (1971).
68. F. H. Dorer and S. N. Johnson, *J. Phys. Chem.* **75**, 3651 (1971).
69. H. W. Chang, D. W. Setser, and M. J. Perona, *J. Phys. Chem.* **75**, 2070 (1971).
70. G. B. Porter and B. T. Connelly, *J. Chem. Phys.* **33**, 81 (1960).
71. E. W. Schlag, S. J. Yao, and H. von Weyssenhoff, *J. Chem. Phys.* **50**, 732 (1969).
72. A. N. Strachan, R. K. Boyd, and K. O. Kutschke, *Can. J. Chem.* **42**, 1345 (1964).
73. R. Atkinson and B. A. Thrush, *Chem. Phys. Lett.* **3**, 684 (1969).
74. R. Atkinson and B. A. Thrush, *Proc. Roy. Soc.* **A316**, 131 (1970).
75. B. S. Rabinovitch, H. F. Carroll, J. D. Rynbrandt, J. H. Georgakakos, B. A. Thrush, and R. Atkinson, *J. Phys. Chem.* **75**, 3376 (1971).
76. S. E. Schwartz and H. S. Johnston, *J. Chem. Phys.* **51**, 1286 (1969).
77. L. F. Keyser, S. Z. Levine, and F. Kaufman, *J. Chem. Phys.* **54**, 355 (1971).
78. J. E. Selwyn and J. I. Steinfeld, *Chem. Phys. Lett.* **4**, 217 (1969).
79. "Selected Values of Physical and Thermodynamic Properties of Hydrocarbons and Related Compounds." Carnegie Press, Pittsburgh, Pennsylvania, 1953.
80. JANAF Thermochemical Tables, 2nd Ed. NSRDS-NBS 37. Washington, D.C., 1971.
81. A. G. Gaydon, "Dissociation Energies and Spectra of Diatomic Molecules," 3rd ed. Chapman and Hall, London, 1968.
82. C. T. Mortimer, "Reaction Heats and Bond Strengths." Pergamon, Oxford, 1962.
83. B. de B. Darwent, "Bond Dissociation Energies in Simple Molecules." NSRDS-NBS no. 31, Washington, D.C., 1970.
84. S. W. Benson, "Thermochemical Kinetics." Wiley, New York, 1968.
85. J. A. Kerr, *Chem. Rev.* **66**, 465 (1966).
86. J. C. Hassler and D. W. Setser, *J. Chem. Phys.* **45**, 3246 (1966).
87. R. L. Johnson and D. W. Setser, *J. Phys. Chem.* **71**, 4366 (1967).
88. D. W. Setser and B. S. Rabinovitch, *Can. J. Chem.* **40**, 1425 (1962).
89. D. W. Setser, B. S. Rabinovitch, and J. W. Simons, *J. Chem. Phys.* **40**, 1751 (1964).
90. H. O. Pritchard, R. G. Sowden, and A. F. Trotman-Dickenson, *Proc. Roy. Soc.* (*London*) **A217**, 563 (1953).
91. E. W. Schlag and B. S. Rabinovitch, *J. Amer. Chem. Soc.* **82**, 5996 (1960).
92. A. T. Blades, *Can. J. Chem.* **39**, 1401 (1961).

93. W. E. Falconer, T. E. Hunter, and A. F. Trotman-Dickenson, *J. Chem. Soc.* 609 (1961).
94. M. C. Flowers and H. M. Frey, *Proc. Roy. Soc.* (*London*) **A257**, 122 (1960).
95. M. C. Flowers and H. M. Frey, *Proc. Roy. Soc.* (*London*) **A260**, 424 (1961).
96. M. C. Flowers and H. M. Frey, *J. Chem. Soc.* 3953 (1959).
97. C. W. Larson, B. S. Rabinovitch, and D. C. Tardy, *J. Chem. Phys.* **47**, 4570 (1967).
98. C. W. Larson, B. S. Rabinovitch, and D. C. Tardy, *J. Chem. Phys.* **49**, 299 (1968).
99. G. O. Pritchard, M. Venugopalan, and T. F. Graham, *J. Phys. Chem.* **68**, 1786 (1964).
100. J. A. Kerr, A. W. Kirk, B. V. O'Grady, D. C. Phillips, and A. F. Trotman-Dickenson, *Discuss. Faraday Soc.* **44**, 263 (1967).
101. J. A. Kerr and D. M. Timlin, *Trans. Faraday Soc.* **67**, 1376 (1971).
102. G. O. Pritchard and R. L. Thommarson, *J. Phys. Chem.* **71**, 1674 (1967).
103. G. O. Pritchard and M. J. Perona, *Int. J. Chem. Kinet.* **1**, 509 (1969).
104. H. W. Chang, N. L. Craig, and D. W. Setser, *J. Phys. Chem.* **76**, 954 (1972).
105. M. Day and A. F. Trotman-Dickenson, *J. Chem. Soc. A* 233 (1969).
106. A. F. Trotman-Dickenson and G. S. Milne, "Tables of Bimolecular Gas Reactions." NSRDS-NBS No. 9, Washington, D.C., 1967.
107. S. W. Benson and G. R. Haugen, *J. Amer. Chem. Soc.* **87**, 4036 (1965).
108. G. Herzberg, "Molecular Spectra and Molecular Structure," Vol. III, Electronic Spectra and Electronic Structure of Polyatomic Molecules. Van Nostrand Reinhold, Princeton, New Jersey, 1966.
109. B. A. Lombos, P. Sauvageau, and C. Sandorfy, *J. Mol. Spectrosc.* **24**, 253 (1967).
110. E. N. Lassettre, A. Skerbele, and M. A. Dillon, *J. Chem. Phys.* **49**, 2382 (1968).
111. D. L. Bunker, *J. Chem. Phys.* **57**, 332 (1972).
112. K. Evans and S. A. Rice, *Chem. Phys. Lett.* **14**, 8 (1972).
113. M. Abramowitz and I. A. Stegun, "Handbook of Mathematical Functions." NBS Appl. Ser. No. 55, Washington, D.C., 1964.
114. S. J. Riley and D. R. Herschbach, *J. Chem. Phys.* **58**, 27 (1973).
115. J. M. Parson and Y. T. Lee, *J. Chem. Phys.* **56**, 4658 (1972).
116. R. Kushner and F. S. Rowland, *J. Phys. Chem.* **75**, 3771 (1971).
117. F. B. Growcock, W. L. Hase, and J. W. Simons, *Int. J. Chem. Kinet.* **5**, 77 (1973).
118. A. Ben-Shaul, R. D. Levine, and R. B. Bernstein, *Chem. Phys. Lett.* **15**, 160 (1972).
119. M. G. Topor and R. W. Carr, *J. Chem. Phys.* **58**, 757 (1973).
120. F. H. Dorer, *J. Phys. Chem.* **77**, 954 (1973).

CHAPTER 10

Fragmentation of Highly Excited Species: Theory of Mass Spectra

The theory of mass spectra, when originally put forward by Eyring and collaborators (*1*),[1] meant just that, a theoretical account of the fragmentation pattern produced by electron impact in a mass spectrometer. The title of this chapter will be used here in a somewhat wider sense as a catch-all term for all high-energy unimolecular decompositions of neutral or charged species produced by any one of the numerous techniques suitable for high excitation, such as electron impact (*2*), charge exchange (*3*), field ionization (*4*), radiolysis (*5*), photoionization (*6*), or vacuum-ultraviolet photolysis (*7*), or by a hot-atom technique using nuclear-recoil ^3H (*8*, p. 108 ff.) or ^{18}F (*9*). Reviews of the various techniques are given in (*2–9*) (see also the beginning of Chapter 9).

In all of these high-energy processes, despite widely different methods of excitation, the decay of the primary excited species should be calculable by the general method of Part I if the assumptions of the statistical theory are satisfied. It turns out that in a good many cases they are, although this requires some extrapolation from the fairly circumscribed treatment of Part I; for this reason, all these processes are considered together under the aegis of the theory of mass spectra.

Important features The common difference that distinguishes the present group of excited systems from those considered in Chapters 8 and 9 is the

[1] The reference list for this chapter can be found on p. 337.

generally much higher level of excitation (> 10 eV) relative to ground state. The consequences are twofold: The reactants are in various stages of vibrational and electronic excitation and often, though not necessarily, are ions, which may or may not behave like similar neutral molecules; and the reactants are sufficiently excited to give rise not only to several parallel channels of reaction, but also to a sequence of consecutive reactions in each parallel channel. This complicates the kinetics and at the same time poses again the problem of energy partition among fragments, previously only touched upon in Chapter 9, Section 6. Since systems involving charged species are always studied at low pressure to avoid the complication of ion–molecule reactions, we have total absence of collisions, and hence of temperature, in contrast with the systems of Chapters 8 and 9.

High excitation energy means that in a polyatomic molecule or ion, a very large number of excited electronic states become energetically accessible and therefore populated. The theory of mass spectra provides a means and a rationalization for dealing with this multitude of excited electronic states. The assumption is made that following excitation, all the electronic excitation energy is rapidly degraded into the internal energy of just one low-lying state (perhaps just the ground state of the neutral or the ion, respectively—a sort of internal conversion), where it is randomized among the internal degrees of freedom; as a result, decomposition from this low-lying state is calculable by the general method of Part I. The theory thus extends the postulate of energy randomization, considered in Chapter 2, Section 3, to include electronic energy as well, and provides arguments in support of this contention.

This chapter is devoted exclusively to unimolecular reactions of positive ions because these are the systems that have been most widely studied both experimentally and theoretically. In comparison, there have been almost no calculations done on the fragmentation of highly excited neutral systems, no doubt due in part to the difficulty connected with obtaining the energy distribution of the excited reactant, particularly in the case of nuclear-recoil reactions. Apart from the energy distribution problem, however, conclusions arrived at from the study of charged systems should be directly transferable to comparable neutral systems excited to energies not exceeding their first ionization potential.

Plan of chapter Section 1 sketches the overall characteristics, both theoretical and experimental, of unimolecular reactions of positive ions. The kinetics of consecutive-parallel reactions is worked out in Section 2, and the statistical aspects of the problem of energy partition among the fragments is considered here in some detail. In order to obtain some information about the initial preparation of excited ions, the process of ion production by the various techniques is discussed in Section 3. The following section, Section 4, examines available information about the potential energy surfaces of polyatomic ions. The basic assumption of the statistical theory is extended in Section 5 to include the randomization of electronic energy and the evidence for and against it. Section 6 discusses the energy distribution functions applicable to the various excitation processes and the reduction of experimental data, while Section 7 deals with problems particular to the calculations of primary radiolytic yields. The connection between experimental observables and theoretical parameters is established in Section 8, and, finally, to close the chapter, Section 9 treats special effects associated with rotational and translational energy.

1. Unimolecular reactions of positive ions

The theory of mass spectra is basically a recipe for applying the general method of Part I to the fragmentation of highly excited species in general, and to the fragmentation of highly excited ions in particular. The theory asserts that there is no fundamental difference between the reactions of neutral and charged molecules, the only determining factor being total internal excitation energy. Hence similar neutral and charged molecules at similar levels of excitation should yield similar fragments, while differences in fragmentation behavior, such as there are, should be only a matter of detail.

The important proviso is, of course, that such a resemblance between the reactions of similar neutral and charged species will exist only to the extent that the decomposition of both shall satisfy the basic assumptions underlying the method of Part I. We cannot expect this to be universally true, but numerous actual examples exist.

For example, in the vacuum-UV photolysis of ethane (*10*), the (neutral) ethane molecule is sufficiently excited to undergo the series of parallel reactions

$$\begin{aligned} C_2H_6 + h\nu &\longrightarrow C_2H_4 + H_2 \\ &\longrightarrow CH_4 + CH_2 \\ &\longrightarrow C_2H_5 + H \\ &\longrightarrow CH_3 + CH_3 \end{aligned} \qquad (10\text{-}1)$$

Of the fragments, C_2H_4 and C_2H_5 contain sufficient energy to decompose further into $C_2H_2 + H_2$ and $C_2H_4 + H$, respectively. In thermal systems, energy is barely sufficient to open the C—C cleavage channel, whereas here we have evidence also for the energetically more expensive C—H cleavage and for the molecular elimination of a lower alkane (CH_4 in this case). This type of reactions is characteristic of all highly excited (neutral) alkanes (*11*, Chapter 3).

Similarly, analysis of fragments found in the mass spectrum of ethane shows that a simplified decomposition scheme of the excited ethane ion is (*12*)

$$\begin{aligned} C_2H_6^+ &\longrightarrow CH_2CH_2^+ + H_2 \\ &\longrightarrow CH_3CH^+ + H_2 \\ &\longrightarrow C_2H_5^+ + H \\ &\longrightarrow CH_3^+ + CH_3 \end{aligned} \qquad (10\text{-}2)$$

where fragments CH_3CH^+ and $C_2H_5^+$ decompose further into $CHCH^+ + H_2$ and $C_2H_3^+ + H_2$, respectively. Ethane ions produced by charge exchange or photoionization decompose in substantially the same fashion. The similarity between the schemes (10-1) and (10-2) will certainly not be lost on the reader, the small difference being easily accounted for by the slightly different energetics of ionic reactions as compared with those of their neutral counterparts.

Experimental and theoretical aspects The experimental observables in such systems, whether involving neutral or charged species, are normally obtained by analysis after a predetermined reaction time, and represent the fractional abundance of each of the fragments in the products. Since a given observed fragment need not be the primary fragment but can be the end result of a long chain of events, it is necessary to retrace these events back to the primary excitation, i.e., the full kinetic scheme must be worked out, rate constants assigned, amount of energy carried off by second-generation fragments determined, and so on, before any comparison between theory and experiment is possible. For practical reasons, most of the experimental work done along these lines has concerned the fragmentation of charged species (ions) because analysis is then a relatively simple affair using a mass spectrometer, but of course the same general principles apply to the study of neutral systems.

The greater ease with which ions can be manipulated and detected promises that ionic systems will provide, eventually, a more fundamental test of unimolecular rate theory than is possible with neutral systems. For one thing, low operating pressures in experiments involving ions permit their fragmentation to be observed directly and not merely in competition with collisional deactivation, as is the case with chemically activated systems discussed in Chapter 9. The experimental technique for direct lifetime measurements on a specific reaction is complex, but results are beginning to appear, as will be discussed. The presence of a charge makes it possible to determine, or to manipulate, the kinetic energy of ions by an applied electric field; some of these results are discussed in Section 9.

On a more fundamental level, the high internal energies at which some of the mass spectral systems can be prepared and studied make these systems potentially a more or less limiting test case for the statistical (RRKM) unimolecular rate theory; one might even argue that treating very highly excited systems by this theory represents an extrapolation beyond the energy region for which the theory was originally meant to apply. This has to do with the fundamental assumption of complete energy randomization prior to decomposition. As explained in Chapter 2, this assumption depends on an unspecified (in detail, anyway) mechanism of internal energy transfer, assumed to be much more rapid than the reaction itself. While rapid, this process obviously can take only a finite time, and therefore when internal energy of the decomposing species becomes sufficiently high, reaction becomes faster than intramolecular energy transfer. In other words, the assumption of complete energy randomization must inevitably fail when internal energy becomes high enough (assuming the system in question does decompose statistically and not via predissociation, for example). There is no clear indication at this time of what is "high enough"; presumably this will be the case when internal excitation becomes some appreciable fraction of the energy required to dissociate a given molecule into its constituent atoms.

Alternatively, the failure of randomization might be detected at low enough energies provided the time of observation is sufficiently short. The technique of field ionization[2] permits lifetime measurements of ions for intervals as hort as 10^{-11} sec after ionization, which might be sufficiently short to detect any incipient failure of randomization in a molecular ion that is not too small. The interpretation of the measurements is unfortunately complicated by the presence of a distribution of energies in the reacting ions. Since lifetimes observed in the absence of collisions are given by relations that are somewhat different from the usual familiar expression derived for conventional thermal systems, it is useful at this point to go into some detail on these measurements.

Ion lifetimes Let $\mathscr{F}(E)$ be the probability per unit energy that ion M^+ is produced with energy E, so that $\mathscr{F}(E)$ is the energy distribution function of *formed* ions, in the sense of Chapter 9, Section 2, and let $P(\tau) = e^{-k(E)\tau}$ be the probability that M^+ shall have lifetime τ when its excitation energy is E, i.e., the probability that it shall survice in the interval $(0, \tau)$. Here, $k(E)$ is the microcanonical rate constant for the decomposition of M^+ at energy E, and the exponentially decreasing distribution of lifetimes embodies the impli-

[2] See Beckey (*4*). For a brief description of the field ionization technique, see the end of Section 2 of this chapter. A different technique which depends on ionization by electron impact but also permits lifetime measurements at very short intervals after ionization has been described by Karachevtsev and Tal'rose (*13,14*).

cit assumption that the lifetimes are random [cf. Eq. (1-14)]. The probability that an ion of energy E will dissociate in t, $t + dt$ is given by the probability of survival in $(0, t)$ times the probability of dissociation in dt:

$$e^{-k(E)t}k(E)\,dt$$

The number of ions of energy E dissociating in t, $t + dt$ per unit energy range is $d\mathfrak{N}(E)$, given by

$$d\mathfrak{N}(E) = \mathfrak{N}_0(E)e^{-k(E)t}k(E)\,dt \tag{10-3}$$

where $\mathfrak{N}_0(E)$ is the total number (per unit energy range) of ions *of energy E* present originally, i.e., formed at $t = 0$. Using $\mathscr{F}(E)$, we can also write $\mathfrak{N}_0(E)\,dE = \mathfrak{N}_0\mathscr{F}(E)\,dE$, where $\mathfrak{N}_0$ is the total number of ions *of all energies* present at $t = 0$. Then, (10-3) becomes

$$d\mathfrak{N}(E) = \mathfrak{N}_0\mathscr{F}(E)e^{-k(E)t}k(E)\,dt \tag{10-4}$$

Since only $d\mathfrak{N}$, the number of ions of *all* energies dissociating in t, $t + dt$ is measurable, we have to sum the individual contributions $d\mathfrak{N}(E)$ over all energies; thus $d\mathfrak{N} = \int_0^\infty d\mathfrak{N}(E)\,dE$, and

$$d\mathfrak{N} = \mathfrak{N}_0\,dt \int_0^\infty \mathscr{F}(E)k(E)e^{-k(E)t}\,dE \tag{10–5}$$

The rate of disappearance is normally referred to the number of ions (of all energies) actually present at t, rather than present originally at $t = 0$. Since $d\mathfrak{N}(E)/dt = k(E)\mathfrak{N}(E)$, we have by integration with respect to time, for a given energy range dE,

$$\mathfrak{N}(E) = \mathfrak{N}_0(E)e^{-k(E)t} = \mathfrak{N}_0\mathscr{F}(E)e^{-k(E)t} \tag{10-6}$$

and summing over all energy ranges

$$\mathfrak{N} = \mathfrak{N}_0 \int_0^\infty \mathscr{F}(E)e^{-k(E)t}\,dE \tag{10–7}$$

where $\mathfrak{N} = \int_0^\infty \mathfrak{N}(E)\,dE$ is the total number of ions of all energies present at t. Substituting for $\mathfrak{N}_0$ in (10-5), we get [cf. Eq. II-6)]

$$\frac{1}{\mathfrak{N}}\frac{d\mathfrak{N}}{dt} = \frac{\displaystyle\int_0^\infty \mathscr{F}(E)k(E)e^{-k(E)t}\,dE}{\displaystyle\int_0^\infty \mathscr{F}(E)e^{-(E)t}\,dE} = \langle k_t\rangle \tag{10-8}$$

where $\langle k_t\rangle$ is the observed, or average, rate constant. We may look on $\mathscr{F}(E)e^{-k(E)t}$ as a weighting function that expresses the fact that the more energetic ions are less likely to survive the time interval $(0, t)$ and therefore are less likely to be present when the decay is measured at t, $t + dt$.

The average rate constant $\langle k_t \rangle$ is thus a function of t; experimentally, it is found that

$$\langle k_t \rangle = \text{const} \times t^{-\Xi} \tag{10-9}$$

where Ξ is a small number of the order of unity. The time dependence of $\langle k_t \rangle$ is the result of the distribution of excitation energies in M^+, for if the M^+'s were all monoenergetic of energy $E = E_{mono}$, we would have $\mathscr{F}(E_{mono}) = 1$ and $\mathscr{F}(E) = 0$ for all other values of E; then,

$$\langle k_t \rangle = k(E_{mono}) \tag{10-10}$$

In thermal systems, by contrast, the average or observed rate constant is time independent even if the reactant is not monoenergetic, because as a result of collisions a steady state is rapidly set up between the gain and loss of reacting molecules. Since measurements are conventionally made under such steady-state conditions, the measured rate constant is in fact the *steady-state* rate constant (cf. Chapter 8, Section 3 and Exercise 1 of this chapter).

Returning now to lifetime measurements obtained by the field-ionization technique (i.e., in the absence of collisions, and therefore *not* under steady-state conditions), Tenschert and Beckey (*15*, Figure 1), measured $\langle k_t \rangle$ for the reaction

$$C_4H_{10}^{+} \longrightarrow C_3H_7^{+} + CH_3 \tag{10-11}$$

and found that $\log\langle k_t \rangle$ is proportional to $\log t$ down to about 10^{-10} sec, but at shorter times, $\langle k_t \rangle$ is smaller than it should have been. When measurements are made at shorter and shorter times, this is tantamount to sampling progressively more energetic ions. The experimental results on the butane ion mean that its decay rate constant $k(E)$ does not increase indefinitely with E. Assuming statistical decomposition, this could be due either to the failure of the requisite amount of energy reaching the appropriate C—C bond fast enough, which would constitute failure of randomization, or it could mean simply that the excitation process does not produce ions of all energies, in which case the excitation function $\mathscr{F}(E)$ has a cutoff at some $E = E_{max}$, and therefore the rate constant spectrum does not contain $k(E)$'s larger than $k(E_{max})$. The cited authors favor this latter explanation, which is all the more plausible because it is known that the average excitation energy in field ionization is quite low, about 0.5 eV.

At the other end of the energy spectrum, Tatarczyk and v. Zahn (*16*)[3] found that at large t, the average constant $\langle k_t \rangle$ of the reaction

$$C_5H_{11}^{+} \longrightarrow C_3H_7^{+} + C_2H_4 \tag{10-12}$$

[3] The technique utilizes ion storage to permit decay measurements at relatively long times.

does not become indefinitely small. This means that the sample of low-energy rate constants $k(E)$ does not contain any members smaller than some $k(E_{min})$. The energy E_{min} represents of course the threshold energy E_0, below which no decay is possible. This result therefore constitutes the first experimental proof that there indeed exists[4] a threshold rate constant $k(E_0)$.

It is obvious that if $\mathscr{F}(E)$ were known, such lifetime measurements would yield $k(E)$ as a function of energy, including $k(E_0)$, i.e., the microcanonical rate constant at threshold. A more practical approach is to prepare mono-energetic ions (e.g., by charge exchange), which avoids the necessity of knowing, or making assumptions about, $\mathscr{F}(E)$ (cf. Section 8). Thus lifetime measurements on ions hold promise to provide more fundamental information than has been hitherto possible with neutral molecules.

2. Consecutive-parallel reactions: energy partitioning

Seen in the perspective of the theory of mass spectra, the fragmentation of ions involves no new principles, expect insofar as these ions are energetic enough to engage in a number of consecutive, or consecutive-parallel, unimolecular reactions, as, for example, in the scheme (10-2). The theory then simply assumes that the rate of each unimolecular step is calculable by the method of Part I. However, the complex kinetics has two consequences: (1) The relation between an experimental observable (e.g., concentration of parent ion or product ion at time t after excitation) is now a complicated function of time that involves the rate constants of several processes; (2) the rate of a consecutive process cannot be calculated unless it is known how much internal energy is available in the fragment for further decomposition.

Point 1 can be resolved exactly by working out the algebra, but point 2, touched upon in Chapter 9, Section 6, involves a new principle because transition-state theory cannot give information about internal states of products, as discussed in Section 3 of Chapter 4. The present section, as a preliminary to actual application of the general method of Part I, deals with these two ancillary problems. The question of how much internal energy is received by the *parent* is a function of the method of primary excitation and is dealt with in Section 6. For the purpose of this section, the internal energy of the parent is assumed to be fixed and known.

Kinetic scheme The easiest way to treat the system of linear differential equations describing a complex kinetic scheme is by the technique of the

[4] This interpretation assumes that appreciable centrifugal effects are absent, because if the fragmenting ion has a spectrum of angular momenta, there is then also a spectrum of threshold energies, i.e., E_0 becomes E_J, and the lowest value of the latter is $E_J = 0$ for a sufficiently high $\mathscr{J}$ (cf. Chapter 7, Section 1).

Laplace–Carson transform. Since a description of the method and explicit results for specific schemes are available (*17*, Chapter V; *18*), it is sufficient to give here only the result. Suppose we have the following consecutive-parallel scheme, where the parent M^+ is originally prepared with energy E:

$$
\begin{array}{llllll}
 & & \mathsf{A}_{11}^+(E_{11}) \xrightarrow{k_{11}} \mathsf{A}_{21}^+(E_{21}) & \longrightarrow \cdots & \mathsf{A}_{i1}^+(E_{i1}) \xrightarrow{k_{i1}} & \\
 & \nearrow^{k_{01}} & & & & \\
\mathsf{M}^+(E) & \xrightarrow{k_{02}} & \mathsf{A}_{12}^+(E_{12}) \xrightarrow{k_{12}} \mathsf{A}_{22}^+(E_{22}) & \longrightarrow \cdots & \mathsf{A}_{i2}^+(E_{i2}) \xrightarrow{k_{i2}} & \quad (10\text{-}13) \\
 & \searrow_{k_{0j}} & & & & \\
 & & \mathsf{A}_{1j}^+(E_{1j}) \xrightarrow{k_{1j}} \mathsf{A}_{2j}^+(E_{2j}) & \longrightarrow \cdots & \mathsf{A}_{ij}^+(E_{ij}) \xrightarrow{k_{ij}} &
\end{array}
$$

The first subscript i selects the step in the consecutive sequence, and the second subscript j identifies the parallel sequence; M^+ is considered as the common zeroth member of every consecutive sequence. Every reaction produces also a neutral fragment which is not indicated since only the kinetics of the charged species is relevant. The internal energy of every charged species E is indicated in parentheses with a subscript appropriate to that species. The critical energy for decomposition is indicated by the additional subscript 0; thus $E_{0(ij)}$ is the critical energy for decomposition of the ith consecutive member of the jth parallel sequence and $E_{0(0j)}$ designates the critical energy for decomposition of M^+ to yield the first member of the jth parallel sequence (cf. Fig. 10-1). All unimolecular rate constants k_{ij} are energy dependent and are understood to be evaluated at the internal energy of the species bearing the same subscript; thus, for example, k_{11} measures the decay of A_{11}^+ of internal energy E_{11}.

The fractional amount of the first-generation fragment of the jth sequence (A_{1j}^+) present at time t when the excitation energy of M^+ is E at $t = 0$ is

$$\mathfrak{F}_{1j}(E, t) = \frac{k_{0j}(E)}{k_{1j}(E_{1j}) - \sum_j k_{0j}(E)} \left\{ \exp\left[-\sum_j k_{0j}(E)t \right] - \exp[-k_{1j}(E_{1j})t] \right\} \tag{10-14}$$

where the energy dependence of every rate constant is indicated explicitly. In a sequence of consecutive reactions, the critical energy for the process $\mathsf{A}_{1j}^+ \rightarrow \mathsf{A}_{2j}^+$ is of necessity larger than that for the process $\mathsf{M}^+ \rightarrow \mathsf{A}_{1j}^+$, i.e., $E_{0(1j)} > E_{0(0j)}$. Since unimolecular rate constants in general increase sharply with energy (cf. Fig. 7-3), the disparity in threshold energies will insure that the rate constant $k_{1j}(E_{1j})$ will be much smaller than $\sum_j k_{0j}(E)$ in virtually every case. Hence at E high enough for the process $\mathsf{A}_{1j}^+ \rightarrow \mathsf{A}_{2j}^+$ to occur, we will have $k_{1j}(E_{1j}) \ll \sum_j k_{0j}(E)$ and $\exp[-\sum_j k_{0j}(E)t] \approx 0$, so that (10-14) reduces in practice to

$$\mathfrak{F}_{1j}(E, t) \simeq \left[k_{0j}(E) \Big/ \sum_j k_{0j}(E) \right] e^{-k_{1j}(E_{1j})t} \qquad (E > E_{0(1j)}) \tag{10-15}$$

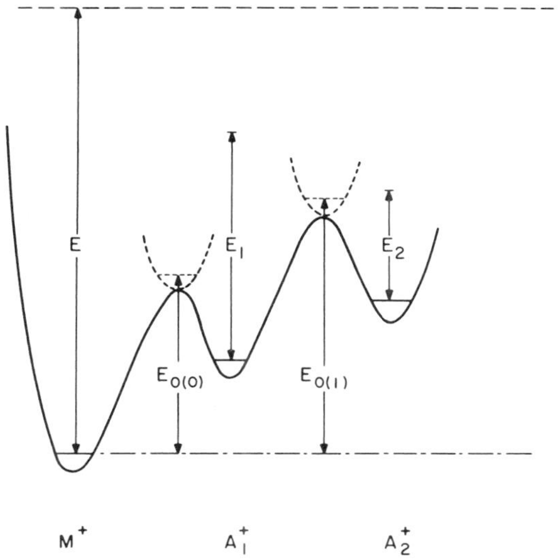

Fig. 10-1. *Schematic energy diagram for a sequence of two consecutive reactions involving positive ions.*

The full reactions are

$$M^+ \rightarrow A_1{}^+ + B_1 \qquad (0)$$
$$A_1{}^+ \rightarrow A_2{}^+ + B_2 \qquad (1)$$

where B_1 and B_2 are neutral fragments. In the general case of parallel-consecutive reactions, all symbols, except M^+ and E, should have a second subscript j identifying the parallel sequence. E_1 and E_2 are internal energies of $A_1{}^+$ and $A_2{}^+$, respectively, and $E_{0(0)}$ and $E_{0(1)}$ are the critical energies for the processes (0) and (1), respectively. All energies are referred to ground state of M^+ as zero. Because of energy partitioning between neutral and charged fragments, $E_1 < (E - E_{0(0)})$ and $E_2 < (E - E_{0(1)})$ (see text). It is assumed in the diagram that there is no formal critical energy for the reverse of processes (0) and (1), i.e. that the forward fragmentation takes place over a type 1 potential energy surface. The reaction coordinate is different for every reaction and therefore the abscissa is not identified. As drawn, fragment $A_2{}^+$ does not have sufficient internal energy to decay further.

Similarly, the fractional amount of the second-generation fragment $A_{2\,j}^+$ present at time t after M^+ has received energy E is

$$\mathfrak{F}_{2j}(E, t) = \frac{k_{0j}(E)k_{1j}(E_{1j})}{[k_{2j}(E_{2j}) - k_{1j}(E_{1j})]\sum_j k_{0j}(E)} \times \{e^{-k_{1j}(E_{1j})t} - e^{-k_{2j}(E_{2j})t}\}, \qquad (E > E_{0(2j)}) \qquad (10\text{-}16)$$

if it is assumed that $\exp[-\sum_j k_{0j}(E)t] \sim 0$, $k_{2j}(E_{2j}) \ll \sum_j k_{0j}(E)$, which will be true if $k_{1j}(E_{1j}) \ll \sum_j k_{0j}(E)$. Furthermore, if $E_{0(2j)}$ is sufficiently larger than $E_{0(1j)}$, the rate constant $k_{2j}(E_{2j})$ will exceed zero only when $k_{1j}(E_{1j})$ is already so large that $e^{-k_{1j}(E_{1j})t} \sim 0$, and (10-16) then reduces to

$$\mathfrak{F}_{2j}(E, t) \approx \frac{k_{0j}(E)k_{1j}(E_{1j})}{[k_{1j}(E_{,1j}) - k_{2j}(E_{2j})] \sum_j k_{0j}(E)} e^{-k_{2j}(E_{2j})t} \tag{10-17}$$

Under these conditions, $\mathfrak{F}_{1j}(E, t)$ of Eq. (10-15) is zero.

This sort of development can be continued until the ith generation fragment, but in practice, the second-generation fragment is probably as far as anyone would want or need to go. The fractional abundance of M^+ at all times t after excitation is

$$\mathfrak{F}_0(E, t) = \exp\left[-\sum k_{0j}(E)t\right] \tag{10-18}$$

The developments inside a sequence of consecutive reactions subsequent to the decomposition of M^+ can be summarized by saying that because of the usual strong dependence of every $k(E)$ on E, combined with the disparity of threshold energies, we have the result that when the excitation energy is high enough for a new process (e.g., $A_{ij}^+ \rightarrow A_{(i+1)j}^+$) to set in, the fractional abundance of the precursor ($A_{(i-1)j}^+$) is already zero; in particular, the abundance of M^+ is negligible when A_{2j}^+ starts appearing. Special cases, where threshold energies are close, or where $k(E)$ does not rise sufficiently with E, require of course consideration of the full expressions.

Let us now concentrate on the abbreviated sequence (cf. Fig. 10-1)

$$M^+ \xrightarrow{k_0} A_1^+ \xrightarrow{k_1} A_2^+ \tag{10-19}$$

where the index j shall be henceforth dropped for greater clarity of notation (except when the k_0's have to be summed over all parallel reactions), and where A_2^+ does not decompose further, i.e., $k_2 = 0$. We then have from (10-15)

$$\mathfrak{F}_1(E, E_1, t) = \left[k_0(E) \Big/ \sum_j k_{0j}(E)\right] e^{-k_1(E_1)t} \tag{10-20}$$

where it is now indicated explicitly that $\mathfrak{F}_1$ depends not only on E and t but also on E_1. Similarly, we have from (10-16), for $k_2 = 0$,

$$\mathfrak{F}_2(E, E_1, t) = \left[k_0(E) \Big/ \sum_j k_{0j}(E)\right](1 - e^{-k_1(E_1)t}) \tag{10-21}$$

where it is again indicated explicitly that $\mathfrak{F}_2$ depends[5] on E, t, and E_1. Observe that assuming random incidence of dissociation (cf. Chapter 1, Section 4), the factor $e^{-k_1(E_1)t}$ can be interpreted as the probability of A_1^+ surviving in the interval $(0, t)$ when its excitation energy is E_1, and therefore

[5] This is true only in the special case when A_2^+ does not decompose further. If it does, $\mathfrak{F}_2$ depends on E_2 as well.

the factor $1 - e^{-k_1(E_1)t}$ in (10-21) is the probability of A_1^+ *not* surviving, i.e., dissociating, in $(0, t)$ when its excitation energy is E_1. Let this probability be designated by $\mathfrak{D}_1(E_1, t)$; then

$$\mathfrak{D}_1(E_1, t) = 1 - e^{-k_1(E_1)t} \tag{10-22}$$

and

$$\mathfrak{F}_1(E, E_1, t) = \left[k_0(E) \Big/ \sum_j k_{0j}(E)\right][1 - \mathfrak{D}_1(E_1, t)] \tag{10-23}$$

$$\mathfrak{F}_2(E, E_1, t) = \left[k_0(E) \Big/ \sum_j k_{0j}(E)\right]\mathfrak{D}_1(E_1, t) \tag{10-24}$$

Energy partition among fragments Unfortunately, the term $e^{-k_1(E_1)t}$ in (10-20)–(10-22) cannot be evaluated as it stands because E_1 is not known. In fact, $E_1 \neq E - E_{0(1)}$, because the full reaction is

$$M^+ \longrightarrow A_1^+ + B_1 \tag{10-25}$$

and most of the difference $E - E_{0(0)}$ is shared between A_1^+ and the neutral fragment B_1; hence all that can be said is that $E_1 \leq (E - E_{0(0)})$. A similar problem exists in connection with E_2 in Eq. (10-16). Since the transition-state theory cannot give information about the partition of energy between fragments, an assumption must be made. In the absence of any information, the simplest assumption that can be made is that the partitioning is statistical (*19*).

The problem of statistical partition of energy was considered may years ago by Kassel (*20*, pp. 101–101), and more recently by Wallenstein and Krauss (*21*), who applied it to mass spectra; a very similar problem appears in connection with transfer of energy on collision in thermal systems (*22,23*). Expressed in the simplest terms, statistical partitioning of energy involves counting the number of ways n quanta of energy can be distributed between two degenerate harmonic oscillators; in this form, it is thus a combinatorial problem of the kind discussed in Chapter 6, Section 2. Let the first oscillator (subscript a) be q_a-fold degenerate and let it contain n_a quanta; similarly, let the second oscillator (subscript b) be q_b-fold degenerate and let it contain n_b quanta, such that $n_a + n_b = n$. The number of ways of distributing n quanta among the two oscillators with n_a quanta in the first and n_b in the second is, by (6-7),

$$W_a(n_a) \times W_b(n_b) = \frac{(n_a + q_a - 1)!}{n_a!\,(q_a - 1)!} \times \frac{(n_b + q_b - 1)}{n_b!\,(q_b - 1)!} \tag{10-26}$$

if the oscillators are independent. Assuming that the oscillators exchange energy freely, the *total* number of ways of distributing $n = n_a + n_b$ quanta among the $q_a + q_b = q$ oscillators is[6]

$$\sum_{k=0}^{n} W_a(k)W_b(n-k) = \sum_{k=0}^{n}\left\{\frac{(k+q_a-1)!}{k!\,(q_a-1)!} \times \frac{(n-k+q_b-1)!}{(n-k)!\,(q_b-1)!}\right\}$$

$$= \frac{(n+q-1)!}{n!\,(q-1)!} = W_{a,b}(n) \tag{10-27}$$

Hence when there are a total of n quanta, the normalized probability of finding n_a quanta in the first oscillator and $n_b = n - n_a$ quanta in the second is

$$P(n_a, n) = W_a(n_a)W_b(n - n_a)/W_{a,b}(n) \tag{10-28}$$

Figure (10-2) shows a histogram of $P(n_a, n)$ versus n_a for the case $q_a = 21$, $q_b = 15$ and for three different values of n. It can be shown that the maximum of each function is located in the vicinity of $n_a = nq_a/(q_a + q_b)$, which represents equipartition. If we identify q_a with the number of vibrational degrees of freedom in A_1^+ and $q_a + q_b$ with the number of vibrational degrees of freedom in M^+, we have the result, used previously in Chapter 9, Section 6, that in the absence of any knowledge about the partitioning of energy, the assumption of free exchange of energy among the degrees of freedom (= "oscillators") of M^+ leads to equipartition as the most probable mode of energy sharing between M^+ and A_1^+. With respect to Eq. (10-25), equipartition thus interpreted would mean that $E_1 = (E - E_{0(0)})q_a/(q_a + q_b)$.

Unfortunately, as pointed out by Wallenstein and Krauss (*21*), the function $P(n_a, n)$ is not at all steep (Fig. 10-2), particularly when n/q is large, so that there is appreciable probability for the fragments to contain energy that is different from equipartition, and therefore must be taken into account. This effect is called "fluctuation."

Before considering fluctuations in more detail, it is convenient to generalize the result (10-28) in terms of an energy-dependent continuous function, i.e., in terms of the densities of states $N(E)$. It can be easily seen from (10-27) that $W_{a,b}(n)$ in the denominator of (10-28) is the discrete convolution (6-33b), while the numerator of (10-28) is merely its summand; hence the equivalent expression in terms of the densities of states can be written directly with the help of Eq. (6-32). If we let x be the energy contained in the fragment of interest (subscript a), and if $\mathscr{E}$ is the total energy available for partition, (10-28) becomes

$$P(x, \mathscr{E}) = N_a(x)N_b(\mathscr{E} - x)\Big/\int_0^{\mathscr{E}} N_a(x)N_b(\mathscr{E} - x)\,dx \tag{10-29}$$

[6] For a proof of this relationship in more general form, see Fry (*24*, p. 41, Eq. (26.2)).

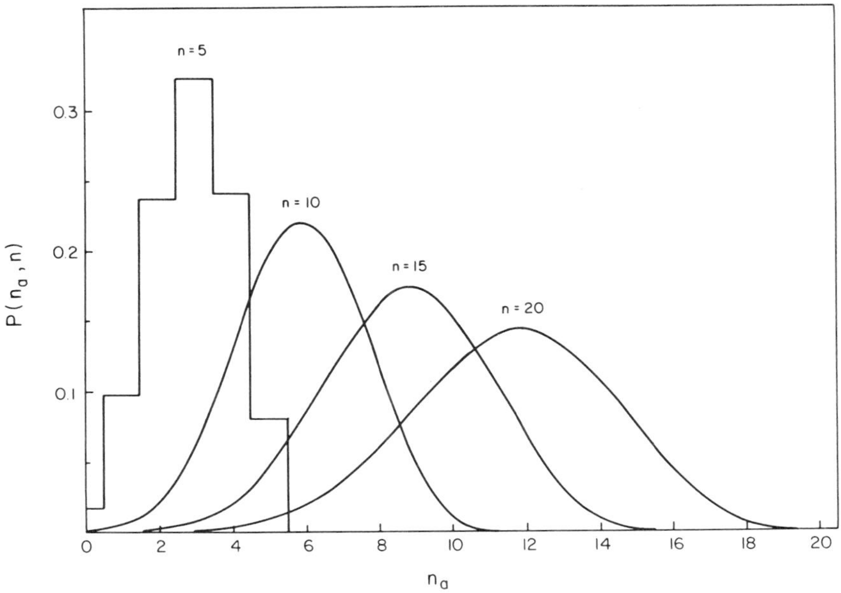

Fig. 10-2. *Plot of probability $P(n_a, n)$ of Eq. (10-28) versus n_a.*
The model consists of oscillator *a*, assumed q_a-fold degenerate, and of oscillator *b*, assumed q_b-fold degenerate. The function $P(n_a, n)$ measures the probability that n_a quanta shall be found in oscillator *a* if the total number of quanta available to both oscillators is n. Calculations were done for $q_a = 21$ and $q_b = 15$ for four different values of n. Since the variable n_a is discontinuous, the function $P(n_a, n)$ should be drawn as a histogram, as shown for $n = 5$. The three other cases ($n = 10$, $n = 15$, and $n = 20$) were drawn as smooth curves for reasons of clarity.

where $P(x, \mathscr{E})$ (dimension: energy^{-1}) is now the probability per unit energy of finding energy x, out of total energy $\mathscr{E}$, in the fragment of interest. The question now arises as to how to interpret $N_b(\mathscr{E} - x)$ and the denominator in (10-29), which is the density of states of the combined system at $\mathscr{E}$.

With reference to the scheme (10-25), the total available energy for partition is $\mathscr{E} = E - E_{0(0)}$, and the configuration, or "combined system" that contains this energy is not M^+ but its transition state (Fig. 10-1). Hence the denominator of (10-29) should be the density of states of the transition state of M^+ at $\mathscr{E}$, i.e., $N^*(\mathscr{E})$ (the asterisk as superscript identifies, as usual, quantities pertaining to the transition state). However, the available energy in the transition state is shared between internal energy and energy of relative translation (cf. Chapter 3, Section 4), so that a measure of the effective number of states in the transition state is $G^*(\mathscr{E})$, the integrated density

(=total number of states). Thus $G^*(\mathscr{E})$ being the denominator of $P(x, \mathscr{E})$ we now require that its numerator be the integrand of

$$G^*(\mathscr{E}) = \int_0^{\mathscr{E}} N_a(x) G_b^*(\mathscr{E} - x)\, dx \tag{10-30}$$

where $G_b^*(\mathscr{E} - x)$ is the integrated density, at $\mathscr{E} - x$, of a configuration having the internal degrees of freedom of the transition state, *less* those of the fragment A_1^+. The reason it is $G_b^*(\mathscr{E} - x)$ and not $N_b^*(\mathscr{E} - x)$ that appears in (10-30) is that, by Eq. (6-35), the integrated density of a combined system is given by the convolution of the density of one part and the integrated density of the other part. Hence (see also Exercise 4)

$$P(x, \mathscr{E}) = N_a(x) G_b^*(\mathscr{E} - x)/G^*(\mathscr{E}) \tag{10-31}$$

The probability $P(x, \mathscr{E})$ was used in the form (10-31) by Lin and Rabinovitch (*25*).[7] It has a maximum somewhat on the *low* side of equipartition, i.e., the most probable distribution is where A_1^+ has somewhat less energy than one would calculate from the ratio of the number of degrees of freedom in A_1^+ and in M^+, particularly if A_1^+ is small compared with M^+ [cf. Table I in (*25*)].

The function (10-31) is close in concept to the function $F(E)$ of Chapter 9 [see, in particular, text accompanying Eq. (9-28)], and shares with it the disadvantage that it is model dependent, since it involves properties of the transition state. It also ignores that part of the available energy $\mathscr{E} - x$ which may not be distributable because of conservation requirements. In most calculations done in mass spectrometry, $P(x, \mathscr{E})$ has been used in the form (10-29), with $N_a(x)$ having the same significance as in (10-31) but with $N_b(\mathscr{E} - x)$ interpreted (*26*) as the density of states of the neutral fragment B_1 [Eq. (10-25)] at energy $\mathscr{E} - x$. The problem with this formulation is that the sum of the number of degrees of freedom in A_1^+ and in B_1 does not add up to the number of degrees of freedom in M^+ (or its transition state) because there are also degrees of freedom of relative motion. Hence the entire energy $\mathscr{E} - x$ is not available to the internal energy of the combined system $A_1^+ + B_1$, as implied by the denominator in (10-29). Equation (10-29) would then predict, for instance, that no fluctuation is possible if one fragment is an atom, because the atom has no internal degrees of freedom. In fact, fluctuation is possible in such a case because some energy is shared between, say, internal degrees of freedom of the large fragment and the degrees of freedom of relative motion of both fragments. This is really

[7] In the text of this reference, $G_b^*(\mathscr{E} - x)$ (our notation) is identified as referring to the remainder of the degrees of freedom in the *molecular ion*, although Eq. (23) in ref. (*25*) makes it clear that it actually refers to the remainder of degrees of freedom in the *transition state*.

another way of saying that Eq. (10-29) neglects the entire kinematics of the fragmentation process.

In spirit (though not in actual detail) Eq. (10-29) is akin to the phase space theory of Light (*27*), in which it is assumed that any state of the *fragment* system consistent with conservation of total energy and angular momentum is equally probable. Busch and Wilson (*28*) have called this approach "statistics after dissociation," in contrast with Eq. (10-31), which represents essentially what they call "statistics before dissociation."

However, (10-29) has the advantage that much more is known about the parameters of the neutral fragment B_1 than about those of the "transition-state-minus-A_1^+" moiety which are required for calculating $G_b^*(\mathscr{E} - x)$. For this reason, the form (10-29) of $P(x, \mathscr{E})$ is probably preferable to (10-31), even though neither is entirely correct. In any event, the two forms yield similar results (see Exercise 2).

Whatever the specific form of $P(x, \mathscr{E})$, it is simply the distribution function for the internal excitation energy of A_1^+, analogous to $\mathscr{F}(E)$ introduced in Section 1. The average fractional abundance of A_1^+ is therefore

$$\mathfrak{F}_1(E, t) = \int_0^{E - E_{0(0)}} \mathfrak{F}_1(E, E_1, t) P(E_1, E - E_{0(0)})\, dE_1 \qquad (10\text{-}32)$$

where $\mathfrak{F}_1$ on the left-hand side of (10-32) is now referred to the excitation energy of the parent (M^+), assumed to be monoenergetic. The maximum possible internal energy of A_1^+ is $E - E_{0(0)}$, corresponding to the case when no energy is received by the neutral fragment B_1, and the minimum energy is obviously zero.

Experimental observables Thus $\mathfrak{F}(E, t)$ gives the time dependence of the fractional abundance of A_1^+ that would be observed for a given excitation energy in M^+. However, in conventional mass spectrometry (unlike in the much more sophisticated lifetime measurements mentioned in Section 1), time is not a variable because "time" is the residence time of the ion in the instrument, and therefore is fixed by the geometry of the instrument. The experimental observable is a "signal," i.e., some minimum abundance of a given ion that the instrument is able to detect above background noise. Since the bulk of the information about reactions of ions has come, and still comes, from conventional mass spectrometry, further account will be slanted mainly toward mass spectrometric applications. For this reason, it is useful to pause a little at this point and discuss briefly the nature of the observables in mass spectrometry; a fuller account may be found in books dealing with instrumental aspects.[8]

[8] See, for example, Roboz (*29*). The discussion in this section omits instruments without a magnetic field, e.g., time of flight, quadrupole, radiofrequency, etc.

A mass spectrometer, in the most general sense of the term, is an instrument consisting of a source, an analyzer, and a detector. The source is where excited ions are produced by any one of a number of techniques, and the analyzer is where ions are subjected to a succession of electric and magnetic fields which can be manipulated so as to have only ions of the desired mass to charge ratio m/e reach the detector and produce a signal; this is called mass analysis. The magnetic field is separated from the electrostatic field and from the detector by field-free space. Typical values of ion residence times in the source and various parts of the analyzer are given in Table 10-1.

TABLE 10-1

ION RESIDENCE TIMES IN VARIOUS PARTS OF A MASS SPECTROMETER[a]

		Analyzer			
	Source	Electrostatic field	Field-free space I	Magnetic field	Field-free space II
Ion residence time (μsec)	1–5	1	5	3	5
Total flight time[b]	τ_1	τ_2	τ_3	τ_4	

[a] After W. A. Chupka, *J. Chem. Phys.* **30**, 191 (1959).

[b] Total flight time is the time spent in flight from source entrance to the exit point of the given instrument element.

Since the ions produced in the source are excited, they fragment en route to the detector, and the value of m/e at which they produce a signal depends on where along the flight path the fragmentation takes place. Suppose for simplicity that the parent ion fragments according to $M^+ \rightarrow A_1^+$ with rate constant k. If M^+ decomposes only in the field-free space II (ahead of the detector), the fragment A_1^+ continues on the trajectory of M^+ and therefore produces a signal at m_0/e corresponding to the parent M^+. If M^+ decomposes in the source, the particle traveling through the analyzer is A_1^+ and yields a signal at m_1/e corresponding to the fragment A_1^+. Ions M^+ decomposing in the electrostatic or magnetic fields are essentially lost, but if M^+ decomposes in the field-free region I between the electric and magnetic fields, a signal will appear at $m_\star/e$, where (*30*)

$$m_\star = m_1^2/m_0 \tag{10-33}$$

While m_0 and m_1 are integral under moderate resolution, $m_\star$ is non integral. Ions that give a signal at $m_\star/e$ are called metastable.

We can now calculate [following Chupka (*31*)] the normalized detector signal, for the simple scheme $M^+ \xrightarrow{k} A_1^+$, as a function of the rate constant k, using flight times of Table 10-1. The normalized detector signal is nothing else but the fractional abundance of the various species as "seen" by the detector. For the parent (signal at m_0/e) we have

$$\mathfrak{F}_0 = e^{-k\tau_4}, \qquad \tau_4 = \tau_1 + 9 \times 10^{-6} \quad \text{sec} \tag{10-34}$$

which is the probability of M^+ surviving in $(0, \tau_4)$; any M^+ that decomposes in this time interval is either lost or appears at some other mass. Similarly, we have for the fragment $A_1{}^+$ (signal at m_1/e)

$$\mathfrak{F}_1 = 1 - e^{-k\tau_1} \tag{10-35}$$

which is the probability of M^+ *not* surviving in $(0, \tau_1)$; again, M^+ not surviving, i.e., decomposing, in some other time interval will be either lost or will yield a signal at some other mass. Finally, for the metastable (subscript $\star$) we have[9]

$$\mathfrak{F}_\star = e^{-k\tau_2}[1 - e^{-k(\tau_3 - \tau_2)}]; \qquad \tau_2 = \tau_1 + 10^{-6} \quad \text{sec}$$
$$\tau_3 = \tau_1 + 6 \times 10^{-6} \quad \text{sec} \tag{10-36}$$

where the first factor is the probability of M^+ surviving in $(0, \tau_2)$, and the second factor is the probability of M^+ *not* surviving[10] in (τ_2, τ_3). Since ions can be formed anywhere in the source, τ_1 varies from 1 to 5 μsec; assuming that for equal fractions of ions $\tau_1 = 1, 2, 3, 4, 5$ μsec, Chupka calculated $\mathfrak{F}_0$, $\mathfrak{F}_1$, and $\mathfrak{F}_\star$ as a function of k. The results are shown in Fig. 10-3, which gives the fractional abundances as a function of log k. In practice, of course, k is not an independent variable, but E (excitation energy in M^+) is. Since k

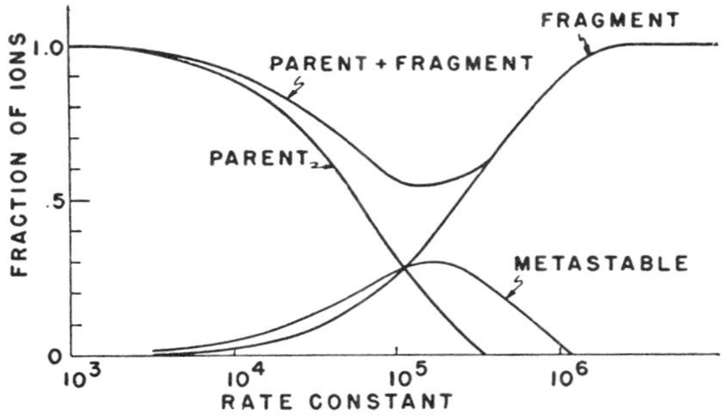

Fig. 10-3. *Fractional abundances as a function of rate constant in the process* $M^+ \rightarrow A_1$. Fractional abundance of parent is $\mathfrak{F}_0$ [Eq. (10-34)], that of fragment is $\mathfrak{F}_1$ [Eq. (10-35)], and that of metastable (3) is $\mathfrak{F}_\star$ [Eq. (10-36)]. From W. A. Chupka, *J. Chem. Phys.* **30**, 191 (1959). Fig. 7.

[9] For more detailed calculations of metastables, see Hills *et al.* (*32*).
[10] Cf. Eq. (1-15) in Chapter 1.

varies strongly with E, roughly as E^n, a plot of $\mathfrak{F}_0$, $\mathfrak{F}_1$, and $\mathfrak{F}_\star$ versus log E would look very similar, and in fact experimental results giving the normalized detector signal as a function of the excitation energy in the parent do look very much like Fig. 10-3. Such a presentation of data is often referred to as the "breakdown graph." Note that $\mathfrak{F}_\star$ provides a rather narrow "window" on processes whose rate constant is located in a small region around 10^5 sec^{-1}.

It may be useful to point out here that every instrument will mass-discriminate to some extent, i.e., the ability of its detector to "see" ions will depend on their mass, and this mass discrimination will be inevitably reflected in the experimental breakdown curve.

Observed fractional abundance We can now return to the more complex scheme (10-19). Now $\mathfrak{F}_1$ [Eq. (10-23)] and $\mathfrak{F}_2$ [Eq. (10-24)] represent the signals that A_1^+ and A_2^+ yield at m_1/e and m_2/e, respectively, and therefore $t = \tau_1$ in both equations. Furthermore, $\mathfrak{F}_1$ and $\mathfrak{F}_2$ depend, through $\mathfrak{D}_1$, on E_1, which is subject to fluctuations, so that the observed normalized signal at m_2/e, as a function of E, is

$$\mathfrak{F}_2(E) = \left[k_0(E)\Big/\sum_j k_{0j}(E)\right]\langle\mathfrak{D}_1(E_1, \tau_1)\rangle_P \tag{10-37}$$

where

$$\langle\mathfrak{D}_1(E_1, \tau_1)\rangle_P = \int_0^{E - E_{0(0)}} (1 - e^{-k_1(E_1)\tau_1})P(E_1, E - E_{0(0)})\, dE_1 \tag{10-38}$$

and $P(E_1, E - E_{0(0)})$ is given by (10-29) or (10-31).

Equation (10-38) can be simplified by noting that because $k_1(E_1)$ rises steeply with E_1, $e^{-k_1(E_1)\tau_1}$ will be virtually zero everywhere except at, and immediately above, threshold; it will be of course unity below threshold. We can therefore write

$$\langle\mathfrak{D}_1(E_1, \tau_1)\rangle_P \approx \int_{E_m}^{E - E_{0(0)}} P(E_1, E - E_{0(0)})\, dE_1 \tag{10-39}$$

where $E_m \geqslant E_{0(1)}$ is some effective lower cutoff on the integral in (10-38) to compensate for $1 - e^{-k_1(E_1)\tau_1}$ not being quite unity in the vicinity of threshold.[11] We may look on E_m as the minimum energy necessary to yield a detectable signal corresponding to $\mathfrak{F}_2$. In practice, $E_m = E_{0(1)}$ is likely to be a good enough approximation, especially if $E_{0(1)}$ is taken as equal to the appearance potential of A_2^+ (cf. Section 8).

[11] Unless, of course, the rate at threshold is already so high that $e^{-k_1(E_1)\tau_1} \sim 0$, in which case $E_m = E_{0(1)}$.

For $\mathfrak{F}_1(E)$ we have, from (10-23),

$$\mathfrak{F}_1(E) = \left[k_0(E)\Big/\sum_j k_{0j}(E)\right]\langle 1 - \mathfrak{D}_1(E_1, \tau_1)\rangle_P \qquad (10\text{-}40)$$

where

$$\langle 1 - \mathfrak{D}_1(E_1, \tau_1)\rangle_P = \int_0^{E - E_{0(0)}} e^{-k_1(E_1)\tau_1} P(E_1, E - E_{0(0)})\, dE_1 \qquad (10\text{-}41)$$

Since the exponential is essentially zero above, but not quite at, $E_{0(1)}$, we may write

$$\langle 1 - \mathfrak{D}_1(E_1, \tau_1)\rangle_P \sim \int_0^{E_m} P(E_1, E - E_{0(0)})\, dE_1 = 1 - \langle \mathfrak{D}(E_1, \tau_1)\rangle_P \qquad (10\text{-}42)$$

where E_m is the same integration limit that appears in (10-39). It is therefore sufficient to calculate only $\langle \mathfrak{D}_1(E_1, \tau_1)\rangle_P$, from which both $\mathfrak{F}_1$ and $\mathfrak{F}_2$ can be obtained.

We can now also calculate the abundance of metastable transitions. For the metastable $M^+ \rightarrow A_1^+$, we have from (10-36) directly[12]

$$\mathfrak{F}_{0\star}(E) = \left[k_0(E)\Big/\sum_j k_{0j}(E)\right] \times \left\{\exp\left[-\sum_j k_{0j}(E)\tau_2\right] - \exp\left[-\sum_j k_{0j}(E)\tau_3\right]\right\} \qquad (10\text{-}43)$$

The fractional abundance of the metastable $A_1^+ \rightarrow A_2^+$ is given similarly by the probability of A_1^+ surviving [Eq. (10-20)] in $(0, \tau_2)$, times the probability of A_1^+ *not* surviving [Eq. (10-21)] in (τ_2, τ_3), i.e.[12]

$$\mathfrak{F}_{1\star}(E) = \left[k_0(E)\Big/\sum_j k_{0j}(E)\right]\langle e^{-k_1(E_1)\tau_2} - e^{-k_1(E_1)\tau_3}\rangle_P \qquad (10\text{-}44)$$

where

$$\langle e^{-k_1(E_1)\tau_2} - e^{-k_1(E_1)\tau_3}\rangle_P = \int_0^{E - E_{0(0)}} \{e^{-k_1(E_1)\tau_2} - e^{-k_1(E_1)\tau_3}\} \times P(E_1, E - E_{0(0)})\, dE_1 \qquad (10\text{-}45)$$

It should be obvious that the exponentials inside the braces in (10-43) and in (10-45) will be different from zero only when $k_0(E)$ or $k_1(E_1)$ is small, i.e., of the order of τ_2^{-1} or τ_3^{-1} or less, as shown in Fig. 10-3. In other words, both $\mathfrak{F}_{0\star}$ and $\mathfrak{F}_{1\star}$ are sensitive functions of E; for this reason, although the integral in (10-45) is likely to be zero over most of the range of integration, the limits of integration over which the integral is finite cannot be specified close enough for arbitrary E.

[12] Cf. also Kropf *et al.* (*33*).

A caveat Most of the developments in this section have concerned the simplified scheme (10-19), but the generalizations to more complex schemes should be obvious. It is also well to point out that these developments are predicated on the assumptions of random incidence of dissociation and of statistical partitioning of energy on dissociation. The former point seems reasonably well established in moderately excited neutral systems, but as discussed in Chapter 9, Section 7, investigation of energy partitioning in (neutral) photochemical systems has revealed some cases that do not follow equipartition. Also some fragmentations (*248*) and metastable transitions (*34*) in charged systems do not follow equipartition. While we have noted some ambiguity about the statistical function $P(x, \mathscr{E})$ used here, it is not yet clear whether the cases of nonequipartition require $P(x, \mathscr{E})$ merely to be interpreted differently or to be discarded altogether. Chances are that a strongly nonstatistical partitioning of energy is merely the result of nonrandom incidence of dissociation,[13] in which case, the statistical theory of unimolecular reactions does not apply anyway.

The problem of energy partition among fragments is complicated further when there is an activation energy for the reverse reaction, as in ionic rearrangement processes, for then the question arises whether this activation energy appears entirely in the relative kinetic energy of fragments ε_t (cf. Section 9), or whether it is shared between ε_t and the total energy available for partition (*248*).

An associated problem—which really constitutes the central assumption of the theory of mass spectra—is the transferability of concepts and results from neutral systems to ionic systems. This will be investigated in the following sections, in particular Sections 4 and 5.

3. Production of ionized species

The primary concern of the theory of mass spectra being the fragmentation of positive molecular ions, it is necessary to consider the excitation process by which these ions are produced in order to obtain some information about the nature and energy content of these ions, that is, about their initial "preparation."

In thermal systems, to use the terminology of ion physics, excitation of the target reactant molecule is accomplished by "bombardment" of the target by other molecules, i.e., by the rather gentle collisions due to thermal motion. When dealing with the production of ions, we still have to start with neutral ground-state molecules, but the excitation process, or "bombardment," must now be considerably more violent because it takes a minimum

[13] Such a conclusion is also suggested by the Monte Carlo calculations of Bunker (*35*).

of about 10 eV to raise a ground-state molecule to its lowest ionized state; this is the so-called lowest or first ionization potential (*36,37,2*). The most convenient bombarding particles for producing positive ions from neutral molecules are photons, electrons, and other ions. We shall start by considering the first two together since they have many features in common.

Electron and photon impact The collision of an isolated molecule with an electron can have several outcomes of varying probability: transfer of translational energy, excitation of vibrational or rotational states of the target molecule, electron exchange or attachment, ionization, or a combination of all of these. These events cause the electron to lose energy and therefore detailed information about molecular excitation can best be obtained by measuring the energy lost by the electron (*38*). Essentially the same excitation processes occur upon a collision of a photon with an isolated molecule (*6*), except that only allowed transitions take place and the photon is annihilated, so that optical excitation is basically a resonance phenomenon.

The complicated events surrounding ionization, or any other electronic transition, can be rationalized by considering the potential energy surfaces for the ground state and the first excited (or ionized) state. Because of the multidimensionality of these surfaces, we have to use as model a diatomic molecule and its ion and extrapolate from there to polyatomic systems, as was done in Chapter 3. It is understood that the potential energy surface or curve refers to the Born–Oppenheimer representation in which nuclear and electronic motions are treated as separable, also assumed in Chapter 3.

Figure 10-4 shows the potential energy curves for a diatomic molecule patterned after the ground-state N_2 and its ion. The probability of excitation to a given vibrational level of the ion can be qualitatively determined by a semiclassical argument, which states that, electronic motion being much more rapid than nuclear motion, electronic excitation will occur only "vertically," that is, while internuclear distance remains unchanged. This is the so-called Franck–Condon principle. The adjective "vertical" refers to the usual plot (Fig. 10-4) of potential energy versus internuclear distance, in which a transition at fixed internuclear distance is represented by a vertical straight line connecting the initial and final states.

Quantum mechanically, (*39*, p. 199), the probability of transition between two vibronic states, characterized by (total) eigenfunctions ψ' and ψ'', is given by $\mathscr{R}^2$, the square of the corresponding matrix element for the electric-dipole transition moment[14]

$$\mathscr{R} = \langle\psi'|\hat{d}|\psi''\rangle \tag{10-46}$$

[14] $\langle\psi'|\hat{d}|\psi''\rangle$ is short-hand notation for the integral $\int \psi'^{*}\hat{d}\psi''\, d\Gamma$, the volume element $d\Gamma$ stretching over both the electronic and nuclear coordinates of the system.

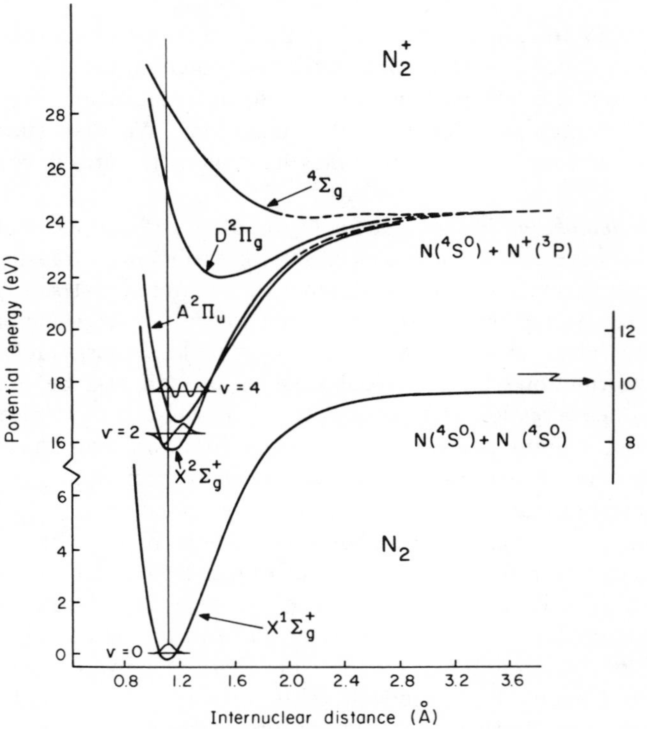

Fig. 10-4. "*Vertical*" *excitation in* N_2.

The vertical line represents a Franck–Condon transition from the $v = 0$ level of ground-state N_2. Among the low vibrational levels of N_2^+, the largest vibrational overlap for this transition occurs roughly for the $v = 2$ level of the ground-state ion (X $^2\Sigma_g{}^+$) and for the $v = 4$ level of the excited state (A $^2\Pi_g$). The form of the harmonic oscillator wave function is shown superimposed on the three vibrational levels concerned. The relative position of the potential energy curve for the weakly bound D $^2\Pi_g$ state of N_2^+ is such that the indicated transition would produce the D state with internal energy exceeding its dissociation limit, resulting in direct dissociation. The $^4\Sigma_g$ state of the ion is repulsive and the indicated transition to that state would cause immediate dissociation.

The *vertical* ionization potential of the ground-state ion, for example, therefore represents a transition to its $v = 2$ vibrational state. The *adiabatic* ionization potential is defined as the transition from the $v = 0$ vibrational state of neutral ground-state molecule to the $v = 0$ vibrational state of the ion, and therefore can only be measured in a vertical process when the vibrational overlap is nonzero for the 0–0 transition; this obviously includes the case when the minima of the two potential energy curves are located at the same internuclear distance. In Figs. 10-4 and 10-7, v signifies the vibrational quantum number. To save vertical space, the energy scale is compressed. The energy scale for the asymptote of the X $^1\Sigma_g{}^+$ state of N_2 is given at the right. The potential energy curves are drawn to scale, in both energy and internuclear distance. Based on the compilation of F. R. Gilmore, *J. Quant. Spectrosc. Radiat. Transfer* **5**, 369 (1965).

where $\hat{d}$ is the electric dipole moment operator. In the Born–Oppenheimer approximation, electronic and nuclear motions are separable, so that[15] $\psi \approx \psi_{el}\psi_{vib}$, where ψ_{el} and ψ_{vib} are the electronic and vibrational parts, respectively, of ψ. Similarly, the operator $\hat{d}$ may be separated into electronic ($\hat{d}_{el}$) and nuclear ($\hat{d}_N$) parts. Then (10-46) becomes

$$\mathscr{R} = \langle \psi'_{el}\psi'_{vib} | \hat{d}_{el} | \psi''_{el}\psi''_{vib} \rangle + \langle \psi'_{el}\psi'_{vib} | \hat{d}_N | \psi''_{el}\psi''_{vib} \rangle \tag{10-47}$$

Since ψ_{el} is not affected by $\hat{d}_N$, the second term has $\langle \psi'_{el} | \psi''_{el} \rangle$ as factor, which is zero because the two eigenfunctions ψ'_{el} and ψ''_{el}, which belong to two different electronic states, are orthogonal. The remaining term in (10-47) can be approximately evaluated by applying the Franck–Condon principle, which amounts to assuming that ψ_{el} varies only very slowly with internuclear distance r. Therefore integrations over the electronic and nuclear coordinates (of which r is the only one that matters because only r affects ψ_{vib}) can be done separately, and the integral involving ψ_{el} can then be replaced by a constant. Thus

$$\mathscr{R} \approx \text{const} \times F_c \tag{10-48}$$

where

$$F_c = \langle \psi'_{vib} | \psi''_{vib} \rangle = \int \psi'_{vib}\psi''_{vib}\, dr \tag{10-49}$$

F_c is the so-called Franck–Condon factor and is given by the vibrational overlap integral; the actual transition probability is then proportional to the square of the overlap integral.

In the harmonic oscillator approximation, ψ_{vib} is given by Hermite polynomials, whose graphical representation as an oscillating curve is well known.[16] The value of the overlap integral, and hence of the transition probability, can be roughly estimated from the extent to which, at fixed internuclear distance, the areas under the curves ψ''_{vib} and ψ'_{vib} overlap; this is shown in Fig. 10-4.

If the molecule possesses symmetry, there are various selection rules that "forbid" certain transitions for reasons of spin and symmetry, but in general these selection rules are much more strict for photon absorption than for low-energy (<100 eV) electron impact because of appreciable spin–orbit coupling produced by electron exchange (incident electron substituted for one present in the target).

A fairly complete theory of excitation by electron or photon impact of a diatomic molecule can be built on the basis of the argument just mentioned

[15] Actually we should have $\psi \approx \psi_{el}\psi_{vib}\psi_{rot}$, but it can be shown (ref. 39) that inclusion of the rotational eigenfunction ψ_{rot} yields only a constant factor in (10-48). For a more refined treatment, see Villarejo (*40*).

[16] See, for example, Kauzmann (*41*, p. 206, Fig. 6-7).

(*42,43*). For polyatomic molecules, the same general principles apply, in particular the Franck–Condon principle, but the details, such as Franck–Condon factors (*44,45*) are much more difficult to work out. The complications arise from the presence of many vibrational degrees of freedom in a polyatomic molecule, many of which are coupled because of anharmonicity, particularly if internal excitation energy is high. The result is that even if the original excitation process has deposited energy only in a few degrees of freedom required by the Franck–Condon principle, this energy rapidly "leaks out" into other vibrational degrees of freedom of the electronic state produced.

Charge exchange A quite different method of excitation is charge exchange, in which[17] an ion A^+, generally atomic, collides with a neutral M to produce ion M^+ and leave a neutral A:

$$A^+ + M \longrightarrow M^+ + A \tag{10-50}$$

For M an atom or diatomic molecule, the theory of the process is reasonably well worked out[18]; it turns out that the cross section (i.e., the probability) for charge exchange is small unless there is resonance, that is, unless the energies on the left side and on the right side of (10-50) match. With reference to M an atom, this means that the energy of the process

$$A^+ + e^- \longrightarrow A \tag{10-51}$$

which is the recombination energy (RE) of A^+, must match one of the ionization potentials (IP) of M, i.e., the energy of the process

$$M \longrightarrow M^+ + e^- \tag{10-52}$$

where M^+ may be any one of the electronic states of the ion. Because the particles colliding in (10-50) are of comparable mass (unlike the case of photon or electron impact), transfer of translational energy is possible, which then enters the energy balance. If the velocity is small, no translational

[17] A somewhat similar process is Penning ionization, where an electronically excited metastable (neutral) atom $(A)^e$ collides with a molecule:

$$(A)^e + M \rightarrow M^+ + A + e^-$$

When the excitation energy of A, the ionization potential of M, and the energy of the ejected electron are known, information can be obtained about the internal excitation of M. So far, the technique has been employed for the determination of electron spectra. See Čermák (*46,47*).

[18] Gurnee and Magee (*48*); Bates (*49*) Chapter 14. Instrumental problems associated with charge exchange experiments are discussed by Friedman and Reuben (*50*). Experimental evidence for near-resonance in charge exchange involving diatomics and triatomics has been reported by Haugh and Bayes (*51*).

energy transfer takes place, and under these conditions, the ion and the neutral interact at relatively large distances because the long-range potential between the two goes as $1/r^2$ when M is an atom and as $1/r^4$ when M is a molecule (cf. Table 7-1 and footnote 7 in Chapter 7).

When M is a polyatomic molecule, the theoretical details cannot be easily worked out, but the requirement of resonance in (10-50) still holds. However, the presence of internal degrees of freedom in M makes the matching of energies somewhat easier, for it is then sufficient that M have an ionization potential merely in the vicinity of RE of A, because any difference RE − IP, if not too large, may be accommodated as internal excitation of the particular ionic electronic state produced.

A qualitative argument Inasmuch as photon, electron, or ion impact involving polyatomic targets is not at present amenable to rigorous treatment, we must resort to qualitative arguments of a general nature, along the lines of a discussion given by Lorquet (*52*), in order to arrive at some picture of the sort of ionic states that might be produced by the various processes of excitation.

Molecular symmetry is not likely to play a very large role in the ionization of a sufficiently large molecule because of the concomitant strong vibrational excitation and the ensuing high anharmonicity; hence symmetry arguments ("correlation rules") cannot be used to predict likely molecular states upon ionization or dissociation. Immediate dissociation upon ionization will occur only if excitation is to a purely repulsive energy hypersurface. Spin conservation is a selection rule generally obeyed in photoexcitation (*53*) but not if the molecular ion contains a heavy atom or if the excitation is by slow electrons. "Slow" electron in this context means one having energy close to threshold for ionization; the incident electron becomes embedded in the target and two electrons are ejected, leading to large spin–orbit coupling and consequent nonconservation of spin (*54*, p. 152). In mass spectrometry, 70-eV electrons are commonly used for ionization, which is considerably in excess of the 10–15 eV threshold usual for ionization. This probably represents an intermediate case because the electrons are not really fast enough to behave like photons. Thus the spin conservation rule in mass spectra may work for small molecules and not for large ones. On the assumption that linear symmetry is preserved throughout the ionization and dissociation of acetylene (a rather small molecule). Fiquet–Fayard (*55*)[19] could account for some difficulties in the mass spectrum of acetylene by invoking the spin

[19] See also Fiquet-Fayard and Guyon (*56*), where the fragmentation of H_2O^+ and H_2S^+ is considered from the same point of view.

conservation rule.[20] In charge exchange, the situation is not clear: ion–molecule interaction at large distances (i.e., at low velocities) would tend to favor conservation of spin because the two particles interact while still distinct systems, but some experimental observations seem to contradict this (*3*).

Autoionization The number of excited states of a neutral polyatomic molecule is large because there are many electrons in many orbitals to excite, and the number of possible combinations is therefore considerable; in principle, their number is infinite. The number of states of a polyatomic ion is even larger because the removal of an electron from any one of the orbitals can be accompanied by simultaneous excitation of the remaining electrons, thus increasing the number of possible combinations. Connected with the existence of such a large number of excited states in the neutral molecule and its ion is the phenomenon of autoionization. Among the excited states of the neutral molecule may be some that will be energetically degenerate with one or more ionic states, that is, a neutral molecule will contain energy in excess of its first ionization potentials. Such states have been called "superexcited" by Platzman (*58*). If the potential energy surfaces for the energetically degenerate neutral and ionic states cross, as is likely, the superexcited state may autoionize, i.e., ionize by a radiationless transition; this is a resonance process. However, such a potential energy surface crossing is slow, roughly within 10^{-13}–10^{-12} sec, which amounts to the period of one or more vibrations. By contrast, direct ionization, which we have considered so far, is essentially immediate (10^{-15}–10^{-16} sec). The excited (neutral) states that autoionize may be those in which two or more electrons are excited, or in which only one electron is excited but there is at the same time appreciable vibrational excitation. This latter case appears to be the dominant one in polyatomic molecules (*59*); the decay into an ion plus electron is caused by the exchange of energy between electronic and vibrational motion, which of course constitutes a breakdown of the Born–Oppenheimer approximation. The autoionization of doubly excited electronic configurations can be treated by the configuration interaction method (*60*) and appears to account for some states of the N_2^+ ion observed in photoelectron spectroscopy (*61*).

High-resolution electron impact data (*62*, Chapter 9) on rare gas atoms and diatomic molecules show that ionization takes place mostly by autoionization (*63–65*). There is also some evidence that in photon or electron impact ionization of large molecules, a substantial fraction (maybe as high as one-half) of the ions produced are due to autoionization, the remainder

[20] McDowell (*57*) used symmetry arguments to deduce states of CH_3^+ formed in the dissociation upon electron impact of the derivatives of methane.

being due to the direct process (*66*). This of course considerably complicates the physics of the ionization process in polyatomic molecules. In general it is only the photoexcitation or photoionization of a small molecule that leads to well-defined excited or ionized states; in fact, most of the detailed knowledge of excited and ionized states of molecules comes from photoelectron spectroscopy (*67,68*) of small molecules. Unfortunately, the transitions observed are often confined to one-electron transitions in the outer shell, and the much more numerous transitions involving the removal of one electron and simultaneous excitation of an inner-shell electron, leading to quartet states of the ion, give rise to weak bands in photoelectron spectroscopy (*53*).

Field ionization A method of production and excitation of ions that is completely different from the preceding is field ionization, mentioned previously in Section 1. In this technique, molecules are adsorbed on a metallic tip and are ionized by the presence of a very high electric field. Ions desorbed from the surface may be dissociated by the electric field, apparently as a gas-phase process, and mass analysis of the fragments then gives rise to the so-called field ionization mass spectrum. This technique has come into prominence in recent years largely by the work of Beckey and collaborators (*4,15,69*). The polarizing effect of the inhomogeneous electric field forces a displacement of the electrons in the ion, so that the dissociation energy of a bond is reduced to a point where fragmentation occurs. The theory has been worked out for diatomic ions (*70*), but the treatment of polyatomic ions[21] is much less straightforward; it seems, however, that dissociation probably occurs mostly from the ground state of the ion.

4. Potential energy surfaces of polyatomic ions

In the previous chapter, we tried to deduce some information about the likely molecular states of ions produced by the various methods of preparation. We shall now turn our attention to the molecular properties of these ionic states, particularly insofar as they have a bearing on the subsequent fragmentation.

Charge density The removal of an electron upon ionization, particularly from a valence orbital, affects the charge density and hence bond strengths in the newly formed ion, and these in turn determine how the ion will fragment. It therefore seems reasonable to assume that calculation of charge densities in ground-state ions should reveal enough about the structure of molecular ions to permit a deduction as to the likely fragmentation pattern.[22]

[21] For polyatomic ions see Lorquet and Hall (*71*).

[22] See Santoro and Spadaccini (*72*) and reply by Lorquet and Hirota (*73*). For arguments in a similar vein, see Spiteller and Spiteller-Friedman (*74*).

In the terminology of Chapter 3, then, charge density should be a measure of critical energy for fragmentation. This is an attractive possibility because charge densities can be estimated relatively easily by molecular orbitral theory.[23] Calculations show a reasonable correlation between charge densities and observed fragmentation of branched (*81*) and cyclic (*82*) alkanes. In the case of large linear alkanes (*81*), charge density is found to be roughly uniform along the C—C chain, suggesting that fragmentation should be nearly random, which is actually the case (*83*). In small linear alkanes, the charge density on the central C—C bond is significantly higher than at the ends of the chain, and yet the fragmentation pattern is highly asymmetric. In particular, the abundance of the CH_3^+ ion in the mass spectrum of ethane is very low despite high charge density on the C—C bond.

A rather extreme case is offered by the aliphatic amines, where in the ground-state ion, most of the charge density is located on the C—N bond (*84,85*), and yet it is the adjacent C—C bond that dissociates because 60–70% of the fragments in the mass spectrum of aliphatic amines belong to the $m/e = 30$ peak, which under high resolution turns out (*86–88*) to have the gross formula CNH_4^+; deuteration studies (*89*) further reveal that the structure is actually $CH_2NH_2^+$. Conventional wisdom would suggest that it should be the electron localized in the lone-pair orbital on the nitrogen atom that should be removed first on ionization because in molecules containing a heteroatom, the nonbonding orbital is usually the highest filled orbital. Obviously in this case, the removal of even a nonbonding electron has far-reaching consequences for the entire molecule,[24] so that considering the orbital as essentially located on the nitrogen is a poor approximation.[25]

A somewhat related approach, originally proposed by Bader (*93–95*) and later used by Lorquet (*96*) and elaborated by Salem (*97,98*), considers the manner in which electron density in a molecule changes in the course of nuclear vibrations. By means of second-order perturbation theory, it can be shown that the most favorable dissociation process, i.e., one requiring least activation energy, is determined by the lowest-lying excited state whose symmetry corresponds to that of a possible reaction coordinate. The pre-

[23] A nonstatistical theory based on calculated charge densities has been given by Lester (*75*). Cf. also Alexandru (*76*) and Ovchinnikov (*77,78*). For a nonstatistical theory of fragmentation based on calculated bond orders, see Lorquet *et al.* (*79,80*). While in several instances these theories may successfully account for the nature of the fragments produced on ionization, they are much less successful in treating the energy dependence of such fragmentation.

[24] Detailed calculations by Cade *et al.* (*90*) on diatomic molecules also show that electronic excitation and ionization produce very substantial changes in electronic charge distribution and chemical binding.

[25] Cf. discussion on localized versus delocalized orbitals in Daudel and Pullman (*91*, p. 78ff). See also Newton *et al.* (*92*, Section VI).

dictions are in general in agreement with the symmetry rules of Woodward and Hoffmann (*99*), but since the whole treatment assumes only small vibrations, it may, at best, determine only *incipient* reaction paths, and so its relevance to reactions of vibrationally excited species is not certain.

Energy surfaces For all these reasons, Leclerc and Lorquet (*100*) calculated the full potential energy surface for the $C_2H_5NH_2^+$ ion by the extended Hückel theory. The interesting result (Fig. 10-5) is that in the neighborhood of the equilibrium internuclear configuration, i.e., at the bottom of the potential trough, equipotential contours are oriented along the C—N axis, showing the C—N bond as having the smaller force constant, i.e., as being "weaker" than the C—C bond. As the energy is increased, however, the equipotential contour lines begin to orient themselves in the C—C direction, and at still higher energies, the potential energy contour lines show clearly the C—C bond dissociating at energies ($\sim$1 eV) well below those necessary to dissociate the C—N bond ($\sim$5 eV). This shows clearly that electronic charge distribution at one particular fixed internuclear distance in the ground state of the ion is insufficient to provide information about the whole dissociation process.

Indications are, therefore, that a full understanding of the ionization and fragmentation process would require the knowledge of the complete manifold of excited states of both the neutral and ionized species. It was mentioned in Chapter 3 that the calculation of a full potential energy surface just for one (ground) state of a molecule is a task arduous enough, so that calculations of a whole manifold of excited states is feasible only for small molecules. Unfortunately, these are the ones that are the least interesting insofar as the statistical theory of fragmentation is concerned because there are not enough low-lying excited states and too few internal degrees of freedom for appreciable randomization of energy to take place prior to dissociation (more on this in Section 5); in fact, dissociation in small molecular ions probably occurs mostly by predissociation, i.e., via a nonadiabatic transition to a repulsive state. With an improved method of calculation,[26] Lorquet and collaborators (*101,102*) have calculated the potential energy surfaces for the ground state and several excited states of the H_2O^+ ion. Although this is a small molecule, the results are interesting enough to be given here in some detail to show the often unsuspected complexity in the dissociation of even a simple system like H_2O^+. Table 10-2 gives a summary of the symmetric and asymmetric dissociation process for ground-state H_2O^+ and two of its excited states.[27]

[26] CNDO/2 with configuration interaction for the ground state and LCAO SCF with a basis of Gaussian-lobe atomic orbitals for the excited states.

[27] For a schematic correlation diagram, see Fiquet-Fayard and Guyon (*56*, Fig. 1).

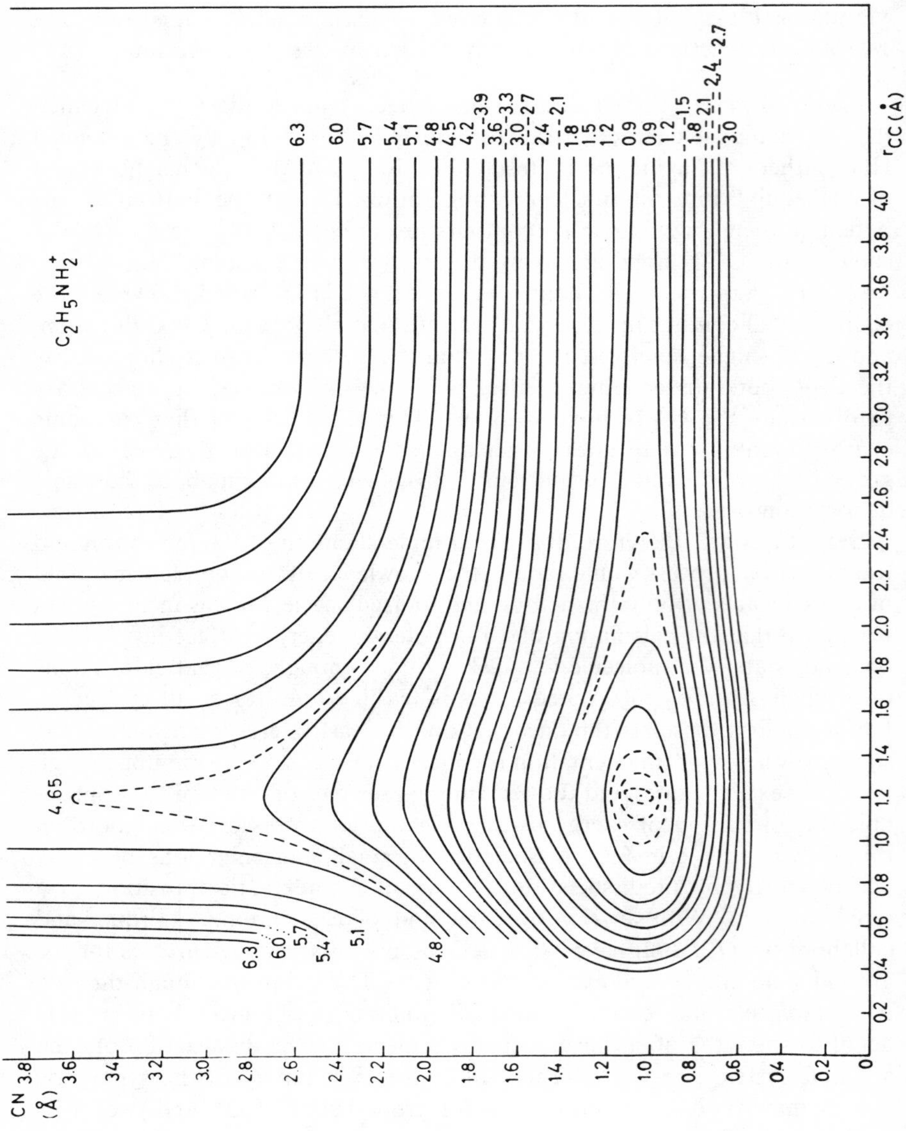
C2H5NH2+
rCC (Å)
CN (Å)
4.65

TABLE 10-2[a]

FRAGMENTATION OF H_2O^+

Process	Dissociation energy (eV)
A. Symmetric (point group C_{2v})	
H_2O^+ ($\tilde{X}\,^2B_1$) → H_2^+ ($^2\Sigma_g^+$) + O (^3P)	> 11.5
H_2O^+ ($\tilde{A}\,^2A_1$) → H_2 ($^1\Sigma_g^+$) + O^+ (^2D)	> 7.5
H_2O^+ ($\tilde{B}\,^2B_2$) → H_2^+ ($^2\Sigma_g^+$) + O (^3P)	> 5.5
B. Asymmetric (point group C_s)	
(i) H_2O^+ ($\tilde{X}\,^2A''$) → H (^2S) + OH^+ ($^3\Sigma^-$)	> 5.25
(ii) H_2O^+ ($\tilde{A}\,^2A'$) → H^+ (^1S) + OH ($^2\Pi$)	> 3.75
(iii) H_2O^+ ($\tilde{B}\,^2A'$) → H (^2S) + OH^+ ($^1\Delta$)	> 3.25

[a] From J. C. Leclerc, Ph.D. Thesis, Université de Liège, 1970. Calculations of parts of the potential energy surface for the process H_2O → H (^2S) + OH ($^2\Pi$) have been reported by K. J. Miller, S. R. Mielczarek, and M. Krauss, *J. Chem. Phys.* **51**, 26 (1969).

The potential energy surfaces show that all the symmetric dissociations proceed by a contraction of the H—O—H angle, accompanied by a symmetric elongation of the two bonds until H_2 or H_2^+ is formed. The asymmetric dissociations show more variety. The potential energy surface for the process B(i) shows a low barrier for angular inversion; dissociation occurs with equal ease when the H—O—H angle is anywhere between 120 and 140°. The potential energy surface for process B(ii) reveals two possible paths for fragmentation, energetically equally favorable, at an H—O—H angle of 180°, and another at about 75°. Finally, the potential energy surface for process B(iii) indicates that as the molecular ion begins to dissociate, the H—O—H angle opens and dissociation takes place when it reaches about 110°.

Segments of the potential energy surface for the dissociation of CH_5^+ into H_2 + CH_3^+ have been calculated by Pedley and collaborators (*103*). Their results are particularly interesting for the reverse reaction H_2 + CH_3^+

Fig. 10-5. *Potential energy surface of* $C_2H_5NH_2^+$ *as a function of the* C—*N and* C—C *distances.*

Energy in electron volts, distance in angstroms. Solid lines are 0.3 eV apart, dashed lines inside potential minimum are at 0.05, 0.1, and 0.2 eV. Near the potential minimum of the ion, the C—N bond is the weakest (ellipse at 0.05 eV, oriented along the C—N axis; force constant of bond 3.5 × 10^5 dyn cm^{-1}), whereas at 0.3 eV, the C—C bond becomes the weaker of the two (first ellipse drawn in solid line, oriented along the C—C axis; force constant of the bond 4.6 × 10^5 dyn cm^{-1}). This behavior, however, seems to be sensitive to the quality of the wave function (J. C. Lorquet, private communication). From J. C. Leclerc, Ph.D. Thesis, Université de Liège, 1970.

$\rightarrow CH_5^+$, for they find that if the hydrogen approaches along the C_3 axis of CH_3^+, then no activation energy is required, however if H_2 approaches the planar CH_3^+ along its C_2 axis, the activation energy is high. This suggests that the customary assumption made in the theory of mass spectra, that there is no activation energy for the reverse of the unimolecular dissociation, may impose a symmetry restriction on the process. It is too early to speculate if this is likely to be a general rule.

Other information For more complex systems, information is less complete because of the expense of elaborate calculations. Segments of the potential energy surface for the doublet and quartet states of the $C_2H_4^+$ ion that have been calculated (*104,105*) show that the ground state of the ion, unlike the ground-state neutral molecule, is twisted, but with a potential energy function for the angle of twist that has a very low minimum, meaning that the molecular ion is easily deformed from its equilibrium configuration. The geometries of the various configurations investigated are shown in Fig. 10-6, and Table 10-3 gives the principal calculated properties of the doublet and quartet ionic states; the ground state of the neutral molecule is included for comparison. The numbers in Table 10-3 have only semiquantitative meaning, but they show clearly a trend toward larger intermolecular distances in the ions compared with the ground state of the neutral molecule. Table 10-3 suggests that hydrogen scrambling in the doublet state of the ion might occur through the intermediate of the asymmetric configuration, which is only 0.8 eV above ground state of the doublet. In the quartet state of the ion, the intermediate in the scrambling process might be the asymmetrically bridged form, which is 0.5 eV above the ground state of the quartet.

Interesting insights into the conformations of ions can be obtained by noting that ions are electron-deficient systems and their electron populations bear some resemblance to the better-known (neutral) boron compounds, which are likewise electron deficient. Thus, using the correlation rules of Walsh (*106*), Lorquet (*107*) suggests that in the ethane ion $C_2H_6^+$, there may be a very low barrier separating the conventional staggered umbrella configuration (point group D_{3d}) from a diborane-like bridged structure (point group D_{2h}) [similar to Fig. 10-6(d)]. This might explain hydrogen scrambling observed on deuteration: the fragments of $CH_3CD_3^+$ contain not only the expected ions CH_3^+ and CD_3^+ but also CH_2D^+ and CHD_2^+. A more detailed analysis (*108*) shows that the analogy with boron compounds should not be pushed too far, because of basic differences between boron and carbon orbitals; thus unlike B_2H_6, the bound state of $C_2H_6^{+}{}^{+}$ is still of the umbrella configuration (D_{3d} symmetry), although at larger C—C distances, the bridged conformation (D_{2h} symmetry) is favored. New *ab-initio* calculations (*250*) on $C_2H_6^+$ show that the ground state of the ion is $^2A_{1g}$ of D_{3d} symmetry with a

TABLE 10-3[a]

GEOMETRY AND ENERGY OF SEVERAL STATES OF THE $C_2H_4^+$ ION AND OF GROUND-STATE C_2H_4

Species	Configuration	Geometric parameters	Energy (eV)
Neutral molecule, ground state	Planar	∢CCH = 122°; R(CH) = 1.37 Å; R(CC) = 1.61 Å, θ = 0°	0
Ion, doublet state	Twisted	∢CCH ≐ 120°; R(CH) = 1.38 Å; R(CC) = 1.69 Å; θ = 37°	11.44
	Planar	∢CCH = 120°; R(CH) = 1.38 Å; R(CC) = 1.70 Å; θ = 0°	11.54
	Perpendicular	∢CCH = 120°; R(CH) = 1.38 Å; R(CC) = 1.68 Å; θ = 90°	12.34
	Bridged	$R(C_1H_2)$ = 1.37 Å; $R(C_1H_1)$ = 1.78 Å; R(CC) = 1.53 Å	15.25
	Doubly bridged	R(CC) = 1.7 Å; R(CH) = 1.7 Å	16.70
	Asymmetric	R(CC) = 1.69 Å; $R(C_1H)$ = 1.37 Å; $R(C_2H)$ = 1.39 Å; ∢C_2C_1H = 180°; ∢C_1C_2H = 104°	12.23
	Asymmetrically bridged	R(CC) = 1.64 Å; $R(CH_1)$ = 1.66 Å; $R(CH_2)$ = 1.39 Å; $R(H_1H_4)$ = 2.44 Å; ∢$C_2C_1H_2$ = 135°	13.76
Ion, quartet state	Planar	∢CCH = 128°; R(CH) = 1.41 Å; R(CC) = 1.72 Å; θ = 0°	18.20
	Perpendicular	∢CCH = 104°; R(CH) = 1.38 Å; R(CC) = 1.8 Å; θ = 90°	17.57
	Bridged	$R(C_1H_2)$ = 1.36 Å; $R(C_1H_1)$ = 1.59 Å; R(CC) = 1.86 Å	19.32
	Doubly bridged	R(CC) = 1.62 Å; R(CH) = 1.80 Å	21.98
	Asymmetric	R(CC) = 1.80 Å; $R(C_1H)$ = 1.41 Å; $R(C_2H)$ = 1.39 Å; ∢C_2C_1H = 180°; ∢C_1C_2H = 94°	18.90
	Asymmetrically bridged	R(CC) = 1.87 Å; $R(CH_1)$ = 1.60 Å; $R(CH_2)$ = 1.38 Å; $R(H_1H_4)$ = 2.56 Å; ∢$C_2C_1H_2$ = 128°	18.07

[a] From A. J. Lorquet, *J. Phys. Chem.* **74**, 895 (1970). Angle θ is the angle of twist about the C—C bond. R is the internuclear distance.

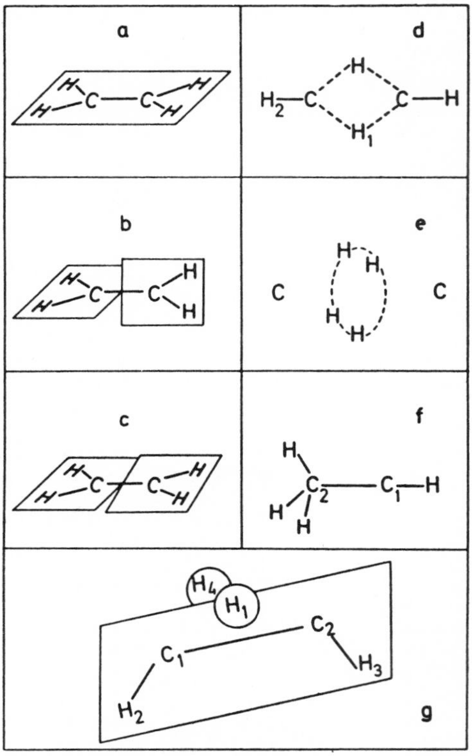

Fig. 10-6. *Possible geometric configurations of the* $C_2H_4^+$ *ion.* (a) Planar, (b) perpendicular, (c) twisted, (d) bridged, (e) doubly bridged, (f) asymmetric, (g) asymmetric bridged. From A. J. Lorquet, *J. Phys. Chem.* **74,** 895 (1970).

larger C—C distance than in the neutral ground state molecule, since it is a C—C σ-bonding electron that is removed on ionization.

An interesting case is the $C_3H_7^+$ ion, which seems to be a very stable species because it is a very abundant fragment in the mass spectra of most large aliphatic hydrocarbons, whether normal or branched chain. It has been suggested (*109*) that the ion, formed originally with the propyl or *iso*-propyl structure, depending on its origin, probably rapidly isomerizes into a cyclopropane configuration. Calculations (*110*) have shown that the cyclopropane configuration of $C_3H_7^+$ is indeed the more stable. The reverse is true, however, of the $C_3H_5^+$ ion, which appears to be more stable in the allyl configuration, rather than as the cyclopropyl cation (*111*). See, however, (*259*).

A summary Such is the extent of available information. It is difficult, and maybe perilous, to try to generalize about potential energy surfaces of polyatomic ions on the basis of the sketchy information collected here. It seems clear, however, that these potential energy surfaces have the same general topological features as the potential energy surfaces for neutral systems discussed in Chapter 3, that is, a dip or bowl in the vicinity of the stable bound configuration, and another dip or bowl representing the dissociated state of the system, the stable and dissociated states separated by an energy col. We can also expect that these surfaces will fall in one of three categories, namely types 1, 2, and 3 as defined in Chapter 3, Section 1, the first two representing fragmentation without and with activation energy for the reverse reaction, respectively, and the third, geometric or structural isomerization. The more common (because more easily detectable) process is probably fragmentation and it produces in general two fragments, one charged and one neutral (cf. Table 10-2, for example); the reverse reaction is thus an ion–molecule reaction, which in general has no formal activation energy (*112*, p. *101*), and therefore we can expect potential energy surfaces for ion fragmentation to be essentially of type 1. (More on this in Section 8.)

Since the methods used in the calculation of potential energy surfaces get progressively worse as the internuclear distance increases, the published data on $C_2H_5NH_2^+$ and H_2O^+, for instance, do not give reliable information about the topology in the vicinity of the energy col (saddle point), but this is perhaps not a major drawback. Reliable information is thus confined to the vicinity of the most stable configuration of the ions. With the knowledge that there are probably many exceptions, the general trends might be summarized as follows.

1. Compared with their neutral stable counterparts, polyatomic ions often have quite different geometries, larger interatomic distances, and show much less resistance to distortion or change of configuration.
2. As a result, potential energy surfaces of ions are flatter and the minima shallower than for comparable neutral molecules.
3. The relative flatness of the potential energy surfaces of ions may make energetically possible several modes of fragmentation, i.e., the energy barrier separating bound and dissociated configurations may have roughly the same height for more than one process—cf. the asymmetric dissociation of the $\tilde{A}\ ^2A'$ state of H_2O^+.

5. Randomization of electronic energy

Suppose that we have prepared a collection of monoenergetic excited ions M^+ of specified total energy *E*. Some, or perhaps most, of the energy is electronic excitation energy, and the remainder is internal (mostly vibrational) energy. In the sense of Chapter 5, we are talking about the energy contained in the pertinent degrees of freedom, except that now electronic degrees of freedom are also considered pertinent. Arguments presented in Section 2 of this chapter show that most of the excitation processes, and electron impact in particular, would produce our collection of M^+ in a multitude of electronic states, each state having its own potential energy surface and its own mode of decomposition. Very likely, at least some of these potential energy surfaces will intersect or come close together, so that if the ion has time to execute one or more vibrations, there exists a possibility of radiationless transitions or intersystem crossings among a number of these surfaces. It is therefore plausible that in this manner the electronic energy of some of the higher electronically excited states will eventually appear as internal (vibrational) energy of a number of lower-lying electronic states. We assume tacitly that none of the electronic states are repulsive, and moreover, that electronic predissociation does not occur, and we suppose, in fact, that at this point in time after the initial excitation, the ion M^+ is still intact and still contains total energy E, although it may now be distributed differently.

Basic assumption In order to make the eventual fragmentation of M^+, containing specified energy E, amenable to treatment by the general method of Part I, which was primarily devised for more conventional kinetics where decomposition involves only one electronic state throughout, Eyring and collaborators (*1*) have made the assumption that the energy of *all* accessible[28] electronic states of M^+ is converted into the internal energy of its *ground* electronic state, where it is randomized among its internal degrees of freedom. The decomposition of M^+ thus proceeds on the potential energy surface for the ground electronic state of the ion. In order that the method of Part I may apply, it is sufficient that decomposition occur on the potential energy surface of a single electronic state, which need not be the ground state; the reasons Eyring *et al.* chose the ground state (characterized by the first ionization potential) is that at the time, this was the only state of the ion about which any information was available. The advantage of this approach is obvious: The difficult problem of which electronic states are populated upon excitation, and how they interact and evolve prior to decomposition, is sidestepped and replaced by a statistical argument, not unlike the one used in Chapter 2, Section 3 to circumvent the dynamics of vibrational excitation. Basically, therefore, the assumption of energy randomization is extended in this way to include electronic as well as internal (vibrational) energy.

Thus if M^+ is prepared with excitation energy of, say, 14 eV relative to ground-state M, and if the first (adiabatic) ionization potential of M is 11 eV,

[28] Meaning all states such that the sum of electronic plus internal energy is equal to the prescribed total E.

for instance, the assumption is that after a while, we shall find only ground-state M^+ ions having 3 eV internal excitation energy, randomly distributed. If this energy is equal to or larger than the critical energy for decomposition into the given reaction channel, the ion will fragment with a rate calculable from Eq. (4-34).

Theoretical evidence What is the evidence for the conversion of electronic energy into vibrational energy of the ground electronic state of the ion? The plausibility of this assumption depends, first of all, on the close spacing of electronic states, since the higher their density (i.e., the higher their number per unit energy interval), the higher the likelihood that their potential energy surfaces will cross at some point, thus providing a means for the assumed "cascading" of electronic energy into the ion ground state or some other low-lying state.

The reason why a crossing is necessary is, in classical terms, again the Franck–Condon principle, which prohibits (1) any large-scale direct conversion of electronic energy into the kinetic energy of the nuclei (vibrations), and (2) any appreciable change in the positions of the nuclei in the course of an electronic transition. The mechanism of the conversion of electronic into vibrational energy can be best visualized with reference to Fig. 10-7. Suppose (diatomic) ion M^+ is prepared in the $v = 2$ vibrational level of state A, which is nearly isoenergetic with the $v = 10$ vibrational level of state B. A radiationless transition ("crossing") from A to B requires a

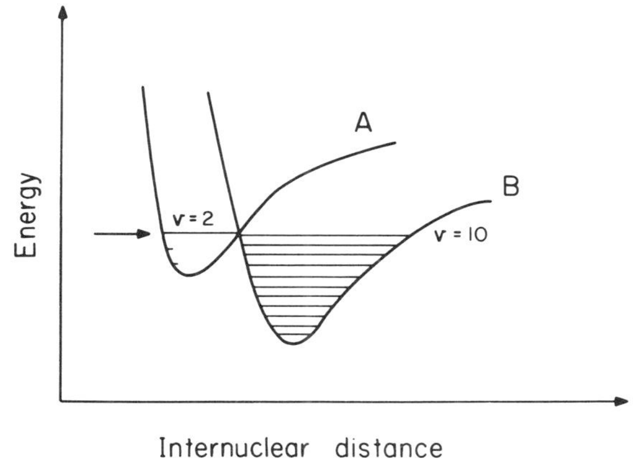

Fig. 10-7. *Intersystem crossing between states A and B of diatomic ion* M^+ *(schematic).* Arrow indicates that M^+ is originally prepared in the $v = 2$ level of state A. This level is nearly isoenergetic with the $v = 10$ level of state B, so that the transition $A \rightarrow B$ occurs with high probability if there are no symmetry restrictions.

perturbation of the system, which depends not only on the size of the energy difference[29] between the two unperturbed levels of A and of B, but also on the square of the matrix element

$$\mathscr{P} = \langle \psi^{\star a} | \hat{P} | \psi^{b} \rangle \tag{10-53}$$

where $\hat{P}$ is the required perturbation operator, which need not be specified further. As with operator $\hat{d}$ in Section 2, the operator $\hat{P}$ can be split into electronic and nuclear parts, so that (10-53), in complete analogy with (10-48), can be written

$$\mathscr{P} \approx \text{const} \times \langle \psi^{a}_{\text{vib}} | \psi^{b}_{\text{vib}} \rangle \tag{10-54}$$

Thus the matrix element $\mathscr{P}$, and hence the probability of crossing from A to B, will be large only for appreciable vibrational overlap, i.e., when the potential energy curves for A and for B cross, or at least come close, as drawn in Fig. 10-7. There are, in addition, various symmetry restrictions, including spin, which are relaxed when appreciable spin–orbit coupling exists (*39*, p. 285). If the vibrational levels of A and B do not match, the probability of crossing from A to B will be small for an isolated diatomic molecule. In an isolated polyatomic molecule, however, vibrational anharmonicity would quickly redistribute vibrational energy from the originally excited level into other vibrational degrees of freedom, thus ensuring that some vibrational levels should match at the intersection of potential energy surfaces of A and of B. An additional factor likely to aid in matching the energy of vibrational levels is the presence of thermal energy in the excited ion, for the experiments are generally performed at room temperature or above. There is therefore a distribution of internal energies in the ground-state neutral molecule to start with, which upon excitation and ionization is carried over into the excited ion (cf. Section 8). Hence the latter is not produced in one particular vibrational level, as implied in Fig. 10-7, but actually in several levels, which then makes the matching of levels between different electronic states easier. It is clear that if the potential energy curve (or surface) of state B crosses that of a state C, a further transition from B to C can occur, and so on, until the difference in electronic energy between the originally prepared excited state and the ground state is found as vibrational energy of the ground electronic state.

In their original paper, Eyring and collaborators presented a simple

[29] Let us specify that we consider here an *isolated*, electronically excited species (which happens to be charged in the present instance). Most texts on photochemistry consider the problem of radiationless conversion of electronic to vibrational energy only with reference to excited species in *condensed* media, where intermolecular collisions rapidly degrade vibrational energy of a given electronic state to its ground vibrational level; therefore transitions are assumed to be possible only from the zeroth level of state A to some isoenergetic vibrational level of state B. Cf. Fig. 4.11 in Turro (*113*).

calculation showing that the average spacing between electronic states of the propane ion should be of the order of 1 mV. In retrospect, and with the benefit of hindsight, this now appears as much too optimistic. Table 10-4 shows the experimentally determined and calculated ionization potentials of propane, where it may be seen that the spacing is of the order of volts rather than millivolts. However, both the experimental and calculated ionization potentials correspond only to the removal of a valence electron, whereas there are many more ionization potentials of comparable energy that involve simultaneous excitation of one or more remaining electrons to vacant orbitals, especially in the high-energy range. Such ionization potentials have not been calculated, and their experimental determination by photoelectron spectroscopy would be difficult, since they would appear as weak and

TABLE 10-4

EXPERIMENTAL AND CALCULATED IONIZATION POTENTIALS OF PROPANE (eV)

Experimental[a]	
Turner[b]	11.07; 13.17; 15.17; 15.70; (18.57); (20.26)
Dewar[c]	11.06; 13.22; 15.19
Inghram[d]	10.94; 13.0; 14.8; 18.3; 18.95
Calculated	
Lorquet[e]	11.21; 11.72; 12.24; 12.64; 16.60; 20.38; 26.22
Lipscomb[f]	11.66; 12.03; 13.12; 14.06; 15.21
Dewar[c,g]	10.83; 11.35; 12.67; 12.84; 13.91; 14.87; 19.71; 23.95; 30.63
Murrell[h]	11.83; 12.37; 12.69; 14.32; 14.67; 15.99; 17.07; 21.38; 24.49; 27.55

[a] Dubious values in parentheses.

[b] M. I. Al-Joboury and D. W. Turner, *J. Chem. Soc.* 4434 (1964) (photoelectron spectroscopy).

[c] M. J. S. Dewar and S. D. Worley, *J. Chem. Phys.* **50**, 654 (1969) (photoelectron spectroscopy).

[d] R. Stockbauer and M. G. Inghram, *J. Phys. Chem.* **54**, 2242 (1971) (threshold electron spectroscopy).

[e] J. C. Lorquet, *Mol. Phys.* **9**, 101 (1965).

[f] M. D. Newton, F. P. Boer and W. N. Lipscomb, *J. Amer. Chem. Soc.* **88**, 2367 (1966). Only ionization potentials for more strongly bound electrons given.

[g] There appears to be some difficulty regarding the nature of the highest occupied orbital in normal paraffins. Dewar's calculations show this orbital to be a C—C bonding orbital, but on the basis of the photoelectron spectrum of ethane, A. D. Baker, C. Baker, C. R. Brundle and D. W. Turner, *Int. J. Mass Spectrom. Ion Phys.* **1**, 285 (1968) conclude that it is more likely a C—H orbital.

[h] J. N. Murrell and W. Schmidt, *JCS Faraday II* **68**, 1709 (1972). Ionization from inner shell orbitals is not included.

broad bands. Thus the density of electronic states of propane is certainly much higher than shown in Table 10-4 in the high-energy range, where it might reach the order of 1 mV, but not in the neighborhood of the ground state (say, up to about 1 eV), where both calculation and experiment should give a fairly good representation of the reality.

Very interesting calculations have been done by Leclerc (*102*) on the methyl and ethyl amine ions. These calculations include not only ionization potentials due to the removal of a valence electron, leading to doublet states of the ion, but also ionization potentials involving simultaneous excitation of one or more of the remaining electrons, leading to quartet states of the ion. The results are shown in Fig. 10-8 together with experimental results obtained by photoelectron specroscopy. The amines, like compounds containing the carbonyl group, and in common with most compounds containing a heteroatom, are a somewhat special case in that the highest occupied orbital in the ground-state neutral molecule is a nonbonding orbital which has a low ionization potential relative to electrons in the other bonding orbitals.[30] There is therefore a gap between the first and second ionization potentials in these compounds, which is not found in hydrocarbons, for instance. The calculated results show rather close-spaced ionization potentials in both methyl and ethyl amine above this gap, with an average spacing of $\sim$0.15 eV for methyl amine and $\sim$0.1 eV for the ethyl amine. Since the method of calculation gave the same ionization potentials for some doublet and quartet states, the actual density of electronic states is probably slightly higher than the number of ionization potentials would indicate.

It is therefore clear that in the larger ions, there is indeed a density of electronic states appreciable enough to lend plausibility to the assumption of numerous crossings of the potential energy surfaces, permitting a "cascade" of electronic energy into a low-lying state. The electronic state in which the cascade is likely to end ("terminal state") is, in the case of amines, not the ground-state ion but probably its first excited state because of the relatively large energy gap between the two which would presumably make the intersystem crossing difficult.[31] Another case seems to be methyl alcohol, a compound likewise containing a heteroatom, where photoionization experiments (*115*) indicate, at about 2 eV above threshold, the existence of an excited electronic state of the ion that does not decompose, presumably because of slow conversion to the ground state.

[30] See, for example, Turro (*113*, p. 12 ff.). For calculated energies of similar formaldehyde orbitals, see Buenker and Peyerimhoff (*114*).

[31] However, the gap might well decrease as the ion evolves on the potential energy hypersurface, e.g., as one or more bond lengths increase (cf. *100*, Fig. 5), and might even result in the actual crossing of potential energy hypersurfaces of the ground electronic state and some electronically excited state (cf. *100*, Fig. 6).

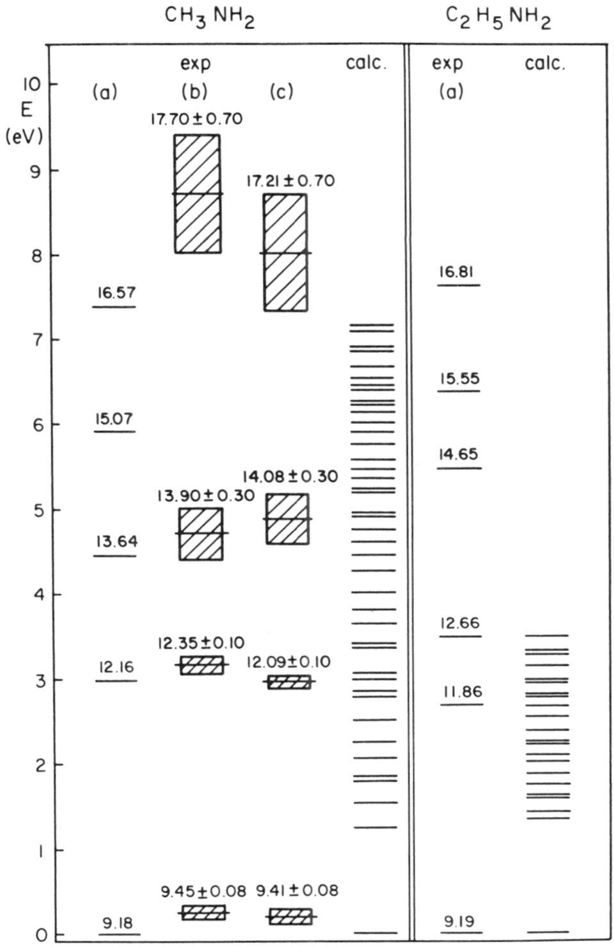

Fig. 10-8. *Experimental and calculated ionization potentials of methyl and ethyl amines.*

Experimental results: (a) M. I. Al-Joboury and D. W. Turner, *J. Chem. Soc.* 4434 (1964). (b) J. Collin, in "IX. Colloquium Spectroscopicum Internationale," p. 596. Publications du GAMS, Lyon 1961. (c) J. Collin, Thèse d'Agrégation de l'Enseignement Supérieur, Université de Liège, 1966. Calculations are those of J. C. Leclerc. From J. C. Leclerc, Ph.D. Thesis, Université de Liège, 1970, Fig. 1-14. Energy scale on left refers to the ground state of the ion as the zero of energy.

It has been suggested (*116*) that benzene might be another case where the terminal state is not the ground state, and where, in fact, there might be several terminal states. Neutral benzene has a doubly degenerate highest occupied orbital, separated by a gap from the others (*117*). This degeneracy is removed on substitution, so that the gap between the first and second ionization potentials in benzene should be smaller in substituted benzenes. It is therefore interesting to compare fragmentation data on benzene with those on toluene, for instance. Experimental data on toluene (*118*) indicate that some fragmentation occurs from at least two states, but undisclosed calculations (*119*, p. 59) are said to account well for the fragmentation of both benzene and toluene by assuming that decomposition occurs only from the ground state.

Evidence has been presented recently (*120*) that acetylene ions produced by charge exchange above 16 eV are in an excited state which persists long enough for the ion to engage in ion–molecule reactions typical of an excited ion. This seems to be another case where the ground and first excited electronic states are separated by a gap that is too large for efficient conversion of electronic energy into ground-state internal energy.

A potentially more direct evidence could be provided by actual lifetime measurements on some particular metastable transition. Thus Coggeshall (*121*) [see also (*122*)] measured transition intensity versus residence time for the metastable transition (10-11) and obtained three breaks in the curve which he interpreted as three different lifetimes due to three different isolated electronic states. The problem with these measurements is discussed at some length in a review by Rosenstock (*123*), who points out that Coggeshall's conclusions are unwarranted because determining lifetime distributions from intensity versus time plots requires data of impossibly high precision. Newer experimental data taken with a special apparatus (*124*) show a continuous distribution of lifetimes for the butane transition (and several others as well), suggesting that only one electronic state is involved having a continuous distribution of internal energies and hence of lifetimes. This conclusion forms the basis of the treatment given in Section 1 discussing the field-ionization results on the same metastable transition.

One other condition for the conversion of electronic excitation energy into the internal energy of the ground-state ion is that there be no strongly repulsive excited states, because they would cause immediate decomposition before much of the electronic energy is degraded. This condition is likely to be fulfilled if orbitals from which the electron is removed are spread over the whole molecule rather than localized just on one particular bond, for the result of ionization is then a small weakening of all bonds rather than a drastic weakening of just one particular bond, causing the electronic state to be strongly repulsive for this particular nuclear motion. We have mentioned previously theoretical indications that the positive hole (left by the departing electron) does in fact extend over the whole molecule.

On the contrary (and perhaps unreasonable) assumption that the electron is removed from localized orbitals, which then determine the nature of the

TABLE 10-5[a]

Molecule C_2H_{2n+2}	Relative number of C—H bonds	Relative abundance of $C_nH_{2n+1}^+$ ions
Propane	0.8	0.8
n-Butane	0.77	0.08
n-Heptane	0.73	0.00002

[a] From Ref. (*1*).

subsequent fragmentation, there should be a direct relation between the relative abundance of a given species of fragment ion and the relative number of bonds in the parent compound available for the given fragmentation. Eyring and collaborators (*1*) gave the results shown in Table 10-5 relating to C—H fragmentation in the mass spectra of several hydrocarbons.

The lack of any correlation between the relative number of C—H bonds and the relative number of ions due to C—H scission effectively argues against nonstatistical bond fission from localized orbitals.

Nonstatistical case If the assumption of electronic energy conversion into vibrational energy of the ground (or some other terminal) electronic state is not made, then we have a nonstatistical case that requires a detailed description of the relaxation process, including consideration of all the selection rules in case of photon impact, and this in turn supposes detailed knowledge of the electronic states involved. Such a calculation has been done, using the so-called multiphonon theory (*125*), for a few selected, narrowly defined nondissociative photoinduced radiationless transitions in (neutral) excited benzene[32]; even in such a narrowly circumscribed case, the solution of the problem is formidable enough. Of more direct interest in the present context is the photoisomerization of (neutral) stilbene. The reaction takes place in solution and involves a complicated mechanism whereby the system, initially prepared optically in a trans singlet level, eventually drops into a twisted triplet which further decays into a cis ground state (*128*). Looked upon as a prototype internal torsion problem, this is a special case of a breakdown of the Born–Oppenheimer approximation since internal torsion gives rise to a new mode of motion of the nuclei; the multiphon theory then suggests that in a stilbene-like case, most of the electronic energy should go into internal torsion rather than the other vibrational degrees of freedom (*129*).

It is clear that nonstatistical conversion of electronic energy is at present rigorously solvable only in a few selected cases. A less rigorous, and hence more widely applicable, solution is possible if all the details of the inter-

[32] Gelbart *et al.* (*126*). For theoretical treatment of nonradiative decay of a single vibronic level in a large isolated molecule, see Nitzan and Jortner (*127*).

action between electronic and vibrational degrees of freedom are neglected and instead attention is focused merely on how much electronic energy reaches different parts of the molecule, the assumption being that a particular bond will break when "enough" energy reaches it. In spirit, therefore, if not in detail, such a theory is fairly close to a nonstatistical theory based on the molecular-orbital approach (cf. footnote 23). There are several possibilities. One is a stochastic approach, involving the shuffling of energy quanta among coupled "cells," discussed in Chapter 2, Section 5, where every cell is coupled to every other cell. On this model, a strong bond can break before a weak bond if the strong bond is more strongly coupled. If the molecule is seen as a linear chain (a model for a polymer or a solid), we have a linear array of "bonds" and interactions are then restricted to adjacent "bond"; this is the "nearest-neighbor" model of Magee and collaborators (*130,131*). On this model, by contrast, it is the weakly coupled "bond" that dissociates first because it traps the energy longest as it is spreading along the chain from the initial point of excitation. Detailed analysis of the linear chain model has been given by Levine (*132*).

Somewhat similar is the exciton model of Partridge (*133*). He sees essentially two kinds of excitations in a linear alkane: one localized in C—H bonds attached to a common carbon bond and the other migrating along the principal C—C chain of the molecule. The theory therefore predicts mostly H or H_2 elimination and only a small probability for a C—C scission, because the energy does not remain localized long enough for a such a scission to occur.

Experimental evidence In practical terms, a nonstatistical distribution of electronic energy implies that the fragmentation pattern of an excited ion (or of any other electronically excited particle) should become dependent on initial preparation. This is a point that can be checked experimentally, although only in a gross fashion. For one thing, the three principal methods of preparation, electron impact, charge exchange, and, to some extent, photoionization, produce mostly an ill-defined mixture of excited states of the ion; the other difficulty is the nature of experimental observables, which makes it frequently impossible to measure the effect of initial preparation alone.

An overall test of the effect of initial preparation is possible by comparing the breakdown curves of a given ion prepared by two different methods. Thus Chupka and Berkowitz (*134*) obtained by photoionization breakdown curves for the ethane, propane, and butane ions which turned out to be in reasonably good agreement with breakdown curves obtained by charge exchange. This comparison, although limited to about 3 eV above ionization threshold, is significant because it does not suffer from the problems of interpretation that plague electron impact breakdown curves (cf. Section 6).

Still another overall test is provided by the general observation that ion thermal energy is effective in bringing about dissociation (Section 6).

Another experimental observable perhaps best suited for checking the effect of initial preparation is the so-called metastable transition in a mass spectrometer, described in more detail in Section 2. It represents essentially a sampling of fragmentation events that take place in the 10^{-5}–10^{-7} sec range, depending on the particular geometry of the instrument used.

One possible way of testing is to observe the same metastable transitions using two different methods of excitation. The following two metastable transitions compete in the fragmentation of the butane ion:

$$C_4H_{10}^+ \longrightarrow C_3H_7^+ + CH_3 \tag{10-55a}$$

$$C_4H_{10}^+ \longrightarrow C_3H_6^+ + CH_4 \tag{10-55b}$$

Chupka and Lindholm (*135*) have measured the ratio of ion intensities corresponding to process (10-55a) relative to (10-55b). The ratio was 1.03–1.10 for $C_4H_{10}^+$ generated by electron impact and 1.18 by photoionization. The two results are thus equal within experimental error, suggesting that the collection of excited states of $C_4H_{10}^+$, although probably initially quite different in electron impact and in photoionization, nevertheless has no detectable effect on the relative importance of these two decay channels of the ion.

The general idea behind another type of experiment is similar to that exploited in experiments designed to check the randomization hypothesis in neutral systems, described previously in Chapter 2, Section 4: preparation of the same excited ion M^+ but of varying chemical ancestry, and detection of a difference in decay behavior, if any. Numerous examples are available, all involving excitation by electron impact. Rosenstock *et al.* (*136*) measured the ratio of intensities due to the competing processes

$$C_6H_{13}^+ \longrightarrow C_4H_9^+ + C_2H_4 \tag{10-56a}$$

$$C_6H_{13}^+ \longrightarrow C_3H_7^+ + C_3H_6 \tag{10-56b}$$

which occur in the mass spectra of normal aliphatic hydrocarbons containing more than seven carbons. The ion $C_6H_{13}^+$ is thus not the parent ion but a daughter ion formed by the loss of $(CH_2)_{n-7}CH_3$ from C_nH_{2n+2}. Various precursors were used for $C_6H_{13}^+$, from *n*-heptane to *n*-tetradecane, and also *n*-hexyl bromide and di-*n*-hexyl ether. At 30 eV, the ratio of intensities was the same for all compounds within 3%.

The absence of an effect on the ratio of intensities means that the $C_6H_{13}^+$ ions formed from different precursors were indeed similar in structure and internal energy distribution. However, the experiment actually measures the end result of the following chain of events subsequent to ionization: (i) distribution of excitation energy in the parent $C_nH_{2n+2}^+$; (ii) partitioning of energy between $C_6H_{13}^+$ and $(CH_2)_{n-7}(CH_3)$; (iii) distribution of excitation energy in $C_6H_{13}^+$. While the experiment provides direct information only on step (iii), the inference is that unless steps (i) and (ii) proceed in a similar fashion for all parents, step (iii) would be different for different parents. Hence the absence of the effect of initial preparation suggests, in a gross way, a very nonspecific, i.e., statistical, intramolecular energy transfer, as well as nonspecific energy partitioning on fragmentation.[33]

In another series of experiments, Khodadadi *et al.* (*138*) measured the kinetic energy released in a series of similar metastable transitions (e.g., $C_2H_4^+ \rightarrow C_2H_2^+ + H_2$) occurring in the electron impact mass spectra of a variety of hydrocarbons (e.g., ethylene, ethane, propane, *n*-butane, neopentane, *n*-heptane), and found that for a given metastable transition, the kinetic energy released is always the same, independent of the nature of the parent precursor. In this case, experiment reflects not only the overall result of steps (i)–(iii), but also, indirectly, the shape of the potential energy surface for the metastable transition, because the kinetic energy release depends on it (cf. Section 9). Again, the absence of any specific effect suggests largely uniform, i.e., statistical, behavior.

A prominent fragment ion in the mass spectra of many benzyl compounds is C_7H_7, which[34] undergoes the metastable transition

$$\underset{m/e\,=\,91}{C_7H_7^+} \longrightarrow \underset{m/e\,=\,65}{C_5H_5^+} + C_2H_2 \qquad (m_\star/e = 46.5) \tag{10-57}$$

giving rise to a metastable peak at $m_\star/e = 46.5$. Jennings and Futrell (*141*) studied the metastable transition (10-57) in benzylchloride, benzyl bromide, toluene, and benzylalcohol by measuring the ratio of ion currents at $m_\star/e =$ 46.5 and at $m/e = 65$, the latter corresponding to $C_5H_5^+$. They found that the ratio of ion currents $i[m_\star/e]/i[m/e = 65]$ is inversely proportional to $E - E_0$, the energy in excess of threshold (E_0) of $C_7H_7^+$, as calculated from the heat of formation of $C_7H_7^+$ using thermodynamic data. The interpretation of the experiment assumes that the fragment $C_7H_7^+$ is produced with a spread of internal energies. The more energetic ones decompose faster (if there is no predissociation), before they reach the field-free region of the mass spectrometer, and the fragment $C_5H_5^+$ is therefore recorded at its

[33] On the other hand, variation in the intensity ratio does not necessarily mean absence of randomization in the parent. Cf. Occolowitz (*137*).

[34] There are reasons to believe that the structure of C_7H_7 is tropyllium (seven carbon cycle) rather than benzyl. Cf. Grubb and Meyerson (*139*). Implications of the tropyllium ion for the randomization hypothesis have been discussed by V. Hanuš and Z. Dolejšek (*140*).

"normal" position at $m/e = 65$. The less energetic ones decompose during the flight across the field-free region and are recorded as a metastable peak at $m_\star/e = 46.5$. As the excitation energy of $C_7H_7^+$ rises, fewer less energetic ions are formed, so that the fraction of $C_5H_5^+$ ions recorded at $m/e = 65$ rises. Since, as in the previous examples, $C_7H_7^+$ is a daughter ion, the experiments only measure the overall result of steps (i)–(iii); however, the absence of any specific effects attributable to the nature of the parent precursor suggests statistical behavior throughout steps (i)–(iii).

There is evidence that in the mass spectrum of toluene, the metastable transition (10-57) is followed by the *consecutive* metastable transition

$$\underset{m/e\,=\,65}{C_5H_5^+} \longrightarrow \underset{m/e\,=\,39}{C_3H_3^+} + C_2H_2 \qquad (10\text{-}58)$$

Hills *et al.* (*32*) used an experimental mass spectrometer arrangement such that the transitions (10-57) and (10-58) could be observed at two different times, once in a field-free region between source and electric sector (region α), and an instant later, as a delayed process, in a field-free region between electric sector and magnetic sector (region β). The fractional abundance of the transitions (10-57) and (10-58) could be successfully calculated as a function of energy transferred to $C_7H_7^+$ by the statistical theory, assuming randomization at every stage and statistical partitioning[35] between neutral and charged fragments. Since the observed fractional abundance in regions α and β is quite sensitive to the rates of the consecutive processes, this system provides a fairly sensitive test of the theory, although it was necessary to make an assumption about the form of the energy transfer function and about the transition states involved.

In the mass spectra of primary and secondary alcohols, the daughter ion $C_2H_5O^+$ gives rise to two competing metastable transitions

$$C_2H_5O^+ \longrightarrow H_3O^+ + C_2H_2 \qquad (m_\star/e = 8.0) \qquad (10\text{-}59a)$$

$$\underset{m/e\,=\,45}{C_2H_5O^+} \longrightarrow CHO^+ + CH_4 \qquad (m_\star/e = 18.7) \qquad (10\text{-}59b)$$

In agreement with the findings of Rosenstock *et al.* (*136*) on the competing transitions (10-56) in hydrocarbons, McLafferty and Pike (*142*) found in the homologous series from C_3 to C_7 that the ratio of ion intensities $i[m_\star/e = 8.0]/i[m_\star/e = 18.7]$ is independent of the parent precursor of $C_2H_5O^+$. They also found that the ratio of intensities $i[m/e = 45]/i[m_\star/e = 8.0 \text{ or } 18.7]$ is proportional to the number of degrees of freedom in the parent (the so-called "degree-of-freedom effect"). In a somewhat similar series of experiments on

[35] $P(x, \mathscr{E})$ in the form (10-29) was used, with $N_b(\mathscr{E} - x)$ interpreted as the density of states of the neutral fragment.

benzoyl compounds, Cooks and Williams (*143*) measured the metastable transition of the benzoyl ion

$$\underset{m/e\,=\,105}{C_6H_5CO^+} \longrightarrow \underset{m/e\,=\,77}{C_6H_5^+} + CO \tag{10-60}$$

and found that the ratio of intensities $i[m/e = 105]/i[m/e = 77]$ increases with the size of the parent precursor.

Inasmuch as $C_2H_5O^+$ and $C_6H_5CO^+$ are both daughter ions, the same problem arises as before concerning the deductions that can be made about the parent. If we assume, as seems reasonable, that in each series of experiments, every parent received the same amount of energy on electron impact, the experimental data show that the partitioning of energy on fragmentation of the parent must have been more or less statisical, because the larger the parent, the larger the neutral fragment and the larger the energy carried away by this neutral fragment on break up of the parent. In other words, the larger the parent, the smaller the energy retained by the daughter ion (which is the same size for all parents), and consequently the larger the fraction of $C_2H_5O^+$ or $C_6H_5CO^+$ ions that do *not* have enough energy to decompose in the field-free region of the mass spectrometer. A statistical partitioning of energy strongly suggests (but does not prove) statistical distribution of energy in the parent as well.

These qualitative deductions have been substantiated by more elaborate calculations of Lin and Rabinovitch (*25*) on the $C_2H_5O^+$ ion, in which they assumed that all the energy of the parent in excess of threshold was available for randomization (cf. Section 2).

Failure of randomization A number of classes of compounds have been reported in the past where the statistical theory of fragmentation was claimed to be inapplicable.[36] What was measured and compared with calculations was the energy dependence of the breakdown pattern of a class of compounds, not just one or more metastable transitions, so that the comparison of theory and experiment measured the overall agreement, not just one particular assumption of the theory. It turned out eventually that the lack of success of these calculations was due to the use of the classical approximation for the density of states, which is in error by several orders of magnitude even at higher energies (*147*). Nevertheless, the theory, even in its very approximate form, proved of heuristic value in providing a deeper insight into the meaning of experimental results (*31*).

There are a few more recent examples where the failure of the statistical theory is not due to crude approximations used in the calculations. One

[36] Friedman *et al.* (*144*) (alcohols); Chupka and Berkowitz (*145*) (amines); Steiner *et al.* (*146*) (alkanes). These results are discussed in Rosenstock (*123*).

such instance is the work of Lifshitz and Long (*148*), who calculated the mass spectra of C_2F_6 and C_3F_8 and concluded that the statistical theory did not apply because they felt they had to use unreasonably low vibrational frequencies in their models for the transition state in order to obtain agreement with experiment. The problem here is that the parent ions disappear much faster than can be accounted for by calculation. The authors therefore concluded that $C_2F_6^+$ and $C_3F_8^+$ probably decompose from repulsive states; however, it is just possible that their activation energies, obtained from appearance potentials, are too high for one reason or another, which of course would make the statistical rate too low. Evidence for possible non-statistical distribution in internal energies, obtained in the measurement of translational energies of fragments, is discussed in Section 9.

6. Determination of energy distribution function

The last two sections have provided justification for using the general method of Part I for calculating the individual rate constants $k_{ij}(E_{ij})$ of an ionic fragmentation process. It was shown in Section 2 how, *at a given E* (E is the excitation energy[37] in the parent), the experimental observables are related to the $k_{ij}(E_{ij})$'s of the individual processes in a complex reaction scheme. Since every actual excitation process will produce a parent ion M^+ with a distribution of E's, all that remains to be done in order to connect calculated results with observables in an actual experiment is to determine the distribution function for the excitation energy in the parent and average over this function the result calculated for a monoenergetic parent. The nature of this distribution function will now be considered.

Thermal energy Inasmuch as experiments are not normally carried out at absolute zero, the neutral molecule M, prior to ionization and excitation, contains thermal energy which then appears in the molecular ion M^+, superimposed on its internal excitation energy. The situation is similar to that in photoexcitation, discussed in Chapter 9, Section 6, except that here the excitation process is in general not monochromatic. Let $\mathscr{D}(y)$ represent the probability per unit energy that excitation deposits energy y in the molecular ion; here, y represents energy in excess of the first (lowest) adiabatic ionization potential and $\mathscr{D}(y)$ (dimension: energy^{-1}) may be called the energy deposition function. If we make the approximation (cf. Chapter 9, Section 6) that the excitation process does not distort the distribution of thermal energies originally present in the neutral M, the same distribution

[37] All energies are understood to refer to the ground vibrational state of the parent ion M^+ as zero.

will therefore appear in M^+; in other words, it will be the thermal distribution $P(E)_e$ of Eq. (8-13) shifted by y along the energy axis, which we shall denote $P(E - y)_e$ [cf. Eq. (9-94) and Fig. 9-7], where E is the internal energy (see footnote 37) of M^+. The probability (per unit energy) $\mathscr{F}(E)$ of finding energy E in M^+ is obtained by folding $\mathscr{D}(y)$ into $P(E - y)_e$, i.e., by the convolution integral:

$$\mathscr{F}(E) = \int_0^E \mathscr{D}(y)P(E - y)_e \, dy \qquad (10\text{-}61)$$

where $\mathscr{F}(E)$ (dimension: energy^{-1}) is the desired distribution function for internal energy E of M^+ referred to the first adiabatic ionization potential as zero.

It is well to recall that E is the total *randomizable* energy of the ion, effective in bringing about dissociation, and the convolution (10-61) implies that thermal energy adds to this pool of randomizable energy. Experimentally, it is known that parent peak intensity in mass spectra decreases on rise of temperature (*149*) due to increased rate of decomposition. More detailed considerations (*150*) of the role of thermal energy in photoionization show that indeed most, if not all, of the thermal energy is effective in the dissociation of ions.[38]

The averaged (observable) fractional abundance of fragment ij is then

$$\langle \mathfrak{F}_{ij}(E) \rangle_{\mathscr{F}} = \int_0^\infty \mathfrak{F}_{ij}(E)\mathscr{F}(E) \, dE \qquad (10\text{-}62)$$

where $\mathfrak{F}_{ij}(E)$ is given by the appropriate equation of Section 2. The approximation of using the undistorted thermal distribution $P(E - y)_e$ in (10-61) is probably justified only for low vibrational levels of M but it is difficult to say by now much it would be distorted for higher levels. In any event, provided the temperature is not too high, the thermal factor will be small, and the convolution (10-61) will be rarely justified by the accuracy of the data.

Charge exchange The energy deposition function will obviously have a different form for different excitation processes. The simplest case is charge exchange; since it is a resonance process (Section 3), we can write[37]

$$\begin{aligned} \mathscr{D}(y) &= \text{const} \qquad \text{when} \quad y = \text{RE} \\ &= 0 \qquad\qquad \text{otherwise} \end{aligned} \qquad (10\text{-}63)$$

where RE is the recombination energy of the incident ion referred to the ionization potential of M as zero. Equation (10-63) contains the assumption

[38] Such effectiveness of thermal energy indicates that dissociation cannot be occurring via electronic predissociation by a repulsive state.

that no part of y appears as internal excitation of the incident ion after neutralization and that the incident ion is in a state with a well-separated RE. These conditions are generally satisfied for rare gas ions, but with polyatomic ions (e.g., COS^+) the possibility should not be overlooked that after charge exchange has taken place, the internal degrees of freedom of the incident particle might carry off some of the energy. Some ions have several states with closely spaced RE's, and it then becomes difficult to specify RE precisely. A table of RE's for a large number of ions has been compiled by Lindholm (*3*).

Charge exchange experiments are usually done at room temperature, where $P(E)_e$ is quite peaked (Fig. 9-3), so that it is sufficient to replace Eq. (10-61) by (see footnote 37)

$$\begin{aligned}\mathscr{F}(E)(\text{charge exchange}) &= \text{const} \qquad \text{for} \quad E = RE + \langle E \rangle \\ &= 0 \qquad \text{otherwise}\end{aligned} \tag{10-64}$$

where $\langle E \rangle$ is the average thermal energy of M [Eq. (8-88)]. Unless M is very large, $\langle E \rangle$ is small at room temperature and can be neglected with respect to RE, particularly if RE is known only with modest precision. The actual value of the constant in (10-64) is immaterial as long as we are interested only in the breakdown graph (which gives the *fractional* abundance of a given fragment at a given E), for the constant then drops out.

Electron and photon impact For excitation by photon and electron impact, $\mathscr{D}(y)$ is not a simple delta function even if the incident photons or electrons are monoenergetic, because the "collision complex," formed by the incident particle plus neutral target, breaks up by forming one (photon impact) or two (electron impact) electrons, in addition to the ion M^+, and each of these can carry off some energy. Thus if $\mathscr{V}$ is the energy of the incident particle, referred to the ionization potential of M^+ as zero, then E, the internal energy of M^+, can be anywhere between zero and $\mathscr{V}$.

For electron impact, numerous empirical or semiempirical expressions for $\mathscr{D}(y)$ have been used. In the early work, it was assumed that

$$\begin{aligned}\mathscr{D}(y)(\text{electron impact}) &= \text{const} \qquad \text{for} \quad 0 \leqslant y \leqslant y_{\max} \\ &= 0 \qquad \text{otherwise}\end{aligned} \tag{10-65}$$

The function is normalized between zero and $y_{\max}$, with $y_{\max}$ representing a more or less arbitrary cutoff (cf. Fig. 10-9). Later, Stevenson (*151*; *152*, Chapter IV, pp. 167 ff.) proposed a modified one-parameter Gaussian form

$$\mathscr{D}(y)(\text{electron impact}) = \frac{\pi}{2\langle y \rangle} \exp\left[-\frac{\pi}{4}\left(\frac{y}{\langle y \rangle}\right)^2\right] \tag{10-66}$$

where $\langle y \rangle = \int_0^\infty y\mathscr{D}(y)\,dy$. This function is normalized to unity between zero and infinity, but most of the area under the curve is confined between zero and $\langle y \rangle$. Howe and Williams (*153*) took the parabolic form

$$\mathscr{D}(y)(\text{electron impact}) = 6(\mathscr{V} - y)y/\mathscr{V}^3 \tag{10-67}$$

This function is normalized to unity between zero and $\mathscr{V}$. All these functions are based on various simplifying assumptions, and therefore are at best only very approximate; hence they also neglect the distribution of electron energies that exists in electrons emitted from a hot filament and treat the electrons as if all had the same energy $\mathscr{V}$. Because of all the approximations, thermal energy may also be neglected, with the result that the convolution (10-61) becomes unnecessary and the given forms of $\mathscr{D}(y)$ have been considered to represent $\mathscr{F}(E)$ itself by taking $y = E$. Other forms of $\mathscr{D}(y)$, discussed later, are likewise not accurate enough to make the convolution (10-61) worthwhile, and therefore no distinction shall be made henceforth between $\mathscr{F}(E)$ and $\mathscr{D}(E)$.

If the problem is approached on a more fundamental level, we can observe that the probability of excitation to a given *state* of M^+ (transition probability) is given by the appropriate Franck–Condon factor; coupled with a (mostly) assumed form of the threshold law [Eq. (10-68)], this would give $\mathscr{F}(E)$. However, we want the probability of excitation to a given *energy range*, and in anything but a diatomic M^+, the problem becomes unmanageable because of the very large number of states involved (of the order of 10^{10}) when the ion is polyatomic and the energy high, as is usually the case.

Much effort has therefore been expended on obtaining $\mathscr{F}(E)$ from experiment. Theoretical considerations (*154–156*)[39] show that for a transition[40] having energy E, $\mathscr{F}(E)$ should be a simple polynomial of the form

$$\mathscr{F}(E) = \Upsilon(\mathscr{V} - E)^{\Lambda - 1} \qquad (\mathscr{V} \geqslant E \geqslant 0) \tag{10-68}$$

Here, $\mathscr{V}$ is the (nominal) energy of the incident electron or photon, the exponent Λ gives the number of electrons that leave the "collision complex," and Υ is a scaling factor proportional to the transition probability, in general different for every vibronic state. For electron impact, $\Lambda = 2$ and for photon impact, $\Lambda = 1$. The total ion current $i(z)$ observed at particle impacting energy z (now considered as variable) is given by the convolution

$$i(z) = \int_0^\infty M(x - z)\mathscr{F}(x)\,dx \tag{10-69}$$

[39] For electron impact, see also Rudge and Seaton (*157*) and Rudge (*158*).

[40] The transition alluded to is a transition from a specified (electronic, vibrational, and rotational) state of M to a specified (electronic, vibrational, and rotational) state of M^+; hence thermal distribution is ignored. Inclusion of thermal effects would modify somewhat the idealized case represented by Eq. (10-68). Cf. Guyon and Berkowitz (*159*).

where $M(x - z)$ is an instrument function that accounts for the spread of energies in the electron beam (in case of electron impact) or for the width of slits (in the case of photon impact), as well as for the response of the detector. It can be shown (*160*) that if $\mathscr{F}(x)$ has the form (10-68), the Λth derivative of $i(z)$ with respect to z is proportional to Υ. Hence to within a normalization factor,

$$\begin{aligned} \mathscr{F}(E)(\text{photoionization}) &\propto [di(z)/dz]_{z=E} \\ \mathscr{F}(E)(\text{electron impact}) &\propto (\mathscr{V} - E)[d^2i(z)/dz^2]_{z=E} \end{aligned} \tag{10-70}$$

In mass spectrometry, $\mathscr{V} = 70$ eV is commonly used, so that $\mathscr{V} \gg E$, and therefore in such cases, $\mathscr{F}(E)$ for electron impact is simply given by the second-derivative ionization efficiency curve, multiplied by the appropriate normalization factor.

The ionization efficiency referred to here in Eqs. (10-70) is the *total* ionization efficiency. As pointed out by Chupka (*31*), the first- (or second-) derivative ionization efficiency curve *for a given species*, divided by the appropriate derivative of the total ionization efficiency (i.e., appropriately normalized), is equivalent to the normalized breakdown graph. In practice, therefore, it is more convenient to convert experimental data to the normalized breakdown graph in the manner described and then compare calculated and "experimental" breakdown graphs, rather than average calculated data over $\mathscr{F}(E)$ and then compare the result with the raw experimental data. This procedure ignores, however, the presence of thermal energy, which is inevitably included in every "experimental" breakdown graph but not in one calculated by the described procedure. For room-temperature data of highest accuracy, or if the temperature is high, one should therefore compare "experimental" breakdown graphs with temperature-averaged calculated breakdown graphs obtained by calculating the fractional abundance of fragment ij at nominal energy E as given from

$$\langle \mathfrak{F}_{ij}(E) \rangle_{P_e} = \int_0^\infty \mathfrak{F}_{ij}(E + x) P(x)_e \, dx \tag{10-71}$$

Except in unusual cases, Eq. (10-71) will result in practice merely in the shift of the energy scale of the breakdown graph by $\langle E \rangle$, the average thermal energy[41] of the neutral parent M, because $P(E)_e$ is a peaked function at room temperature [cf. Eq. (10-64)].

It is nevertheless instructive to consider the actual form of $\mathscr{F}(E)$ as obtained from (10-70) because it can be compared with $\mathscr{F}(E)$ obtained in other ways. In the ideal case (infinite resolution), the derivatives (first or second, for photon and electron impact, respectively) should be a succession

[41] For calculated breakdown graphs for propane and butane ions at several temperatures, see Vestal *et al.* (*26*). The differences due to temperature are minimal.

of delta functions centered on the onset energy of each transition. Instrumental factors do not allow transitions to individual states to be resolved, in which case the first- or second-derivative curve as a function of energy is an envelope of all the unresolved delta functions. This will be particularly true of electron impact results, where electrons are usually produced by emission from a hot filament and therefore have a broad distribution of energies characteristic of the temperature of the filament. Because photons can be produced with a smaller spread of energies, the first-derivative photoionization efficiency curves generally shows more detail than the second-derivative electron impact ionization efficiency, but overall for a given M^+, they should show similar structure. However, because some transitions are likely to be disallowed for electrons, but not for photons, we can expect in general less structure in the second-derivative electron impact ionization efficiency curves.

Figures 10-9 and 10-10 show a comparison of second-derivative electron impact (*161*) and first-derivative photon impact (*134*) ionization efficiency

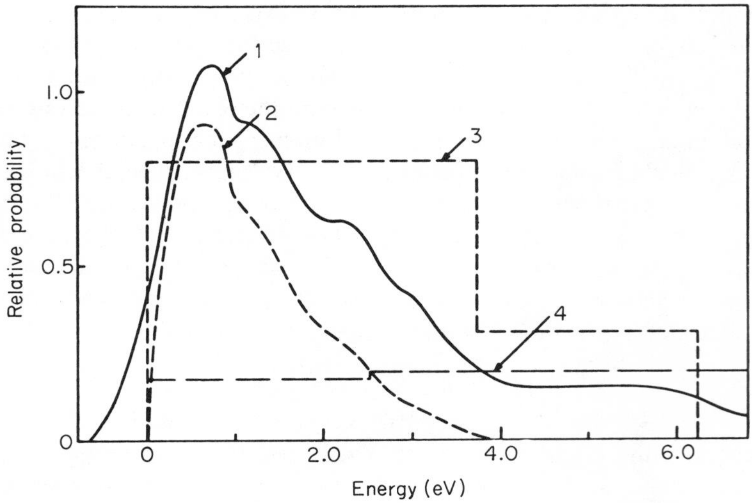

Fig. 10-9. *$\mathscr{F}(E)$ for excitation of propane by electron impact.*
Energy referred to the ground state of the ion as zero. Curve 1 is the second derivative of ionization efficiency [Eq. (10-70)] for $\mathscr{V} \gg E$; curve 2 is the same curve but for $\mathscr{V} = 4$ eV above ionization potential. Rectangular distributions are those used in early calculations of breakdown curves of propane by (curve 3) A. Kropf, E. M. Eyring, A. L. Wahrhaftig and H. Eyring, *J. Chem. Phys.* **32**, 149 (1960), and (curve 4) H. M. Rosenstock, M. B. Wallenstein, A. L. Wahrhaftig and H. Eyring, *Proc. Nat. Acad. Sci. U.S.* **38**, 667 (1952). From W. A. Chupkå and M. Kaminsky, *J. Chem. Phys.* **35**, 1991 (1961).

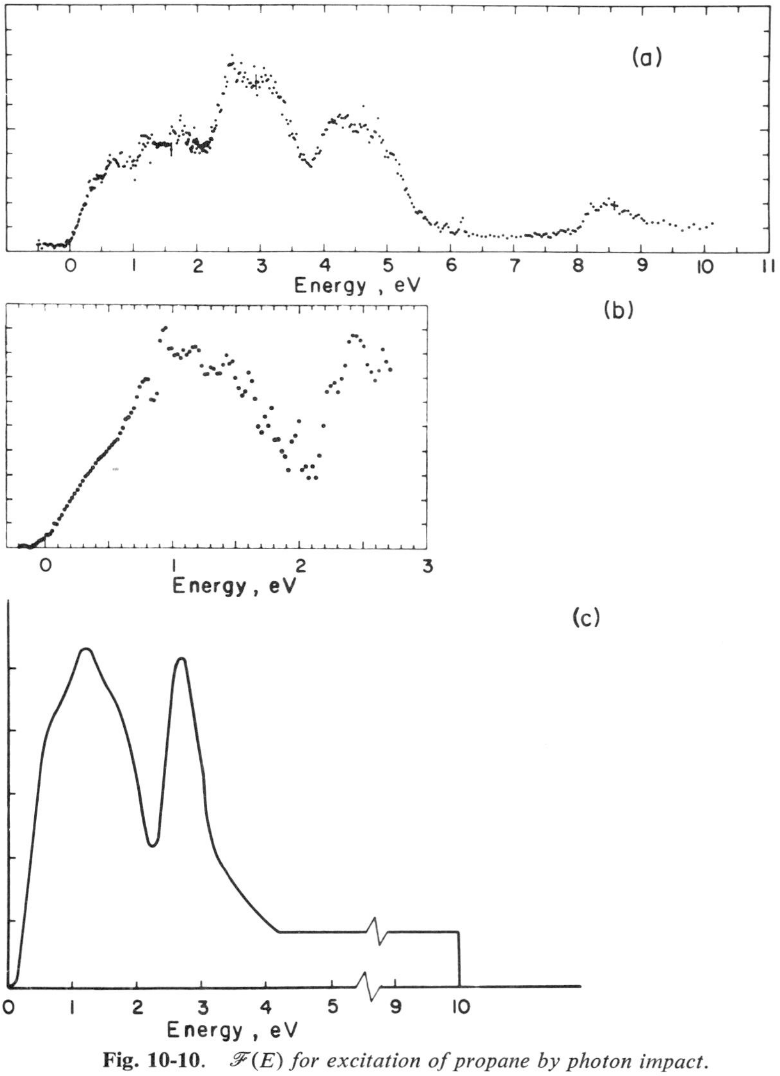

Fig. 10-10. *ℱ(E) for excitation of propane by photon impact.*

The ordinate is essentially ℱ(E) in arbitrary units, not normalized. Energy referred to the ground state of the ion as zero. (a) As obtained from threshold electron spectroscopy by R. Stockbauer and M. G. Inghram, *J. Chem. Phys.* **54**, 2242 (1971), Fig. 3. The curve gives the relative Franck–Condon factor, which for photon impact is equivalent to ℱ(E) if Eq. (10-68) applies. (b) As obtained from the first derivative of total photoionization efficiency by W. A. Chupka and J. Berkowitz, *J. Chem. Phys.* **47**, 2921 (1967), Fig. 11. (c) Used by M. Vestal and J. H. Futrell [*J. Chem. Phys.* **52**, 978 (1970), Fig. 2] in calculating the mass spectrum of propane. The energy scales of (a) and (c) are the same, that of (b) is about twice as large.

curves for the propane ion, the workhorse system in the theory of mass spectra. Superimposed on these curves are shapes of $\mathscr{F}(E)$ actually used in calculations[42] of the mass spectrum of propane. The agreement between the two figures is only moderate, showing that the theoretical basis of the treatment is only approximately true. In fact, doubts exist not only about the validity of Eq. (10-68) for large $(E - \mathscr{V})$ but also about the form of the power law itself for electron impact (*162,163*), and about a possible contribution from autoionizing states which would presumably require a Λ between 1 and 2. In purely instrumental terms, taking a second derivative of an experimental quantity tends to magnify out of all proportion any short-term instability of the instrument.

A differential experimental technique, in which autoionization is suppressed, is threshold electron spectroscopy (*40*), which allows direct determination of relative Franck–Condon factors. It should give results equivalent to the first derivative of total ionization efficiency in photoionization if the photoionization cross section is constant above threshold and no autoionization occurs. Figure 10-10 shows data of Stockbauer and Inghram (*164*) on propane which differ somewhat from those of Chupka and Berkowitz (*134*) because of variation in cross section and contribution from autoionization. Nevertheless, the difference is not large; in addition, breakdown graphs obtained from the first derivative of photoionization efficiency are in general in good agreement with those obtained by charge exchange (*134*), which establishes the validity of Eq. (10-68) for photoionization, at least not too far from ionization threshold.

Combined with mass analysis, threshold electron spectroscopy yields the breakdown graph directly, independent of cross-section variations or autoionization. The internal energy of the parent is precisely known since only ions produced in coincidence with electrons of zero kinetic energy are detected. The technique thus avoids the pitfalls of conventional photoionization and charge exchange experiments, and is therefore superior to both techniques. Stockbauer (*249*) has recently used threshold electron spectroscopy to obtain breakdown graphs for the methane and ethane ions; these new results essentially confirm earlier charge exchange and photoionization data, except for a few details.

This leaves only the distribution function for electron impact still in doubt. Recently, Meisels and collaborators (*165*) proposed a function based on the optical approximation, which treats the electron as a quasiphoton. Then, the probability that the energy loss x suffered by the incident electron shall

[42] Because of the complicated shapes, averaging over $\mathscr{F}(E)$ must be done by piecewise integration over linear portions, if there are any, or by point-by-point numerical integration for other shapes.

result in the formation of an ion pair is proportional to

$$\sigma(x)/x \tag{10-71a}$$

where $\sigma(x)$ is the total cross section for photoionization, obtainable from photoelectron spectra. The energy loss is divided between the kinetic energy of the ejected electron and the ionization and internal energies of the ion, and the actual energy deposition function $\mathscr{F}(E)$ is obtained by folding the energy-loss function (10-71a) into the appropriate photoelectron spectrum. The shape of $\mathscr{F}(E)$ so obtained is quite similar to the threshold electron spectroscopy function shown in Fig. 10-10(a), and when available breakdown graphs of methane, ethane, and ethylene are averaged over this $\mathscr{F}(E)$, reasonably good 70-eV mass spectra are obtained.

In summary, then, the distribution function $\mathscr{F}(E)$ is not really needed, except in electron impact, because properly normalized charge exchange results yield the breakdown graph directly, and photoionization with mass analysis yields the breakdown graph after differentiation of the ionization efficiency curve of every species and normalization. It is only in mass spectra that $\mathscr{F}(E)$ is needed to correlate theory with experiment, but unfortunately the mass spectrum is quite insensitive to the actual form of $\mathscr{F}(E)$. It is fairly certain that $\mathscr{F}(E)$ for 70-eV electron impact should rise after ionization threshold, go through a maximum, and vanish somewhere beyond 20 eV or so because there is usually little change in the mass spectrum about 20 or 30 eV. For this reason, $\mathscr{F}(E)$ of Eqs. (10-66) and (10-67) is probably not quite as bad as it looks. In any event, it seems that conventional 70-eV electron impact with hot-filament-produced electrons is a rather ill-defined system that probably does not merit a great expenditure of effort.

7. Primary yield in radiolysis

When ionizing radiation (e.g., X rays, gamma rays, or any high-energy particle) passes through gaseous matter, it is well known (*166*, Chapter 2) that essentially all the chemical effect arises from slow (<100 eV) secondary electrons. Hence the primary act in radiolysis is mostly excitation by electron impact, and this results in the formation of excited ions. It was recognized early (*167–169*) that there should therefore exist a connection between the primary process in radiolysis and the fragmentation pattern observed in conventional 70-eV electron impact mass spectra, despite a different distribution of electron energies in radiolysis, since it is known that the mass spectrum is relatively independent of electron energy above ~ 30 eV. Hence if it is possible to calculate the mass spectrum of a given substance, then, provided the basic assumptions of the theory of mass spectra still hold, it should also be possible to calculate the primary yield in its radiolysis—which is the reason why radiolysis is discussed in a chapter dealing with the theory of mass spectra.

The major difference between radiolytic and mass spectrometric systems is pressure, i.e., the presence of collisions in the former and their absence in the latter: Radiolysis is usually studied at atmospheric pressure, while mass spectrometers are run at 10^{-8} atm or so. The consequence is that the similarity between the two systems is confined to the interval preceding the first collision in the radiolytic system, i.e., some 10^{-10} sec at atmospheric pressure. Hence, to this approximation, the primary yield of ions in the radiolysis of a gas should be given (*170*) by its mass spectrum at 10^{-10} sec.

Secondary events Unfortunately, primary radiolytic yields are not observable experimentally, and therefore there is no direct check on the calculated primary yields. What is experimentally observable is the total radiolytic yield (and perhaps also the yield of some intermediates), which refers to neutral substances, inasmuch as the final products of radiolysis are all neutral. Clearly, ions produced in the primary event of radiolysis must collide, mostly with neutral molecules, which constitute the bulk of the gas phase; thus one of the secondary events in radiolysis are ion–molecule reactions, of which there is usually quite a variety.[43] Electron impact, or ion–electron recombination, may also produce highly excited neutral molecules, which then fragment and produce radicals that can react further. Thus the total observable radiolytic yield is removed by several steps from the primary event, and this of course introduces numerous elements of uncertainty if one wishes to compare calculated primary yields with experimentally observed total yields. For this reason, radiolytic systems are not suitable for testing unimolecular rate theory. In fact, it is rather the reverse: Assuming the validity of unimolecular rate theory and of its basic assumptions, one can use the theory to obtain a foothold on the basis on which the mechanism of radiolysis can be quantitatively interpreted. Some of the principles involved in the application of the theory of mass spectra to radiation chemical yields have been discussed by Prášil (*172*).

It would be outside the scope of this chapter to discuss ion–molecule reactions and other secondary phenomena in the radiolysis of gases, and therefore the discussion shall not extend beyond primary yields. If we subscribe to the view that the primary yield of ions in radiolysis is given by the mass spectrum at 10^{-10} sec, then we are left, at this point, only with the determination of the energy deposition[44] function $\mathscr{F}(E)$, the basic theory of mass spectra having been dealt with in Sections 1–5 of this chapter.

[43] For a discussion of ionic reactions in the radiolysis of ethylene, see Meisels (*171*).

[44] Because of all the approximations, the thermal energy convolution (10-61) can be safely neglected.

Energy deposition function Wincel (*173*), who calculated yields in the radiolysis of methane and ethane, used $\mathscr{F}(E)$ of the from (10-66) for methane and a simple rectangular distribution for ethane. Prášil (*174*) calculated yields in the radiolysis of methanol, methane, ethane, and propane using a variety of rectangular distributions. In this study, the theory of mass spectra was used to calculate the fragmentation of not only excited ions, but also of excited neutral molecules (at energies below the first ionization potential), assuming that both excited and neutral molecules fragmented in the same fashion. The function $\mathscr{F}(E)$ therefore extended below, as well as above, the ionization potential. Above the ionization potential, $\mathscr{F}(E)$ was adjusted to give roughly (at 10^{-5} sec) the observed mass spectrum, while below the ionization potential, the adjustment was such as to make $\mathscr{F}(E)$ yield overall an average energy equal roughly to W, the average energy necessary to form an ion pair, which is known from dosimetry.

A more fundamental attack on the problem of the energy deposition function was made by Prášil *et al.* (*175–177*), who derived from the optical approximation the following form for the energy-loss function:

$$\frac{\chi}{N}\cdot\frac{1}{x}\,\frac{df(x)}{dx} \tag{10-72}$$

where χ is the total energy absorbed by all molecules, N is the number of electrons in the molecule being investigated, and $f(x)$ is its optical oscillator strength. Equation (10-72) is analogous to (10-71a) and is probably better justified in the case of radiolysis than in mass spectrometry. In practical terms, Eq. (10-72) means that $\mathscr{F}(E)$ will have a shape very similar to the electronic excitation spectrum of the molecule in question. Such a spectrum for methane was published by Platzman (*178*) and was used by Prášil in his computation of primary yields in the radiolysis of methane as an alternative to a rectangular distribution.[45]

In all of the studies mentioned, reasonable agreement with observed overall radiolytic yields was obtained after combining the computed primary yields with a plausible choice of ion–molecule and radical–molecule reactions. It is difficult to assess the significance of these results because the theory of mass spectra was used in its old, classical formulation. For this reason, it is also not very disturbing that $\mathscr{F}(E)$ of Eq. (10-72) actually gave a poorer mass spectrum of methane than an assumed rectangular distribution. By all odds, the total yield in radiolysis is not very sensitive to any particular form of $\mathscr{F}(E)$ because, for one thing, the total yield represents only an average, and calculations are subject to possible cancellation of errors arising from the simplified treatment of primary and secondary processes.

[45] The entire theory of energy deposition in irradiated matter has been reviewed recently by Santar (*179*).

It has been pointed out (*14,180*) that the ordinary mass spectrum (computed at 10^{-10} sec or some other time) may be a poor approximation to the primary yield because direct studies with fast electrons (~1 keV) seem to indicate a different abundance of fragments. A nonstatistical theory (called the "local" theory) has been proposed by Gur'ev (*181*) in which it is assumed that on electron impact, dissociation occurs in a "collision zone" without any significant redistribution of energy over the rest of the molecule.

8. Theoretical parameters and experimental observables

In suitable experimental systems involving ionization combined with mass analysis, the experimental observable is the breakdown graph (an example of which is shown in Fig. 10-11), i.e., the normalized fractional abundance of fragments as a function of excitation energy. Its theoretical counterpart is the appropriate $\mathfrak{F}_{ij}(E)$ of Section 2 (averaged, if necessary, over the thermal distribution of energies). Since $\mathfrak{F}_{ij}(E)$ is always proportional to $k_{0j}(E)/\sum_j k_{0j}(E)$ [cf. Eqs. (10-20) and (10-21)], it depends on all the $k_{0j}(E)$'s and therefore is not a very sensitive function of any one particular $k_{0j}(E)$.

There is, however, an interesting new method called the "beam technique" (*182,183*), which allows actual lifetime measurements on selected metastable

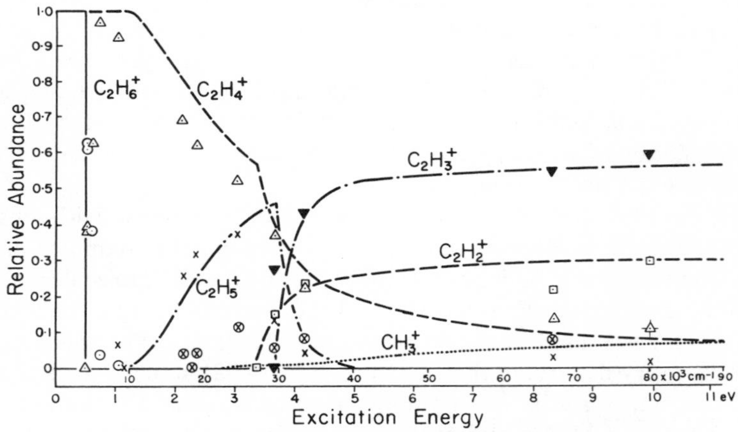

Fig. 10-11. *Breakdown graph for the* $C_2H_6^+$ *ion.*
Points are experimental results of H. von Koch, *Arkiv Fys.* **28**, 559 (1965) obtained by charge exchange. (○) $C_2H_6^+$, (△) $C_2H_4^+$, (×) $C_2H_5^+$, (⊗) CH_3^+, (▼) $C_2H_3^+$, (□) $C_2H_2^+$. Curves were calculated by Z. Prášil and W. Forst, *J. Phys. Chem.* **71**, 3166 (1967). For new experimental results obtained by threshold electron spectroscopy see (*249*, Fig. 10); these results remove the discrepancy for CH_3^+ between theory and the charge exchange data.

transitions. It is beyond the scope of the present chapter to go into experimental details, but suffice it to say that using ionization by charge exchange, Andlauer and Ottinger (*184,185*) were able to measure, over a range of excitation energies in the parent, the rate constant for the decomposition of an essentially monoenergetic benzonitrile molecular ion[46]:

$$C_6H_5CN^+ \longrightarrow C_6H_4^+ + HCN \tag{10-73}$$

Such a measurement has not yet been accomplished even for a neutral reactant. Similar measurements have been done for the metastable transitions $C_6H_6^+ \rightarrow C_4H_4^+ + C_2H_2$ and $C_6H_6^+ \rightarrow C_6H_5^+ + H$. Obviously results of this kind would provide a very direct test of theoretical calculations since all that is needed are the properties of the molecular ion and its transition state; all problems relating to the distribution of excitation energies and complex reaction schemes are eliminated. Calculations have been done on the cited reactions of the benzonitrile and benzene ions by Klots (*247*) with moderate success.

Mechanism of fragmentation The first step in calculating a breakdown graph for an ionic process is to establish the mechanism of the fragmentation. This is by no means an easy task, especially in the case of large parent molecular ions[47] because mass analysis provides an identification of the fragments according to mass but not according to structure. Metastable transitions provide direct evidence for the occurrence of a given reaction[48] and therefore much effort has been expended on the exploitation of metastable transitions for the elucidation of reaction paths in mass spectra.[49] On the other hand, the absence of a metastable transition for a given process may simply mean that the process in question has a rate constant higher than needed for the metastable "window" (i.e., rate in excess of $\sim 10^6$ sec^{-1}). Isotopic labeling and energetics of ionic fragmentation also provide supplemental clues that are helpful in reducing the number of possibilities.[50] A complicating feature of many electron impact spectra are rearrangement processes in which fragmentation is accompanied by scrambling of bonds inside the newly formed

[46] Unfortunately, benzonitrile may not be a very good example of a statistical decomposition.

[47] Several books are devoted entirely or in part to the correlation between mass spectra and molecular structure; for instance: Beynon (*186*); McLafferty (*187*); Budzikiewicz *et al.* (*188*). For a recent review, see Williams (*189*).

[48] A very thorough determination of metastable transitions in ethane has been done by Löhle and Ottinger (*190*), who found 54 metastable transitions in CD_3CH_3 and 19 in C_2H_6.

[49] For a review, see Beynon (*191*).

[50] See, for instance, Natalis (*192*), who used energetic arguments to interpret geometric isomerism in mass spectra.

fragments (*193,194*). Finally, spectroscopic selection rules may also be helpful, in some of the simple cases, to determine the likely states of the fragments.

Theoretical equations and parameters Assuming that the mechanism of the ionic process being studied has been worked out in the form of a consecutive-parallel scheme of the sort given in (10-13), we now have to collect the parameters necessary for calculating the rates of the individual reactions. It was shown in Section 2, however, that in most cases, only the rate constant k_{0j} for decomposition of M^+ into A_{1j}^+ is necessary because the fractional abundances $\mathfrak{F}_{1j}$ and $\mathfrak{F}_{2j}$ of the first and second generation fragments, respectively, can be expressed in terms of the k_{0j}'s and the energy partition function $P(x, \mathscr{E})$ [cf. Eqs. (10-37), (10-38), (10-40), and (10-42)]; the same is true of the metastable abundance $\mathfrak{F}_{0\star}$ [Eq. (10-43)]. We have for $k_{0j}(E)$

$$k_{0j}(E) = \alpha_{0j} G_{0j}^*(E - E_{0(0j)})/hN(E) \tag{10-74}$$

where $N(E)$ is the density of states of M^+ at energy E, $G_{0j}^*(E - E_{0(0j)})$ is the integrated density of the transition state for decomposition of M^+ into A_{1j}^+, $E_{0(0j)}$ is the critical energy for this process, and α_{0j} is its reaction path degeneracy. The function $P(x, \mathscr{E})$ always involves the density of states of the fragment A_{1j}^+ at energy x as one factor, which we shall call $N_{1ja}(x)$; the other factors depend on the form of $P(x, \mathscr{E})$ used. Thus in the form (10-31) we have

$$P(x, E - E_{0(0j)}) = N_{1ja}(x) G_{0jb}^*(E - E_{0(0j)} - x)/G_{0j}^*(E - E_{0(0j)}) \tag{10-75}$$

where $G_{0jb}^*(y)$ is the integrated density for the species having the degrees of freedom of the transition state in $M^+ \rightarrow A_{1j}^+$, *less* those of A_{1j}^+ itself, and $G_{0j}^*(z)$ has been defined earlier. The alternative form of $P(x, \mathscr{E})$ is Eq. (10-29), i.e.

$$P(x, E - E_{0(0j)}) = N_{1ja}(x) N_{1jb}(E - E_{0(0j)} - x) \Big/ \int_0^{E - E_{0(0j)}} N_{1ja}(x) N_{1jb}(E - E_{0(0j)} - x)\, dx \tag{10-76}$$

where $N_{1jb}(y)$ is the density of states of the neutral fragment B_{1j} formed in the reaction $M^+ \rightarrow A_{1j}^+ + B_{1j}$. It is not necessary to actually calculate the convolution integral in the denominator of (10-76) because we see immediately from (6-32) that it is equal to $N_{1jab}(E - E_{0(0j)})$, the density of the combined system (fragment A_{1j}^+ + fragment B_{1j}) at energy $(E - E_{0(0j)})$, and this density can be obtained directly by inverting the product of their partition functions (Chapter 6, p. 106).

The metastable abundance $\mathfrak{F}_{1\star}$ [Eq. (10-44)], and possibly also $\mathfrak{F}_{1j}$ and

$\mathfrak{F}_{2j}$ when the simplifying approximation leading to Eq. (10-42) does not apply, requires the knowledge of $k_{1j}(E_{1j})$, where

$$k_{1j}(E_{1j}) = \alpha_{1j} G_{1j}^*(E_{1j} - E_{0(1j)})/hN_{1ja}(E_{1j}) \tag{10-77}$$

where E_{1j} is the excitation energy in fragment A_{1j}^+, $G_{1j}^*(y)$ is the integrated density of the transition state for the reaction $\mathrm{A}_{1j}^+ \rightarrow \mathrm{A}_{2j}^+$, α_{1j} is its reaction path degeneracy, and $N_{1ja}(z)$ has been defined.

The rate expressions (10-74) and (10-77) ignore tunneling and rotational effects; the latter will be taken up in the next section (Section 9). The extension of the treatment to more complicated schemes where the second generation fragments decompose further should be fairly obvious, but it will no doubt be appreciated that the algebraic complications become formidable.

The following parameters are required for numerical evaluation of expressions (10-74)–(10-77):

1. The critical energies $E_{0(0j)}$ and $E_{0(1j)}$.
2. $N(E)$, the density of states of M^+, at energies from the smallest $E_{0(0j)}$ onward.
3. $N_{1ja}(x)$, the density of states of fragment A_{1j}^{∞}, from $x = E_{0(0j)}$ onward.
4. $N_{1jb}(y)$, the density of states of fragment B_{1j}, from $y = 0$ onward.
5. $G_{0j}^*(z)$, the integrated density of the transition state for decomposition of M^+ into A_{1j}^+, from $z = 0$ onward.
6. $G_{0jb}^*(w)$, if (10-75) for $P(x, \mathscr{E})$ is used, from $w = 0$ onward.
7. $N_{1jab}(x')$, the density of the combined system $(\mathrm{A}_{1j}^+ + \mathrm{B}_{1j})$, at $E - E_{0(0j)}$.
8. $G_{1j}^*(y)$, defined in Eq. (10-77), from $y = 0$ onward.
9. The reaction path degeneracies α_{0j}.

Critical energies and appearance potentials If every reaction in the ionic decomposition scheme being calculated occurs over a type 1 potential energy surface (no activation energy for the back reaction), the critical energies $E_{0(ij)}$ (parameter 1) are obtainable, in principle, from the appearance potentials $\mathrm{AP}(\mathrm{A}_{i+1,j}^+)$ of fragments $\mathrm{A}_{i+1,j}^+$. The process measured by AP is[51]

$$\mathrm{M} \longrightarrow \mathrm{A}^+ + \mathrm{B} + e^- \tag{10-78}$$

for which we have

$$\mathrm{AP}(\mathrm{A}^+) = \Delta H_f^0(\mathrm{A}^+) + \Delta H_f^0(\mathrm{B}) - \Delta H_f^0(\mathrm{M}) + \Delta E' \tag{10-79}$$

where $\Delta H_f^0(X)$ is the enthalpy of formation of species X in its standard state

[51] It is understood that in the case of electron impact, an additional electron appears on both sides of the equation.

at 0°K [cf. Eq. (9-95)], and $\Delta E'$ is excess internal or kinetic energy of the fragments. Since we must refer all energies to the ground state of M^+ as zero, we also need the ionization potential IP of M, i.e., the threshold energy for the process (see footnote 51)

$$M \rightarrow M^+ + e^- \tag{10-80}$$

for which

$$IP(M^+) = \Delta H_f^0(M^+) - \Delta H_f^0(M) + \Delta E' \tag{10-81}$$

where $\Delta E'$ is the excess internal energy of M^+.

The problems associated with converting AP's into actual critical energies are multiple.[52] First of all, there is the question of interpretation of the experimental appearance potential data,[53] which is particularly complex, for instrumental reasons, in electron impact, and to a lesser extent in photon impact. Second, Eq. (10-79) contains the indeterminate term $\Delta E'$ which arises, in part, from the presence of thermal energy in the reactants and products, but also, more importantly, from possible internal (vibrational) excitation of the products above the thermal level, and from possible channeling of excitation energy into the kinetic energy of the fragments.[54] The purely thermal part of $\Delta E'$ is small and calculable, but the other parts may be large and are not calculable, except in a few simple cases.[55] It is generally believed that if the potential energy surface is of type 1, the kinetic energy of the fragments will be small, but the vibrational (or indeed electronic) excitation of the fragments is usually uncertain. A related problem afflicts the ionization potential IP [process (10-80)] needed to refer AP to the appropriate zero of energy. Both photon and electron impact are "vertical" processes which may lead to formation of M^+ in an excited vibrational state depending on the relative positions of the potential energy curves for M and M^+ (cf. Fig. 10-4); it is this excess vibrational energy, plus the difference in thermal energies, that is measured by $\Delta E'$. In optical spectroscopy, on the other hand, it is possible to measure IP(M^+) such that $\Delta E' = 0$, which is the "true" IP, against which IP's obtained by other methods can be compared. Usually photoionization IP's are quite close to the spectroscopic values, but electron impact IP's are often higher, if the transition for which $\Delta E' = 0$ is not sufficiently probable.

The third problem affecting AP's is the limited time scale of the mass

[52] For reviews on the general subject of appearance and ionization potentials, see (*2*) and Krauss and Dibeler (*195*); C. A. McDowell (*196*); Collin (*197*).

[53] See, for example, Collin (*197*, p. 138); Beckey and Comes (*198*).

[54] We exclude, however, the case where the kinetic energy of the fragments is due to dissociation of the parent via a transition to a *repulsive* state of the ion.

[55] See, for example, the classical case of $CO^+ \rightarrow C^+ + O$ calculated by Hagstrum and Tate (*199*).

spectrometer. As indicated in Section 2, fragments A_{ij}^+ are detected as ion current at m_{ij}/e, only if A_{ij}^+ is formed in the source, i.e., within a very short interval of time (cf. Table 10-1). Thus the onset of the ion current at m_{ij}/e, i.e., the AP(A_{ij}^+), basically represents the energy at which the process $M^+ \rightarrow A_{ij}^+$ becomes sufficiently rapid for A_{ij}^+ to be detected as such. From Fig. 10-3, we see that the requirement is that the rate constant for fragmentation to first generation fragments be roughly of the order of 10^7 sec^{-1} or higher. Hence unless $k_{0j}(E_{0(0j)}) > 10^7$ sec^{-1}, i.e., unless the rate constant at threshold is already sufficiently high, the AP(A_{1j}^+) will *not* be a measure of $E_{0(0j)}$ but will include whatever excess energy is required to make $k_{0j}(E)$ large enough. The difference between the true threshold energy and the energy measured by AP has been called the "kinetic shift" by Chupka (*31*). The kinetic shift will be particularly important if $k_{0j}(E)$ is small at threshold and if it rises only slowly with energy. A similar problem exists with AP's of second generation fragments A_{2j}^+ (*21*), except that it is further aggravated by the fluctuation of energy in A_{1j}^+ with the result that AP(A_{2j}^+) usually does not show a sharp onset and that part of the energy measured as AP may have actually gone into the excitation of the neutral fragment B_{1j}.

The result is then that the AP's are likely to be overestimates of unknown magnitude of the corresponding critical energies. Nevertheless, a body of data on AP and IP now exists from a wide variety of sources that shows a certain amount of self-consistency, so that the "good" values of AP are probably not too far from thermodynamic enthalpies, perhaps within 0.3 eV or better. This is not the sort of accuracy one would hope to have for critical energies, but this is the best that is available at present. Overall, however, the situation does not justify using the critical energies as adjustable parameters except insofar as "adjustment" is confined to within the estimated accuracy of the AP's used.

Type 1 versus type 2 reactions Note that the use of AP's for critical energies is based on the assumption that the ionic reactions in question do not have an activation energy for reaction in the opposite sense. If there *is* an appreciable activation energy for the reverse reaction, the problem becomes vastly more complicated. However, it seems that only an ionic rearrangement process would fall into this category, and in any event, the activation energy for the reverse process may be smaller than the uncertainty in the critical energy for the forward process as determined from the AP. Note also that because of the relatively large error inherent in the critical energies $E_{0(ij)}$ obtained in the manner described, we have neglected the fact that, strictly speaking, they are model dependent because they include the zero-point energy of the transition state of the reaction in question.

A juxtaposition of analogous reactions of neutral and charged species reveals that beside simple bond-breaking reactions (e.g., $C_2H_6 \rightarrow 2CH_3$ and $C_2H_6^+ \rightarrow CH_3^+ + CH_3$), which in both ions and neutrals occur over a type 1 potential energy surface, there is also complex bond fission (e.g., $C_2H_5Cl \rightarrow C_2H_4 + HCl$ and $C_2H_6^+ \rightarrow C_2H_4^+ + H_2$) which in *neutral* species always occurs over a type 2 potential energy surface since it requires an activation energy for the reverse reaction. In an *ionic* fragmentation, however, even if complex bond fission is involved, the reverse reaction is always an ion–molecule reaction which, as mentioned, usually requires no measurable activation energy. There is an appreciable difference between simple and complex bond fission (cf. Chapter 11, Sections 2 and 3), which should be reflected in the properties of the potential energy surface, and yet, seemingly this is not so in the case of ionic fragmentation. This contrast between the fragmentations of neutral and ionic species is probably only apparent: Since the potential energy surfaces of ions are in general quite flat (Section 4), it is likely that an ionic type 2 potential energy surface will show only a small activation energy for the reverse reaction, which will thus escape detection and be lost in experimental error. This error, as pointed out in connection with appearance potentials, may be fairly large.

The conclusion is therefore that the absence of measurable activation energy in ion–molecule reactions is not a sensitive enough criterion to decide whether or not the ionic fragmentation process in question occurs over a type 1 or type 2 potential energy surface; presumably, the same is true of type 3 processes (isomerizations). We shall assign to a given ionic fragmentation the same type of potential energy surface over which the analogous reaction of its neutral counterpart occurs.

Molecular parameters of ions Parameters 2–4 refer to the density of states of stable species, which requires as input vibrational frequencies and also moments of inertia if rotations are involved. These general aspects have been discussed in Chapter 8, Section 7; in the present instance, the particular problem is that the species to which parameters 2 and 3 refer are charged and the requisite information is not normally available. We saw in Section 4 that, as a rule, potential energy surfaces for ions are probably flatter than in the case of similar neutral molecules, which suggests lower vibrational frequencies and higher moments of inertia. However, this general statement is hardly sufficient to allow us to make a definite assignment in any one particular case, and therefore it is best to assume that molecular parameters of a ground-state ion (vibrational frequencies, moments of inertia, and the like) are the same as those of its neutral counterpart unless definite information to the contrary is available.

Table 10-6 summarizes the somewhat fragmentary data available at present for a few polyatomic molecules of interest in connection with the theory of mass spectra. In the case of the methyl halides, the frequencies in the ion and in the neutral molecule are not very different, but this is really a somewhat special case in that the electron removed on ionization comes from a nonbonding orbital localized on the heteroatom. The case of ethylene in Table 10-6 shows a considerable variation in frequencies, depending on the state of the ion. Quite likely, this is typical, not only in ions, but in neutral molecules as well: Vibrational frequencies depend strongly on the electronic state, which is not too unexpected since the geometry of excited states is often quite different from that of the ground state. Hence the

TABLE 10-6

SOME FUNDAMENTAL FREQUENCIES OF NEUTRAL MOLECULES AND THEIR IONS (in cm^{-1})

A. The methyl halides[a]

Species	Vibration	Neutral molecule $\tilde{X}\,{}^1A_1$	Positive ion state ${}^2E_{3/2}$	${}^2E_{1/2}$
CH_3I	ν_1	2953	3031	3053
	ν_2	1250	1224	1235
CH_3Br	ν_2	1305	—	1272
	ν_4	3056	3112	—
	ν_6	953	871	—
CH_3Cl	ν_2	1355	—	1514(?)
	ν_3	732	871	—
	ν_5	1488	2748(?)	—

B. Ethylene[b]

Vibration	Neutral molecule ${}^1A_{1g}$	Ion state $\tilde{X}$	$\tilde{A}$	$\tilde{B}$	$\tilde{C}$
ν_1	3026	—	2900	1700	1240
ν_2	1623	1230	1300	—	—
ν_3	1342	1340	800	—	—
ν_4	1027	430	—	—	—

[a] From J. L. Ragle, I. A. Stenhouse, D. C. Frost and C. A. McDowell, *J. Chem. Phys.* **53**, 178 (1970). ν_1: CH_3 sym. stretch; ν_2: CH_3 def.; ν_3: C—X stretch; ν_4: CH_3 deg. stretch; ν_5: CH_3 deg. deform.; ν_6: CH_3 rock.

[b] From D. W. Turner, C. Baker, A. D. Baker and C. R. Brundle, "Molecular Photoelectron Spectroscopy," p. 167. Wiley (Interscience), New York, 1970. ν_1: sym. C—H stretch; ν_2: sym. C—C stretch; ν_3: C—H def.; ν_4: CH_2 twist.

assumption that vibrational frequencies of ions are more or less the same as those of their ground-state neutral counterparts is likely to be particularly poor when the geometry of the state of the ion from which decomposition is assumed to occur ("terminal state") is very different from that of the ground-state neutral molecule. As mentioned in Section 4, the ground state ethane ion seems to have the same symmetry as ground state neutral ethane, but with a larger C—C distance; this leads to a reduced ν_3 (C—C stretching frequency) in the ion [$\nu_3 \sim 470 \text{ cm}^{-1}$, calculated in (*250*)], compared with neutral C_2H_6 ($\nu_3 \sim 990 \text{ cm}^{-1}$, observed).

The Jahn–Teller effect[56] may operate in many instances and cause the ion to have less symmetry than its neutral counterpart. For instance, whereas CH_4 is of T_d symmetry, CH_4^+ appears to be of D_{2d} symmetry, although the C—H distance in the ion is only about 0.01 Å larger than in CH_4 (*201*).

Parameters 5 and 6 refer to an ionic transition state, or a part thereof, respectively. Detailed information about transition states is meager even in the case of neutral species, and therefore the best course to adopt in the present instance is to suppose that for ionic species, the same principles will apply as in the case of neutral species. To this approximation, the gross properties of the transition state will be governed only by the type of potential energy surface over which the reaction takes place, while the detailed properties will be essentially those of the transition state in a similar reaction involving neutral species. This approximation justifies treating all transition states together (cf. Chapter 11) according to the type of potential energy surface but regardless of the charge (or indeed, regardless of the electronic state) of the species to which they refer. Such a procedure is made all the more necessary by the absence of any direct relation between a mass spectral experimental observable and a transition-state parameter, with the possible exception of the appearance potential AP, which could, in principle, yield some clues about the zero-point energy of the transition state if only it could be determined with sufficient accuracy.

Calculations There is one computational aspect, common to all highly excited species, that concerns parameters 2 and 3 and, to a lesser extent, parameters 4–6: As a result of the large excitation energy that can be deposited in the parent M^+ by most excitation processes, anharmonic effects are likely to be important in such highly excited M^+'s, and possibly also in their fragments. Hence at high energies, the function calculated in the harmonic oscillator approximation [$G^*(E)$ in the case of the transition state] will be an underestimate of the actual density function. Since, as we have seen, vibrational frequencies in ions are probably somewhat lower than in similar neutral molecules, this underestimate is further increased in the case of ions

[56] See, for example, King (*200*, p. 410).

by using frequencies of their neutral counterparts in the density-of-states calculations. This underestimate will be more important for $N(E)$ than for $G^*(E - E_0)$ because of the larger energy at which $N(E)$ is calculated; since $N(E)$ appears in the denominator in the expression for the rate constant, the result, when the energy is not too high, will be an *overestimate* of the rate constant, assuming of course it is not (accidentally) outweighed by the error in the critical energy. Unfortunately, the assignment of anharmonic parameters exceeds our present knowledge of molecular properties.[57] Calculations on a model system using Morse oscillators and harmonic oscillators with a cutoff were done by Forst and Prášil (*203*).

Recent (1965 and later) applications of the theory of mass spectra include calculations of the breakdown curves of the ions of methane (*204*), ethane (*12*), propane (*205,206*), cyclobutane (*207*), C_6 to C_8 alkanes (*208*), phenanthrene (*209*), and tetrahydronaphthalene (*210*), and of product distribution in two ion–molecule reactions known to proceed with the formation of a long-lived complex (*211*), $C_2H_4^+ + C_2H_4 \rightarrow \cdots$ and $C_2H_4^+ + C_2H_2 \rightarrow \cdots$. All of these calculations give reasonable agreement with experiment, showing the usefulness of the theory, although in one instance (*208*), it was necessary to adjust somewhat the critical energies E_0. These calculations were all done with the use of one of the better approximations for the density and sum of states, except for the two very large molecules, where the classical approximation was necessary for reasons of cost. Earlier claims (*251,252*) that the fragmentation of the methane and ethane ions does not obey the statistical theory of mass spectra, now appear to have been based merely on an instrumental artifact (*249*). For calculations on ions produced by surface ionization, see (*258*).

9. Kinetic energy of fragments and rotational effects

In the fragmentation of ions, one fragment bears a charge, and therefore its kinetic energy[58] can be measured mass spectrometrically either by applying a retarding potential or by exploiting the focusing properties of ions with kinetic energies.[59] Kinetic energy measurements in neutral systems would be vastly more difficult, and therefore the fragmentation of ions provides a convenient system on which to examine one aspect of unimolecular dissociations not otherwise easily accessible to measurement. Besides this fundamental interest, the kinetic energy of fragment ions is of importance in providing

[57] For some of the problems associated with the assignment of anharmonic parameters, see Prášil (*202*). Cf. also Chapter 6, Section 9.

[58] Let us specify clearly that kinetic energy of *relative* motion is meant here.

[59] For instrumental aspects of kinetic energy measurements, see Stanton and Monahan (*212*); Taubert (*213*); Durup and Heitz (*214*); Barber *et al.* (*215*); Franklin *et al.* (*216*) and references therein, Weinstein (*254*) and Franklin and Sharma (*255*).

indirect information about the shape of the potential energy surface over which the dissociation is taking place and in assessing the significance of appearance potentials (cf. Section 8).

Theory of kinetic energy release A release of appreciable kinetic energy on fragmentation can arise in several ways: (i) as a result of transition to a repulsive state of the singly charged ion, a case mentioned previously (footnote 54), or (ii) in the decomposition of a doubly charged ion (*217–220*). The two cases cannot always be easily distinguished,[60] but this is immaterial in the present context because in the discussion that follows we shall specifically *exclude* cases (i) and (ii) because they represent basically a nonstatistical mechanism of decomposition, to which the theory of mass spectra cannot be expected to apply. We are interested here only in a third way: (iii) relative kinetic energy of fragments released as a result of statistical partitioning of energy between the internal degrees of freedom of the transition state and the reaction coordinate. This is the same kinetic energy ε_t that appears in Eq. (4-9) and several other equations in Chapter 4.

In those cases where the statistical theory can apply, it is generally believed that if the potential energy surface for decomposition is of type 2, fragments will be formed with higher than normal translational energy. Evidence that this view is essentially correct comes from calculations on potential energy surfaces of bimolecular reaction done by Polanyi and co-workers (*222,223*). They find that if the potential barrier is located in the exit valley of the potential energy surface ("surface II" or "repulsive surface" in their nomenclature), most reagent energy becomes product translation. Such is the case with all type 1 and type 2 potential energy surfaces in unimolecular reactions, if we take the view that the surfaces shown in Figs. 3-1 and 3-2 are at all typical. In the contrary case ("surface I" or "attractive surface" of Polanyi), the corresponding unimolecular potential energy surface would have to have an unusual feature indeed, in that the stable configuration of the reactant would have to be located in an elongated bowl or dip and the exit valley would have to be at nearly a right angle to the long axis of the dip or bowl. Polanyi also finds that his "surface II" exists mostly for strongly endothermic reactions, and this is of course the case of all electronically adiabatic unimolecular dissociations.

These considerations do not say whether the entire activation energy for the back reaction (over a type 2 potential energy surface) will appear as excess kinetic energy of fragments formed in the forward process. Appreciable

[60] Appell *et al.* (*221*) ascribe the high kinetic energy of CH_3^+ produced in $C_3H_8^+ \rightarrow CH_3^+ + C_2H_5$ to cause (i), while H. Ehrhardt, in the discussion following the presentation of this paper, showed that it was more likely due to the formation of CH_3^+ from $C_3H_8^{++}$. For a more recent study of case (i), see (*253*).

activation energy for the back reaction exists only for ionic rearrangement processes, which are difficult to deal with (see end of Section 2). The vast majority of processes of decomposition of ions are simple fragmentations, where the activation energy for the back reaction is small, as was assumed throughout this chapter. In the following, we shall therefore confine our attention to fragmentations not involving activation energy for the reverse process.

Within the framework of the transition-state theory, a simple account of the kinetic energy release in fragmentation can be given along the lines of a treatment given originally by Taubert (*224*) and Klots (*225*) [see also (*226*)]. The validity of this approach depends on the transition state being one in which the fragments already begin separating from each other, i.e., where the top of the potential energy barrier is located at a fairly large interatomic distance. Such transition states are termed "loose" (Chapter 11, Section 2) and are generally associated with type 1 potential energy surfaces.

The object is to calculate the *average* relative kinetic energy of fragments, since this is the quantity that can be related to an experimental observable (after transformation from laboratory coordinates to center-of-mass coordinates). The situation where fragments are formed with relative kinetic energy ε_t can be looked upon as one particular reaction channel. Assuming that no redistribution of energy occurs once the reaction system has left the transition-state configuration in the direction of products, the fractional yield for the reaction channel with specified relative kinetic energy ε_t is

$$P(\varepsilon_t, E - E_0) = k(E, \varepsilon_t) \Big/ \int_0^{E-E_0} k(E, \varepsilon_t)\, d\varepsilon_t \tag{10-82}$$

where $P(\varepsilon_t, E - E_0)$ is the probability per unit energy that the relative kinetic energy of the fragments shall be ε_t when total available energy is $E - E_0$. On using Eqs. (4-10) and (4-18), we find that Eq. (10-82) yields

$$\begin{aligned} P(\varepsilon_t, E - E_0) &= N^*(E - E_0 - \varepsilon_t) \Big/ \int_0^{E-E_0} N^*(E - E_0 - \varepsilon_t)\, d\varepsilon_t \\ &= N^*(E - E_0 - \varepsilon_t)/G^*(E - E_0) \end{aligned} \tag{10-83}$$

where the notation of Eq. (4-20) is used for the integral in the denominator. The average relative translation (kinetic) energy is then

$$\langle \varepsilon_t \rangle = [1/G^*(E - E_0)] \int_0^{E-E_0} \varepsilon_t N^*(E - E_0 - \varepsilon_t)\, d\varepsilon_t \tag{10-84}$$

This equation can be integrated by parts if we take advantage of the fact that $N(E)$ is the derivative of $G(E)$ (Chapter 6, Section 1); thus

$N^*(E - E_0 - \varepsilon_t) = dG^*(E - E_0 - \varepsilon_t)/d(E - E_0 - \varepsilon_t) = -dG^*(E - E_0 - \varepsilon_t)/d\varepsilon_t$; the result is then simply[61] (with $y = E - E_0 - \varepsilon_t$)

$$\langle \varepsilon_t \rangle = \left[\int_0^{E - E_0} G^*(y)\, dy \right] \Big/ G^*(E - E_0) \tag{10-85}$$

This equation was obtained somewhat less directly by Lau and Lin (*227*) and by much more circuitous argument by LeRoy and Conway (*228*). It is not necessary to perform the tedious integration in the numerator of (10-85), for by applying the integration theorem of the Laplace transformation [Eq. (6-27)] twice in succession, we see immediately that

$$\int_0^{E - E_0} G(y)\, dy = \mathfrak{L}^{-1}\{\ell(s)/s^2\} \tag{10-86}$$

Hence instead of having to generate, for the purpose of numerical integration, $G^*(y)$ at a number of y values between $y = 0$ and $y = E - E_0$, we can obtain the integral in the numerator of (10-85) directly by simply taking $k = 2$ in the steepest-descent formula (6-89) or (6-91). Note also that both the numerator and denominator of (10-85) refer to the transition state, i.e., to the same collection of oscillators and rotors, which further simplifies the calculations. Calculations based on (10-86) have been reported (*227*). The result (10-85) is applicable to monoenergetic reactant ion. In an actual case, there will be a distribution of energies, and therefore the observable average relative translational energy will be

$$\langle\langle \varepsilon_t \rangle\rangle = \int_0^{\infty} \langle \varepsilon_t \rangle \mathscr{F}(E)\, dE \tag{10-87}$$

Taubert (*224*) used a rectangular distribution for $\mathscr{F}(E)$, and Klots (*225*) used $\mathscr{F}(E)$ of the form (10-70) to calculate from (10-87) average kinetic energies of (positive) fragment ions in a number of fragmentations produced by impact of 75-eV electrons. In a later study, Haney and Franklin (*229*) measured $\langle \varepsilon_t \rangle$ over a small range above the appearance potential of the fragment, and by a short extrapolation, obtained $\langle \varepsilon_t \rangle$ *at* the appearance potential, which they then compared with $\langle \varepsilon_t \rangle$ calculated from (10-84). The assumption made was that at the appearance potential AP, the decomposing ion is essentially monoenergetic, the excess energy ("kinetic shift") being $E - E_0 = \text{AP} - \Delta H_0$, where ΔH_0 is the dissociation energy obtained from thermodynamic data. The same procedure was later (*230*) applied to kinetic

[61] Since $G^*(y)$ here is a smooth-function approximation, the development leading to Eq. (10-85) makes use of the result $G^*(0) = 0$, although the exact result is $G^*(0) = 1$ (cf. legend to Fig. 6-1).

energies of negative ions produced by dissociative electron attachment processes, i.e., fragmentations of the type

$$M + e^- \rightarrow A^- + B \tag{10-88}$$

where A^- is a negative ion.

In these studies, the classical [Eq. (6-55)] or semiclassical [Eqs. (6-56) and (6-57)] approximation for the density function was used, and thus in order to obtain agreement between measured and calculated average translational energies, it was necessary to assume that the number of oscillators in the *parent* ion was less[62] than the actual number[63]; in other words, the calculated $\langle \varepsilon_t \rangle$ was too small. In a later, more detailed study, Spotz *et al.* (*232*) recalculated most of the previous results using better methods for the enumeration of states, including direct count, and, taking all the vibrations in the parent as pertinent degrees of freedom, obtained quite good agreement between calculated and observed values of $\langle \varepsilon_t \rangle$ (for $\langle \varepsilon_t \rangle$ between 3–4 kcal mole^{-1}), except at high observed $\langle \varepsilon_t \rangle$ (above 10 kcal mole^{-1}). The agreement was improved if it was assumed that not just one but three coordinates were transformed into internal translation in the transition state, i.e., by taking a three-dimensional reaction coördinate.

Nonstatistical effects The fact that the calculated $\langle \varepsilon_t \rangle$ always seems to come out too small, suggesting a smaller than expected number of pertinent degrees of freedom in the parent ion when $\langle \varepsilon_t \rangle$ is large, has led LeRoy[64] (*233,234*) to suggest that this is due to incomplete randomization of energy in the parent. This is of course a possibility, although, the dissociative electron attachment (10-88) probably occurs via a repulsive state, so that the transition-state theory cannot be expected to apply; yet the observed $\langle \varepsilon_t \rangle$ in these cases is not much different from that observed in the more familiar fragmentations of positive ions where, in most instances, transition-state theory does apply (the conspicuous exceptions are $CF_4 + e^- \rightarrow CF_3 + F^-$ and $BF_3 + e^- \rightarrow F^- + BF_3$ or $F_2^- + BF_2$, where the measured $\langle \varepsilon_t \rangle$ is $\sim$20 kcal mole^{-1}).

In the same connection, it is useful to cite Henglein (*236*), who has observed that in (positive) ion–molecule reactions, which are the reverse of the processes

[62] A similar result was obtained by Beynon *et al.* (*231*), who measured and calculated the kinetic energy released in a metastable decomposition. See (*256*) for thermochemical aspects.

[63] Since Eq. (10-83) refers to the density of states in the transition state, the number of vibrational degrees of freedom (or of pertinent degrees of freedom in general) to be used in (10-83) is always *one less* than whatever their number is assumed to be in the parent ion.

[64] It seems that incomplete randomization of energy occurs also in the "persistent complex" $C_4H_8^+$ formed in the ion–molecule reaction $C_2H_4^+ + C_2H_4$. Cf. Lee *et al.* (*235*).

considered earlier, there is a transition from complex (transition state in our terminology) formation to spectator stripping when the translational energy of the incident ion exceeds about 3 eV. This he interprets from the standpoint of unimolecular theory as meaning that at higher velocities, there is increasing difficulty in converting the translational energy into the internal energy of the "complex," whose lifetime now becomes too short for statistical redistribution to be efficient, and the products therefore are no longer scattered isotropically; instead, the impulsive spectator-stripping mechanism takes over.

Rotational effects Actually, the question of possible nonstatistical effects in these dissociations cannot be adequately discussed without dealing first with the contribution of rotational degrees of freedom to the rate constant. Nothing has been said so far about the decomposition of ions under the influence of an effective (vibrational plus rotational) potential because this is a subject that has always been, and probably can be, neglected when dealing with breakdown curves, which are fairly insensitive to a good deal of the detail, but assumes much larger importance in connection with more detailed measurements, like those of translational energies, and of threshold behavior in general (*182*).

The general principle involved is well-known: As in all unimolecular reactions, dissociations of ions must proceed with conservation of angular momentum. Therefore overall rotations (quantum number $\mathscr{J}$) will be adiabatic, as discussed in Chapter 7, and will have the effect of reducing the critical energy, i.e., increasing the rate. The individual rate constant k_{ij} in the scheme (10-13) will become $k_{ij}(E, \mathscr{J})$, where $\mathscr{J}$ is the angular momentum quantum number of the parent; the explicit form of this $\mathscr{J}$-dependent k_{ij} is given by Eq. (7-1). Other rotational degrees of freedom may function as pertinent degrees of freedom that contribute to the density of states.

The excitation process will leave the parent ion in a number of angular momentum states, and therefore the question arises as to what will be its rotational distribution function. Given the large mass of the nuclei with respect to electrons, we can expect that the removal of one electron, i.e., ionization, will not have much effect on the angular momentum of the nuclei, so that the excited ion should possess a distribution of rotational energies characteristic of the temperature of the neutral ground-state molecule from which it was produced. The distribution in the ion will be therefore the thermal equilibrium distribution $P(\mathscr{J})_e$ of Eq. (8-14), at least in the case of photoionization. In electron impact ionization, however, it is doubtful that the rotational distribution in the ion will be simply $P(\mathscr{J})_e$ because it is known (*42*, p. 726 ff.; *237*) that slow electrons (~ 10 eV) are capable of exciting molecular rotations, particularly if the target molecule possesses a permanent dipole.

Hence ions produced by electron impact will probably have a rotational distribution that may be appreciably distorted from $P(\mathscr{J})_e$. In addition, the relevant temperature, in the case of a conventional hot-filament electron gun, will be the temperature existing in the source, usually around 200–300°C.

These considerations apply to the decomposition of parent to first generation fragment. If the fragment decomposes further, the partitioning of angular momentum among the first generation fragments must be considered. Because of conservation laws, such partitioning cannot be statistical, and therefore in the few cases where the rotational distribution in fragments was determined, some extreme nonthermal distributions were observed.[65]

Chupka (*241*) [see also (*159*)] has studied the photoionization threshold of the process

$$CH_4^+ \rightarrow CH_3^+ + H \tag{10-89}$$

at room temperature and found that the thermal rotational energy of the methane ion did indeed contribute to the lowering (actually a smearing-out because of the thermal distribution of rotational quantum numbers) of the critical energy, in agreement with Eq. (7-13) if a thermal average value of $\mathscr{J}(\mathscr{J} + 1)$ is used (about 40 in the case cited). This experiment therefore lends credence to the concept of adiabatic rotations and their effect on the critical energy.

The relevance of adiabatic rotations, or, more precisely, of the effective potential, to the discussion of translational energies of fragments in type 1 dissociations, is due, first, to the appearance of an activation energy for the reverse reaction (which we called V_{eff} in Chapter 7) in the effective potential (cf. Fig. 7-1). Therefore when decompositions under an effective potential are considered, the total kinetic energy released is larger by V_{eff}, given by Eq. (7-12). The second consequence of the effective potential is that the critical energy E_0 becomes $\mathscr{J}$ dependent, which we indicate, following previous practice, by writing E_J for E_0. Hence $\langle \varepsilon_t \rangle$ of Eq. (10-85) also becomes $\mathscr{J}$ dependent. We have then

$$\langle \varepsilon_t(\mathscr{J}) \rangle = \left[\int_0^{E - E_J} G^*(y)\, dy \right] \Big/ G^*(E - E_J) \tag{10-90}$$

[65] A well-studied case is the photodissociation $H_2O + h\nu \rightarrow H + OH\,(^2\Sigma^+)$. Cf. Carrington (*238*). The dissociation seems to occur from a linear excited state of H_2O, so that, by the Franck–Condon principle, this state is formed with a considerable bending torque; as a result, the OH is "spun off" with a very nonthermal rotational distribution. For calculations of the rotational distribution in this system, see Horie and Kasuga (*239*); Pechukas *et al.* (*240*). Ground state H_2O is of course bent.

and the observable average translational energy is obtained by averaging (10-90) over $\mathcal{F}(E, \mathcal{J})$, i.e.,[66]

$$\langle\!\langle \varepsilon_t \rangle\!\rangle_{\mathcal{F}(E,\mathcal{J})} = \int_0^\infty \int_0^\infty \langle \varepsilon_t(\mathcal{J}) \rangle \mathcal{F}(E, \mathcal{J})\, dE\, d\mathcal{J} \tag{10-91}$$

If the distribution of the $\mathcal{J}$'s is thermal, then the $\mathcal{J}$ part of $\mathcal{F}(E, \mathcal{J})$ is simply $P(\mathcal{J})_e$ [Eq. (8-14)], provided $\mathcal{F}(E, \mathcal{J}) = \mathcal{F}'(E) \times P(\mathcal{J})_e$. The total kinetic energy release is[67]

$$(\mathrm{KE})_{\mathrm{total}} = \langle\!\langle \varepsilon_t(\mathcal{J}) \rangle\!\rangle_{\mathcal{F}(E,\mathcal{J})} + \langle V_{\mathrm{eff}} \rangle \tag{10-92}$$

assuming that all of $\langle V_{\mathrm{eff}} \rangle$ is used to increase the kinetic energy of the fragments and no part of it goes into their internal excitation. While probable, this is by no means certain, and therefore (10-92) gives only an upper limit for $(\mathrm{KE})_{\mathrm{total}}$. [See (*226*) when the distribution of $\mathcal{J}$'s is nonthermal.

Evaluation of (10-90) and (10-91) must be done numerically if an accurate approximation for $N^*(E)$ and $G^*(E)$ is used, but we can get some idea of the order of magnitude of $(\mathrm{KE})_{\mathrm{total}}$ if we use $\langle \varepsilon_t(\mathcal{J}) \rangle \simeq (E - E_J)/\mathsf{n}$ (cf. Exercise 3a); taking Eq. (7-13) for E_J, it is then easy to show that

$$(\mathrm{KE})_{\mathrm{total}} = \langle\!\langle \varepsilon_t \rangle\!\rangle + (1/\mathsf{n})[kT + (\mathsf{n} - 1)\langle V_{\mathrm{eff}} \rangle] \tag{10-93}$$

where n is the number of pertinent degrees of freedom (assumed to be vibrational) in the ion. If we use $\mathscr{C}$ of Table 7-2 for the process (10-89), and take $I = 5.33 \times 10^{-40}$ g cm^2 for the moment of inertia of CH_4^+, then $\langle V_{\mathrm{eff}} \rangle$ at 600°K, roughly the relevant temperature in electron impact experiments, is only 40 cal mole^{-1}. Thus (10-93) shows that even if the effective potential is considered, kinetic energy release is only some small 150 cal mole^{-1} larger (for $\mathsf{n} = 10$) than under purely vibrational potential, whereas it would have to be several kilocalories per mole larger in order to bring experimental and calculated translational energies into agreement. While the experimental values of the translational energies were obtained with electron impact ionization, and therefore the rotational distribution was probably not thermal, or perhaps not that corresponding to 600°K, Eq. (10-93) would give the correct answer only if the rotational temperature of CH_4^+ [i.e., T in Eq. (10-93)] were several thousand degrees, which seems far-fetched.

Implications for transition state We must therefore conclude that in those cases where the observed $\langle \varepsilon_t \rangle$ exceeds the value calculated from (10-85) or (10-91), the cause is more in the structure of the transition state itself than in the neglect of the centrifugal potential. The suggestion, already mentioned,

[66] As in Eq. (8-41), the infinite upper limit in (10-91) with respect to $\mathcal{J}$ is really inappropriate but is left here for simplicity.

[67] $\langle V_{\mathrm{eff}} \rangle$ is V_{eff} of Eq. (7-12) but with a thermal average substituted for $\mathcal{J}(\mathcal{J} + 1)$. For a two-dimensional rotor, $\langle \mathcal{J}(\mathcal{J} + 1) \rangle_{P(\mathcal{J})_e} = 2IkT/\hbar^2$.

that perhaps in these cases we are dealing with a three-dimensional reaction coordinate, means, insofar as the kinematics of the dissociation process is concerned, that one vibrational degree of freedom is transformed into relative translation of the fragments, and two other vibrational degrees of freedom of skeletal bending become the orbital rotation of the fragments. This is an extreme case of a "loose" transition state, one which has essentially become products.

How a "loose" transition state leads to an increase in the kinetic energy of the fragments can easily be seen by considering the classical result for $\langle \varepsilon_t \rangle$. With a conventional one-dimensional reaction coordinate, classically, $\langle \varepsilon_t \rangle = (E - E_0)/\mathsf{v}$ (Exercise 3), where v is the total number of vibrational degrees of freedom in the *parent* (not the transition state). If the reaction coordinate has two more dimensions, the classical result is then $\langle \varepsilon_t \rangle = (E - E_0)/(\mathsf{v} - 2)$, which of course means higher translational energy. But the same result will be obtained for a conventional one-dimensional reaction coordinate if it is assumed that four vibrations in the parent become rotations in the transition state. The reason is that, insofar as the exponent of the energy term in the density function expression is concerned (i.e., the denominator of the expression for $\langle \varepsilon_t \rangle$), two rotations count for one vibration, so that if four vibrations are replaced by four rotations, the effect is the same as if two vibrations were deleted (e.g., became intrinsic or orbital motion of the fragments).

Physically, the appearance of rotations and the disappearance of vibrations in the transition state signifies incipient fragmentation, and therefore only vibrational[68] degrees of freedom can be considered representative of a still-intact molecular framework. A reduction in the number of vibrational degrees of freedom in the transition state amounts to the molecule behaving as if it were smaller than it actually is, which is usually presumed to be the sign of a failure of energy randomization. In the present context, however, it is not really a failure of randomization that would be involved; rather, we have randomization over degrees of freedom that are, in part, almost those of the fragments. It appears therefore that "failure of randomization" is, to a degree, a question of interpretation.

An alternative treatment, which borrows from statistical theory of nuclear reactions (*242*), views a unimolecular fragmentation proceeding via a "loose" transition state in terms of the cross section for the reverse process.[69] This

[68] We exclude considerably "softened" vibrations which, as shown in Chapter 8, Section 7, are really equivalent to rotations.

[69] This point of view has been elaborated by Knewstubb (*243*). Note that he uses the term "transmission coefficient" in a somewhat different sense from that used in Chapter 4, Section 4: His meaning of the term refers essentially to E_J, i.e., the lowering of the critical energy due to the centrifugal potential. For his treatment of the ethane ion fragmentation, see (*257*).

approach, first used by Klots (*225*), is basically equivalent to transition-state theory (*244,245*), but it circumvents the ambiguity concerning the properties of the transition state, since rate is expressed in terms of the properties of the products which, unlike those of the transition state, are accessible to observation. An updated version (*246,247*) has been applied by Klots to the process (10-89) and to the related reactions $CH_4^+ \rightarrow CH_2^+ + H_2$ and $CH_3^+ \rightarrow CH_2^+ + H$. The calculated results are little different from those of the "orthodox" transition-state theory but were obtained without the need of estimating transition-state parameters, a worthwhile advantage. Cf. Exercise 4 in Chapter 6.

Exercises

1. Show that if lifetime measurements are made over an (infinitely) long time, the observed average rate constant over this time interval is then given by an equation similar to Eq. (9-39). *Hint:* Use Eq. (1-17) for $\langle \tau \rangle$ and then take $\langle k \rangle = \langle \tau \rangle^{-1}$. In this connection, "infinite" means in practice any time interval of the order of minutes. This exercise shows the connection between lifetime measurements and conventional measurements in thermal systems.

2. Using the classical form of $N(E)$ and $G(E)$, show for the process (10-25) that:
(a) $P(x, \mathscr{E})$ of Eq. (10-29) has a maximum at

$$x_{\max} = (\mathsf{v}_a - 1)\mathscr{E}/(\mathsf{v}_a + \mathsf{v}_b - 2)$$

where v_a is the number of vibrational degrees of freedom in A_1^+ and v_b is the number of vibrational degrees of freedom in B_1.
(b) $P(x, \mathscr{E})$ of Eq. (10-31) has a maximum at

$$x_{\max} = (\mathsf{v}_a - 1)\mathscr{E}/(\mathsf{v}_M - 2)$$

where v_M is the number of vibrational degrees of freedom in M^+. The assumption is that only vibrational degrees of freedom are involved in M^+, A_1^+, and B_1.

3. Let n be the number of pertinent (vibrational) degrees of freedom in the molecular ion. (a) Using the semiclassical approximations (6-61) and (6-62) for $N^*(E)$ and $G^*(E)$, show that Eq. (10-84) gives for $\langle \varepsilon_t \rangle$

$$\langle \varepsilon_t \rangle = \frac{E' + aE_z}{\mathsf{n}} - \left(\frac{aE_z}{E' + aE_z}\right)^{\mathsf{n}-1} \left(E' + \frac{aE_z}{\mathsf{n}}\right)$$

[cf. (*224*) and (*229*)], where $E' = E - E_0$. Compare with the result obtainable from Eq. (10-85). (b) Using the fully classical form of $\langle \varepsilon_t \rangle$

(i.e., taking $a = 0$ in the equation just given), show that

$$\langle\!\langle \varepsilon_t \rangle\!\rangle_{P(E_0,E)_{e,norm}} = kT$$

taking for $P(E_0, E)_{e,norm}$ the thermal distribution $P(E)_e$ [Eq. (8-13)] in its classical form, normalized in (E_0, ∞) [cf. Chapter 8, Section 5 and Fig. 9-4(a)]. *Hint:* Use $\Gamma(a, z)$ of Chapter 1, Exercise 2 in the expanded form

$$\Gamma(a, z) = z^{a-1}e^{-z}\left[1 + \frac{a-1}{z} + \frac{(a-1)(a-2)}{z^2} + \cdots\right]$$

4. Let j represent a particular reaction product channel, and let $k(j, E)$ represent the unimolecular rate constant for decomposition into channel j when total energy available to reactant is E. Show that the probability (10-31) is merely a special case of the more general relation

$$P(j, E) = k(j, E) \Big/ \sum_j k(j, E)$$

where $P(j, E)$ is the probability of reaction occuring into channel j when total available reactant energy is E.

References

1. H. M. Rosenstock, M. B. Wallenstein, A. L. Wahrhaftig, and H. Eyring, *Proc. Nat. Acad. Sci. U.S.* **38**, 667 (1952).
2. F. H. Field and J. L. Franklin, "Electron Impact Phenomena," 2nd ed. Academic Press, New York and London, 1970.
3. E. Lindholm, *Advan. Chem.* **58**, 1 (1966).
4. H. D. Beckey, "Field Ionization Mass Spectrometry." Pergamon, Oxford, 1971.
5. G. G. Meisels, *in* "Fundamental Processes in Radiation Chemistry" (P. Ausloos, ed.) Chapter 6, pp. 347 ff. Wiley (Interscience), New York, 1968.
6. G. V. Marr, "Photoionization Processes in Gases." Academic Press, New York and London, 1967.
7. P. J. Ausloos and S. G. Lias, *Annu. Rev. Phys. Chem.* **22**, 85 (1971).
8. F. S. Rowland, *in* "Molecular Beams and Reaction Kinetics" (C. Schlier, ed.), Course XLIV, Italian Physical Society. Academic Press, New York and London, 1970.
9. K. A. Krohn, N. J. Parks, and J. W. Root, *J. Chem. Phys.* **55**, 5785 (1971).
10. S. G. Lias, G. J. Collin, R. E. Rebbert, and P. Ausloos, *J. Chem. Phys.* **52**, 1841 (1970).
11. L. W. Sieck, *in* "Fundamental Processes in Radiation Chemistry" (P. Ausloos, ed.). Wiley (Interscience), New York, 1968.
12. Z. Prášil and W. Forst, *J. Phys. Chem.* **71**, 3166 (1967).
13. G. V. Karachevtsev and V. L. Tal'rose, *Kinet. Kataliz* **8**, 1 (1967) (English transl.).
14. G. V. Karachevtsev and V. L. Tal'rose, *Kinet. Kataliz* **8**, 447 (1967) (English transl.).
15. G. Tenschert and H. D. Beckey, *Int. J. Mass. Spectrom. Ion Phys.* **7**, 97 (1971), Fig. 1.

16. J. Tatarczyk and U. von Zahn, *Z. Naturforsch.* **20a**, 1708 (1965).
17. N. M. Rodiguin and E. N. Rodiguina, "Consecutive Chemical Reactions." Van Nostrand Reinhold, Princeton, New Jersey, 1964.
18. A. E. R. Westman and D. B. DeLury, *Can. J. Chem.* **34**, 1134 (1956).
19. H. M. Rosenstock, Ph.D. Thesis, Univ. of Utah (1952).
20. L. S. Kassel, "The Kinetics of Homogeneous Gas Reactions." Chem. Catalog, New York, 1932.
21. M. B. Wallenstein and M. Krauss, *J. Chem. Phys.* **34**, 929 (1961).
22. M. Hoare, *Mol. Phys.* **4**, 465 (1961).
23. R. V. Serauskas and E. W. Schlag, *J. Chem. Phys.* **42**, 3009 (1965).
24. T. C. Fry, "Probability and its Engineering Uses," 2nd ed. Van Nostrand Reinhold, Princeton, New Jersey, 1965.
25. Y. N. Lin and B. S. Rabinovitch, *J. Phys. Chem.* **74**, 1769 (1970).
26. M. Vestal, A. L. Wahrhaftig, and W. H. Johnston, Rep. ARL 62-426, Aeronautical Res. Lab., Office of Aerospace Res., USAF (1962).
27. P. Pechukas, J. C. Light, and C. Rankin, *J. Chem. Phys.* **44**, 794 (1966).
28. G. F. Busch and K. R. Wilson, *J. Chem. Phys.* **56**, 3626 (1972).
29. J. Roboz, "Introduction to Mass Spectrometry: Instrumentation and Techniques." Wiley (Interscience), New York, 1968.
30. J. A. Hipple, R. E. Fox, and E. U. Condon, *Phys. Rev.* **69**, 347 (1946).
31. W. A. Chupka, *J. Chem. Phys.* **30**, 191 (1959).
32. L. P. Hills, J. H. Futrell, and A. L. Wahrhaftig, *J. Chem. Phys.* **51**, 5255 (1969).
33. A. Kropf, E. M. Eyring, A. L. Wahrhaftig, and H. Eyring, *J. Chem. Phys.* **32**, 149 (1960).
34. U. Löhle and C. Ottinger, *Int. J. Mass Spectrom. Ion Phys.* **5**, 265 (1970).
35. D. L. Bunker, *J. Chem. Phys.* **40**, 1946 (1964).
36. J. L. Franklin, J. G. Dillard, H. M. Rosenstock, J. T. Herron, K. Draxl, and F. H. Field, "Ionization Potentials, Appearance Potentials and Heats of Formation of Gaseous Positive Ions." NSRDS-NBS No. 26, Washington, D.C., 1969.
37. V. I. Vedeneyev, L. V. Gurvich, V. N. Kondratiev, V. A. Medvedev, and Y. L. Frankevich, "Bond Energies, Ionization Potentials and Electron Affinities." St. Martin's Press, New York, 1966.
38. S. Trajmar, J. K. Rice, and A. Kuppermann, *Advan. Chem. Phys.* **18**, 15 (1970).
39. G. Herzberg, "Molecular Spectra and Molecular Structure," Vol. II, Spectra of Diatomic Molecules. Van Nostrand Reinhold, Princeton, New Jersey, 1950.
40. D. Villarejo, *J. Chem. Phys.* **48**, 4014 (1968).
41. W. Kauzmann, "Quantum Chemistry." Academic Press, New York and London, 1957.
42. H. S. W. Massey, E. H. S. Burhop, and H. B. B. Gilbody, "Electronic and Ionic Impact Phenomena," 2nd ed., Vol. II, Electron Collisions with Molecules and Photoionization. Oxford Univ. Press, London and New York, 1969.
43. M. Misakian, J. C. Pearl, and M. J. Mumma, *J. Chem. Phys.* **57**, 1891 (1972).
44. T. E. Sharp and H. M. Rosenstock, *J. Chem. Phys.* **41**, 3453 (1964).
45. R. Botter, V. H. Dibeler, J. A. Walker, and H. M. Rosenstock, *J. Chem. Phys.* **44**, 1271 (1966).
46. V. Čermák, *Coll. Czech. Chem. Commun.* **33**, 2739 (1968).
47. V. Čermák, *Advan. Mass Spectrom.* **4**, 697 (1968).
48. E. F. Gurnee and J. L. Magee, *J. Chem. Phys.* **26**, 1237 (1957).
49. D. R. Bates, *in* "Atomic and Molecular Processes" (D. R. Bates, ed.). Academic Press, New York and London, 1962.

50. L. Friedman and B. G. Reuben, *Advan. Chem. Phys.* **19**, 33 (1971).
51. M. J. Haugh and K. D. Bayes, *Phys. Rev.* **A2**, 1778 (1970.)
52. J. C. Lorquet, *Mol. Phys.* **10**, 493 (1966).
53. J. C. Lorquet and C. Cadet, *Int. J. Mass Spectrom. Ion Phys.* **7**, 245 (1971).
54. H. Eyring, J. Walter, and G. E. Kimball, "Quantum Chemistry." Wiley, New York, 1944.
55. F. Fiquet-Fayard, *J. Chim. Phys.* **64**, 320 (1967).
56. F. Fiquet-Fayard and P. M. Guyon, *Mol. Phys.* **11**, 17 (1966).
57. C. A. McDowell, *Trans. Faraday Soc.* **50**, 423 (1954).
58. R. L. Platzman, *Radiat. Res.* **17**, 419 (1962).
59. R. S. Berry, *J. Chem. Phys.* **45**, 1228 (1966).
60. J. N. Beardsley, *J. Phys. B* **1**, 349 (1968).
61. J. Berkowitz and W. A. Chupka, *J. Chem. Phys.* **51**, 2341 (1969).
62. L. Kerwin, P. Marmet, and J. D. Carette, *in* "Case Studies in Atomic Collision Physics" (E. W. McDaniel and R. R. C. McDowell, eds.). Vol. 1. North-Holland Publ., Amsterdam, 1969.
63. E. Bolduc, J. J. Quemener, and P. Marmet, *Can. J. Phys.* **49**, 3095 (1971).
64. E. Bolduc, J. J. Quemener, and P. Marmet, *J. Chem. Phys.* **57**, 1957 (1972).
65. P. Marmet, private communication (1971).
66. H. Ehrhardt, F. Linder, and T. Tekaat, *Advan. Mass Spectrom.* **4**, 705 (1968).
67. D. W. Turner, A. D. Baker, and C. R. Brundle, "Photoelectron Spectroscopy." Wiley (Interscience), New York, 1970.
68. D. W. Turner, *Ann. Rev. Phys. Chem.* **21**, 107 (1970).
69. K. Levsen and H. D. Beckey, *Int. J. Mass Spectrom. Ion Phys.* **9**, 51, 63 (1972).
70. J. R. Hiskes, *Phys. Rev.* **122**, 1207 (1961).
71. J. C. Lorquet and G. G. Hall, *Mol. Phys.* **9**, 29 (1965).
72. V. Santoro and G. Spadaccini, *J. Phys. Chem.* **73**, 462 (1969).
73. J. C. Lorquet and K. Hirota, *J. Phys. Chem.* **73**, 463 (1969).
74. G. Spiteller and M. Spiteller-Friedman, *Ann. Chem.* **690**, 1 (1965).
75. G. R. Lester, *Advan. Mass Spectrom.* **1**, 287 (1959).
76. G. Alexandru, *Int. J. Mass Spectrom. Ion Phys.* **4**, 1 (1970).
77. A. A. Ovchinnikov, *Theoret. Exp. Chem.* **1**, 44 (1965) (Engl. transl.).
78. A. A. Ovchinnikov, *Theoret. Exp. Chem.* **2**, 8 (1966) (Engl. transl.).
79. J. C. Lorquet, *J. Chim. Phys.* **57**, 1085 (1960).
80. L. d'Or, J. C. Lorquet, and J. Momigny, *Bull. Classe Sci. Acad. Roy. Belg.* **46**, 650 (1960).
81. J. C. Lorquet, *Mol. Phys.* **9**, 101 (1965).
82. K. Hirota and Y. Niwa, *J. Phys. Chem.* **72**, 5 (1968).
83. S. Meyerson, *J. Chem. Phys.* **42**, 2181 (1965).
84. J. C. Lorquet, A. J. Lorquet, and J. C. Leclerc, *Advan. Mass Spectrom.* **4**, 569 (1968).
85. K. Hirota, I. Fujita, M. Yamamoto, and Y. Niwa, *J. Phys. Chem.* **74**, 410 (1970).
86. J. Collin, M. J. Franskin, and D. Hyatt, *Bull. Soc. Roy. Sci. Liège* **5–6**, 318 (1967).
87. J. Collin, M. J. Franskin, and D. Hyatt, *Bull. Soc. Roy. Sci. Liège* **11–12**, 744 (1966).
88. J. Collin and M. J. Franskin, *Bull. Soc. Roy. Sic. Liège* **3–4**, 267, 285 (1966).
89. J. Collin, *Bull. Soc. Chim. Belg.* **67**, 549 (1958).
90. P. E. Cade, R. F. W. Bader, and J. Pelletier, *J. Chem. Phys.* **54**, 3517 (1971).
91. R. Daudel and A. Pullman, eds., "Aspects de la chimie quantique contemporaine." Editions du CNRS, Paris, 1971.

92. M. D. Newton, F. P. Boer, and W. N. Lipscomb, *J. Amer. Chem. Soc.* **88**, 2367 (1966).
93. R. F. W. Bader, *Mol. Phys.* **3**, 137 (1960).
94. R. F. W. Bader, *Can. J. Chem.* **40**, 1164 (1962).
95. R. F. W. Bader and A. D. Bandrauk, *J. Chem. Phys.* **49**, 1666 (1968).
96. J. C. Lorquet, *Mol. Phys.* **10**, 489 (1966).
97. L. Salem, *Chem. Phys. Lett.* **3**, 99 (1969).
98. L. Salem and J. S. Wright, *J. Amer. Chem. Soc.* **91**, 5947 (1969).
99. R. B. Woodward and R. Hoffman, "The Conservation of Orbital Symmetry." Verlag Chemie—Academic Press, New York, 1970.
100. J. C. Leclerc and J. C. Lorquet, *J. Phys. Chem.* **71**, 787 (1967).
101. J. C. Lorquet, A. J. Lorquet, J. C. Leclerc, and C. Cadet, *Advan. Mass Spectrom.* **5**, 43 (1971).
102. J. C. Leclerc, Ph.D. Thesis, Univ. de Liège (1970).
103. M. F. Guest, J. N. Murrell, and J. B. Pedley, *Mol. Phys.* **20**, 81 (1971).
104. A. J. Lorquet and J. C. Lorquet, *J. Chem. Phys.* **49**, 4955 (1968).
105. A. J. Lorquet, *J. Phys. Chem.* **74**, 895 (1970).
106. A. D. Walsh, *J. Chem. Soc.* 2325 (1953).
107. J. C. Lorquet, *Discuss. Faraday Soc.* **35**, 83 (1963).
108. S. D. Peyerimhoff and R. J. Buenker, *J. Chem. Phys.* **49**, 312 (1968).
109. P. N. Rylander and S. Meyerson, *J. Amer. Chem. Soc.* **78**, 5799 (1956).
110. J. D. Petke and L. J. L. Whitten, *J. Amer. Chem. Soc.* **90**, 3338 (1968).
111. S. D. Peyerimhoff and R. J. Buenker, *J. Chem. Phys.* **51**, 2528 (1969).
112. M. Henchman, *in* "Ion-Molecule Reactions" (J. L. Franklin, ed.), Vol. 1. Plenum Press, New York, 1972.
113. N. J. Turro, "Molecular Photochemistry." Benjamin, New York, 1965.
114. R. J. Buenker and S. D. Peyerimhoff, *J. Chem. Phys.* **53**, 1368 (1970).
115. K. M. A. Refaey and W. A. Chupka, *J, Chem. Phys.* **48**, 5205 (1968).
116. H. M. Rosenstock and M. Krauss, *Advan. Mass Spectrom.* **2**, 257 (1964).
117. M. J. S. Dewar, and S. D. Worley, *J. Chem. Phys.* **50**, 654 (1969).
118. F. Meyer and A. G. Harrison, *J. Chem. Phys.* **43**, 1778 (1965).
119. M. L. Vestal, *in* "Fundamental Processes in Radiation Chemistry" (P. Ausloos, ed.). Wiley (Interscience), New York, 1968.
120. I. Szabo and P. J. Derrick, *Int. J. Mass Spectrom. Ion Phys.* **7**, 55 (1971).
121. N. D. Coggeshall, *J. Chem. Phys.* **37**, 2167 (1962).
122. J. C. Schug, *J. Chem. Phys.* **40**, 1283 (1964).
123. H. M. Rosenstock, *Advan. Mass Spectrom.* **4**, 253 (1968).
124. C. Ottinger, *Z. Naturforsch.* **22a**, 20 (1967).
125. K. F. Freed and J. Jortner, *J. Chem. Phys.* **52**, 6272 (1970).
126. W. M. Gelbart, K. G. Spears, K. F. Freed, J. Jortner, and S. A. Rice, *Chem. Phys. Lett.* **6**, 345 (1970).
127. A. Nitzan and J. Jortner, *J. Chem. Phys.* **58**, 1355 (1971).
128. W. M. Gelbart and S. A. Rice, *J. Chem. Phys.* **50**, 4775 (1969).
129. W. M. Gelbart, K. F. Freed, and S. A. Rice, *J. Chem. Phys.* **52**, 2460 (1970).
130. J. C. Lorquet, S. G. ElKolmoss, and J. L. Magee, *J. Chem. Phys.* **37**, 1991 (1962).
131. K. Funabashi and J. L. Magee, *in* "Actions chimiques et biologiques des radiations" (M. Haïssinsky, ed.), Huitième Série. Masson, Paris, 1965.
132. R. D. Levine, *J. Chem. Phys.* **44**, 2035 (1966).
133. R. H. Partridge, *J. Chem. Phys.* **52**, 2485, 2501 (1970).

134. W. A. Chupka and J. Berkowitz, *J. Chem. Phys.* **47**, 2921 (1967).
135. W. A. Chupka and E. Lindholm, *Arkiv. Fys.* **25**, 349 (1963).
136. H. M. Rosenstock, V. H. Dibeler, and F. N. Harllee, *J. Chem. Phys.* **40**, 591 (1964).
137. J. L. Occolowitz, *J. Amer. Chem. Soc.* **91**, 5202 (1969).
138. G. Khodadadi, R. Botter, and H. M. Rosenstock, *Int. J. Mass Spectrom. Ion Phys.* **3**, 397 (1969).
139. H. M. Grubb and S. Meyerson, *in* "Mass Spectrometry of Organic Ions" (F. W. McLafferty, ed.). Academic Press, New York and London, 1963.
140. V. Hanuš and Z. Dolejšek, *Coll. Czech. Chem. Commun.* **28**, 652 (1963).
141. K. R. Jennings and J. H. Futrell, *J. Chem. Phys.* **44**, 4315 (1966).
142. F. W. McLafferty and W. T. Pike, *J. Amer. Chem. Soc.* **89**, 595 (1967).
143. R. G. Cooks and D. H. Williams, *Chem. Commun.* 627 (1968).
144. L. Friedman, F. A. Long, and M. Wolfsberg, *J. Chem. Phys.* **27**, 613 (1957).
145. W. A. Chupka and J. Berkowitz, *J. Chem. Phys.* **32**, 1546 (1960) (amines).
146. B. Steiner, C. F. Giese, and M. G. Inghram, *J. Chem. Phys.* **34**, 189 (1961) (alkanes).
147. W. Forst, *Chem. Rev.* **71**, 339 (1971), Table V.
148. C. Lifshitz and F. A. Long, *J. Phys. Chem.* **69**, 3746 (1965).
149. H. Ehrhardt and O. Osberghaus, *Z. Naturforsch.* **15a**, 575 (1960).
150. W. A. Chupka, *J. Chem. Phys.* **54**, 1936 (1971).
151. D. P. Stevenson, *Radiat. Res.* **10**, 610 (1959).
152. D. P. Stevenson and D. O. Schissler, *in* "Actions chimiques et biologiques des radiations" (M. Haïssinsky, ed.), 5ème Sér. Masson, Paris, 1961.
153. I. Howe and D. H. Williams, *J. Amer. Chem. Soc.* **90**, 5461 (1968).
154. E. P. Wigner, *Phys. Rev.* **73**, 1002 (1948).
155. G. H. Wannier, *Phys. Rev.* **100**, 1180 (1956).
156. S. Geltman, *Phys. Rev.* **102**, 771 (1956).
157. M. R. H. Rudge and M. J. Seaton, *Proc. Roy. Soc. (London)* **A283**, 262 (1965).
158. M. R. H. Rudge, *Rev. Mod. Phys.* **40**, 564 (1968).
159. P. M. Guyon and J. Berkowitz, *J. Chem. Phys.* **54**, 1814 (1971).
160. J. D. Morrison, *J. Chem. Phys.* **21**, 1767 (1953).
161. W. A. Chupka and M. Kaminsky, *J. Chem. Phys.* **35**, 1991 (1961).
162. L. J. Kieffer and G. H. Dunn, *Rev. Mod. Phys.* **38**, 1 (1966).
163. P. Marchand, C. Paquet, and P. Marmet, *Phys. Rev.* **180**, 123 (1969).
164. R. Stockbauer and M. G. Inghram, *J. Chem. Phys.* **54**, 2242 (1971).
165. G. G. Meisels, C. T. Chen, G. B. Giessner, and R. H. Emmel, *J. Chem. Phys.* **56**, 793 (1972).
166. L. G. Christophorou, "Atomic and Molecular Radiation Physics." Wiley (Interscience), New York, 1972.
167. J. H. Futrell, *J. Amer. Chem. Soc.* **81**, 5921 (1959).
168. J. H. Futrell, *J. Phys. Chem.* **64**, 1634 (1960).
169. L. M. Dorfman and M. C. Sauer, *J. Chem. Phys.* **32**, 1886 (1960).
170. J. H. Futrell, *J. Chem. Phys.* **35**, 353 (1961).
171. G. G. Meisels, *Advan. Chem.* **58**, 243 (1966).
172. Z. Prášil, *Coll. Czech. Chem. Commun.* **31**, 3252 (1966).
173. H. Wincel, *Nukleonika* **7**, 25 (1962).
174. Z. Prášil, *Coll. Czech. Chem. Commun.* **32**, 3105 (1967).
175. Z. Prášil, K. Vacek, and J. Bednář, *Coll. Czech. Chem. Commun.* **30**, 2693 (1965).
176. I. Santar, *Coll. Czech. Chem. Commun.* **33**, 1 (1968).

177. I. Santar, *Coll. Czech. Chem. Commun.* **34**, 1, 311 (1969).
178. R. L. Platzman, *Vortex* **23**, 1 (1962).
179. I. Santar, *Proc. 10th Czech. Annu. Meeting Radiat. Chem., Prague* **1**, 51 (1971).
180. N. N. Tunitskii, S. E. Kapriyanov, and A. A. Perov, *Izv. Akad. Nauk SSR* (*Chem.*) 1945 (1962) (English Transl.).
181. M. V. Gur'ev, *Dokl. Akad. Nauk SSSR* **136**, 856 (1961) (English transl.).
182. C. Ottinger, *Z. Naturforsch.* **22a**, 20 (1967).
183. I. Hertel and C. Ottinger, *Z. Naturforsch.* **22a**, 40, 1141 (1967).
184. B. Andlauer and C. Ottinger, *J. Chem. Phys.* **55**, 1471 (1971).
185. B. Andlauer and C. Ottinger, *Z. Naturforsch.* **27a**, 293 (1972).
186. J. H. Beynon, "Mass Spectrometry and its Applications to Organic Chemistry." Elsevier, Amsterdam, 1960.
187. F. W. McLafferty, *in* "Topics in Organic Mass Spectrometry" (A. L. Burlingame, ed.). p. 223. Wiley (Interscience), New York, 1970.
188. H. Budzikiewicz, C. Djerassi, and D. H. Williams, "Mass Spectrometry of Organic Compounds." Holden-Day, San Francisco, California, 1967.
189. D. H. Williams, *Advan. Mass Spectrom.* **5**, 569 (1971).
190. U. Löhle and C. Ottinger, *J. Chem. Phys.* **51**, 3097 (1969).
191. J. H. Beynon, *Advan. Mass Spectrom.* **4**, 123 (1968).
192. P. Natalis, *in* "Mass Spectrometry" (E. I. Reed, ed.). p. 379. Academic Press, New York, 1965.
193. S. Meyerson and A. W. Weitkamp, *Org. Mass Spectrom.* **1**, 659 (1968).
194. R. G. Cooks, *Org. Mass Spectrom.* **2**, 481 (1969).
195. M. Krauss and V. H. Dibeler, *in* "Mass Spectrometry of Organic Ions" (F. W. McLafferty, ed.). p. 117. Academic Press, New York, 1963.
196. C. A. McDowell, *in* "Mass Spectrometry" (C. A. McDowell, ed.). p. 506. McGraw-Hill, New York, 1963.
197. J. E. Collin, *in* "Mass Spectrometry" (R. I. Reed, ed.). p. 201. Academic Press, New York, 1965.
198. H. D. Beckey and F. J. Comes, *in* "Topics in Organic Mass Spectrometry" (A. L. Burlingame, ed.). p. 1. Wiley (Interscience), New York, 1970.
199. H. D. Hagstrum and J. T. Tate, *Phys. Rev.* **59**, 354 (1941).
200. G. W. King, "Spectroscopy and Molecular Structure." Holt, New York, 1964.
201. F. A. Grimm and J. Godoy, *Chem. Phys. Lett.* **6**, 336 (1970).
202. Z. Prášil, *Advan. Mass Spectrom.* **5**, 53 (1971).
203. W. Forst and Z. Prášil, *J. Chem. Phys.* **53**, 3065 (1970).
204. L. P. Hills, M. L. Vestal, and J. H. Futrell, *J. Chem. Phys.* **54**, 3834 (1971).
205. M. L. Vestal, *J. Chem. Phys.* **43**, 1356 (1965).
206. M. Vestal and J. H. Futrell, *J. Chem. Phys.* **52**, 978 (1970).
207. C. Lifshitz and T. O. Tiernan, *J. Chem. Phys.* **55**, 3555 (1971).
208. J. C. Tou, L. P. Hills, and A. L. Wahrhaftig, *J. Chem. Phys.* **45**, 2129 (1966).
209. P. Nounou, *Advan. Mass. Spectrom.* **4**, 551 (1968).
210. H. Wincel, Z. Kecki, and S. Minc, *Proc. 2nd Tihany Symp. Radiat. Chem.* (J. Dobo and P. Hedvig, eds.). Publ. House of the Hungarian Acad. of Sci., Budapest, 1967.
211. S. E. Butrill, Jr., *J. Chem. Phys.* **52**, 6174 (1970).
212. H. E. Stanton and J. E. Monahan, *J. Chem. Phys.* **41**, 3694 (1964).
213. R. Taubert, *Z. Naturforsch.* **19a**, 484 (1964).
214. J. Durup and F. Heitz, *J. Chim. Phys.* **61**, 470 (1964).
215. M. Barber, K. R. Jennings, and R. Rhodes, *Z. Naturforsch.* **22a**, 15 (1967).

216. J. L. Franklin, P. M. Hierl, and D. A. Whan, *J. Chem. Phys.* **47**, 3148 (1967).
217. R. Fuchs and R. Taubert, *Z. Naturforsch.* **19a**, 1181 (1964).
218. R. Fuchs and R. Taubert, *Z. Naturforsch.* **20a**, 823 (1965).
219. M. Barber and K. R. Jenkins, *Z. Naturforsch.* **24a**, 143 (1969).
220. R. Fuchs, *Z. Naturforsch.* **21a**, 2069 (1966).
221. J. Appell, J. Durup, and F. Heitz, *Advan. Mass Spectrom.* **3**, 457 (1966).
222. M. H. Mok and J. C. Polanyi, *J. Chem. Phys.* **53**, 4588 (1970).
223. J. C. Polanyi, *Accounts Chem. Res.* **5**, 161 (1972).
224. R. Taubert, *Z. Naturforsch.* **19a**, 911 (1964).
225. C. E. Klots, *J. Chem. Phys.* **41**, 117 (1964).
226. S. A. Safron, N. D. Weinstein, D. R. Herschbach, and J. C. Tully, *Chem. Phys. Lett.* **12**, 564 (1972).
227. K. H. Lau and S. H. Lin, *J. Phys. Chem.* **75**, 2458 (1971).
228. R. L. LeRoy and D. C. Conway, *J. Chem. Phys.* **56**, 5199 (1972).
229. M. H. Haney and J. L. Franklin, *J. Chem. Phys.* **48**, 4093 (1968).
230. J. U. DeCorpo, B. A. Bafus, and J. L. Franklin, *J. Chem. Phys.* **54**, 1592 (1971).
231. J. H. Beynon, J. A. Hopkinson, and G. R. Lester, *Int. J. Mass Spectrom.* **1**, 343 (1968).
232. E. L. Spotz, W. A. Seitz, and J. L. Franklin, *J. Chem. Phys.* **51**, 5142 (1969).
233. D. J. LeRoy, *J. Chem. Phys.* **53**, 846 (1970).
234. D. J. LeRoy, *J. Chem. Phys.* **55**, 1476 (1971).
235. A. Lee, R. L. LeRoy, Z. Herman, R. Wolfgang, and J. C. Tully, *Chem. Phys. Lett.* **12**, 569 (1972).
236. A. Henglein, *J. Chem. Phys.* **53**, 458 (1970).
237. K. Takayanagi and Y. Itikawa, *Advan. At. Mol. Phys.* **6**, 105 (1970).
238. T. Carrington, *J. Chem. Phys.* **41**, 2012 (1964).
239. T. Horie and T. Kasuga, *J. Chem. Phys.* **40**, 1683 (1964).
240. P. Pechukas, J. C. Light, and C. Rankin, *J. Chem. Phys.* **44**, 794 (1966).
241. W. A. Chupka, *J. Chem. Phys.* **48**, 2337 (1968).
242. V. Weisskopf, *Phys. Rev.* **52**, 295 (1937).
243. P. F. Knewstubb, *Int. J. Mass Spectrom. Ion Phys.* **6**, 217 (1971).
244. K. Morokuma, B. C. Eu, and M. Karplus, *J. Chem. Phys.* **51**, 5193 (1969).
245. D. G. Truhlar, *J. Chem. Phys.* **53**, 2041 (1970).
246. C. E. Klots, *J. Phys. Chem.* **75**, 1527 (1971).
247. C. E. Klots, *Z. Naturforsch.* **27a**, 553 (1972).
248. R. G. Cooks, D. W. Setser, K. Jennings and S. Jones, *Int. J. Mass Spectrom. Ion Phys.* **7**, 493 (1971).
249. R. Stockbauer, *J. Chem. Phys.* **58**, 3800 (1973). The author is grateful to Dr. Stockbauer for a preprint.
250. W. A. Lathan, L. A. Curtiss and J. A. Pople, *Mol. Phys.* **22**, 1081 (1972).
251. B. Brehm and E. von Puttkamer, *Z. Naturforsch.* **22a**, 8 (1967).
252. E. von Puttkamer, *Z. Naturforsch.* **25a**, 1062 (1970).
253. E. G. Jones, J. H. Beynon and R. G. Cooks, *J. Chem. Phys.* **57**, 3207 (1972).
254. N. D. Weinstein, *J. Chem. Phys.* **58**, 408 (1973).
255. J. L. Franklin and D. K. S. Sharma, *J. Chem. Phys.* **58**, 409 (1973).
256. E. G. Jones, J. H. Beynon and R. G. Cooks, *J. Chem. Phys.* **57**, 2652 (1972).
257. P. F. Knewstubb, *JCS Faraday II* **68**, 1196 (1972).
258. N. D. Potekhina, *Theoret. Exp. Chem.* **6** 276 (1970) (English transl.)
259. L. Radom, J. A. Pople, V. Buss, and P. von R. Schleyer, *J. Amer. Chem. Soc.* **94**, 311 (1972).

CHAPTER 11

Transition States in Unimolecular Reactions

In the early days of transition-state theory, it was felt that the full potential energy surface would eventually become known for every reaction, and that from the various dimensions and curvatures of the surface in the vicinity of the saddle point, all vibrational frequencies and moments of inertia of the transition state would be calculable in a straightforward manner by matrix technique. It was in this sense that the theory was viewed as an "absolute" theory. Unfortunately, this promise has not been realized because potential energy surface calculations turned out to be much more arduous than anticipated, and even now the potential energy surface for the simplest of all reactions, $H + H_2 \rightarrow H_2 + H$, is still in some doubt. The picture is not much brighter for unimolecular potential energy surfaces, as we have seen in Chapter 3 and Chapter 10, Section 4.

Consequently, much empiricism had to be used, and is still being used, to formulate the structure of transition states in unimolecular reactions, but this writer feels that, in an effort to make calculated results agree with experiment in specific isolated cases, there has been altogether too much tampering with the various parameters of the transition state. Some authors (*1,2*)[1] now think that the predictive power of the statistical theory of unimolecular reactions is uncertain because of the large number of adjustable

[1] The reference list for this chapter can be found on p. 389.

parameters. Actually, the formulation of transition-state structures can be put on a rational basis to a considerable extent; then the range of adjustment of the remaining undetermined parameters is usually found to be fairly small, if the simple rules outlined in Section 1 are followed.

The ultimate test of transition-state structure assignment is of course the agreement (or the lack of it) between the calculated and measured rates. This test involves, however, *all* the assumptions of the theory, not just the assumptions regarding the transition state. In some instances, the isotope effect on rates provides a useful additional test that can be more closely identified with the properties of the transition state. Whatever the test, the transition-state models are inspired by, and tested on, kinetic data, and therefore the empiricism that enters into their formulation should come, as much as possible, from a field outside chemical kinetics if the application of the theory is not to be merely a demonstration of the kineticist's skill at the art of adjusting parameters.

The range of adjustment, such as there is after these precautions, is further reduced if the transition-state structure is based on the treatment of a whole *collection* of data on a given type of unimolecular reaction involving similar reactant molecules, members of a homologous series, and/or those which are isotopically substituted. The discussion of transition states given in this chapter is therefore slanted, as a far as possible, in favor of work where one basic transition state has been demonstrated to account for detailed experimental rate data, thermal as well as nonthermal, on groups of homologous compounds, both normal and isotopically substituted. The emphasis is therefore on depth rather than breadth.[2]

Plan of chapter Section 1 discusses general principles that are independent of reaction type. Included here are also those aspects of the isotope effect that have a bearing on the transition state and its structure. Sections 2–4 discuss transition states for selected reactions of types 1, 2, and 3, respectively.

1. General remarks; isotope effect

In specific terms, we are concerned in this chapter with model-dependent parameters that enter the basic equation (7-1):

$$k(E, \mathscr{J}) = \alpha G^*(E - E_J)/hN(E) \tag{11-1}$$

that is, everything in the numerator of (11-1): the reaction path degeneracy α because it depends on the spatial conformation of atoms in the transition state (Chapter 4, Section 5); the integrated density of states $G^*(z)$ because it

[2] A broad survey may be found in Bamford and Tipper (*3*). See also Robinson and Holbrook (*5*).

depends on the vibrational frequencies (and/or moments of inertia if rotations are involved) in the transition state; the critical energy E_0 because it is a function of the zero-point energy of the transition state (Chapter 8, Section 7); and E_J because it depends on the interatomic distances in the transition state. Thus a rather detailed knowledge of the transition state is necessary.

Such "knowledge" is of a rather special kind because the transition state does not "exist" in the normal sense of the word, and its properties are not accessible to experimental observation, although in principle they are calculable. Thus if the potential energy surface for a reaction of chemical interest were known in sufficient detail, the properties of the transition state would be calculable rigorously; however, this is not the case in practice because extensive potential energy surface calculations for polyatomic reactants are not feasible at present. A useful procedure is to note that the transition state, as we insisted on in Chapter 4, Sections 2 and 3, is merely a convenient device to circumvent the problems connected with determining the volume of phase space relevant to the decomposition process. This volume can be described and determined on the basis of diverse postulates, some of which can be related to experimental observables.

Essence of transition state formulation We shall adopt the point of view, which we believe to preserve the essence of the heuristic value of transition-state formulation, that the transition state is basically similar to a comparable stable molecular system, except in degrees of freedom involving atoms more or less directly concerned in the chemical transformation. In these degrees of freedom, the transition state is considered to represent a structure intermediate between reactants and products, and the usual rules relating geometry to observable properties of stable molecules are assumed to apply. Formally, this can be expressed by treating a transition state consisting of N atoms as a problem in $3N - 3$ degrees of freedom,[3] separable[4] in a small region near the top of the potential energy saddle point; the internal potential and kinetic energies can then be expressed in terms of normal coordinates with normal mode frequencies, in quite the same way as in the case of a stable molecule, except that one vibrational degree of freedom is associated with displacement from a potential *maximum* and gives rise to an imaginary normal mode frequendy and hence a negative force constant (*4*, p. 115 ff.).

[3] Since they have no effect on the rate, we eliminate immediately three degrees of freedom due to translation of the center of mass.

[4] The separability of the degree of freedom that ultimately becomes the reaction coordinate is a *necessary* assumption (cf. Chapter 4, Section 1); the assumption that the other degrees of freedom are also separable (i.e., are normal modes) is made merely for convenience to reduce the problem of calculating $G^*(z)$ to the simpler problem of calculating the integrated density of states for a collection of independent oscillators. This has been discussed previously in Chapter 5, Section 4.

The imaginary frequency[5] is the property characteristic of any unstable structure located at the top of a potential energy barrier, and therefore is also characteristic of the transition state.

A somewhat different formulation has been given by Bunker and Pattengill (*12*) [see also (*13*)], who describe the volume in phase space relevant to the decomposition process in terms of a transition state ("critical configuration" may be a better term, to avoid confusion) characterized by a minimum in the local density of states along a reaction coordinate q, such that (assuming $\mathscr{J} = 0$ for simplicity)

$$\partial N^*(E - E_0)/\partial q = 0$$

Note that here we abandon the assumption that the reaction coordinate is a separable degree of freedom, for the reaction coordinate is now coupled to the other degrees of freedom. This formulation is inspired by Bunker's results on the computer simulation of actual reactions (cf. Chapter 2, Section 5), which reveal that the randomness of molecular lifetimes—the basic postulate of RRKM theory—supposes as a prerequisite the existence of a "bottleneck" in phase space. Evaluation of the local minimum of states along a reaction coordinate requires of course the knowledge of the coupling between the reaction coordinate and the other degrees of freedom; in short, closer attention must be paid to molecular dynamics. This considerably complicates the calculation of the rate constant, so that so far this formulation has been applied only to the dissociation of quasilinear molecules (*12–15*), where q is simply the relevant internuclear distance. We shall therefore stick to the more traditional formulation, where the transition state is assumed to be localized at the extremum of the potential; this is perhaps more crude, but also at the same time more widely (and more easily) applicable.

The reaction coordinate, in the traditional formulation, corresponds to the normal mode coordinate with imaginary frequency. Since the ultimate purpose of the present treatment is to obtain for the transition state the density of states of its normal modes having real (positive or zero) frequencies, the normal mode coordinate of imaginary frequency may be ignored since it has already been considered explicitly in the treatment leading to the basic equation (11-1).

Within these very general guidelines, it is usually possible to assume some sort of structure for the transition state, and once this is done, one could then calculate the normal mode vibrational frequencies of the transition state by the F–G matrix technique (*16*, Chapter 4). Since the complexity of these calculations increases rapidly with molecular size, this procedure is

[5] For further discussion on the subject of imaginary frequency, see Johnston (*49*, p. 84–87); Weston (*6*); Laidler (*7*); LeRoy (*8*); Yao and Zwolinski (*9*); Ridley *et al.* (*10*); Yao and Zwolinski (*11*).

probably worthwhile only in cases where only few atoms are involved,[6] for while the calculations of the frequencies are rigorous, the result is of course no better than the assumptions that went into the postulated transition state. It is therefore necessary to consider, in addition to the general guidelines, more specific constraints imposed on the structure of transition states by appropriate experimental observations. The remainder of this chapter is devoted essentially to this task.

Isotope effect: the product rule Useful general constraints on the structure of transition states can be obtained when the isotope effect is considered.[7] The separation of electronic and nuclear motion (Born–Oppenheimer approximation) has the result that the potential energy surface for a given reaction is independent of isotopic substitution;[8] however, the masses of the nuclei do affect the vibrational frequencies and the moments of inertia of both reactant and transition state, and through these affect the rate. Additionally, isotopic substitution influences the rate through the mass dependence of the transmission coefficient $\kappa(E)$ (Chapter 4, Section 4), and through a purely statistical effect on the reaction path degeneracy. For example, whereas $\alpha = 4$ for the process $CH_4 \rightarrow CH_3 + H$ (Chapter 4, Section 5), we have $\alpha = 1$ for $CH_3D \rightarrow CH_3 + D$ and $\alpha = 3$ for $CH_3D \rightarrow CH_2D + H$.

We shall consider here only those aspects of isotopic substitution which have a bearing on the structure of the transition state, i.e., the effect on vibrational frequencies and moments of inertia. In stable molecules, the product of the vibrational normal mode frequencies ω in the "normal" (unprimed symbols) and isotopically substituted molecule (primed symbols) are related by the Teller–Redlich product rule (*16*, p. 184, Eq. (8)]:

$$\prod_{k=1}^{3N-6} \frac{\omega_k'}{\omega_k} = \prod_{i=1}^{N} \left(\frac{m_i}{m_i'}\right)^{3/2} \left(\frac{M'}{M}\right)^{3/2} \left(\frac{I_a' I_b' I_c'}{I_a I_b I_c}\right)^{1/2} \tag{11-2}$$

where m_i are the masses of the atoms in the molecule (N is the total number of atoms), M is the total mass of the molecule, and I_a, I_b, and I_c are the

[6] For an example, see Johnston *et al.* (*17*).

[7] For a survey oriented toward bimolecular reaction, see Weston (*18*).

[8] The full Hamiltonian for a system of several nuclei and electrons contains the kinetic energy term

$$-\tfrac{1}{2}\hbar^2 \textstyle\sum_i (1/m_i)\nabla_i^2 - (\hbar^2/2m_e) \sum_j \nabla_j^2$$

where m_i is the mass of nucleus i and m_e the mass of the electron; the index i refers to nuclei and the index j to electrons. In the Born–Oppenheimer approximation, the nuclear kinetic energy terms are neglected because $m_i \gg m_e$, and therefore the solution of the stationary-state Schrödinger equation (i.e., the potential energy surface) contains no dependence on the masses of the nuclei. Cf. Daudel *et al.* (*19*, p. 306 ff.).

moments of inertia about the axes *a*, *b*, and *c*, respectively. The product rule may also be applied only to a subset of vibrational frequencies of the molecule, provided they all belong to the same symmetry species; in such a case, the factor M'/M in (11-2) appears only if the set comprises a nongenuine vibration resulting in translation, and the moment-of-inertia ratio for a given axis appears only if there is a nongenuine vibration resulting in rotation about the given axis. The requisite information can be obtained from the character table for each species [see also (*20*, p. 232)]. It is clear that the application of the product rule in this form requires the complete specification of the symmetry properties of the molecule and its vibrations.

The product rule is a generalization of the well-known result that the frequency of a harmonic oscillator is inversely proportional to the square root of the mass. For those frequencies on the left-hand side of (11-2) that represent a nongenuine vibration resulting in translation (subscript t), we have therefore (in the limit of vanishing forces) $\omega_t'/\omega_t = (M/M')^{1/2}$, and for those frequencies resulting in rotation (subscript r), $\omega_r'/\omega_r = (I/I')^{1/2}$, where I is the moment of inertia with respect to the appropriate axis. Equation (11-2) then follows.

Following the rule that the transition state is a "molecule like any other," except for one vibrational degree of freedom with imaginary frequency, the product rule (11-2) will apply to all the frequencies of the transition state[9] or to a subset of a given symmetry species. The product rule thus provides a definite relation between the vibrational frequencies of the "normal" and the isotopically substituted transition state. Since the molecular parameters of the "normal" and the isotopically substituted *reactant* are not subject to adjustment (i.e., are known from nonkinetic information and cannot, or should not, be tampered with), the frequencies of the transition state must be so chosen as to yield the experimentally observed kinetic isotope effect and, simultaneously, satisfy the product rule (11-2); this simultaneous requirement reduces considerably the amount of adjustment that can be made in the transition state.

Let us now take a closer look at the kinetic isotope effect[10] and how it depends on the nature of the transition state. There are two kinds of isotope effect: the primary effect,[11] when the isotopically substituted atom is involved

[9] For the transition state, the left-hand side of (11-2) becomes

$$\frac{\omega^{*\prime}}{\omega^*}\prod_{k=1}^{3N-7}\frac{\omega_k'}{\omega_k}$$

Here, ω^* is the imaginary frequency, proportional to $1/\mu^*$, where μ^* is the effective mass in the reaction coordinate (cf. Chapter 4, Section 1).

[10] For a recent general review, see Collins and Bowman (*21*).

[11] For details, see O'Ferrall and Kouba (*22*).

in a bond that is made or broken in the course of the reaction, and the secondary effect,[12] when no isotopically substituted atoms are involved in bonds made or broken. In chemical kinetic practice, deuterium labeling is the most common isotopic substitution, and therefore the discussion will be confined to the hydrogen–deuterium isotope effect. The subscript H will identify quantities referring to hydrogen-containing (or "normal") species and the subscript D will refer to deuterated species. We shall first consider the isotope effect on the rate constant for a monoenergetic reactant and then introduce whatever modifications are necessary due to the presence of a distribution of energies.[13]

Primary isotope effect As an example of the primary isotope effect (sometimes referred to, in addition, as the intermolecular effect) we may take the reaction

$$\begin{aligned} C_2H_5 &\xrightarrow{k_H} C_2H_4 + H \\ C_2H_4D &\xrightarrow{k_D} C_2H_4 + D \end{aligned} \tag{11-3}$$

If we compare the rate constants at the same excitation energy E in the reactant, we have

$$\frac{k_H(E)}{k_D(E)} = \frac{\alpha_H G_H^*(E - E_{0(H)})N_D(E)}{\alpha_D G_D^*(E - E_{0(D)})N_H(E)} \tag{11-4}$$

The $G^*(y)$ functions for the transition states in the two reactions (11-3) do not include the C—H or C—D stretch, which becomes the reaction coordinate, but they depend nevertheless on isotopic substitution through the bending modes; hence, *at the same energy*, $G_D^*(y) > G_H^*(y)$ because the isotopically substituted bending mode has a lower frequency. There is also an isotope effect in the reactant molecule [cf. Fig. 11-1(a)]: Deuteration reduces the zero-point energy of the molecule (but probably much less so that of the transition state), and therefore affects the critical energy; hence $E_{0(D)} > E_{0(H)}$, which tends to reverse the inequality of the $G^*(y)$ functions, so that, overall, $G_H^*(E - E_{0(H)})/G_D^*(E - E_{0(D)}) \gtrsim 1$. The frequency of the C—H stretch (ν_H) or C—D stretch (ν_D), which becomes the reaction coordinate in the transition state, contributes to the density of states of the molecule. Since $\nu_D < \nu_H$, $N_D(E)/N_H(E) > 1$; in addition, since isotope substitution reduces the symmetry of the molecule (and of its transition state), we can expect[14] in general

[12] For a review, see Halevi (*23*).

[13] A review of isotope effects in unimolecular reactions can be found in Rabinovitch and Setser (*24*).

[14] If we assume that only a methyl C—H bond can be broken in (11-3) and that the D atom is located on a methyl group, we have $\alpha_H = 3$ and $\alpha_D = 1$ in this example. The fundamental frequencies ν are used here since these are known directly from the analysis of vibrational spectra.

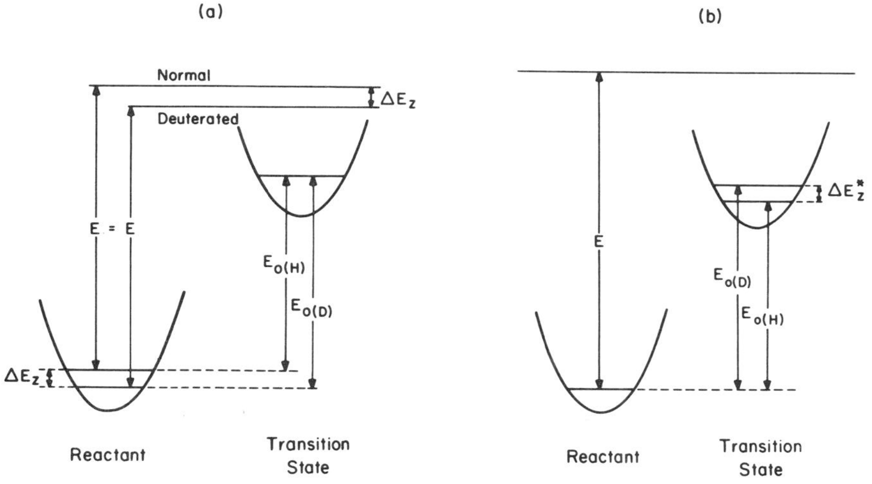

Fig. 11-1. *Schematic representation of energy relationships in the primary isotope effect.* Subscripts (H) and (D) identify ordinary and deuterated species, respectively. Asterisk refers to transition state; E_0 is the critical energy for reaction, E_z is the zero-point energy. (a) The primary *inter*molecular isotope effect. The same excitation energy E in reactant yields excitation energies in the transition state differing roughly by $\Delta E_z = E_{z(\text{H})} - E_{z(\text{D})}$. (b) The primary *intra*molecular isotope effect. At the same excitation energy E in the reactant, excitation energies in the transition state differ by $\Delta E_z^* = E_{z(\text{H})}^* - E_{z(\text{D})}^*$.

$\alpha_{\text{H}} > \alpha_{\text{D}}$. The result is then that

$$k_{\text{H}}(E)/k_{\text{D}}(E) > 1 \qquad (11\text{-}5)$$

i.e., the deuterated reactant will decompose with a smaller rate constant, when rates are compared at the same excitation energy in the reactant. Equation (11-5) represents the "normal" kinetic isotope effect. As excitation energy increases, the ratio $G_{\text{H}}^*(E - E_{0(\text{H})})/G_{\text{D}}^*(E - E_{0(\text{D})})$ tends to unity when $E \gg E_{0(\text{H})}$ or $E_{0(\text{D})}$, and therefore $k_{\text{H}}(E)/k_{\text{D}}(E)$ will decrease with increasing energy (see Exercise 1). Model calculations on the system (11-3) showing this trend were done by Rabinovitch and Setser (*24*, Table VIII).

A special case of the primary isotope effect (also called the "intramolecular" effect) arises when two parallel processes compete, e.g.,

$$C_2H_4D \xrightarrow{k_{\text{H}}} C_2H_3D + H \qquad (11\text{-}6a)$$

$$C_2H_4D \xrightarrow{k_{\text{D}}} C_2H_4 + D \qquad (11\text{-}6b)$$

Note here that the labeling H or D always refers to the *elimination* of H or D, respectively; thus $G_{\text{H}}^*(E)$ refers to the transition state of (11-6a), which contains one C—D stretch, although it is labeled H, and $G_{\text{D}}^*(E)$ refers to the transition state of (11-6b), containing only C—H bonds, since here the C—D stretch is the reaction coordinate. Equation (11-4) still applies, except that $N_D(E) = N_H(E)$, but the transition states are now different for the H and D labels, respectively, and so are the critical energies [Fig. 11-1(b)]. We have $G_H^*(y) > G_D^*(y)$ because the H label refers to thc transition state comprising the C—D stretch of lower frequency, causing the integrated density to be higher; therefore we also have $E_{0(H)} < E_{0(D)}$. The result is that $G_H^*(E - E_{0(H)})/G_D^*(E - E_{0(D)}) > 1$, $\alpha_H/\alpha_D > 1$, and again $k_H(E)/k_D(E) > 1$. Model calculations on the system (11-6) are available (*24*, Table X).

Thus for a monoenergetic reactant (excluding the effect of α), the primary *inter*molecular isotope effect is due mostly to the effect of the isotopic substitution on the properties of the reactant, whereas the primary *intra*molecular isotope effect is due mostly to the effect of isotopic substitution on the properties of the transition state. These properties cannot be adjusted independently for reactions (11-6a) and (11-6b) but must be related through the Teller–Redlich product rule.

Secondary isotope effect As an example of the secondary isotope effect, we may take

$$CH_3CH_3 \xrightarrow{k_H} CH_3 + CH_3 \qquad (11\text{-}7a)$$

$$CD_3CD_3 \xrightarrow{k_D} CD_3 + CD_3 \qquad (11\text{-}7b)$$

where the reaction coordinate in both cases is the C—C stretch, not isotopically substituted. Eq. (11-4) applies again; in this special case, $\alpha_H = \alpha_D = 1$, but $N_D(E) \gg N_H(E)$ and $E_{0(H)} \approx E_{0(D)}$ since both reactant and transition state have similar zero-point energies, both being isotopically substituted to the same degree. Thus $G_H^*(E - E_{0(H)})/G_D^*(E - E_{0(D)}) < 1$, but because this ratio is evaluated at a lower energy than $N_D(E)/N_H(E)$, the latter ratio is more important, and therefore $k_H(E)/k_D(E)$ is larger, and often much larger, than unity. Like the primary effects, the secondary isotope effect also decreases with increasing energy. Calculation on a model system may be found in (*24*, Table VII).

It should be obvious that in more heavily isotopically substituted molecules, both the primary and secondary isotope effects may appear; in such a case, the isotope effect will depend on the properties of the transition state whether or not the primary effect is intermolecular or intramolecular.

Isotope effects are also observed in the decomposition of ions[15] and the general principles involved are of course the same as in the case of neutral molecules. Isotope effects have been calculated by the theory of mass spectra for the decomposition of the ions of methane (*26*), ethane (*27,28*), propane (*29–31*), and toluene (*32,178*).

[15] For the isotope effect upon kinetic energy release in the fragmentation of metastables, see Bertrand *et al.* (*25*).

When centrifugal effects are considered, E_0 becomes E_J, the two being related by Eq. (7-13):

$$E_J = E_0 + V_{\text{eff}}(r_m) - E_r(\mathscr{J}) \tag{11-8}$$

The last two terms are both a function of $1/\mu_{AB}$, where μ_{AB} is the reduced mass of the reactant, seen as a quasidiatom $\mathscr{AB}$ (Chapter 7, Section 1), and therefore both terms depend on isotopic substitution. Replacing hydrogen by deuterium diminishes the difference $V_{\text{eff}}(r_m) - E_r(\mathscr{J})$ because $V_{\text{eff}}(r_m)$ depends on $(1/\mu_{AB})^{n/(n-2)}$ (n is the power law of the attractive part of the potential), whereas $E_r(\mathscr{J})$ depends on $1/\mu_{AB}$. The critical energy E_0, on the other hand, generally increases when deuterium replaces hydrogen in the reactant (Fig. 11-1). Overall, therefore, centrifugal effects would tend to diminish the isotope effect[16] in monoenergetic reactant having specified angular momentum quantum number $\mathscr{J}$.

Isotope effect and the distribution of energies The presence of a distribution of energies in the reactant molecule will modify the size of the isotope effect relative to the monoenergetic reactant, particularly if the energy distribution function is significantly different for the isotopically substituted reactant. In thermal reactions, the equilibrium distribution function $P(E)_e$ [Eq. (8-13)] gives somewhat more weight to higher energies in an isotopically substituted molecule relative to the "normal" molecule,[17] which tends to diminish the inequality (11-5), i.e., the isotope effect is smaller than it would be if $P(E)_e$ were independent of isotopic substitution.

The same is true of the distribution function $F(E)$ [Eq. (9-26)] for chemical activation, as of course one would expect in view of Eqs. (9-35) and (9-36), except that the difference between $F(E)$ for a "normal" and an isotopically substituted species is smaller because of the lower temperature at which the experiments are generally done [cf. (*34*, Fig. 3)]. Because the starting point of $F(E)$ depends on E_0^{ar}, which is itself dependent on isotopic substitution, it is difficult to determine a priori how $F(E)$ will modify the isotopic effect relative to the monoenergetic reactant.

In the case of ions, the function $\mathscr{F}(E)$, which measures the probability of forming an ion with internal energy E from a ground-state neutral molecule, will depend on isotopic substitution if an appreciable fraction of the ions are formed by autoionization from superexicited states (*35*). The reason is that while the formation of the superexcited molecule by photon or electron

[16] Tunneling through a centrifugal barrier might be expected to lead to a large isotope effect because of the mass dependence of the transmission coefficient $\kappa(E)$ [Eq. (4-32)]. In fact, it does not, as shown by Klots (*33*) because the smaller transmission coefficient for the heavier isotope is compensated by a lower $V_{\text{eff}}(r_m)$.

[17] This is the case also with $P(\mathscr{J})_e$; cf. Exercise 2.

impact is presumably isotope independent, the subsequent ionization is due to an exchange of energy between electronic and vibrational motions (cf. Chapter 10, Section 3), which is mass dependent, and thus gives rise to the isotope effect in autoionization. Experiment shows an isotope effect in ionization by electron impact (*36*) but not in photoionization (*37*), in agreement with the general belief that autoionization is important in electron but less so in photon impact.

The isotope effect on the averaged thermal rate constant k_{uni} can be obtained in a simple closed form at the two pressure limits. At the high-pressure limit, we start with Eq. (8-23); then,

$$k'_{\infty(\text{H})}/k'_{\infty(\text{D})} = (\alpha_{\text{H}} Q_{\text{H}}^* Q_{\text{D}}/\alpha_{\text{D}} Q_{\text{D}}^* Q_{\text{H}}) e^{-\Delta E_0/kT} \tag{11-9}$$

where $\Delta E_0 = E_{0(\text{H})} - E_{0(\text{D})}$. Note that the centrifugal correction factor f_∞ [Eqs. (8-56) and (8-59)] is mass independent and cancels out. Theoretical investigations (*38–40*) of the isotope effect are usually confined to analysis of equations of the form (11-9), applicable to thermal systems. Table 11-1 shows the approximate size of the various factors in (11-9), except for the reaction path degeneracy, for which we always have $\alpha_{\text{H}}/\alpha_{\text{D}} \geqslant 1$. The combined effect of the various factors very nearly cancels and we have in most instances $k'_{\infty(\text{H})}/k'_{\infty(\text{D})} \approx 1$. Since for secondary isotope effect $E_{0(\text{H})} \approx E_{0(\text{D})}$, Eq. (11-9) shows that an appreciable temperature dependence of the ratio $k'_{\infty(\text{H})}/k'_{\infty(\text{D})}$ indicates an important *primary* isotope effect.

At the low-pressure limit, the applicable equation is Eq. (8-43), assuming equilibrium theory; then,

$$k_{0(\text{H})}/k_{0(\text{D})} = (\mu_{\text{D}}/\mu_{\text{H}})^{1/2} \left[Q_{\text{D}} \int_{E_{0(\text{H})}}^{\infty} N_{\text{H}}(E) e^{-E/kT}\, dE \right] \Big/ \left[Q_{\text{H}} \int_{E_{0(\text{D})}}^{\infty} N_{\text{D}}(E) e^{-E/kT}\, dE \right] \tag{11-10}$$

where μ_{D} and μ_{H} are the reduced masses of the isotopically substituted and of the "normal" collision partners, respectively; the ratio $(\mu_{\text{D}}/\mu_{\text{H}})^{1/2}$ arises by cancellation of common factors in the ratio of collision frequencies Z [cf. Eq. (8-100)]. Isotopic substitution affects both the integrand and the integration limit[18] of the integrals in (11-10), with the result that the ratio of integrals is much less than unity. Although $\mu_{\text{D}}/\mu_{\text{H}} > 1$ and $Q_{\text{D}}/Q_{\text{H}} > 1$ (Table 11-1), the ratio of the integrals in (11-10) is generally sufficiently small to outweigh the combined effects of the other two factors in (11-10), with the result that $k_{0(\text{H})}/k_{0(\text{D})} < 1$, i.e., there is an *inverse* isotope effect at the low-pressure limit of a thermal reaction. These considerations have ignored the low-pressure centrifugal correction factor f_0 [Eq. (8-44)] which, in the case of type 2 or type 3 reactions, depends on isotopic substitution through the critical energy E_0 (in the case of primary isotope effect), and on

[18] For a pure secondary effect, however, we have $E_{0(\text{H})} = E_{0(\text{D})}$.

TABLE 11-1

MAGNITUDES OF VARIOUS FACTORS IN EQ. (11-9)

Ratio	Isotope effect: Primary: Intermolecular	Isotope effect: Primary: Intramolecular	Isotope effect: Secondary
Q_H^*/Q_D^*	1	>1	<1
Q_D/Q_H	>1	1	>1
$e^{-\Delta E_0/kT}$	<1	<1	~1

the molecule density function $N(E)$ [Eq. (8-62)], and also on molecular mass through the upper integration limit z_0 in case of type 1 reactions [Eqs. (8-68) and (8-69)]. No calculations of the isotope effect on f_0 have been reported so far, but it can be reasonably assumed that the effect will be small since f_0 itself is small.

We thus have the general result that in thermal unimolecular reactions, the isotope effect is pressure dependent[19] and on decrease of pressure, should change from "normal" to inverse. This prediction has been experimentally verified in all the cases studied so far: the cyclopropane–cyclopropane-d_6 system (*43,44*) the cyclobutane–cyclobutane-d_8 system (*45*), the methyl–methyl-d_1 (*46*), and methyl-d_3 (*47*) as well as the ethyl–ethyl-d_5 (*48*) isocyanide systems. The decline in k_H/k_D with pressure is generally less than a factor of ten, from about 1.5 at $p \to \infty$ to about 0.2 at $p \to 0$. Theoretical calculations (*41,43,46–48*) reproduce the experimental trends quite well.

In chemically activated systems k_H/k_D *increases* markedly as pressure decreases (*167*). The source of this pressure dependence is the cascade deactivation (Chapter 9, Section 5), not so much because of an isotope effect on the transition probability, but because of the isotope effect on the microcanonical rate constant, for which $k_H(E)/k_D(E)$ is always larger than unity.

Information about transition state parameters We may now summarize briefly the discussion of the isotope effect by saying that the constraints imposed by the isotope effect are such that the rate constant for the isotopically

[19] Experimentally, we have in thermal reactions at any pressure, $k_H/k_D = (A_H/A_D) \exp[E_{a(D)} - E_{a(H)}]/kT$, where A and E_a are the *experimental* preexponential factors and activation energies, respectively (do not confuse $E_{a(H)}$ and $E_{a(D)}$ with the critical energies $E_{0(H)}$ and $E_{0(D)}$ which are pressure- and temperature-independent intrinsic parameters) It then turns out that the difference in experimental activation energies $E_{a(H)} - E_{a(H)}$ is a complicated function of pressure. Cf. (*41*); a more general discussion can be found in Vogel and Stern (*42*).

substituted species must be calculable from theoretical expressions for the "normal" species by modifying the parameters of the latter in accordance with the Teller–Redlich product rule. Thus once the transition state of the "normal" species is set up, the amount of modifications that can be made in the transition state to account for the rate of the isotopically substituted species is limited.

Since the isotope effect governs only the relationship between transition states for "normal" and isotopically substituted species, it is still necessary to consider ways how the transition state for the "normal" species is set up in the first place. This will be done in the following sections of this chapter, in which the transition states are discussed according to the type of the potential energy surface over which the reaction is occurring.

Two general remarks are in order, however, before the discussion gets down to specifics: First, as shown in Chapter 8, Section 7, thermal unimolecular reactions at the high-pressure limit provide a direct link with thermally averaged properties of the transition state. In particular, the thermal high-pressure preexponential factor A_∞ is related to the entropy of activation [Eq. (8-114)], and this is a more sensitive reflection of the properties of the transition state than the absolute magnitude of the calculated rate constant, the only test available in nonthermal systems. For this reason, much of the available information on transition states comes from thermal systems or is tested on thermal systems and on the magnitude of A_∞ (cf. Chapter 9, Section 3). Consequently, most of the discussion in the following sections is based on results obtained in thermal systems. Unfortunately, thermal data for the same reaction obtained from different sources sometimes disagree, due to experimental complications, surface effects, and other factors, and in addition, the quoted values of A_∞ and $E_{a\infty}$ (the experimental high-pressure activation energy) may not correspond to the actual high-pressure limit because of extrapolation problems [cf. Chapter 8, Section 7]. For all these reasons, numerical values of A_∞, even if based on "good" thermal data (summarized in the following sections in Tables 11-2–11-4), should be taken with a grain of salt and considered only as a rough guide, suitable at most for formulating semiquantitative arguments. For a review of preexponential factors, see (*179*).

The second general remark concerns the course to adopt when, as is frequently the case, there is insufficient information about a given degree of freedom in the transition state. If it is a degree of freedom that clearly is not directly involved in the chemical transformation (e.g., a C—H stretch in the transition state for the reaction $C_2H_6 \rightarrow 2CH_3$), then it should be assigned the same parameters (vibrational frequency, moment of inertia, and the like) as in the reactant. If, however, it is a degree of freedom that is involved in the chemical transformation directly [e.g., the C—Č stretch in Eq. (11-3)], and if, within the guidelines set out in this chapter, no information has emerged about its parameters in the transition state, the rule should be followed that such a degree of freedom shall be assigned parameters intermediate between those of the reactant (C_2H_5) and the product (C_2H_4), rather than be arbitrarily adjusted. If followed consistently, this procedure would eliminate much of the arbitrariness in the assignment of transition-state parameters that has led to doubts about the predictive value of transition-state theory (*1*).

2. Transition states for type 1 reactions

These reactions occur over a type 1 potential energy surface (Chapter 3, Section 1) and involve simple bond rupture yielding two radicals: schematically, $\mathcal{AB} \rightarrow \mathcal{A} + \mathcal{B}$, where $\mathcal{A}$ and $\mathcal{B}$ are in general polyatomic fragments.

Kinetic evidence The circumstantial kinetic evidence from thermal reactions that can be used as inspiration for formulating a structure for the transition state is the following (listed roughly in order of decreasing pertinence).

(a) At the high-pressure limit in thermal systems, most of these reactions are characterized by an "abnormally" high A factor ($\log A_\infty \approx 17$), which in the thermodynamic interpretation of reaction rates, means a high entropy of activation, i.e., one that is an appreciable fraction of the total entropy change in the reaction. A high entropy of activation implies a transition state which is more like the final state (products) than like the reactants. For example, in the dissociation of ethane (Table 11-2), the total entropy change in $C_2H_6 \rightarrow 2CH_3$ is 38 eu mole^{-1} at 298°K, while the experimental pre-exponential factor is equivalent to an entropy of activation of about 13 eu.

(b) Since $E_{a\infty} \approx \Delta H^0$ (Table 11-2), the reverse reaction of association of two radicals should proceed essentially without activation energy and on every[20] collision. Thermodynamic interpretation of these collision rates leads to the same conclusions (*50*) about the structure of the transition state as in (a), as of course it should because by microscopic reversibility, the transition state must be the same for the forward and reverse reactions. Similarly (*50*), in the case of the decomposition of charged species, the

[20] This ignores a statistical factor due to the necesssity of proper orientation of electronic spins of the radicals in order that they may yield a singlet product molecule on collision. For example, ground-state CH_3 radicals are doublets, and therefore only one-fourth of the CH_3–CH_3 collisions lead to singlet ethane; thus the rate constant for the recombination of methyls is one-fourth the collision frequency Z [Eq. (8-100)]. At room temperature, $Z/4$ works out to $\sim 10^{10}$ liters mole^{-1} sec^{-1} for the recombination of most alkyl radicals, and this is roughly the rate constant observed experimentally [cf. (*49*, p. 152, Table 9.2)].

TABLE 11-2

THERMAL PARAMETERS FOR SOME TYPE 1 REACTIONS

Reaction	Ref.	$\log A_\infty$ (sec^{-1})	T (°K)	$E_{a\infty}$ (kcal $mole^{-1}$)	ΔS_* (eu $mole^{-1}$)	ΔS^0_{298} (eu $mole^{-1}$)	ΔH^0_{298} (kcal $mole^{-1}$)
$CH_4 \rightarrow CH_3\cdot + H$	*a*	14.6	1400	103	3.2	29.4	104
$C_2H_6 \rightarrow 2CH_3\cdot$	*b*	16.0	900	86.0	10.5	37.9	88.2
	c	16.3	850	88.0	12.0	—	—
	d	17.4	850	91.7	17.0	—	—
$C_2F_6 \rightarrow 2CF_3\cdot$	*e*	17.63	1450	94.4	17.0	45.4	91.8
$C_2H_6N_2 \rightarrow CH_3\cdot + N_2CH_3\cdot$	*f*	17.3	550	55.4	17.4	39.6	52.5
$C_2D_6N_2 \rightarrow CD_3\cdot + N_2CD_3\cdot$	*g*	15.5	550	50.7	9.2	—	—
$C_2F_6N_2 \rightarrow CF_3\cdot + N_2CF_3\cdot$	*h*	16.2	600	55.2	12.2	44.3	56.0
iso-C_4H_{10} $\rightarrow CH_3\cdot +$ *iso*-$C_3H_7\cdot$	*i*	17.8	800	82.5	19.0	—	—

A_∞ and $E_{a\infty}$ are experimental parameters determined from $k_\infty = A_\infty e^{-E_{a\infty}/RT}$; T is the mean experimental temperature in °K, ΔS_* is defined as $R \ln(A_\infty h/ekT)$, evaluated at the mean experimental temperature (see text in Section 2), ΔS^0_{298} is the standard entropy change for the indicated reaction at 298°K, referred to standard state of ideal gas at 1 atm pressure, ΔH^0_{298} is the standard enthalpy change, similarly defined; 1 eu = 1 cal deg^{-1} $mole^{-1}$.

References for experimental results:

[a] H. B. Palmer, J. Lahaye and K. C. Hou, *J. Phys. Chem.* **72**, 348 (1968);
[b] M. C. Lin and M. H. Back, *Can. J. Chem.* **44**, 2357 (1966);
[c] A. B. Trenwith, *Trans. Faraday Soc.* **62**, 1538 (1966);
[d] C. P. Quinn, *Proc. Roy. Soc.* (*London*) **A275**, 190 (1963);
[e] E. Tschuikow-Roux, *J. Chem. Phys.* **49**, 3115 (1968);
[f] W. Forst and O. K. Rice, *Can. J. Chem.* **41**, 562 (1963);
[g] D. R. Chang and O. K. Rice, *Int. J. Chem. Kinet.* **1**, 171 (1969);
[h] E. Leventhal, C. R. Simonds and C. Steel, *Can. J. Chem.* **40**, 930 (1962);
[i] R. S. Konar, R. M. Marshall and J. H. Purnell, *Trans. Faraday Soc.* **64**, 405 (1968).

Parameters of reference (*g*) are probably preferable to those of reference (*f*). Thermodynamic data from S. W. Benson and H. E. O'Neal, "Kinetic Data on Gas Phase Unimolecular Reactions," NSRDS-NBS 21, Washington, D.C. 1970.

reverse reaction is an ion–molecule reaction, which is known to proceed without activation energy (Chapter 10, Section 4).

(c) The forward and reverse rates are related through the equilibrium constant. From the reverse rate, it can be calculated, and is found experimentally, that the forward rate of dissociation is quite high, if allowance is made for the activation energy. To account for this high rate theoretically, a high density of states in the transition state is needed.

Table 11-2 summarizes kinetic and thermodynamic parameters of some of the better-known type 1 dissociations. The kinetic parameters A_∞, $E_{a\infty}$, and also ΔS_* refer to the mean experimental temperature (given under T), whereas the standard overall entropy change (ΔS°_{298}) and enthalpy change (ΔH°_{298}) in the reaction refer to 298°K. ΔS_* is defined as $R \ln(A_\infty h/ekT)$ (in eu mole^{-1}), so that in order to convert it to the entropy of activation ΔS^* defined by (8-114), $R \ln(\alpha f_\infty)$ would have to be subtracted from the value of ΔS_* in Table 11-2. The reason for using ΔS_* is that this quantity, apart from universal constants, contains only the purely experimental parameters A_∞ and T, whereas ΔS^* contains, in addition, the experimentally inaccessible parameters α and f_∞.

It is safer to base general semiquantitative deductions about transition states on ΔS_* rather than on ΔS^* because α is a model-dependent parameter and f_∞ is not known directly from experiment. In any one specific case, however, actual rate calculations require a transition-state model, which is then fitted to ΔS^* and not merely to ΔS_*. The temperature difference between ΔS_* and ΔS°, and between $E_{a\infty}$ and ΔH°, which are calculated at T and at 298°K, respectively, is not significant if only semiquantitative comparisons are intended. More extensive compilations of experimental data are available on type 1 dissociations[21] and the reverse radical associations (*49*, p. 152, Table 9-2) and ion–molecule reactions (*54*).

Theoretical considerations For nonkinetic information, we turn to the potential energy surface (Fig. 3-1). The only potential energy barrier is the rotational barrier, and therefore the transition state must be located on top of it, since it is the only feature of the potential energy surface between reactant and products where the transition state can be defined within the "traditional" formalism discussed in Section 1. We shall focus attention on the *location* of the barrier, since its height and its effect on rate have already been considered in Chapter 7.

We have seen before [Eq. (7-11)] that for a potential that depends on internuclear distance as $1/r^{n}$ at large distances, the maximum of the rotational barrier is located at r_m given by

$$r_m = [n\mathscr{C}\mu_{AB}/\mathscr{J}(\mathscr{J}+1)\hbar^2]^{1/(n-2)} = (n\mathscr{C}/2r_e{}^2E_r)^{1/(n-2)} \qquad (11\text{-}11)$$

where $E_r = \mathscr{J}(\mathscr{J}+1)\hbar^2/2\mu_{AB}r_e{}^2$ is the rotational energy of the quasi-diatomic molecule $\mathscr{AB}$ as it starts decomposing; r_e is the equilibrium distance between the centers of gravity of the (future) fragments $\mathscr{A}$ and $\mathscr{B}$ inside the stable molecule $\mathscr{AB}$. In systems with thermal distribution of

[21] Frey and Walsh (*51*). Benson's book (*52*) discusses at length the thermochemistry of activation. Benson and O'Neal (*53*) is a compilation of data on thermal reactions of all types.

rotational energies, the average rotational energy of the diatom $\mathscr{AB}$ is $\langle E_r \rangle = kT$, which may be used to replace E_r, so that

$$\langle r_m \rangle = (n\mathscr{C}/2r_e^2 kT)^{1/(n-2)} \tag{11-12}$$

For a reaction between neutral species, $n = 6$; if we write $\mathscr{C} \simeq D_e r_e^6$ [Eq. (7-9)], we get

$$\langle r_m \rangle / r_e = (3D_e/kT)^{1/4} \tag{11-13}$$

The dissociation $\mathscr{AB} \rightarrow \mathscr{A} + \mathscr{B}$ typically involves the rupture of a C—C bond, for which $D_e \simeq 85$ kcal mole^{-1}; at 600°K, a fairly usual temperature, $kT = 1.2$ kcal mole^{-1}, so that

$$\langle r_m \rangle \simeq 3.8 r_e \tag{11-14}$$

For a reaction involving dissociation into one charged and one neutral species, for example, $C_2H_6^+ \rightarrow CH_3^+ + CH_3$, we have $n = 4$, and if the distribution of rotational energies is thermal, as is likely (cf. Chapter 10, Section 9), we can again replace E_r by kT in (11-11), so that

$$\langle r_m \rangle / r_e = (2D_e/kT)^{1/2} \tag{11-15}$$

The dissociation energy D_e can be replaced by the appearance potential AP of the charged fragment, referred to the ionization potential of the parent ion as zero of energy. In the previous example, AP(CH_3^+) $\simeq$ 2.3 eV, which leads to

$$\langle r_m \rangle \simeq 9.4 r_e \tag{11-16}$$

if we take $kT = 1.2$ kcal mole^{-1} as before. For interaction involving a charged fragment with a neutral fragment which has a dipole moment, the attractive constant $\mathscr{C}$, and hence also $\langle r_m \rangle / r_e$, is even larger.

From Eq. (11-14), we see that in the dissociation of a neutral molecule into two neutral fragments, the top of the rotational barrier corresponds to the breaking bond being extended to about four times its normal length;[22] this is therefore also the amount of bond extension in the transition state, since by definition, the transition state is located at the top of the centrifugal barrier. Similarly, Eq. (11-16) shows that in the dissociation of a charged molecule into one charged and one neutral fragment, the relevant bond extension in the transition state is about ten times the normal bond length.

It is useful now to introduce into the discussion the concept of bond order,

[22] Similar results were obtained by Tschuikow-Roux (*55*). Actually, in the case of thermal reactions, it would be somewhat more consistent to calculate $\langle r_m \rangle$ from $f_\infty^{1/2}$ [Eq. (8-56)] rather than from (11-12), since it is the overall rate and not E_r which should be averaged over $P(\mathscr{J})_e$. For $D_e = 85$ kcal mole^{-1}, $kT = 1.2$ kcal mole^{-1}, the result is (for $n = 6$) $f_\infty^{1/2} = 2.65$, which is somewhat smaller than the result of Eq. (11-14).

originally due to Pauling,[23] which extends to fractional bonds the well-known notion that a double bond is shorter than a single bond and a triple bond is shorter than a double bond. If r_s is the single-bond length, the bond length for bond of order x is, according to Pauling,

$$r_x = r_s - 0.6 \log_{10} x \qquad (r_x, r_s \text{ in Å}) \tag{11-17}$$

If we take $r_s = 1.5$ Å, the normal C—C bond length in (neutral) ethane, and identify $\langle r_m \rangle$ of Eq. (11-14) with r_x, Eq. (11-17) yields $x = 3 \times 10^{-8}$ for the C—C bond order in the transition state for the dissociation of ethane into two methyls, which is essentially a bond order of zero. It can be shown by semiempirical arguments[24] in the somewhat special case of a linear triatomic molecule X—H—Y, with bond order x between X—H and bond order y between H—Y, that the bending force constant f_ϕ depends on a power of the product xy. Therefore as either bond order goes to zero, so does f_ϕ. In the more general case of a polyatomic molecule $\mathscr{A}\mathscr{B}$, the dependence of bending force constants on bond order cannot be so neatly demonstrated,[25] but it seems reasonable to assume that a considerable reduction of bond order of one of the bonds involved should similarly lead to a considerable reduction of the bending force constant.

For example, a considerable increase in the C—C distance (i.e., a reduction of the C—C bond order) in (neutral) ethane would presumably reduce appreciably the force constants of the two bending vibrations around 800 cm^{-1} and of the two CH_3 rocking vibrations around 1200 cm^{-1}.

A very small bending constant signifies that the molecule has practically separated into two independent fragments $\mathscr{A}$ and $\mathscr{B}$ having each roughly the same properties (vibrational frequencies, moments of inertia) as if they were completely free. Such a transition state resembles of course the final state much more than it does the initial state, and this is quite in line with the prediction made from the entropy of activation in thermal reactions of type 1 as mentioned earlier.

A triatomic molecule is a simple enough case where the decrease in the bending force constant can be taken into account more explicitly. This was done by Troe *et al.* (*15,61,63*) in the case of NO_2, a bent triatomic molecule.

[23] Pauling (*56*). See also (*49*, p. 80 ff.). However, in his book "The Nature of the Chemical Bond" (*57*, p. 239, Eq. (7-5)), Pauling gives Eq. (11-17) in the form $r_x = r_s - 0.71 \log_{10} x$.

[24] This is the so-called bond-energy bond-order method (BEBO) of Johnston (*49*, Appendix E, Eq. (E-32)).

[25] The force constant for bending depends on the second derivative of potential with respect to angle of bending, and could be, in principle, calculated rigorously if the potential energy surface were sufficiently known. For more details on the bending motions in triatomic molecules, see Thorson and Nakagawa (*58*); Brown and Parr (*59*); and, in particular, Brown and Charles (*60*).

If the coupling of the bending vibration with the other degrees of freedom is considered, somewhat in the manner of Bunker and Pattengill (*12*), it turns out that the local minimum in the density of states of the critical configuration is located at a smaller distance than the centrifugal maximum, and is only weakly dependent on $\mathscr{J}$. Thus it appears that although rotational energy reduces the critical energy, the centrifugal maximum is not actually the principal factor determining the location of the transition state. The same conclusion emerges from more detailed consideration of angular momentum conservation (ref. *24* in Chapter 5). At present, these results are not readily transferable to more complex cases, so that we shall have to make do with the very simplified independent fragment model of the transition state if we wish to treat all the cases of chemical interest.

This picture of the transition state consisting of two (almost) independent fragments $\mathscr{A}$ and $\mathscr{B}$ is a very crude one, since in the majority of cases, the actual amount of the reduction in the bending force constants cannot be readily calculated. However, even if it were calculable, the amplitude of vibrations with very low force constant would be so large[26] that the usual small-vibration theory could not be used for obtaining the frequencies of the transition state. In the absence of any definite information, we can assume that the bending force constant is zero, i.e., that the bending vibrations become free rotations, or tumblings, of the two fragments.

We wish to recall at this point that the calculation of rate by Eq. (11-1) calls for the density of states of the active degrees of freedom, and the understanding was made in Chapter 5, Section 2 that all vibrations will be taken as a pertinent or active degree of freedom. In the present case, the free rotations represent an approximate way of handling low-frequency vibrations, and therefore must also be considered as pertinent degrees of freedom. This does not conflict too much with angular momentum conservation because the conditions is merely that $\boldsymbol{M}_{\mathscr{A}\mathscr{B}}$, the angular momentum of the quasi-diatomic molecule $\mathscr{A}\mathscr{B}$, be equal to the vector sum of the (intrinsic) angular momenta of the two fragments ($\boldsymbol{M}_{\mathscr{A}}$, $\boldsymbol{M}_{\mathscr{B}}$), and of the orbital angular momentum $\boldsymbol{L}$ associated with the movement of $\mathscr{A}$ and $\mathscr{B}$ around their common center of gravity in such transition state (boldface symbols represent vectors):

$$\boldsymbol{M}_{\mathscr{A}\mathscr{B}} = \boldsymbol{M}_{\mathscr{A}} + \boldsymbol{M}_{\mathscr{B}} + \boldsymbol{L} \tag{11-18}$$

There is thus a restriction on the magnitude that $\mathscr{J}_{\mathscr{A}}$ and $\mathscr{J}_{\mathscr{B}}$ can take; and hence also on the corresponding rotational energy. Since it is not easy to calculate the density of only partially active states, it is a fair approximation to treat rotations associated with $\boldsymbol{M}_{\mathscr{A}}$ and $\boldsymbol{M}_{\mathscr{B}}$ as pertinent degrees of freedom in the usual sense, i.e., without any restriction. It is clear however that den-

[26] Classically, the Boltzmann average of the vibrational amplitude (A_ϕ, in radians) for angular vibration is given by $A_\phi = (2\pi kT/f_\phi)^{1/2}$. Cf. (*49*, p. 92, Eq. (6-23)).

sities of states and, ultimately, rates calculated in this fashion will represent an *upper limit.*

Loose transition state, model A The transition state in which fragments are assumed to tumble freely, and their rotations become pertinent degrees of freedom included in the integrated density of states of the transition state, is usually referred to as "loose"; it was first introduced by Gorin (*4,62*) and later discussed by Rice (*64*). We shall refer to it as "loose, model A," since another model will be discussed further below. To date, a number of radical associations (*54,65*), ion–molecule reactions (*54*), and unimolecular dissociations, thermal and nonthermal, have been successfully interpreted in terms of the model A for the loose transition state. In particular, among reactions of small molecules, we may cite the unimolecular dissociations of methane (*24*) ($CH_4 \rightarrow CH_3 + H$), methyl iodide (*66,67*) ($CH_3I \rightarrow CH_3 + I$), ethane (*68*) ($C_2H_6 \rightarrow 2CH_3$), ethane ion (*69*) ($C_2H_6^+ \rightarrow CH_3^+ + CH_3$), hexafluoroethane (*70*)[27] ($C_2F_6 \rightarrow 2CF_3$), azomethane (*71*) ($C_2H_6N_2 \rightarrow CH_3 + N_2CH_3$), acetaldehyde (*72*) ($CH_3CHO \rightarrow CH_3 + CHO$), propane (*24*) ($C_3H_8 \rightarrow CH_3 + C_2H_5$), butane (*24*) ($C_4H_{10} \rightarrow 2C_2H_5$), and nitrogen pentoxide (*73*) ($N_2O_5 \rightarrow NO_2 + NO_3$).

In the case of the dissociation of ethane into two methyls, for example, which may be considered the prototype of a type 1 dissociation, the free tumblings of the methyls lead to the appearance of five free rotations in the transition state (two per methyl plus one arising from what was formerly internal torsion in the molecule). It follows from Eq. (8-115) that one rotation (subscript r = 1) then contributes, in a thermal system, the entropy (per molecule)

$$S_{r=1} = k(\ln Q_{r=1} + \tfrac{1}{2}) \tag{11-19}$$

where $Q_{r=1}$ is the partition function for a one-dimensional rotor, as given by Eq. (6-39), including its symmetry number since σ (cf. Chapter 6, Section 2) is not included via the reaction path degeneracy α in the entropy of activation ΔS^* calculated from Eq. (8-114). The approximate moment of inertia for each rotation can be obtained with the help of Eq. (11-13) and the known geometry of the methyl radical. The remaining degrees of freedom in the loose transition state are then vibrational (in the case of ethane, these would be the vibrations of two free methyl radicals[28]), the entropy of which can be obtained directly from tables of thermodynamic functions of the harmonic oscillator. Some small adjustment in the frequencies or moments of inertia (preferably) may be necessary in order to arrive at the correct total entropy

[27] An algebraic error here leads to the wrong sign for the centrifugal effect.

[28] The vibrational frequencies of the methyl radical have been calculated by Johnston (*74*). For an experimental determination, see Lester *et al.* (*75*).

of activation ΔS^*. Each rotation will be found to make a much larger contribution to the entropy of the transition state than would the molecular vibration it replaces,[29] as of course one would expect, since the loose transition model is designed to account for high entropy of activation [point (a) at the beginning of this section].

The integrated density of states for the transition state, $G^*_{vr}(E)$, can then be calculated on the basis of the vibrational frequencies so determined, and on the basis of the appropriate number of free rotors (five in the case of ethane), whose states are included in $G^*_{vr}(E)$ since for simplicity they are assumed to exchange energy freely with the other degrees of freedom in the transition state, as stated at the end of the preceding paragraph. Every rotation included in $G^*_{vr}(E)$ increases $G^*_{vr}(E)$, and hence also the calculated rate constant, in accord with point (c).

In nonthermal systems, the absence of temperature means that experimental results cannot be interpreted in terms of the entropy of activation. However, one can still calculate the entropy of activation from the postulated transition state, and then, by comparison with a similar thermal system for which ΔS^* is known, determine whether the nonthermal transition state is reasonable.

In the dissociation of charged species, which of course is always nonthermal, a similar procedure may be used. As Eq. (11-16) shows, the separation of the fragments (one neutral, one charged) in the transition state is even larger than in the case where both fragments are neutral, so that the transition state is even "looser," i.e., still closer to the configuration of the products. It is then feasible to interpret the dissociation in terms of the cross section for the reverse process (see Chapter 10, Section 9 and Chapter 4, Exercise 2).

In physical terms, the tumbling of the fragments in the loose transition state means that their relative orientation does not matter insofar as the reverse association is concerned, which must be the case if the association is to proceed on every collision [point (b)]. Note that the fragments are assumed to rotate freely in the transition state but not freely enough to separate. What holds them together is the balance between the centrifugal force of the rotating fragments and the attractive force derived from the attractive part of the interfragment potential. This is the physical meaning of the condition $dV_{\text{eff}}/dr = 0$ which defines r_m [Eq. (7-11)]. If the centrifugal force is expressed in terms of the kinetic energy of translation of the colliding fragments, r_m may be interpreted as defining the cross section for the association process; this has been used in calculating cross sections for ion–molecule reactions.[30]

[29] For example, we find from (11-19) that at 300°C, a free rotation for which $\sigma = 3$ $3\hbar^2/2I \simeq 5\ \text{cm}^{-1}$ (case of free methyl rotation) contributes an entropy of $\sim$4.3 eu mole^{-1}, whereas a torsional vibration at 300 cm^{-1} would contribute $\sim$2.6 eu mole^{-1}.

[30] Gioumousis and Stevenson (*76*). The influence of rotations on the reverse association of *neutral* fragments has been considered by Benson (*77*). Cf. Chapter 7, Exercise 1.

The problem of reaction path degeneracy in type 1 reactions is relatively simple. In the dissociation of ethane into two methyls, $\alpha = 1$ if the internal torsion is taken as a free rotation in both the molecule and transition state, as seems preferable, and $\alpha = 2$ under the same conditions in the dissociation of ethane ion because the positive charge can be carried by either methyl. Similarly, $\alpha = 2$ in the dissociation of symmetric azo compounds if the decomposition takes place as written in Table 11-2, since in a two-fragment dissociation, the alkyl radical can split off from either end of the molecule.

Loose transition state, model B As we have mentioned, model A for the loose transition state represents a sort of upper limit on "looseness." Accordingly, there is no thermal reaction on record where the model gave an entropy of activation that was too low, but claims have been made (*78*) that it may give an entropy of activation that is too high. An alternative way of treating vibrations of low force constant—which is the basis of looseness in transition states—is to treat them simply as low-frequency (or "softened") vibrations; this approach has been championed in particular by Steel and Laidler (*79*). We shall refer to such loose transition states with low-frequency vibrations as "loose, model B."

The advantage of model B is that the frequencies can be manipulated with impunity to yield the proper entropy of activation, since the force constant is not known and cannot be calculated (at least not by small-vibration theory). Thus the degree of looseness of the transition state can be conveniently varied. By contrast, the only way the contribution of free rotations to the entropy of the transition state can be manipulated is through their number (i.e., by choosing how many low-frequency bendings are to be taken as free rotations) and/or their moments of inertia. However, as shown in Chapter 8, Section 7, the number of rotations can be determined from kinetic data only if the moments of inertia are known, and these, in turn, can be determined only if the number of rotations is known. Thus model A of the loose transition state demands, in practice, a more or less arbitrary decision about the number of rotations or their moments of inertia, or both, whereas model B demands only the more or less arbitrary specification of the vibrational frequency; thus model B is somewhat less ambiguous. In addition, this model may seem conceptually more pleasing in that it does not involve the perhaps disturbing notion that the bits and pieces of the transition state are not properly bonded together and actually rotate, a behavior unknown in stable molecules. Additionally, the more orthodox molecular framework in this model makes it somewhat easier to decide on the likely overall symmetry of the transition state, which has a bearing on the pertinence of external (overall) rotational degrees of freedom (cf. Chapter 5, Section 3).

Type 1 unimolecular dissociations that have been successfully treated, in varying degrees of rigor, using model B of the loose transition state include the decomposition of methyl halides (*67*) ($CH_3X \rightarrow CH_3 + X$, X = I, Br), ethane (*78,80*) ($C_2H_6 \rightarrow 2CH_3$), azomethane-d_6 (*81*) ($C_2D_6N_2 \rightarrow CD_3 + N_2CD_3$), ethyl bromide (*82*) ($C_2H_5Br \rightarrow C_2H_5 + Br$), ethyl iodide (*83*) ($C_2H_5I \rightarrow C_2H_5 + I$), ethane ion (*69*) ($C_2H_6{}^+ \rightarrow C_2H_5{}^+ + H$), propane ion (*29*) ($C_3H_8{}^+ \rightarrow C_2H_5{}^+ + CH_3$ and $C_3H_8{}^+ \rightarrow C_3H_7{}^+ + H$) and its deuterated analogs, butane (*84*) ($C_4H_{10} \rightarrow 2C_2H_5$), and hydrazine (*85*) ($N_2H_4 \rightarrow 2NH_2$), as well as transition states for reactions of the type $CH_2X + CH_2X \rightarrow C_2H_4X_2$ (X = Cl, Br, F, or H in one or both radicals) which are needed for the calculation of the distribution function $F(E)$ [Eq. (9-26)] in various chemically activated haloethanes (*82,86–88*).

The list includes several reactions that have also been treated by model A of the loose transition state. This is not too surprising because, at least in thermal reactions, the two models lead to the same high-pressure rate constant k_∞ [Eq. (8-23)] if the partition function for one "softened" vibration ($Q^*_{v=1}$) of model B is equal to the partition function ($Q^*_{r=1}$) for one equivalent free rotation of model A (asterisk refers to transition state):

$$Q^*_{v=1} = Q^*_{r=1}$$

or

$$[1 - \exp(-h\nu^*/kT)]^{-1} = (8\pi^2 I^* kT)^{1/2}/h \tag{11-20}$$

from which the equivalent moment of inertia I^* can be determined if ν^* is known, and vice versa. Since entropy depends on the logarithmic derivative of Q with respect to temperature [Eq. (8-115)], the equality in Eq. (11-20) does not imply equality of the corresponding vibrational and rotational entropies because the two partition functions have each a different temperature dependence. Application of Eq. (8-115) to $Q^*_{v=1}$ gives

$$S_{v=1} = k[(\ln Q^*_{v=1}) + (h\nu^*/kT)Q^*_{v=1} \exp(-h\nu^*/kT)] \tag{11-21}$$

which is different from (11-19) if we let $Q^*_{v=1} = Q^*_{r=1}$. Similarly, if Eq. (11-20) is satisfied, it does not imply that the microcanonical rate constant $k(E)$ will be equivalent for models A and B of the loose transition state; in fact, the energy dependence of $k(E)$ will be different for the two models (*89*). It is only because of the nature of averaging in thermal systems that models A and B give the same k_∞ if (11-20) is satisfied.[31] Only if the energy dependence of $k(E)$ were known in detail would it be possible to make a distinction between models A and B.

Model B, when applied to the prototype dissociation of a molecule into two fragments (e.g., ethane dissociating into two methyls), considers the transition state as involving vibrational degrees of freedom only; these are, in essence, the vibrational degrees of freedom with frequencies of the two

[31] For details, see Forst (*90*, p. 75 ff.).

separate fragments, one internal torsion, which can also be approximated by a free internal rotation, and four low-frequency bending vibrations, in lieu of the free rotations of model A. There is no universal agreement as to how low these bending vibrations should be. In the ethane molecule, there are two skeletal bendings around 800 cm^{-1} and methyl waggings around 1200 cm^{-1}; in the ethane transition state, these are generally lowered to around 100 cm^{-1}, and are down to about 50 cm^{-1} if there are heavy substituents like I or Br. In thermal systems, the amount of reduction in the frequency of the bending vibrations is determined primarily by entropy considerations, and in nonthermal systems, these bending frequencies are reduced by a more or less arbitrary factor varying from three[32] to ten (*84*).

The suggestion has been made by Benson (*91*) that loose transition states may involve ionic binding[33] because "softened" vibrations or free rotations are difficult to fit into the framework of normal covalent bonds. Thus in the case of the dissociation of ethane into two methyls, the transition state might be the structure $CH_3^-CH_3^+$, with both electrons of the about-to-be-broken C—C bond on the negatively charged methyl. Alternatively, the transition state might be represented as

$$CH_2^- \cdots H^+CH_3 \longleftrightarrow H_2C\text{—}H\cdots CH_3 \qquad (11\text{-}22)$$

In the structure on the right, one of the methyls is bonded via a three-center bond. This type of bonding has been suggested by Rice (*64,81*) to account for the low-frequency bending vibrations in the transition state for the decomposition of azomethane.

A summary In summary, then, experimental data show greatly reduced bonding in transition states for type 1 reactions of neutral (and some charged) species, without, however, pinpointing precisely which degrees of freedom are affected and by how much. Because this type of bonding is not found in normal stable molecules, analogy with similar stable molecules is not of much help in formulating these transition states. The ambiguity of the evidence accounts for the variety of structures that have been suggested to deal

[32] From entropy considerations, Benson (*52*) favors a reduction of the bending frequencies by a factor of three for elimination of groups like CH_3, NH_2, OH, and C_2H_5, except that in ethane dissociation, he favors a factor of six. In the dissociations $CH_3X \rightarrow CH_3 + X$ (X = I, Br) he favors a reduction by a factor of five; this factor was also used in Hassler and Setser (*67*).

[33] An ionic transition state has been invoked by Hassler and Setser (*67*) to account for the reaction of diazomethane with hydrogen halides, $CH_2N_2 + HX \rightarrow CH_3X + N_2$ (X = Cl, Br, I), where two bonds are broken and two bonds are formed, and yet the activation energy is low.

with the reduced bonding, structures which are all to an extent arbitrary and basically equivalent. The author favors at present the model using "softened" vibrations (model B) as slightly preferable to others in that it requires fewest assumptions in dealing with thermal and nonthermal reactions of the sort discussed in the previous chapters.

No systematic and unified theoretical treatment of all type 1 reactions has appeared so far, so that we lack a coherent overall picture. Critical analysis by Waage and Rabinovitch (*80*) of the ethane–methyl radical system, which is the most extensively investigated reaction in both the forward and reverse directions, reveals an inconsistency in that the dissociation into two methyls seems to require a somewhat tighter transition state than the reverse association.[34] A similar discrepancy is indicated for the dissociation of higher alkanes and the reverse radical associations (*93*).

3. Transition states for type 2 reactions

These reactions involve either (i) simple bond fission, e.g.,

$$C_2H_5\cdot \longrightarrow C_2H_4 + H\cdot \tag{11-23}$$

or (ii) complex bond fission, e.g.,

$$C_2H_5Cl \longrightarrow C_2H_4 + HCl \tag{11-24}$$

In either case, there occurs some internal reorganization of the reactant in the course of the reaction, and therefore there must be an activation energy for the reverse reaction [generally larger in case (ii) than in case (i)]; hence these reactions occur over type 2 potential energy surfaces, at least when neutral reactants are involved. Simple and complex bond fission occurs also in the decomposition of ions, e.g., simple bond fission:

$$C_2H_5{}^{+}\cdot \longrightarrow C_2H_4{}^{+} + H \tag{11-25}$$

and complex bond fission:

$$C_2H_4{}^{+} \longrightarrow C_2H_2{}^{+} + H_2 \tag{11-26}$$

except that here the activation energy for the reverse reaction is negligible, and the potential energy surface is therefore presumably of type 1. We shall assume that transition states for simple and complex bond fission in charged species nevertheless are similar to their neutral type 2 counterparts (cf. Chapter 10, Section 8).

[34] However, this view is not shared by Kobrinsky *et al.* (*92*), who find that the loose model A accounts well for the cross-combination rate of CH_3 and CF_3 radicals. According to Hase (*14*), the inconsistency between calculated forward and reverse rates in the ethane–methyl system is removed if the minimum density criterion is used for the critical configuration.

TABLE 11-3

THERMAL PARAMETERS FOR SOME TYPE 2 REACTIONS

Reaction	Ref.	$\log A_\infty$ (sec^{-1})	T (°K)	$E_{a\infty}$ (kcal $mole^{-1}$)	ΔS_* (eu $mole^{-1}$)	ΔS°_{298} (eu $mole^{-1}$)	ΔH°_{298} (kcal $mole^{-1}$)
(1) Butene-1 → $CH_3\cdot + C_3H_5\cdot$	*a*	12.7	950	59.1	−4.7	34.2	72
	b	13.9	850	69.5	1.0	—	—
(2) $C_2H_5\cdot \rightarrow C_2H_4 + H\cdot$	*c*	13.6	850	38.0	−0.4	20.3	38.9
	d	14.4	750	40.9	3.5	—	—
(3) $C_2H_5Cl \rightarrow C_2H_4 + HCl$	*e*	13.51	725	56.6	−0.5	31.1	16.7
n-$C_3H_7Cl \rightarrow C_3H_6 + HCl$	*e*	13.50	725	55.1	−0.5	32.8	14.0
$C_2H_5Br \rightarrow C_2H_4 + HBr$	*f*	13.45	650	53.9	−0.5	30.9	19.9
(4) Cyclobutene → butadiene	*g*	13.4	425	32.9	0.1	—	~6

A_∞ and $E_{a\infty}$ are experimental parameters determined from $k_\infty = A_\infty e^{-E_{a\infty}/kT}$; T is the mean experimental temperature in °K, ΔS_* is defined as $R \ln(A_\infty h/ekT)$, evaluated at the mean experimental temperature (see text in Section 2), ΔS°_{298} is the standard entropy change for the indicated reaction at 298°K, referred to standard state of ideal gas at 1 atm pressure, ΔH°_{298} is the standard enthalpy change, similarly defined; 1 eu = 1 cal deg^{-1} $mole^{-1}$.

References for experimental results:

[a] J. A. Kerr, R. Spencer and A. F. Trotman-Dickenson, *J. Chem. Soc.* 6652 (1965);
[b] M. P. Halstead and C. P. Quinn, *Trans. Faraday Soc.* **64**, 103 (1968);
[c] M. C. Lin and M. H. Back, *Can. J. Chem.* **44**, 2357 (1966);
[d] L. F. Loucks and K. J. Laidler, *Can. J. Chem.* **45**, 2795 (1967);
[e] H. Hartmann, H. G. Bosche and H. Heydtmann, *Z. physik. Chem.* (*N.F.*) **42**, 329 (1964);
[f] P. J. Thomas, *J. Chem. Soc.* 1192 (1959);
[g] R. W. Carr, Jr. and W. D. Walters, *J. Phys. Chem.* **69**, 1073 (1965).

Thermodynamic data from S. W. Benson and H. E. O'Neal, "Kinetic Data on Gas Phase Unimolecular reactions," NSRDS-NBS 21, Washington, D.C. 1970, except for reaction (4), where ΔH°_{298} is based on an estimate of C. S. Elliott and H. M. Frey, *Trans. Faraday Soc.* **62**, 895 (1966).

Table 11-3 summarizes thermal data on several type 2 reactions. Rate parameters for reactions (1) and (2) had to be extracted from a complex scheme because these reactions produce radicals, and therefore the parameters are less certain than for reactions (3) and (4), which occur essentially as written. More extensive summaries of experimental data on type 2 reactions of hydrocarbons (*51,53*) and alkyl halides (*53*)[35] are available.

Circumstantial kinetic evidence from thermal systems that throws some light on transition states in type 2 reactions is the following.

[35] Thermal data for other alkyl halides have been collected by Maccoll (*94*) and Benson (*52*, Table 3.5).

(a) There is indeed an activation energy for the reverse reaction, as a comparison between $E_{a\infty}$ and ΔH° in Table 11-3 shows. Reactions (1) and (2) in Table 11-3 are possibly exceptions in that ΔH° seems to be about equal to $E_{a\infty}$; however, like all other reactions in the table, reaction (1) involves some internal reorganization because the allyl radical $C_3H_5\cdot$ is resonance stabilized.

(b) The preexponential factors are all in the vicinity of 10^{13} sec^{-1}, which corresponds in most cases to a small negative entropy of activation. Since the overall entropy change in these reactions has a large [except for reaction (4)] *positive* value, this suggests that the transition states are closer to reactants than to products, and possibly (in the case of alkyl halides) are even more highly structured than the reactants themselves.

Simple bond fission We shall take up simple bond fission first. The example (11-23) represents an H split-off; in larger radicals, there may also be alkyl split-off, e.g.,

$$\underset{sec\text{-butyl radical}}{C_4H_9\cdot} \longrightarrow CH_3\cdot + \underset{\text{propylene}}{C_3H_6} \tag{11-27}$$

a well-known chemically activated system, quoted frequently in Chapter 9. Since in both H-atom and alkyl split-off the other product is an olefin, there is reason to believe that an incipient double bond will be forming in the transition state, and therefore transition states for this type of fission are commonly represented by a semirigid model. The activation energy for the reverse of (11-27) is larger (*95*, p. 43) ($\sim$8 kcal $mole^{-1}$) than for the reverse of (11-23) ($\sim$2 kcal $mole^{-1}$) and therefore the transition state for H split-off is probably looser than for alkyl split-off.

There is no universal agreement as to how tight or how loose the transition state for H split-off should be. Rabinovitch and collaborators[36] use a model consisting of the frequencies of the product olefin plus two C—H bending vibrations at $\sim$300 cm^{-1} (i.e., reduced by a factor of about five from their usual value in the radical). In a looser version of this model, the C—H bendings are further decreased to 150 cm^{-1}. This olefin-based model accounts reasonably well for the decomposition of thermally (*78*) and chemically activated (*96*) C_2H_5 and its deuterated analogs and also for the transition state in the reaction (*97*) H + butene which produces the chemically activated butyl radicals; this transition state is necessary for calculating the chemical activation distribution function $F(E)$ [Eq. (9-26)]. In the similar decomposition of chemically activated trichloroethyl radical (*98*) $CHCl_2CHCl \rightarrow C_2H_2Cl_2 + Cl$, the olefin-based model was also satisfactory, with the two additional C—Cl bends reduced by a factor of eight from their usual value in the radical.

[36] See (*24*) for a review. For ethyl radical decomposition, see also (*180*).

However, another model is possible for the transition state in the H (or Cl) split-off, one based on the frequencies of the *radical* (after elimination of the C—H or C—Cl stretch that becomes the reaction coordinate), in which the radical C—C torsion and C—C stretching frequencies are sometimes (*98*)[37] but not always (*1*) increased, to account for the incipient double bond formation, and two C—Cl bends (in the chloroethyl radical) are reduced eightfold. This model was used by Knox and collaborators (*1*) in the calculation of transition states in the decomposition of all the chemically activated chloroethyl radicals, from monochloroethyl to pentachloroethyl.[38]

The frequencies in the transition state for alkyl split-off, e.g., in the reaction (11-27) and in similar reactions of higher alkyl radicals (*101*), are usually based on those of the reactant radical. One C—C stretch becomes the reaction coordinate; in the transition-state model for reaction (11-27) used by Rabinovitch and co-workers (*97*), bending frequencies of the radical which belong to the separating fragments were lowered to one-half their value (the radical frequencies to be lowered at the CH_2 wagging at 1304 cm^{-1}, two CH_3 rocking modes at 1148 and 959 cm^{-1}, and one skeletal bend at 365 cm^{-1}), and the other C—C stretch and torsion frequencies were increased to a value corresponding to a bond order of 1.5, i.e., from 1050 to 1350 cm^{-1} and from 125 to 250 cm^{-1}, respectively, to account for the double bond forming in the future propylene product.

Transition states for simple bond fission in olefins are likewise generally based on the reactant olefin (*102*) since these reactions may be looked upon as intermediate between type 1 and type 2. The following numbers refer to bond fission in butene-1 [reaction (1) in Table 11-3], which is a fairly typical example. The reaction coordinate is a C—C stretch (around 900 cm^{-1} in the reactant, deleted in the transition state); another C—C stretch (at ~850 cm^{-1}) is increased to ~1200 cm^{-1} to account for 1.5 bond order due to allylic resonance in the transition state. For the same reason, the C—C torsion in butene-1 (at ~1000 cm^{-1}) is lowered to ~600 cm^{-1}. In order to build some looseness in the transition state, the following lowerings in the transition state were made relative to reactant: CH_2 rock, from ~750 cm^{-1} to ~200 cm^{-1}; CH_3 wag, from ~1200 cm^{-1} to ~300 cm^{-1}; and a skeletal

[37] The increase in the C—C stretch and torsion frequencies is made to correspond to bond order 1.5, i.e., a bond intermediate between a single and a double bond.

[38] Huy *et al.* (*99*) have successfully calculated the decomposition of the trichloroethyl radical by leaving the radical frequencies unchanged in the transition state (after elimination of the reaction coordinate). Similarly, Dees and Setser (*100*) calculated the rate in the chemically activated system $CH_2CO\ (^3A_2) \rightarrow CH_2\ (^3\Sigma_g^-) + CO\ (^1\Sigma_g^+)$ by taking the vibrational frequencies of the transition state to be the same as those of the ketene reactant, with the C—C stretch frequency deleted as reaction coordinate. This procedure is recommended when the critical energy is not known with certainty and more elaborate calculations are not warranted.

bend from $\sim 300\ cm^{-1}$ to $\sim 50\ cm^{-1}$. More or less similar assignments were used in transition states for the decomposition of butane derivatives (*103*) and higher olefins (*102*). It is significant that a loose type 1 transition state for these bond fission reactions in chemically activated olefins does not account for experimental data (*102*), showing that these are indeed type 2 reactions with a tighter transition state.

Reaction path degeneracy in these bond fission reactions is generally straightforward: it is unity for process (11-27) or reaction (1) in Table 11-3, but will be higher if the reactant molecule contains several methyl groups, any one of which can split off. Similarly, the process (11-23) has $\alpha = 3$, if it is assumed that the product hydrogen atom can arise from any one of the three hydrogens on the methyl (see footnote 14).

It should be clear that transition states for H or alkyl split-off in alkyl radicals, or for bond fission in olefins, represent no fundamental departure from transition states for type 1 reactions discussed in the previous section; in fact, there are various degrees of "looseness" or tightness," depending on how much internal structural reorganization of the reactant is taking place in the course of the reaction. This aspect has been discussed by Hay (*104*) from the point of view of the simple Hückel molecular orbital theory. Loose transition states involve simply a stretching, and ultimately, the rupture of a σ bond into two radicals, without much significant structural reorganization. If the product radicals have a strong tendency to exist in a π configuration, a change from a tetrahedral to planar configuration takes place, necessitating a change of hybridization from sp^3 to sp^2, plus the formation of a π bond, which leads to a stiffening in the transition state, the extent of which may vary from case to case. A correlation exists between the delocalization energy of the product radicals and the entropy of activation of the corresponding thermal unimolecular dissociation (i.e., the thermal preexponential factor A_∞) (*104*, Fig. 1), and between the π-electron energy of the radicals and the preexponential factor for their thermal bimolecular association (*104*, Fig. 2). However, some of the experimental data cited by Hay in support of his argument is suspect.

Complex bond fission The example (11-24) for complex bond fission is one of the most extensively studied reactions, in both thermal (*105–110*)[39] and chemically activated (*115*) systems and in varying degrees of deuterium substitution (-d_0, -d_3, and, -d_5). In addition, chemical activation data are

[39] A survey of thermal results for HF elimination from alkyl fluorides has been given by Kerr and Timlin (*111*, Table VI). Data on HF and HCl elimination from α-fluoromethanol and α-chloromethanol, respectively, have been reported by Lin (*112,113*) and by Brus and Lin (*114*).

available for: HBr and HF elimination from bromoethane (*82*)[40] and fluoroethane (*88*) respectively; HX (X = Cl, Br, F) elimination from 1,2-substituted dichloroethane (-d_0 and -d_4) (*34*), dibromoethane (*82*)[40], bromochloroethane (*82*) and difluoroethane (*117*), respectively; HCl elimination from 1,1-dichloroethane (*118*), 1,1,2-trichloroethane (*118*), 1,3-dichloropropane (*119*), 1,4-dichlorobutane (*119*), and 1-chloropropane (*119*); and HF elimination from 1,1,1-trifluoroethane (*88,120*).

Setser and collaborators, who produced practically all of the chemical activation results, have made a systematic theoretical analysis of these reactions and have shown that essentially the same transition state accounts for all the results regardless of whether the system is thermally or chemically activated and regardless of the nature of the substrate molecule and the halogen it contains. They visualize the reaction as proceeding by the simultaneous rearrangement of four bonds: the C—Cl and C—H bonds are breaking and the H—Cl and the C—C double bonds are forming; thus

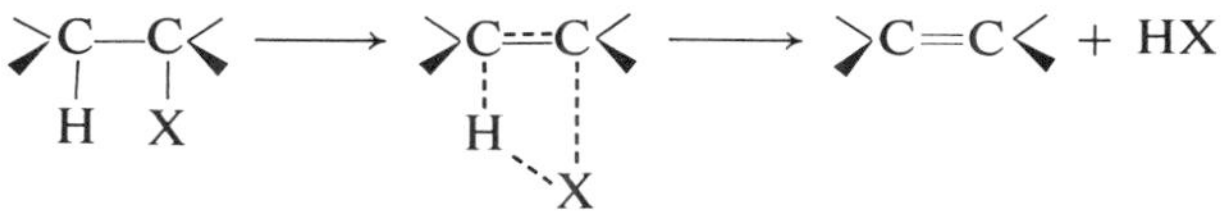

$$\begin{array}{c}{>}\!\!{\underset{\underset{H}{|}}{C}}\!\!-\!\!{\underset{\underset{X}{|}}{C}}\!\!{<} \longrightarrow {>}\!\!{\underset{\underset{H}{\vdots}}{C}}\!\!\cdots\!\!{\underset{\underset{\ddots X}{\vdots}}{C}}\!\!{<} \longrightarrow {>}C{=}C{<} + HX \end{array} \qquad (11\text{-}28)$$

The reaction is thus a four-center process, and in setting up the transition state, attention is focused on the in-plane frequencies of the C—C—X—H ring, assumed to be planar, which contains the atoms directly concerned in the chemical transformation.

Tritium hot-atom experiments (*121*) on various deuterated chloroethanes have shown rather conclusively that the elimination of HX is an α, β process. In more heavily substituted haloethanes, e.g. 1,1,1-trifluoro-2-chloroethane (*116*) and 1,1,2-trichloroethane-1-d_1 (*157*), however, there also occurs HX elimination via an α, α process. Data on the α, α eliminations are still sparse and the process will not be discussed here; presumably it occurs via a *three-center* transition state.

In order to investigate systematically the variation of transition-state parameters, Setser and collaborators have used as guiding principle the assumption that throughout the reaction, the total bond order in the C—C—X—H ring remains constant at three. One possibility, for example, would be to assign a bond order of 1.5 to the C—C bond (a double bond in the process of formation) and bond order of 0.5 to the other three bonds (single bonds forming or breaking). Other, less symmetric assignments of bond orders are possible, and were also considered by Setser and collaborators. The assigned bond order is then related to bond length r_x by Pauling's

---

[40] This reference also includes a summary of thermal data on HBr elimination from bromoethane.

rule [Eq. (11-17)] and from the bond length there can be obtained the corresponding force constant by means of a semiempirical relation due to Johnston (*122*)[41]:

$$\log f_2 = (a - r_x)/b \qquad (11\text{-}29)$$

where f_2 is force constant and a and b are constants for a given pair of atoms. Finally, these force constants are used in an F–G matrix to obtain the in-plane ring frequencies in the transition state. There are five normal modes corresponding to the in-plane motions, of which the one with the lowest frequency (generally below 50 cm^{-1}) is taken as the reaction coordinate.[42] The out-of-plane C—H frequencies[43] were assigned by analogy with cyclobutane and ethylene. There remains the ring-puckering out-of-plane frequency, which was adjusted to fit the data, and was generally set at 400 cm^{-1}. Internal torsion in the transition state was treated as a low-frequency vibration. In transition states for the decomposition of larger molecules (chloropropanes and chlorobutane), the frequencies of the group attached to the ring were assumed to be the same as in the corresponding chloroalkane.

Vibrational frequencies in isotopically substituted molecules were derived from those of the normal molecules by the application of the Teller–Redlich product rule (11-2); the moments of inertia, necessary for calculating the moment-of-inertia ratio that appears in (11-2), are given by the molecular geometry in the transition state, which depends on bond length, and, through Eq. (11-17), ultimately on the assigned bond order.

Since thermal data (*105–109*) show an appreciable temperature dependence of $k_{\infty(H)}/k_{\infty(D)}$, there is an important primary[44] isotope effect in reactions of the type (11-28), and Setser and collaborators therefore had to consider asymmetric assignments of bond orders in the transition state. It turns out that the model that accounts best for all the available data, i.e., chemical activation, thermal activation, and hydrogen–deuterium kinetic isotope effect, is a highly asymmetric model of the transition state, with the ring hydrogen weakly bonded to both carbon and chlorine (bond order 0.2 each) and a fairly strongly bonded chlorine to carbon (bond order 0.8); by difference, the C—C bond order then comes out to be 1.8. This same model also accounts successfully for HBr and HF elimination from bromoethanes and fluoro-

[41] A similar useful relation is Badger's rule as modified by Herschbach and Laurie [Eq. (6-11)]. The f_2 referred to in (11-29) is the *quadratic* force constant.

[42] For details, see Hassler (*123*, p. 78a), and (82, Table V).

[43] In the case of the C_2H_5Cl transition state, these are the C—H stretch (four each, around 3000 cm^{-1}), CH_2 deformation (four each, around 1350 cm^{-1}), CH_2 twist (two each, around 900 cm^{-1}), and CH_2 rock (two each, around 900 cm^{-1}).

[44] However, the *overall* observed isotope effect is a complicated combination of both primary and secondary effects.

ethanes.[45] Reaction path degeneracy has been considered in Chapter 4, Section 5.

The assumption of constant bond order during the course of a reaction is the central idea of the BEBO method (*49*) used successfully by Johnston to account for bimolecular exchange reactions, and the procedure used by Setser and collaborators may be considered an interesting adaptation of it to a particular type of unimolecular dissociation.

However, relations (11-17) and (11-29) are only semiempirical and therefore not unique (see footnote 23), and perhaps also are not particularly suited to very low bond orders, so that the transition-state model postulated by Setser and collaborators is likewise not unique, of which they are well aware (*34*). For example, O'Neal and Benson (*124*) assumed a transition state without a fully developed ring structure ("loose cycle") and with a partially ionic character;[46] it appears, however, that this model does not fit the observed isotope effect (*34*). Nevertheless, O'Neal and Benson's model is not radically different from that of Setser and collaborators and because it assigns only simple partial bond orders ($\frac{1}{2}$ or $\frac{3}{2}$), it may be useful for treating reaction systems where available experimental information is limited. Others (*126–128*), notably Maccoll (*94*), have considered the transition state in (11-28) to be entirely ionic.

Another approach was used by Heydtmann (*110*), who, while retaining a cyclical transition state for reaction (11-28), argued against conservation of total bond order in the reaction because of the high activation energy. Recent results on energy partitioning among the products of HF elimination from CH_3CF_3 show (*120*) that at the top of the barrier separating reactants from products, the C—C double bond is probably not fully formed, i.e., the C—C bond order in the transition state is closer to one rather than close to two as assumed by Setser and collaborators. In practical terms, reduction of the C—C bond order from 1.8 to 1 means that the C—C stretching frequency in the transition state has to be reduced from $\sim 1400\ cm^{-1}$ to $\sim 1000\ cm^{-1}$, which will have negligible influence on the calculated results but destroys the principle of conservation of total bond order. In recent work, Setser and collaborators use bond orders of 1.1, 0.8, 0.1, and 0.1, for the C—C, C—X, X—H, and H—C bonds, respectively (total bond order 2.1).

Chemical activation rate constants, thermal *A* factors, and thermal rate constant fall-off with pressure all depend only weakly on the various assumptions regarding the transition state, and therefore all the different models of

[45] In more recent work (*115,117,119*) this model was modified slightly to bond orders 1.9, 0.9, 0.1, and 0.1 for the C—C, C—X, X—H, and H—C bonds, respectively.

[46] This transition-state model was also used by Kerr and Timlin (*125*) to treat HF elimination from 1,2-difluoroethane.

the transition state may account equally well for a given segment of the data. The interest in the procedure set up by Setser and collaborators, arbitrary though it may be, is that transition-state parameters can be varied systematically and that in arriving at the final model they considered the total available data.

A four-center transition state similar to (11-28) but much less elaborately optimized has been used also in H_2 elimination reactions from the ethane ion (*69*) ($C_2H_6^+ \rightarrow C_2H_4^+ + H_2$) and from the propane ion (*29*) ($C_3H_8^+ \rightarrow C_3H_6^+ + H_2$).

Cyclobutene Reaction (4) in Table 11-3, the isomerization of cyclobutene to butadiene, which is typical of isomerization reactions of cyclobutenes (*51*, Table XI), yields only one product and should therefore be a type 3 reaction. However, comparison with thermal parameters of legitimate structural isomerizations listed in Table 11-4 shows that cyclobutene isomerizes with an abnormally low activation energy ($\sim$35 kcal mole^{-1} versus $\sim$65 kcal mole^{-1} for cyclopropanes and cyclobutane) and a low preexponential factor, i.e., a low entropy of activation.

The low entropy of activation suggests that the reaction probably involves a puckering of the cyclobutene ring and simultaneous stretching of the C—C bond opposite the double bond, which, if it takes place as a concerted process, would explain the low activation energy. The transition state is thus akin to that of the process (11-28) and therefore different from that of other type 3 isomerizations. Elliott and Frey (*129*) have done calculations on the thermal isomerization of three cyclobutenes, assuming either a type 2 or type 3 transition state. The calculated results (entropy of activation, pressure fall-off) could not distinguish between the two kinds of transition states, but a type 2 transition state seems indicated because the isomerizations are highly stereospecific. Similarly, calculations on the very similar thermal isomerization of 3,3,4,4-tetrafluorocyclobutene were insensitive to transition state assignments (*168*).

Because of its stereospecificity, the isomerization of cyclobutene has always intrigued the theoretical chemist. The latest self-consistent field-configuration interaction calculations (*130,131*) show that the reaction must proceed via a stepwise mechanism in which the rotation of the methylene groups, in either the conrotatory or disrotatory mode, occurs only after an expanded cyclobutene structure has been formed, and then continues to completion before further significant increase in terminal C—C bond distance occurs. Calculations therefore seem to rule out the concerted process but do not really explain the low activation energy. Some unusual aspects of the potential energy surface for this process have been pointed out by Dewar and Kirschner (*169*), and by Rastelli *et al.* (*170*).

4. Transition states for type 3 reactions

Type 3 reactions, as mentioned in Chapter 3, are isomerizations. If the product contains only a different geometric arrangement of the same bonds as in the reactant, the isomerization is said to be geometric, whereas if the product contains atoms of the reactant bonded together differently, implying that existing bonds have been broken and new bonds have been made in the course of the reaction, the isomerization is said to be structural.

Product(s) of a reaction contain (in the high-pressure limit of a thermal reaction) the energy difference $E_{a\infty} - \Delta H^\circ$, which in type 1 and type 2 reactions is partitioned among at least *two* particles, most of it in the form of vibrational energy and some of it in the form of relative translational energy (cf. Chapter 10, Section 9). In an isomerization, the single product particle contains the entire energy difference $E_{a\infty} - \Delta H^\circ$, and therefore can easily revert back to the reactant, especially if ΔH° is positive, i.e., if the energy of activation for the reverse reaction is low. Lin and Laidler (*132*) have applied the necessary correction for back reaction to the calculated rates of isomerization of *cis*-butene-2, cyclobutene, and cyclopropane. As might have been expected from the size of ΔH° in these reactions (Tables 11-3 and 11-4), the correction for the back reaction is important only for the first two reactions and is negligible in the case of the (structural) isomerization of cyclopropane.

Geometric isomerization Geometric isomerizations generally involve merely a change from a *cis* conformation to a trans conformation (or vice versa), and the simplest case, at least conceptually, is the cis–trans isomerization of an open-chain (noncyclical) compound where the change is accomplished by a 180° rotation about a bond. The reaction coordinate is the angle of torsion and the activation energy is the height of the rotational barrier separating the two isomers. A very simple example is the cis–trans isomerization of nitrous acid, for which Hisatsune (*133,134*) has calculated the thermal rate [i.e., k_∞, Eq. (8-23)] using data obtained from infrared spectroscopy. The *trans* isomer is known to be planar, with the structure

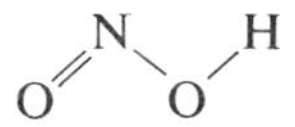

and the cis isomer is very likely nonplanar with the O—H group twisted out of the O═N—O plane; the isomerization therefore involves torsion about a single bond. The rotational barrier maximum in this case is at 86° from the cis minimum, and Hisatsune therefore took the structural parameters of the transition state to be essentially intermediate between those of the cis and trans acids. He also calculated the quantum mechanical tunneling correction

TABLE 11-4

THERMAL PARAMETERS FOR SOME TYPE 3 REACTIONS

Reaction	Ref.	$\log A_\infty$ (sec^{-1})	$T(°K)$	$E_{a\infty}$ (kcal mole^{-1})	ΔS_* (eu mole^{-1})	ΔS°_{298} (eu mole^{-1})	ΔH°_{298} (kcal mole^{-1})
cis → *trans* Isomerizations							
(I) *cis*-CDH=CHD → *trans*	a	13	750	65	−2.9	0	0
trans-CHCl=CHCl → *cis*	b	12.7	800	55.3	−4.4	0	−1.0
cis-CH$_3$CH=CHCH$_3$ → *trans*	c	13.8	700	62.8	0.9	−1.2	−1.0
cis-CF$_3$CF=CFCF$_3$ → *trans*	d	13.6	700	56	0.02	0.5	−0.8
(II) *trans*-1,2-Dideuterocyclopropane → *cis*	e	16.0	750	64.2	10.9	1.4	0
cis-1,2-Dimethylcyclopropane → *trans*	f	15.5	700	59.4	8.7	−2.6	−1.0
cis-1,2-Dimethylcyclobutane → *trans*	g	14.8	700	60.1	5.5	−1.2	−1.0
Structural isomerizations							
Propane → propylene	h	15.2	750	65	7.2	6.8	−7.9
trans-1,2-Dideuterocyclopropane → propylene-d_2	i	15.12	750	65.4	6.8	6.0	−7.9
Methylcyclopropane → isobutylene	j	14.6	750	66	4.5	3.3	−9.4
1,1-Dimethylcyclopropane → 3-methyl-1-butene	k	14.8	750	62.6	5.4	8.8	−5.0
→ 2-methyl-2-butene		14.8	750	62.6	5.4	10.2	−8.3
1,2-Dimethylcyclopropane → 2-methyl-1-butene	l	13.9	700	62	1.4	6.7	−8.3
→ 2-methyl-2-butene		14.0	700	62.3	1.8	5.9	−9.9
Cyclobutane → 2C$_2$H$_4$	m	15.6	730	62.5	9.1	41.5	18.7
CH$_3$NC → CH$_3$CN	n	13.6	500	38.4	0.7	−0.4	−16.7
C$_2$H$_5$NC → C$_2$H$_5$CN	o	13.9	500	38.4	2.0	—	—

A_∞ and $E_{a\infty}$ are experimental parameters determined from $k_\infty = A_\infty e^{-E_{a\infty}/RT}$; T is the mean experimental temperature in °K, ΔS_* is defined as $R \ln(A_\infty h/e\mathcal{k}T)$, evaluated at the mean experimental temperature (see text in Section 2), ΔS°_{298} is the standard entropy change for the indicated reaction at 298°K, referred to standard state of ideal gas at 1 atm pressure, ΔH°_{298} is the standard enthalpy change, similarly defined; 1 eu = 1 cal deg^{-1} mole^{-1}.

References for experimental results:

[a] J. E. Douglas, B. S. Rabinovitch and F. S. Looney, *J. Chem. Phys.* **23**, 315 (1955);
[b] L. D. Hawton and G. P. Semeluk, *Can. J. Chem.* **44**, 2143 (1966);
[c] B. S. Rabinovitch and K. W. Michel, *J. Amer. Chem. Soc.* **81**, 5065 (1965);
[d] E. W. Schlag and E. W. Kaiser, Jr., *J. Amer. Chem. Soc.* **87**, 1171 (1965);
[e] E. W. Schlag, B. S. Rabinovitch and K. Wiberg, *J. Chem. Phys.* **28**, 504 (1958);
[f] M. C. Flowers and H. M. Frey, *Proc. Roy. Soc.* (*London*), **A257**, 122 (1960);
[a] H. R. Gerberich and W. D. Walters, *J. Amer. Chem. Soc.* **83**, 4884 (1961);
[h] T. S. Chambers and G. B. Kistiakowsky, *J. Amer. Chem. Soc.* **56**, 399 (1934):
[i] E. W. Schlag and B. S. Rabinovitch, *J. Amer. Chem. Soc.* **82**, 5996 (1960);
[j] D. W. Placzek and B. S. Rabinovitch, *J. Chem. Phys.* **69**, 2141 (1965);
[k] M. C. Flowers and H. M. Frey, *J. Chem. Soc.* 3953 (1959);
[l] M. C. Flowers and H. M. Frey, *Proc. Roy. Soc.* (*London*) **A260**, 424 (1961);
[m] C. T. Genaux, F. Kern and W. D. Walters, *J. Amer. Chem. Soc.* **75**, 6196 (1952);
[n] F. W. Schneider and B. S. Rabinovitch, *J. Amer. Chem. Soc.* **84**, 4215 (1962);
[o] K. M. Maloney and B. S. Rabinovitch, *J. Phys. Chem.* **73**, 1652 (1969).

Thermodynamic data from S. W. Benson and H. E. O'Neal, "Kinetic Data on Gas Phase Unimolecular Reactions," NSRDS-NBS 21, Washington, D.C. 1970.

by fitting an Eckart potential to the rotational barrier. Unfortunately, there are no experimental data to compare his calculations with.

A large group of cis–trans isomerizations for which there exists a good deal of experimental information are isomerizations involving rotation about a double bond. The thermal parameters for a few representative reactions are summarized in Table 11-4 (group I). From the listed values of $E_{a\infty}$ and ΔH°, we see that these reactions involve an appreciable activation energy for both the forward and back reactions. The preexponential factor is in the neighborhood of 10^{13}, which means that, depending on temperature, the entropy of activation ΔS_* may be negative or positive but is always small in absolute value, and the overall entropy change in the reaction ΔS° is likewise small in absolute value, positive or negative. It is not possible to deduce much about the structure of the transition state in these reactions from entropy considerations because the overall entropy change is very likely smaller than the probable error in ΔS_*. The small value of ΔS° does suggest, however, that reactant and product are very similar, as of course we would expect for a cis–trans isomerization.

If entropy of activation does not give much useful information in the present case, activation energy does. In the theoretical interpretation of these cis–trans isomerizations, Lin and Laidler (*135*) have noted that there exists a close relation between the activation energies and π-bond energies[47] (E_π); thus for ethylene, E_π comes out to be ~ 62 kcal mole^{-1}, which is fairly close to the $E_{a\infty}$ values for the reactions of group I in Table 11-4. The activation energies are therefore of the right magnitude necessary to destroy the π bond of the reactant, which suggests that the transition state must correspond to a twisted configuration where the p orbitals of the adjacent carbons are at right angles to each other, with no effective overlap, i.e., no π bonding. Essentially similar interpretation has been given by Benson *et al.* (*136*), except that they see the transition state as a biradical because of the absence of π bonding.

Thus, at least insofar as the angle of twist is concerned, the transition state is a configuration exactly half-way between the configuration of the reactant and that of the product. If we therefore take the angle of twist as the reaction coordinate and assume that, except for the angle of twist, reactant and products are identical, then in the expression for k_∞ [Eq. (8-23)], every term of

[47] The π-bond energy E_π in ethylene, for example, may be defined as $E_\pi(C_2H_4) = D(H{-}C_2H_4^\dagger) - D(H{-}C_2H_4)$, where $D(H{-}C_2H_4^\dagger)$ is the dissociation energy of the H bond in the hypothetical process $C_2H_5 \rightarrow H + C_2H_4^\dagger$ in which $C_2H_4^\dagger$ is an ethylene species *without* a π bond (e.g., some electronically excited state of ethylene), and $D(H{-}C_2H_4)$ is the dissociation energy of the H bond in a similar process where C_2H_4 is formed *with* the usual π bond; thus the difference E_π measures effectively the strength of the π bond.

the partition function ratio Q^*/Q cancels except the partition function for the torsional motion (frequency ν_θ), which of course is absent from Q^* but is present in Q, where it contributes the factor $(1 - e^{h\nu_\theta/kT})^{-1}$. Therefore we should have for the cis–trans isomerization about a double bond

$$k_\infty = \alpha f_\infty(kT/h)(1 - e^{-h\nu_\theta/kT})e^{-E_0/kT} \tag{11-30}$$

where E_0 is the height of the potential barrier separating the *cis* and *trans* isomers.

In *cis*-CDH═CDH, the first reaction listed in Table 11-3, the torsional frequency is (*137*) $\nu_\theta \simeq 990\ \text{cm}^{-1}$, and (11-30) then yields, with[48] $\alpha = 2$ and $f_\infty = 1$ (i.e., ignoring centrifugal effects) the result

$$A_\infty = 2.5 \times 10^{13} \quad \text{sec}^{-1} \tag{11-31}$$

which is of the right magnitude.[49] Fall-off calculations on the isomerization of *cis*-butene-2 were done by Wieder and Marcus (*73*) with rather poor results. Later calculations by Lin and Laidler (*132*), taking into account the back reaction, gave better agreement. However, for many other cis–trans isomerizations, especially when large substituents are present in the molecule (not listed in Table 11-4), Eq. (11-30) yields a much worse agreement with experiment, for reasons that are not entirely clear. The experimental data may be uncertain in these cases, inasmuch as thermal isomerizations are often plagued by side reactions and surface effects and photochemical isomerizations are complicated by the presence of excited electronic states (*138*, Chapter 4, p. 165). Alternatively, the theory may be too crude in that it does not take into account possible interactions between the double bond and other substituents that may be present in the molecule.

Geometric and structural isomerization of cyclopropanes Group II of cis–trans isomerizations in Table 11-4 involves cyclic compounds, and it is immediately obvious that these isomerizations are different from those of group I in that they have significantly higher preexponential factors, and therefore also higher entropies of activation, although the overall entropy and enthalpy changes ($\Delta S°$ and $\Delta H°$, respectively) are quite comparable for both groups. In fact, the thermal parameters for the geometric isomerization of a group II compound are quite similar to the thermal parameters for the

[48] Reaction path degeneracy in cis–trans isomerization is $\alpha = 2$ since presumably the internal twisting motion can take place in either of two directions.

[49] However, if one uses the value $\nu_\theta = 990\ \text{cm}^{-1}$ to calculate (*20*, Eq. II.304, p. 227) the height of the potential barrier separating the two isomers, then, assuming harmonic vibrations and taking the rotational constant A of CDH═CDH as being intermediate between those of C_2H_4 (A = 4.867 cm^{-1}) and C_2D_4 (A = 2.437 cm^{-1}) (*20*, p. 437), the result is $E_0 = 1.68 \times 10^4\ \text{cm}^{-1} = 48\ \text{kcal mole}^{-1}$, which is considerably less than the value of $E_{a\infty}$ listed for this isomerization in Table 11-4.

structural isomerization of the same compound,[50] as was first noted by Rabinovitch *et al.* (*141*) in the case of 1,2-cyclopropane-d_2. In absolute terms, however, the rate of the cis–trans isomerization is faster than for the structural isomerization because the latter has a slightly higher activation energy.

The similarity in the preexponential factors A_∞ for the geometric and structural isomerizations of these cyclic compounds suggests that the transition states must be similar. Inasmuch as the entropy of activation ΔS_* that can be calculated from A_∞ for either geometric or structural isomerization is similar to the overall entropy change in the structural isomerization, the transition states must be akin to the product of the structural isomerization, i.e., some sort of acyclic structure.

Unfortunately, these general observations are compatible with at least two different, though similar, mechanisms. The two basic possibilities in cyclopropane isomerization, first suggested by Chambers and Kistiakowsky (*142*), are (i) a direct or one-step mechanism, involving simultaneous carbon rupture and hydrogen migration, or (ii) a two-step mechanism, where a trimethylene diradical intermediate is first formed by carbon rupture and can then either revert to the original cyclopropane reactant (or its geometric isomer) or yield propylene by hydrogen migration. In the one-step mechanism for the structural isomerization of cyclopropane, for example, the transition state is seen as a structure more or less half-way between cyclopropane and propylene, whereas in the two-step mechanism, the transition state is seen as a structure roughly half-way between the trimethylene biradical and propylene. There is no definite proof for or against either mechanism, mainly because the postulated trimethylene biradical intermediate remains undetectable despite the fact that the isomerization of cyclopropane is one of the most-studied reactions and has played an important part in the development of unimolecular rate theory, particularly the theory of Slater.[51] Calculations by the statistical (RRKM) theory are insensitive to transition-state assignment and therefore either mechanism yields calculated results in reasonably good agreement with experiment.

Rabinovitch and co-workers have systematically (and successfully) interpreted the considerable amount of data produced in his laboratory on various thermally and chemically activated cyclopropanes (cf. Chapter 9, Sections 3–5) in terms of the one-step mechanism but involving transition states that are assumed to be slightly different in the geometric and structural isomerizations.

In the geometric isomerization of 1,2-cyclopropane-d_2, the transition state

[50] Extensive reviews of thermal data on the isomerization of small ring compounds have been given by Frey (*139*) and by Willcott *et al.* (*140*).

[51] For a brief historical review, see the article by Laidler and Loucks in Bamford and Tipper (*3*, p. 1).

is assumed to involve ring expansion (*143*) (but not ring opening) with the concomitant rotation of a CHD group (*144*). The reaction coordinate is a ring mode at $\sim 1000\ cm^{-1}$, which is therefore discarded in the transition state: two CHD, or one CH_2 and one CHD, twisting vibrations near 1200 cm^{-1} in the molecule become hindered rotors in the transition state, with a hindering potential assumed to be ~ 3.5 kcal $mole^{-1}$; the other frequencies are assumed to have the same value in the transition state and in the molecule. The change from twisting vibrations to hindered one-dimensional rotors is done mainly because the geometric isomerization has a higher entropy of activation ΔS_* than the structural isomerization (cf. Table 11-4). Since the ring expansion (arrow) of the cyclopropane ring can occur at any one of three

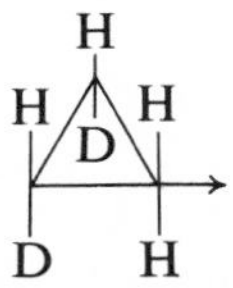

apexes, the reaction path degeneracy is three. One path is associated with two CHD rotors, two paths each are associated with one CHD and one CH_2 rotor; these were averaged in the transition state. In alkyl-substituted cyclopropanes, a similar procedure was used, and frequencies of alkyl groups attached to the ring were assumed to be the same in the molecule and in the transition states. The alkyl group torsion with respect to the cyclopropane ring was taken as a free rotation and the barrier for hindered internal rotation[52] of the CH_2 or CHCH groups was taken as ~ 20 kcal $mole^{-1}$ in the case of the 1,2-dimethyl compound in order to obtain fit with the observed thermal rate. This model accounts well[53] for the rate and kinetic isotope effect in the geometric isomerization of thermally and chemically activated 1,2-cyclopropane-d_2 and 1,2-dimethyl cyclopropane-d_8.

The transition state of Rabinovitch and co-workers[54] for the structural isomerization of cyclopropane is assumed to have the structure of a species between an expanded cyclopropane ring and trimethylene:

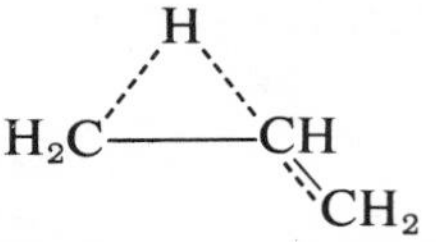

[52] In a more recent study by Rynbrandt and Rabinovitch (*145*) the hindered internal rotation in the transition state for the geometric isomerization of 1,2-dimethyl cyclopropane was replaced by torsional vibrations at 175 and 150 cm^{-1}, and five low CH_3—C bending frequencies in the transition state were lowered from 682 (3) to 490 (3) and from 400 (2) to 233 (2) (degeneracies in parentheses) relative to the old assignment (*146*) as a result of a new vibrational assignment for the molecule.

[53] Simons and Rabinovitch (*146*). See Tables VI–VIII for vibrational assignment.

[54] Schlag and Rabinovitch (*144*). Setser and Rabinovitch (*147*); Appendix II of this paper contains full vibrational assignments in the molecule and transition state.

Broken lines represent bonds of order $\frac{1}{2}$, full lines bonds of order 1 (normal single bonds), and $\overline{- - -}$ represents bonds of order $\frac{3}{2}$. Bond lengths were calculated from bond orders using Eq. (11-17), and the C—C$\overline{- - -}$C angle was taken as 109°, intermediate between its values in propylene and cyclopropane. The C—H—C triangle was assumed to be equilateral, with a H—C—C angle of 53°. A C—H stretch in the molecule at 3000 cm^{-1} was taken as the reaction coordinate and was eliminated in the transition state. The molecule frequencies that were changed in the transition state relative to their value in the molecule are shown in Table 11-5; the assignment

TABLE 11-5

VIBRATIONAL FREQUENCIES OF CYCLOPROPANE THAT ARE ASSUMED TO CHANGE WHEN ITS TRANSITION STATE FOR STRUCTURAL ISOMERIZATION TO PROPYLENE IS FORMED[a]

Molecule		Transition state	
Designation	Frequency (cm^{-1})	Designation	Frequency (cm^{-1})
Ring mode	1029	C—C stretch	900
Ring mode	1029	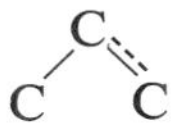 bend	500
Ring mode	1188	C$\overline{- - -}$C stretch	1100
CH_2 rock	741	C$\overline{- - -}CH_2$ twist	500
CH_2 rock	741	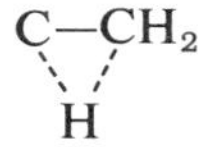 twist	500

[a] Assignment of D. W. Setser and B. S. Rabinovitch, *Can. J. Chem.* **40**, 1425 (1962). Cf. also D. W. Setser, Ph.D. Thesis, University of Washington, 1961. p. 158 ff.

was made with the aid of bond order rules and by analogy with cyclopropane and propylene and estimated frequencies of trimethylene. All other frequencies of the molecule were left unchanged in the transition state. With this model, the reaction path degeneracy is 12 since the hydrogen atom about to be transferred to the future CH_3 group of the product propylene can be any one of four equivalent hydrogens of the cyclopropane, and the CH_2 group receiving the hydrogen can be any one of three equivalent CH_2 groups of the cyclopropane, giving a total of 12 equivalent ways of forming the transition state.

Transition states for the structural isomerization of alkyl-substituted cyclopropanes (*143*, *146–148*) were obtained by making analogous changes in the frequencies of the molecule. Alkyl group torsion with respect to the ring was assumed to be a free rotation in the molecule as well as in the transition state. Substituted cyclopropanes isomerize by parallel paths into several different olefins, but the same transition state was assumed to apply for every path regardless of the geometric isomerism of the reactant. For example, 1,2-dimethyl cyclopropane isomerizes (*146*) into *cis*- and *trans*-pentene-2, 2-methyl butene-2, and 2-methyl butene-1. The total reaction path degeneracy here is eight since, compared with cyclopropane, there is only two-thirds the number of hydrogens available for migration.

In addition to thermal and chemical activation data on the geometric and structural isomerization of a variety of cyclopropanes, these transition states also account for the hydrogen–deuterium isotope effect in the thermal system (*43*,*44*) cyclopropane–cyclopropane-d_6 (structural and geometric isomerization), the hydrogen–fluorine isotope effect in the chemical activation systems (*149*,*150*) of methyl, ethyl, and propyl cyclopropane and their trifluoro analogs (structural isomerization), and for the hydrogen-deuterium isotope effect in the thermal isomerizations (*143*) of methyl cyclopropane and of 1,2-dideutero-3-methyl cyclopropane (both structural and geometric isomerizations).

In cyclopropanes formed by CH_2 addition across a double bond, the transition state for the reaction *forming* the chemically activated reactant was assumed to have the structure[55]

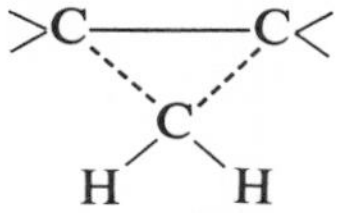

where broken lines again represent bonds of order $\frac{1}{2}$. Frequencies were assigned by analogy with cyclopropane; a cyclopropane ring motion was taken as the reaction coordinate. The CH_2 group was taken to be attached rigidly, with no internal rotation. This transition state is used only for calculating the distribution function $F(E)$ [Eq. (9-26)] and a relatively crude assignment is sufficient.

The two-step mechanism (ii) involving a biradical intermediate has been vigorously advocated by Benson (*154*), who has used it to systematize a large amount of thermal data and the associated thermodynamics of activa-

[55] For details, see Setser (*151*, p. 163). For a model based on MO calculations, see Dobson *et al.* (*152*); Hoffmann *et al.* (*153*).

tion (*156*). Detailed rate calculations were done by Lin and Laidler (*41*) [see also (*158*)], based on the following kinetic scheme that applies at energies above threshold for product formation:

$$\text{molecule} \underset{k_{-2}}{\overset{k_2}{\rightleftharpoons}} \text{biradical} \underset{k_{-3}}{\overset{k_3}{\rightleftharpoons}} \text{product} \tag{11-32}$$

By "product" is meant here the product of the *structural* isomerization; therefore in geometric isomerization, k_3 and k_{-3} are zero, and it can be shown that the thermal high-pressure rate for geometric isomerization is then $k_\infty(\text{geometric}) = \frac{1}{2}k_2$, i.e., it depends only on the rate of the ring-opening process. The factor $\frac{1}{2}$ arises from the biradical having equal probability to reform either geometric isomer. In structural isomerization, k_3 and k_{-3} are not zero, and the resulting kinetic expressions are complex; however, if step 3 is rate determining, k_∞(structural) reduces to the product of k_3 and the equilibrium constant for the ring-opening process. This appears to be the case with cyclopropane and methyl cyclopropane, but in the case of cyclobutane, it seems that $k_{-2} \simeq k_3$. The calculations of the thermal rate of isomerization of these three compounds and some of their deuterated analogs, based on the scheme (11-32), gave fairly good agreement with experiment, but in general proved quite insensitive to the assignment of frequencies in the transition states involved, and therefore, by extension, also insensitive to the fundamental assumption regarding the existence of a biradical intermediate in the scheme (11-32).

The isomerization of cyclopropane has received a good deal of theoretical attention, particularly the geometric isomerization. If the isomerization proceeds via mechanism (ii), the trimethylene biradical intermediate would appear as a secondary potential minimum along the reaction coordinate. In a calculation by the Extended Hückel Method, Hoffmann (*171*) did find along the reaction coordinate a shallow minimum which could be identified with Benson's biradical. Later *ab initio* calculations (*172,173*) have shown, however, that there is in fact *no* secondary potential minimum. Salem and co-workers (*172*), using their own particular optimization procedure (*174*), have traced out part of the potential energy surface for the geometric isomerization, and found that the reaction is essentially a concerted process, involving a conrotatory motion of two CH_2 groups in the initial and final stages of the reaction, and disrotatory motion in the transition state region. The shallow minimum found by Hoffmann could at best be interpreted as lying on a path leading to *optical* isomerization.

These results apply to the geometric isomerization of cyclopropane on the singlet potential energy surface, which is certainly the relevant surface for the thermal reaction. By contrast, on the triplet surface there *is* a well-defined potential minimum, corresponding to the biradical, and this minimum lies *below* the singlet transition state. Thus the thermal reaction could pass through the biradical intermediate, which is actually a lower energy pathway, but only if conditions are favorable for the intersystem crossing singlet → triplet (*175*). In other words, Benson's biradical would have to be a triplet.

We may note in this connection that the structural isomerization of cyclobutane and its derivatives, although apparently proceeding via a mechanism and transition state similar to those of cyclopropane[56], is an exception to our classification scheme, in that the cyclobutanes yield *two* ethylenic fragments[57] and therefore, strictly speaking, are not type 3 reactions. Another exception is the structural isomerization of cyclobutenes, which do yield a single product and therefore should be type 3 reactions, but the transition states appear to be those of type 2 reactions. Hence transition states of cyclobutenes have been dealt with in Section 3.

Structural isomerization of isocyanides   Another type of isomerization that has played an important role in unimolecular rate theory, particularly the theory of energy transfer in thermal reactions at low pressure (cf. Chapter 8, Section 8), is the isomerization isocyanide → cyanide, of which the isomerizations of the methyl and ethyl compounds are the most-studied examples. The work is again due to Rabinovitch and co-workers.

The transition state, as originally formulated by Schneider and Rabinovitch (*160*), for the methyl isocyanide system, was taken to be a cyclic species with three resonance structures:

$$H_3C-\overset{..}{N}\diagdown_{\underset{..}{C}} \qquad H_3C\text{----}\overset{..}{N}\diagdown_{\text{--}C} \qquad H_3\overset{+}{C} \quad \overset{N^-}{\underset{C}{\mathbb{I}}}$$

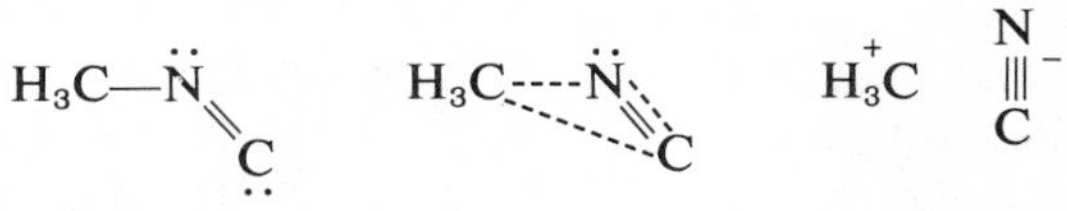

Giving equal weight to each structure, the bond orders are then 0.5 for H_3C-N, 2.5 for $N\!\!=\!\!C$, and 0.16 for $H_3C\cdots C$; from these bond orders the corresponding bond lengths were calculated from Eq. (11-17). Frequency assignments were made with the help of Badger's rule [Eq. (6-121)] and by

---

[56] The potential energy surface for the isomerization cyclobutane → $2C_2H_4$ (see Table 11-4) was investigated by Hoffmann *et al.* (*155*). If the biradical mechanism is valid, the reaction should involve tetramethylene as an intermediate. The authors find that, "if by the term 'intermediate' is meant a true minimum in the many-dimensional potential energy surface, our calculations imply that there is *no such species* intervening between cyclobutane and two ethylenes." However, there appears to be a large flat region in the surface which should be *operationally* indistinguishable from a true minimum.

[57] This is also true of the cyclobutane ion: cyclo-$C_4H_8{}^+ \rightarrow C_2H_4{}^+ + C_2H_4$. Cf. Lifshitz and Tiernan (*159*).

analogy with cyclopropylene. The reaction coordinate is an asymmetric ring deformation

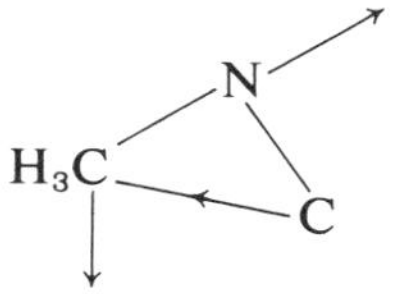

The N≡C stretch at 2161 cm^{-1} in the molecule becomes a ring deformation mode in the transition state at 1990 cm^{-1}, a C—N stretch at 945 cm^{-1} becomes a ring deformation at 600 cm^{-1}, and of the two C—N—C bends at 270 cm^{-1} in the molecule, one becomes the reaction coordinate and the other a twisting motion at 300 or 600 cm^{-1}, depending on the model of the transition state; all other frequencies are assumed to be the same in the molecule and transition state. The transition-state model with the 300 cm^{-1} frequency was judged to give the best overall agreement with experiment.

The same transition-state model was used for calculating the isomerization of ethyl isocyanide (*161*), except for the inclusion of additional vibrational modes due to the replacement of methyl by ethyl. The CH_3 torsion in the ethyl group was taken to be a torsional vibration at 224 cm^{-1}. Again the transition state with the twisting motion in the neighborhood of 300 cm^{-1} gave the best agreement with experiment. Calculations of the hydrogen–deuterium kinetic isotope effect in methyl-d_1 (*46*), methyl-d_3 (*47*), and ethyl-d_5 isocyanides (*48*), and of the C^{13} isotope effect (*162*) in the methyl compound all made successful use of the same transition-state model.

Rabinovitch and collaborators have used a reaction path degeneracy of $\alpha = 3$ for the isomerization of both methyl and ethyl isocyanides. The methyl compound is of C_{3v} symmetry, and therefore $\sigma = 3$, while the transition state is of C_s symmetry, with $\sigma^* = 1$. If one assumes that in this case $\alpha = \sigma/\sigma^*$, we get $\alpha = 3$. However, this result would be true only if the methyl hydrogens were somehow involved in the reaction, which they apparently are not, so that by direct count, $\alpha = 1$. In terms of the discussion of Chapter 4, Section 5, we might say that if the transition state is nonplanar, as is likely, we have a "type b failing" so that $\sigma/\sigma^* \neq \alpha$, and the direct count should take precedence. In the case of the ethyl compound, $\sigma = 1$ and $\sigma^* = 1$, but two different transition states may be recognized (*161*), depending on the relative positions of C≡N, CH_3, and the two hydrogens of CH_2, and one of these transition states can exist in two enantiomeric forms, so that $\alpha = 3$.

It will be noted that this transition state is quite different from that suggested by the calculations of Van Dine and Hoffmann (*163*), whose calculated potential energy surface for the isocyanide → cyanide isomerization is shown in Fig. 3-3. Their transition state would involve appreciable charge separation and such an ionic model apparently could not be fitted to the observed thermal preexponential factor A_∞. Additional experimental evidence that is cited against the model of VanDine and Hoffmann concerns the isomerization of butyl isocyanides (*164*): optically active *sec-butyl* isocyanide isomerizes with retention of configuration, which points to a synchronous process, and no product other than cyclobutylcyanide is obtained in the isomerization of cyclobutylisocyanide, which seems to rule out cyclobutyl cation as an intermediate because of its known propensity to carbon skeleton rearrangement. VanDine and Hoffmann's calculations, which were done by the Extended Hückel Method, also predict the methyl group in the methyl isocyanide to flatten out into a planar configuration in the transition state.

Recent *ab initio* calculations (*176*) modify somewhat these conclusions: the methyl group in the methyl isocyanide remains pyramidal throughout the reaction, including the transition state, and the transition state seems to have much less ionic character than the original estimate of VanDine and Hoffmann. These new calculations are more in line with experimental results on the optically active butyl isocyanides, but the picture of the transition state that emerges is still basically that of VanDine and Hoffmann.

If the potential energy surface of VanDine and Hoffmann implies a transition-state model that does not fit thermal experimental results, one can then wonder to what extent the computer results of Harris and Bunker (*165*) are relevant to the behavior of a real experimental system. The computer results are based on the model of VanDine and Hoffmann and, as mentioned in Chapter 2, Section 5, show that the isocyanide isomerization may be a nonstatistical process. Could it be that these computer results merely reflect the aberrant behavior of an unrealistic model for the reaction? Conversely, perhaps the isocyanide isomerization actually *is* a nonstatistical system, as suggested by the calculations, in which case the transition state of Rabinovitch, even though derived from experiment, would be merely a meaningless construct designed to interpret good experimental results on the basis of a wrong, i.e., statistical, model. Unfortunately, the thermal experimental results on the methyl isocyanide isomerization are now themselves under attack, for Yip and Pritchard (*166*) have found evidence that the isomerization may be free-radical catalyzed. Whether the extent of free-radical catalysis is sufficient to disqualify the isomerization of isocyanides as a well-behaved unimolecular reaction is presently a matter of active discussion; nevertheless, the possibility of a free-radical catalysis does bring into some doubt the usefulness of the isocyanide isomerization for testing unimolecular rate theories.[58]

[58] Recent work (*177*) suggests that the free radical catalysis may be only minor. In any event, the catalysis appears only at high pressures, so that the usefulness of the methyl isocyanide system for energy transfer studies at low pressure (cf. Chapter 8, Section 8) appears to be unimpaired.

Exercises

1. Show that for the primary intermolecular isotope effect, Eq. (11-4) reduces to the expression

$$\frac{k_{\mathrm{H}}(E)}{k_{\mathrm{D}}(E)} = \frac{\alpha_{\mathrm{H}}}{\alpha_{\mathrm{D}}}\left(\frac{\mu_{\mathrm{D}}^*}{\mu_{\mathrm{H}}^*}\right)^{1/2}\left(\frac{E - E_{0(\mathrm{H})}}{E - E_{0(\mathrm{D})}}\right)^{\mathrm{v}-1}$$

if the classical expression (6-55) is used for the density of states. At high energies ($E \to \infty$), this equation reduces further to

$$k_{\mathrm{H}}(E)/k_{\mathrm{D}}(E) = (\alpha_{\mathrm{H}}/\alpha_{\mathrm{D}})(\mu_{\mathrm{D}}^*/\mu_{\mathrm{H}}^*)^{1/2}$$

which gives the *minimum* value of the intermolecular isotope effect.

2. Show that the maximum contribution to $P(\mathscr{J})_{\mathrm{e}}$ [Eq. (8-14)] occurs for

$$\mathscr{J}_{\max} = \tfrac{1}{2}[(2kT/\mathbf{B})^{1/2} - 1]$$

where $\mathbf{B} = h^2/8\pi^2 I_b kT$ is the rotational constant and $I_b = \mu r_{\mathrm{e}}^2$ is the moment of inertia. Isotopic substitution diminishes **B** and therefore increases $\mathscr{J}_{\max}$.

References

1. P. C. Beadle, J. H. Knox, F. Placido, and K. C. Waugh, *Trans. Faraday Soc.* **65**, 1571 (1969).
2. F. Kaufman, *Ann. Rev. Phys. Chem.* **20**, 77 (1969).
3. C. H. Bamford and C. F. H. Tipper, eds., *in* "Comprehensive Chemical Kinetics," Vol. 5. Elsevier, Amsterdam, 1972.
4. S. Glasstone, K. J. Laidler, and H. Eyring, "The Theory of Rate Processes." McGraw-Hill, New York, 1941.
5. P. J. Robinson and K. A. Holbrook, "Unimolecular Reactions," Chapter 7. Wiley (Interscience), New York and London, 1972.
6. R. E. Weston, Jr., *Discuss. Faraday Soc.* **44**, 163 (1967).
7. K. J. Laidler, *Discuss. Faraday Soc.* **44**, 172 (1967).
8. D. J. LeRoy, *Discuss. Faraday Soc.* **44**, 173 (1967).
9. S. J. Yao and B. J. Zwolinski, *J. Phys. Chem.* **72**, 373 (1968).
10. B. A. Ridley, K. A. Quickert, and D. J. LeRoy, *J. Phys. Chem.* **72**, 1845 (1968).
11. S. J. Yao and B. J. Zwolinski, *J. Phys. Chem.* **72**, 1845 (1968).
12. D. L. Bunker and M. Pattengill, *J. Chem. Phys.* **48**, 772 (1968).
13. W. H. Wong and R. A. Marcus, *J. Chem. Phys.* **55**, 5625 (1971).
14. W. L. Hase, *J. Chem. Phys.* **57**, 730 (1972).
15. J. Troe, *in* "Physical Chemistry. An Advanced Treatise" (W. Jost, ed.). Academic Press, New York, 1972 (in press).
16. E. B. Wilson, Jr., J. C. Decius, and P. C. Cross, "Molecular Vibrations. The theory of Infrared and Raman Vibrational Spectra." McGraw-Hill, New York, 1955.
17. H. S. Johnston, W. A. Bonner, and D. J. Wilson, *J. Chem. Phys.* **26**, 1002 (1957).
18. R. E. Weston, Jr., *Science* **158**, 332 (1967).

19. R. Daudel, R. Lefebvre, and C. Moser, "Quantum Chemistry." Wiley (Interscience), New York, 1959.
20. G. Herzberg, "Molecular Spectra and Molecular Structure," Vol. II, Infrared and Raman Spectra of Polyatomic Molecules. Van Nostrand Reinhold, Princeton, New Jersey, 1945.
21. C. J. Collins and N. S. Bowman, eds., "Isotope Effect in Chemical Reactions," ACS Monograph 167. Van Nostrand Reinhold, Princeton, New Jersey, 1971.
22. R. A. M. O'Ferrall and J. Kouba, *J. Chem. Soc. B* 985 (1967).
23. E. A. Halevi, *Progr. Phys. Org. Chem.* **1**, 109 (1963).
24. B. S. Rabinovitch and D. W. Setser, *Advan. Photochem.* **3**, 1 (1964).
25. M. Bertrand, J. H. Beynon, and R. G. Cooks, *Int. J. Mass Spectrom. Ion Phys.* **9**, 346 (1972).
26. L. P. Hills, M. L. Vestal, and J. H. Futrell, *J. Chem. Phys.* **54**, 3834 (1971).
27. Z. Prášil and W. Forst, *J. Phys. Chem.* **71**, 3166 (1967).
28. C. Lifshitz and R. Sternberg, *Int. J. Mass Spectrom. Ion Phys.* **2**, 303 (1969).
29. M. Vestal and J. H. Futrell, *J. Chem. Phys.* **52**, 978 (1970).
30. C. Lifshitz and M. Shapiro, *J. Chem. Phys.* **45**, 4242 (1966).
31. C. Lifshitz and M. Shapiro, *J. Chem. Phys.* **46**, 4912 (1967).
32. I. Howe and F. W. McLafferty, *J. Amer. Chem. Soc.* **93**, 99 (1971).
33. C. E. Klots, *Chem. Phys. Lett.* **10**, 422 (1971).
34. K. Dees and D. W. Setser, *J. Chem. Phys.* **49**, 1193 (1968).
35. R. L. Platzman, *J. Chem. Phys.* **38**, 2775 (1963).
36. S. Meyerson, H. M. Grubb, and R. W. V. Haar, *J. Chem. Phys.* **39**, 1445 (1963).
37. J. C. Person and P. O. Nicole, *J. Chem. Phys.* **49**, 5421 (1968).
38. J. Bigeleisen and M. Wolfsberg, *Advan. Chem. Phys.* **1**, 15 (1958).
39. C. R. Gatz, *J. Chem. Phys.* **44**, 1861 (1966).
40. W. E. Buddenbaum and P. E. Yankwich, *J. Phys. Chem.* **71**, 3136 (1967).
41. M. C. Lin and K. J. Laidler, *Trans. Faraday Soc.* **64**, 927 (1968).
42. P. C. Vogel and M. J. Stern, *J. Chem. Phys.* **54**, 779 (1971).
43. B. S. Rabinovitch, D. W. Setser, and F. W. Schneider, *Can. J. Chem.* **39**, 2609 (1961).
44. B. S. Rabinovitch, P. W. Gilderson, and A. T. Blades, *J. Amer. Chem. Soc.* **86**, 2994 (1964).
45. R. W. Carr, Jr. and W. D. Walters, *J. Amer. Chem. Soc.* **88**, 884 (1966).
46. B. S. Rabinovitch, P. W. Gilderson, and F. W. Schneider, *J. Amer. Chem. Soc.* **87**, 158 (1965).
47. F. W. Schneider and B. S. Rabinovitch, *J. Amer. Chem. Soc.* **85**, 2365 (1963).
48. K. M. Maloney, S. P. Pavlou, and B. S. Rabinovitch, *J. Phys. Chem.* **73**, 2756 (1969).
49. H. S. Johnston, "Gas Phase Reaction Rate Theory." Ronald Press, New York, 1966.
50. T. S. Ree, T. Ree, H. Eyring, and T. Fueno, *J. Chem. Phys.* **36**, 281 (1962).
51. H. M. Frey and R. Walsh, *Chem. Rev.* **69**, 103 (1969).
52. S. W. Benson, "Thermochemical Kinetics." Wiley, New York, 1968.
53. S. W. Benson and H. E. O'Neal, "Kinetic Data on Gas Phase Unimolecular Reactions." NSRDS-NBS 21, Washington, D.C., 1970.
54. J. C. Walton, *J. Phys. Chem.* **71**, 2763 (1967).
55. E. Tschuikow-Roux, *J. Phys. Chem.* **72**, 1009 (1968).
56. L. Pauling, *J. Amer. Chem. Soc.* **69**, 542 (1947).

57. L. Pauling, "The Nature of the Chemical Bond," 3rd ed. Cornell Univ. Press, Itchica, New York, 1960.
58. W. R. Thorson and I. Nakagawa, *J. Chem. Phys.* **33**, 994 (1960).
59. J. E. Brown and R. G. Parr, *J. Chem. Phys.* **54**, 3429 (1971) Appendix.
60. F. B. Brown and N. G. Charles, *J. Chem. Phys.* **55**, 4481 (1971).
61. M. Jungen and J. Troe, *Ber. Bunsenges.* **74**, 2776 (1970).
62. E. Gorin, *Acta Physicochim. URSS* **9**, 691 (1938).
63. H. Gaedtke and J. Troe, *Ber. Bunsenges.* **77**, 24 (1973).
64. O. K. Rice, *J. Phys. Chem.* **65**, 1588 (1961).
65. H. S. Johnston and P. Goldfinger, *J. Chem. Phys.* **37**, 700 (1962).
66. R. A. Marcus and O. K. Rice, *J. Phys. Coll. Chem.* **55**, 894 (1951).
67. J. C. Hassler and D. W. Setser, *J. Amer. Chem. Soc.* **87**, 3793 (1965).
68. D. W. Setser and B. S. Rabinovitch, *J. Chem. Phys.* **40**, 2427 (1964).
69. Z. Prášil and W. Forst, *J. Phys. Chem.* **71**, 3166 (1967).
70. E. Tschuikow-Roux, *J. Chem. Phys.* **49**, 3115 (1968).
71. W. Forst, *J. Chem. Phys.* **44**, 2349 (1966).
72. D. W. Setser, *J. Phys. Chem.* **70**, 826 (1966).
73. G. M. Wieder and R. A. Marcus, *J. Chem. Phys.* **37**, 1835 (1962).
74. H. S. Johnston, *Advan. Chem. Phys.* **3**, 133 (1961).
75. W. Lester, S. Andrews, and G. C. Pimentel, *J. Chem. Phys*, **44**, 2527 (1966).
76. G. Gioumousis and D. P. Stevenson, *J. Chem. Phys.* **29**, 294 (1958).
77. S. W. Benson, *J. Amer. Chem. Soc.* **91**, 2152 (1969).
78. M. C. Lin and K. J. Laidler, *Trans. Faraday Soc.* **64**, 79 (1968).
79. C. Steel and K. J. Laidler, *J. Chem. Phys.* **34**, 1827 (1961).
80. E. V. Waage and B. S. Rabinovitch, *Int. J. Chem. Kinet.* **3**, 105 (1971).
81. D. R. Chang and O. K. Rice, *Int. J. Chem. Kinet.* **1**, 171 (1969).
82. R. L. Johnson and D. W. Setser, *J. Phys. Chem.* **71**, 4366 (1967).
83. J. Yang and D. C. Conway, *J. Chem. Phys.* **43**, 1296 (1965).
84. G. Z. Whitten and B. S. Rabinovitch, *J. Phys. Chem.* **69**, 4348 (1965).
85. D. W. Setser and W. C. Richardson, *Can. J. Chem.* **47**, 2593 (1969).
86. H. W. Chang and D. W. Setser, *J. Amer. Chem. Soc.* **91**, 7648 (1969).
87. J. C. Hassler and D. W. Setser, *J. Chem. Phys.* **45**, 3246 (1966).
88. H. W. Chang, N. L. Craig, and D. W. Setser, *J. Phys. Chem.* **76**, 954 (1972).
89. W. Forst, *J. Phys. Chem.* **76**, 342 (1972), Fig. 3.
90. W. Forst, *in* "Reaction Transition States" (J. E. Dubois, ed.). Gordon and Breach, New York, 1972.
91. S. W. Benson, *Advan. Chem. Phys.* **2**, 1 (1964).
92. P. C. Kobrinsky, G. O. Pritchard, and S. Toby, *J. Phys. Chem.* **75**, 2225 (1971).
93. W. L. Hase, R. L. Johnson, and J. W. Simons, *Int. J. Chem. Kinet.* **4**, 1 (1972).
94. A. Maccoll, *Chem. Rev.* **69**, 33 (1969).
95. A. F. Trotman-Dickenson and G. S. Milne, "Tables of Bimolecular Gas Reactions." NSRDS-NBS 9, Washington, D.C., 1967.
96. J. H. Current and B. S. Rabinovitch, *J. Chem. Phys.* **38**, 783 (1963).
97. B. S. Rabinovitch, R. F. Kubin, and R. E. Harrington, *J. Chem. Phys.* **38**, 405 (1963).
98. D. C. Tardy and B. S. Rabinovitch, *Trans. Faraday Soc.* **64**, 1 (1968).
99. Le-Khac Huy, W. Forst, J. A. Franklin, and G. Huybrechts, *Chem. Phys. Lett.* **3**, 307 (1969).
100. K. Dees and D. W. Setser, *J. Phys. Chem.* **75**, 2240 (1971).

101. M. J. Pearson and B. S. Rabinovitch, *J. Chem. Phys.* **42**, 1624 (1965).
102. F. H. Dorer and B. S. Rabinovitch, *J. Phys. Chem.* **69**, 1952 (1965).
103. G. W. Taylor and J. W. Simons, *J. Chem. Kinet.* **3**, 453 (1971).
104. J. M. Hay, *J. Chem. Soc. B* 1175 (1967).
105. A. T. Blades, P. W. Gilderson, and M. G. H. Wallbridge, *Can. J. Chem.* **40**, 1526 (1962).
106. N. Capon and R. A. Ross, *Trans. Faraday Soc.* **62**, 1560 (1966).
107. K. A. Holbrook and A. R. W. Marsh, *Trans. Faraday Soc.* **63**, 643 (1967).
108. H. Heydtmann and G. W. Völker, *Z. Phys. Chem.* (*Frankfurt*) **55**, 296 (1967).
109. G. W. Völker and H. Heydtmann, *Z. Naturforsch.* **23b**, 1407 (1968).
110. H. Heydtmann, *Ber. Bunsenges.* **72**, 1009 (1968).
111. J. A. Kerr and D. M. Timlin, *Int. J. Chem. Kinet.* **3**, 427 (1971).
112. M. C. Lin, *J. Phys. Chem.* **75**, 3642 (1971).
113. M. C. Lin, *J. Phys. Chem.* **76**, 811, 1425 (1972).
114. L. E. Brus and M. C. Lin, *J. Phys. Chem.* **76**, 1429 (1972).
115. W. G. Clark, D. W. Setser, and K. Dees, *J. Amer. Chem. Soc.* **93**, 5328 (1971).
116. G. E. Millward and E. Tschuikow-Roux, *Int. J. Chem. Kinet.* **4**, 559 (1972).
117. H. W. Chang and D. W. Setser, *J. Amer. Chem. Soc.* **91**, 7648 (1969).
118. J. C. Hassler and D. W. Setser, *J. Chem. Phys.* **45**, 3237 (1966).
119. K. Dees, D. W. Setser, and W. G. Clark, *J. Phys. Chem.* **75**, 2231 (1971).
120. P. N. Clough, J. C. Polanyi, and R. T. Taguchi, *Can. J. Chem.* **48**, 2919 (1970).
121. Y. N. Tang and F. S. Rowland, *J. Amer. Chem. Soc.* **90**, 570 (1968).
122. H. S. Johnston, *J. Amer. Chem. Soc.* **86**, 1643 (1964).
123. J. C. Hassler, Ph.D. Thesis, Kansas State Univ. (1966).
124. H. E. O'Neal and S. W. Benson, *J. Phys. Chem.* **71**, 2903 (1967).
125. J. A. Kerr and D. M. Timlin, *Trans. Faraday Soc.* **67** 1376 (1971).
126. H. dePuy and R. W. King, *Chem. Rev.* **60**, 431 (1960).
127. A. Maccoll, *Advan. Phys. Org. Chem.* **3**, 91 (1965).
128. A. Maccoll and P. J. Thomas, *Progr. Reaction Kinet.* **4**, 119 (1967).
129. C. S. Elliott and H. M. Frey, *Trans. Faraday Soc.* **62**, 895 (1966).
130. K. Hsu, R. J. Buenker, and S. D. Peyerimhoff, *J. Amer. Chem. Soc.* **93**, 2117 (1971).
131. R. J. Buenker, S. D. Peyerimhoff, and K. Hsu, *J. Amer. Chem. Soc.* **93**, 5005 (1971).
132. M. C. Lin and K. J. Laidler, *Trans. Faraday Soc.* **64**, 94 (1968).
133. G. E. McGraw, D. L. Bernitt, and I. C. Hisatsune, *J. Chem. Phys.* **45**, 1392 (1966).
134. I. C. Hisatsune, *J. Phys. Chem.* **72**, 269 (1968).
135. M. C. Lin and K. J. Laidler, *Can. J. Chem.* **46**, 973 (1968).
136. S. W. Benson, D. M. Golden, and K. Egger, *J. Amer. Chem. Soc.* **87**, 468 (1965).
137. R. L. Arnett and B. L. Crawford, Jr., *J. Chem. Phys.* **18**, 118 (1950).
138. R. B. Cundal, *in* "Progress in Reaction Kinetics" (G. Porter ed.), Vol. 2. Pergamon, Oxford, 1964.
139. H. M. Frey, *Advan. Phys. Org. Chem.* **4**, 147 (1966).
140. M. R. Willcott, R. L. Cargill, and A. B. Sears, *Progr. Phys. Org. Chem.* **9**, 25 (1972).
141. B. S. Rabinovitch, E. W. Schlag, and K. Wiberg, *J. Chem. Phys.* **28**, 504 (1958).
142. T. S. Chambers and G. B. Kistiakowsky, *J. Amer. Chem. Soc.* **56**, 399 (1934).
143. D. W. Setser and B. S. Rabinovitch, *J. Amer. Chem. Soc.* **86**, 564 (1964).
144. E. W. Schlag and B. S. Rabinovitch, *J. Amer. Chem. Soc.* **82**, 5996 (1960).
145. J. D. Rynbrandt and B. S. Rabinovitch, *J. Phys. Chem.* **74**, 1679 (1970).
146. J. W. Simons and B. S. Rabinovitch, *J. Phys. Chem.* **68**, 1322 (1964).
147. D. W. Setser and B. S. Rabinovitch, *Can. J. Chem.* **40**, 1425 (1962).

148. G. W. Taylor and J. W. Simons, *Int. J. Chem. Kinet.* **3**, 25 (1971).
149. F. H. Dorer, B. S. Rabinovitch, and D. W. Placzek, *J. Chem. Phys.* **41**, 3995 (1964).
150. F. H. Dorer and B. S. Rabinovitch, *J. Phys. Chem.* **69**, 1973 (1965).
151. D. W. Setser, Ph.D. Thesis, Univ. of Washington (1961).
152. R. C. Dobson, D. M. Hayes, and R. Hoffmann, *J. Amer. Chem. Soc.* **93**, 6188 (1971).
153. R. Hoffmann, D. M. Hayes, and P. S. Skell, *J. Phys. Chem.* **76**, 664 (1972).
154. S. W. Benson, *J. Chem. Phys.* **43**, 34, 521 (1961).
155. R. Hoffmann, S. Swaminathan, G. Odell, and R. Gleiter, *J. Amer. Chem. Soc.* **92**, 7091 (1970).
156. H. E. O'Neal and S. W. Benson, *J. Phys. Chem.* **72**, 1866 (1968).
157. K. C. Kim and D. W. Setser, *J. Phys. Chem.* **76**, 283 (1972).
158. E. Jakubowski, H. S. Sandhu, and O. P. Strausz, *J. Amer. Chem. Soc.* **93**, 2610 (1971).
159. C. Lifshitz and T. O. Tiernan, *J. Chem. Phys.* **55**, 3555 (1971).
160. F. W. Schneider and B. S. Rabinovitch, *J. Amer. Chem. Soc.* **84**, 4215 (1962).
161. K. M. Maloney and B. S. Rabinovitch, *J. Phys. Chem.* **73**, 1652 (1969).
162. J. F. Wettaw and L. B. Sims, *J. Phys. Chem.* **72**, 3440 (1968).
163. G. W. VanDine and R. Hoffmann, *J. Amer. Chem. Soc.* **90**, 3227 (1968).
164. J. Casanova, Jr., N. D. Werner, and R. E. Schuster, *J. Org. Chem.* **31**, 3473 (1966).
165. H. H. Harris and D. L. Bunker, *Chem. Phys. Lett.* **11**, 433 (1971).
166. C. K. Yip and H. O. Pritchard, *Can. J. Chem.* **48**, 2942 (1970).
167. D. W. Setser and E. E. Siefert, *J. Chem. Phys.* **57**, 3613 (1972).
168. H. M. Frey, R. G. Hopkins, and I. C. Vinall, *JCS Faraday I* **68**, 1874 (1972).
169. M. J. S. Dewar and S. Kirschner, *J. Amer. Chem. Soc.* **93**, 4292 (1971).
170. A. Rastelli, A. S. Pozzoli, and G. DelRe, *JCS Perkin II*, 1571 (1972).
171. R. Hoffmann, *J. Amer. Chem. Soc.* **90**, 1475 (1968).
172. J. A. Horsley, Y. Jean, C. Moser, L. Salem, R. M. Stevens, and J. S. Wright, *J. Amer. Chem. Soc.* **94**, 279 (1972); *Pure Appl. Chem. Suppl.* **1**, 197 (1971).
173. P. J. Hay, W. J. Hunt, and W. A. Goddard, *J. Amer. Chem. Soc.* **94**, 638 (1972).
174. L. Salem, *Accounts Chem. Res.* **4**, 322 (1971).
175. L. Salem and C. Rowland, *Angew. Chem. Int. Ed.* **11**, 92 (1972).
176. D. H. Liskow, C. F. Bender, and H. F. Schaefer, *J. Amer. Chem. Soc.* **94**, 5178 (1972).
177. T. Fujimoto, F. M. Wang, and B. S. Rabinovitch, *Can. J. Chem.* **50**, 3251 (1972).
178. R. Dunbar, *J. Amer. Chem. Soc.* **95**, 472 (1973).
179. G. M. Nazin, *Russian Chem. Rev.* **41**, 711 (1972).
180. J. V. Michael and G. N. Suess, *J. Chem. Phys.* **58**, 2807 (1973).

Appendix

Described here are four computer programs for the calculation of $N_v(E)$, $G_v(E)$, $N'_{vr}(E)$ and $G'_{vr}(E)$ by the method of steepest descents (Chapter 7, Sections 8 and 9). It is hoped that the inclusion of actual computer programs will impress on the interested reader how simple the steepest-descent method is, and will encourage him to make use of the method for his own particular applications. The programs were originally written by Z. Prášil and, with minor modifications, have been used extensively in the author's laboratory. They are written in APL and should be readily translatable into other computer languages if one takes into account the right-to-left sequence of operations in APL. The programs were deliberately left in a simple form to make them readily understandable even to the relatively uninitiated, while readers with experience in programming can easily modify them for greater efficiency.

The first three programs represent applications of Eq. (6-89) to various oscillator systems, and therefore have common features which it is convenient to discuss first.

Programs 1, 2, and 3 evaluate, for various values of the energy E, the expression

$$\frac{Q_v(\theta)}{\theta^E(\ln \theta^{-1})^{k+\mathsf{r}/2}[2\pi\theta^2\phi''(\theta)]^{1/2}} \qquad \text{(notation NVR)} \qquad (1)$$

The meaning of NVR depends on the assigned value of $2k + \mathsf{r}$, which is represented by the notation NR. Thus

if NR $= 0$, NVR represents the density of vibrational states $N_v(E)$;
if NR $= 2$, NVR represents the integrated density of vibrational states $G_v(E)$;
if NR $= \mathsf{r}$, NVR represents $N'_{vr}(E)$ for r (classical) rotational degrees of freedom;

if NR = r + 2, NVR represents $G'_{vr}(E)$ for r (classical) rotational degrees of freedom.

Note that to convert $N'_{vr}(E)$ and $G'_{vr}(E)$ into $N_{vr}(E)$ and $G_{vr}(E)$, respectively, both must be multiplied by Q'_r [Eq. (6-42)].

The energy-dependent parameter θ (notation TH) is the root of $\theta\phi'(\theta) = 0$, and is obtained by the Newton–Raphson method. A first, approximate value of θ, say t_1 (notation T1), is used to calculate $f_0 = t_1\phi'(t_1)$ (notation F0), and then a "better" value of θ, say t_2 (notation T2), is calculated from

$$t_2 = t_1 - (f_0/f_1) \tag{2}$$

where $f_1 = [df_0/d\theta]_{\theta=t_1}$ (notation F1). Using t_2, new f_0 and f_1 are calculated, which are then used is Eq. (2) to obtain a still better t_2, and so on. The iteration is continued until $|t_2 - t_1| \leqslant \epsilon$, where ϵ is a small number (notation EPS); then $t_1 = \theta$. The iteration for the first energy value E[1] starts with t_1 equal to some assigned value (usually 0.999, see below), and the iterations at higher E[KK] (KK > 1) start with θ determined for the immediately preceding E[KK − 1].

To start the execution of a program, the workspace must contain:

(1) the vector E($E_1, \ldots, E_k, \ldots, E_{max}$) [in cm^{-1}] which specifies the energies at which the density or integrated density is to be calculated. The smallest value of E must be larger than zero ($E_1 > 0$);

(2) the vector FR($\alpha_1, \ldots, \alpha_n$) [in cm^{-1}], the vector DEG($g_1, \ldots, g_n$), and in nonharmonic systems, the vector X($x_1, \ldots, x_n$), all three of the same length. The elements of these vectors represents, respectively, the frequency, degeneracy, and anharmonicity constant of each member of the collection of independent oscillators for which the density or integrated density is to be calculated;

(3) the number NR (see above); the number T1, usually 0.999; the number EPS, which determines the precision with which θ is to be calculated, usually 10^{-10};

(4) the standard functions DFT and EFT which determine the format of the output. As written, the programs print out the results in columns, the first column giving energy in cm^{-1}, and the second column giving the corresponding density or integrated density. The units of the density are $(cm^{-1})^{-1}$. A third column prints out θ (TH).

Following is a more detailed description of each program.

1. Program DENS: reduced density or integrated density for a collection of harmonic oscillators and rigid free rotors. Here [cf. Eq. (6-90)]

$$Q_v(\theta) = \prod_{i=1}^{v} (1 - \theta^{\alpha_i})^{-g_i} \qquad \text{(notation QV)}$$

where $\alpha_i \equiv h\nu_i$ (or $h\omega_i$) (notation ALFA); then [cf. Eq. (6-92)]

$$\theta^2\phi''(\theta) = \sum_{i=1}^{\mathrm{v}} \frac{g_i\alpha_i^2\theta^{\alpha_i}}{(1-\theta^{\alpha_i})^2} + \frac{k+\mathrm{r}/2}{(\ln\theta^{-1})^2} \qquad \text{(notation TH} \times \text{F1)}$$

The value of θ is the root of the equation [cf. Eq. (6-93)]

$$f_0 = \sum_{i=1}^{\mathrm{v}} \frac{g_i\alpha_i\theta^{\alpha_i}}{1-\theta^{\alpha_i}} + \frac{k+\mathrm{r}/2}{\ln\theta^{-1}} - E = 0 \qquad \text{(notation F0)}$$

and f_1 (notation F1) is the first derivative of f_0 with respect to θ:

$$f_1 = \frac{df_0}{d\theta} = \sum_{i=1}^{\mathrm{v}} \frac{g_i\alpha_i^2\theta^{\alpha_i-1}}{(1-\theta^{\alpha_i})^2} + \frac{k+\mathrm{r}/2}{\theta(\ln\theta^{-1})^2}$$

so that

$$\theta f_1 = \theta^2\phi''(\theta)$$

2. Program CUTOFF: reduced density or integrated density for a collection of rigid free rotors and truncated harmonic oscillators. Here [cf. Eq. (6-117)]

$$Q_{\mathrm{v}}(\theta) = \prod_{i=1}^{\mathrm{v}} \left(\frac{1-\theta^{(m_i+1)\alpha_i}}{1-\theta^{\alpha_i}}\right)^{g_i} \qquad \text{(notation QV)}$$

The value of m_i is the maximum vibrational quantum numbers of the ith vibrational mode calculated from the corresponding anharmonicity constant x_i as *the nearest lower integer* satisfying

$$m_i = \left(\frac{1}{4x_i} - \frac{1}{2}\right) \qquad \text{(notation M)}$$

We have

$$\theta^2\phi''(\theta) = \underbrace{\sum_{i=1}^{\mathrm{v}} \left[\frac{\alpha_i^2\theta^{\alpha_i}}{(1-\theta^{\alpha_i})^2} - \frac{(m_i+1)^2\alpha_i^2\theta^{(m_i+1)\alpha_i}}{(1-\theta^{(m_i+1)\alpha_i})^2}\right] g_i}_{\text{SUM2}} + \frac{k+\mathrm{r}/2}{(\ln\theta^{-1})^2}$$

(notation F1 × TH)

The value of θ is the root of the equation

$$f_0 = \underbrace{\sum_{i=1}^{\mathrm{v}} g_i \left[\frac{\alpha_i\theta^{\alpha_i}}{1-\theta^{\alpha_i}} - \frac{(m_i+1)\alpha_i\theta^{(m_i+1)\alpha_i}}{1-\theta^{(m_i+1)\alpha_i}}\right]}_{\text{SUM1}} + \frac{k+\mathrm{r}/2}{\ln\theta^{-1}} - E = 0$$

(notation F0)

and f_1 (notation F1) is the first derivative of f_0 with respect to θ:

$$f_1 = \frac{df_0}{d\theta} = \sum_{i=1}^{\mathrm{v}} g_i \left[\frac{\alpha_i^2\theta^{\alpha_i-1}}{(1-\theta^{\alpha_i})^2} - \frac{(m_i+1)^2\alpha_i^2\theta^{(m_i+1)\alpha_i-1}}{(1-\theta^{(m_i+1)\alpha_i})^2}\right] + \frac{k+\mathrm{r}/2}{\theta(\ln\theta^{-1})^2}$$

so that

$$\theta f_1 = \theta^2\phi''(\theta)$$

3. Program ANH: reduced density or integrated density for a collection of rigid free rotors and Morse oscillators. Here [cf. Eq. 6-110)]

$$Q_{\mathsf{v}}(\theta) = \prod_{i=1}^{n} \left(\sum_{\mathsf{v}=0}^{\mathsf{v}\,\max} \theta^{\alpha_i \gamma_i} \right)^{g_i} \qquad \text{(notation } \times/\text{S1}\bigstar\text{DEG)}$$

$$\gamma_i = [\mathsf{v}(1 - x_i) - \mathsf{v}^2 x_i] \qquad \text{(notation G)}$$

where v is the vibrational quantum number running from 0 to $\mathsf{v}_{\max}$; $\mathsf{v}_{\max}$ is the nearest lower integer satisfying

$$\mathsf{v}_{\max} = \frac{1}{2}\left(\frac{1}{x_i} - 1\right) \qquad \text{(notation VMAX)}$$

The value of $\theta^2\phi''(\theta)$ is given by

$$\theta^2\phi''(\theta) = \underbrace{\sum_{i=1}^{n} g_i \left[\frac{\sum_{\mathsf{v}=0}^{\mathsf{v}\max} \alpha_i^2 \gamma_i^2 \theta^{\alpha_i \gamma_i}}{\sum_{\mathsf{v}=0}^{\mathsf{v}\max} \theta^{\alpha_i \gamma_i}} - \left(\frac{\sum_{\mathsf{v}=0}^{\mathsf{v}\max} \alpha_i \gamma_i \theta^{\alpha_i \gamma_i}}{\sum_{\mathsf{v}=0}^{\mathsf{v}\max} \theta^{\alpha_i \gamma_i}} \right)^2 \right]}_{\text{SUM2}} + \frac{k + \mathsf{r}/2}{(\ln \theta^{-1})^2}$$

Other notation, as used in the program; is

$$\theta^2\phi''(\theta) = \text{F1} \times \text{TH} \qquad \sum_{\mathsf{v}=0}^{\mathsf{v}\max} \alpha_i \gamma_i \theta^{\alpha_i \gamma_i} = \text{S2}$$

$$\sum_{\mathsf{v}=0}^{\mathsf{v}\max} \theta^{\alpha_i \gamma_i} = \text{S1} \qquad \sum_{\mathsf{v}=0}^{\mathsf{v}\max} \alpha_i^2 \gamma_i^2 \theta^{\alpha_i \gamma_i} = \text{S3}$$

The value of θ is the root of the equation

$$f_0 = \underbrace{\sum_{i=1}^{n} g_i \left[\frac{\sum_{\mathsf{v}=0}^{\mathsf{v}\max} \alpha_i \gamma_i \theta^{\alpha_i \gamma_i}}{\sum_{\mathsf{v}=0}^{\mathsf{v}\max} \theta^{\alpha_i \gamma_i}} \right]}_{\text{SUM1}} + \frac{k + \mathsf{r}/2}{\ln \theta^{-1}} - E = 0 \qquad \text{(notation F0)}$$

and f_1 (notation F1) is the first derivative of f_0 with respect to θ:

$$f_1 = \frac{df_0}{d\theta} = \frac{\text{SUM2}}{\theta} + \frac{k + \mathsf{r}/2}{\theta(\ln \theta^{-1})^2}$$

so that $\theta f_1 = \theta^2\phi''(\theta)$.

N.B. The notation here follows that actually used in the program, and is slightly different from the notation of Chapter 6 (p. 117 ff). In particular, n represents the number of degenerate oscillators, the ith being g_i-fold degenerate; thus the total number of oscillators is $\sum_{i=1}^{n} g_i$. v is the quantum number of the ith g_i-fold degenerate oscillator. Thus v and v_{max} are implicit functions of the index i.

4. Program DERIVG: density of states for a collection of rigid free rotors and truncated harmonic oscillators via the derivative of $G(E)$. This program represents an application of the formula [cf. Eq. (6-95)]

$$G(E)\left\{\ln\theta^{-1} - \frac{1}{\theta^2\phi''(\theta)} - \frac{\theta^3\phi'''(\theta)}{2[\theta^2\phi''(\theta)]^2}\right\} \qquad \text{(notation NVRD)}$$

The first part of the program (up to line 16) calculates $G(E)$ (notation NVR) by operations borrowed directly from CUTOFF, and therefore NR in this program must be either NR = 2 (no rotors), or NR = $\mathsf{r} + 2$ (r rotors). New information required is $\theta^3\phi'''(\theta)$ which can be written in terms of the variables previously defined in CUTOFF as

$$\theta^3\phi''' = -(\text{F1} \times \text{TH}) - \text{SUM2} + \text{PHI3}$$

where the new element PHI3 is defined by

$$\text{PHI3} = \sum_{i=1}^{\mathsf{v}} \left\{\frac{\alpha_i^3(1+\theta^{\alpha_i})\theta^{\alpha_i}}{(1-\theta^{\alpha_i})^3} - \frac{\beta_i^3(1+\theta^{\beta_i})\theta^{\beta_i}}{(1-\theta^{\beta_i})^3}\right\} + \frac{(k+\mathsf{r}/2)(2-\ln\theta^{-1})}{(\ln\theta^{-1})^3}$$

with $\beta_i = (m_i + 1)\alpha_i$. The program prints out θ, $G(E)$, and $N(E)$.

```
      ∇DENS[⎕]∇
    ∇ DENS
[1]    '       DENSITIES OF STATES BY THE METHOD OF STEEPEST DESCENTS
[2]    ''
[3]    ALFA←FR
[4]    KK←1
[5]   LABEL1:F0←(+/((DEG×ALFA×T1*ALFA)÷1-T1*ALFA))+((NR÷2)÷⍟÷T1)-E[KK]
[6]    SUM←+/((DEG×(ALFA*2)×T1*(ALFA-1))÷(1-T1*ALFA)*2)
[7]    F1←SUM+(NR÷2)÷T1×(⍟÷T1)*2
[8]    T2←T1-(F0÷F1)
[9]    →((|T2-T1)≤EPS)/LABEL2
[10]   T1←T2
[11]   →LABEL1
[12]  LABEL2:TH←T1
[13]   QV←×/((1-TH*ALFA)*(-DEG))
[14]   NVR←QV÷((TH*E[KK])×((⍟÷TH)*(NR÷2))×((F1×TH×○2)*0.5))
[15]  ('       '; 10 2 DFT E[KK]; 20 8 EFT NVR;'      ';TH)
[16]   →((KK←KK+1)≤⍴E)/LABEL1
[17]   ''
[18]   '       END'
    ∇
```

```
      ∇CUTOFF[⎕]∇
    ∇ CUTOFF
      '        DENSITIES OF STATES OF A COLLECTION OF HARMONIC OSCILLATORS WITH A CUT-OFF
      ''
      M←⌊((÷4×X)-0.5)
      ALFA←FR
      KK←1
    LABEL1:SUM1←+/(DEG×(((ALFA×T1*ALFA)÷(1-T1*ALFA))-(((M+1)×ALFA×T1*(M+1)×ALFA)÷(1-T1*(M+1)×ALFA))))
      SUM2←+/(DEG×((((ALFA*2)×T1*ALFA)÷((1-T1*ALFA)*2))-(((((M+1)×ALFA)*2)×T1*(M+1)×ALFA)÷((1-T1*(M+1)×ALFA)*2))))
      F0←SUM1+((NR÷2)÷⍟÷T1)-E[KK]
      F1←(SUM2÷T1)+(NR÷2)÷T1×(⍟÷T1)*2
      T2←T1-(F0÷F1)
       →((|T2-T1)≤EPS)/LABEL2
      T1←T2
      →LABEL1
    LABEL2:TH←T1
      QV←×/(((1-TH*(M+1)×ALFA)÷(1-TH*ALFA))*DEG)
      NVR←QV÷((TH*E[KK])×((⍟÷TH)*(NR÷2))×((F1×TH×○2)*0.5))
      ('       '; 10 2 DFT E[KK]; 20 8 EFT NVR;'      ';TH)
       →((KK←KK+1)≤⍴E)/LABEL1
      ''
      '        END'
    ∇
```

```
      ∇ANH[⎕]∇
    ∇ ANH
[1]   '       DENSITIES OF STATES OF A COLLECTION OF ANHARMONIC MORSE OSCILLATORS'
[2]   ''
[3]   ALFA←FR
[4]   S1←0×⍳⍴FR
[5]   VMAX←⌊(0.5×((÷X)-1))
[6]   KK←1
[7]  LABEL1:SUM1←SUM2←0
[8]   I←1
[9]  LABEL2:V←0,⍳VMAX[I]
[10]  G←(V×(1-X[I]))-((V*2)×X[I])
[11]  ZZ←T1*(ALFA[I]×G)
[12]  S1[I]←+/ZZ
[13]  S2←+/ALFA[I]×G×ZZ
[14]  S3←+/((ALFA[I]×G)*2)×ZZ
[15]  SUM1←SUM1+(S2÷S1[I])×DEG[I]
[16]  SUM2←SUM2+((S3÷S1[I])-((S2÷S1[I])*2))×DEG[I]
[17]  →((I←I+1)≤⍴FR)/LABEL2
[18]  F0←SUM1+((NR÷2)÷⍟÷T1)-E[KK]
[19]  F1←(SUM2÷T1)+(NR÷2)÷T1×(⍟÷T1)*2
[20]  T2←T1-(F0÷F1)
[21]  →((|T2-T1)≤EPS)/LABEL3
[22]  T1←(T2×T2<1)+(0.9999×T2>1)
[23]  →LABEL1
[24] LABEL3:TH←T1
[25]  NVR←(×/S1*DEG)÷(TH*E[KK])×((⍟÷TH)*NR÷2)×((F1×TH×○2)*0.5)
[26]  ('      '; 10 2 DFT E[KK]; 20 8 EFT NVR;'      ';TH)
[27]  →((KK←KK+1)≤⍴F)/LABEL1
[28]  '       END'
    ∇
```

```
      ∇DERIVG[⎕]∇
    ∇ DERIVG
      '             DENSITY VIA DERIVATIVE OF  G(E) '
      ''
      M←⌊((÷4×X)-0.5)
      ALFA←FR
      KK←1
    LABEL1:SUM1←+/(DEG×(((ALFA×T1*ALFA)÷(1-T1*ALFA))-(((M+1)×ALFA×T1*(M+1)×ALFA)÷(1-T1*(M+1)×ALFA))))
      SUM2←+/(DEG×((((ALFA*2)×T1*ALFA)÷((1-T1*ALFA)*2))-(((((M+1)×ALFA)*2)×T1*(M+1)×ALFA)÷((1-T1*(M+1)×ALFA)*2))))
      F0←SUM1+((NR÷2)÷⊛÷T1)-E[KK]
      F1←(SUM2÷T1)+(NR÷2)÷T1×(⊛÷T1)*2
      T2←T1-(F0÷F1)
      →((|T2-T1)≤EPS)/LABEL2
      T1←T2
      →LABEL1
    LABEL2:TH←T1
      QV←×/(((1-TH*(M+1)×ALFA)÷(1-TH*ALFA))*DEG)
      NVR←QV÷((TH*E[KK])×((⊛÷TH)*(NR÷2))×((F1×TH×○2)*0.5))
      W1←((ALFA*3)×(1+TH*ALFA)×TH*ALFA)÷((1-TH*ALFA)*3)
      W2←((((M+1)×ALFA)*3)×(1+TH*((M+1)×ALFA))×TH*((M+1)×ALFA))÷((1-TH*((M+1)×ALFA))*3)
      SUM3←+/DEG×(W1-W2)
      PHI3←SUM3+((NR÷2)×(2-⊛÷TH))÷(⊛÷TH)*3
      NVRD←NVR×((⊛÷TH)-(÷(F1×TH))+((PHI3-(F1×TH)+SUM2)÷(2×((F1×TH)*2))))
      RESULTS←NVR,NVRD
      ('            '; 6 0 DFT E[KK]; 15 10 DFT TH; 15 8 EFT RESULTS)
      →((KK←KK+1)≤ρE)/LABEL1
      ''
      '       END'
    ∇
```

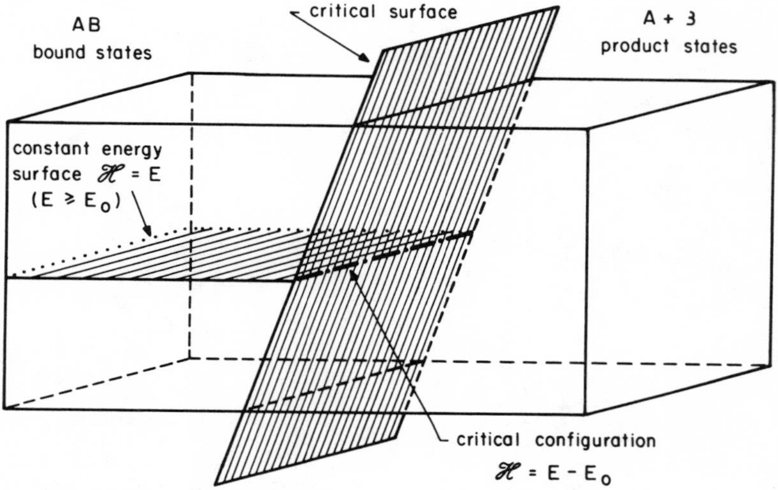

Fig. A-1. *Classical phase space for the decomposition* $\mathscr{A}\mathscr{B} \rightarrow \mathscr{A} + \mathscr{B}$.

This figure is an attempt to help the reader, struggling with Chapter 4, to visualize the phase space for the decomposition of $\mathscr{A}\mathscr{B}$. To make a pictorial representation possible, it is assumed that the phase space of $\mathscr{A}\mathscr{B}$ is three dimensional,[1] and the rectangular parallelepiped shown in the figure is a portion of it in the vicinity of the critical surface. Phase space "volume" is therefore a volume in the usual sense. The critical "surface" is consequently an actual two-dimensional surface (shown as a hatched almost vertical plane) that cuts the total phase space into two regions, one corresponding to bound states $\mathscr{A}\mathscr{B}$, and the other to dissociated states $\mathscr{A} + \mathscr{B}$. The constant energy "surface" is also an actual two-dimensional surface, shown for simplicity as a plane parallel to one of the faces of the parallelepiped; only its part on the bound states side of the critical surface is shown. For every particular value of E, there will be one such surface; we may think of these surfaces as stacked up one on top of another, and the critical surface will then cut through all of them. The intersection of the critical surface and a constant energy surface defines the critical configuration, in this case one dimensional; it is the dot-dashed line. Crudely speaking, decomposition at fixed E would correspond to an object (a dot) moving along the constant energy surface; when it reaches the critical configuration, it "explodes," i.e. forms "products."

[1] This is a grossly oversimplified representation, for an actual phase space can only have an even number of dimensions; moreover, even in a three-dimensional representation, most surfaces would be in general curved.

List of Symbols

In general, script is used for quantum numbers, German capitals for time-dependent quantities, and sans serif capitals for reactants. The asterisk superscript to a symbol always indicates the symbol in question refers to transition state. The list here contains only those transition-state symbols where the presence of the asterisk may lead to ambiguity. Square brackets [ ] always indicate concentration. Vector quantities are in boldface.

a	constant defined locally
a_{xy}	constant in Eq. (6-121)
a_0	radius of first Bohr orbit (0.5292×10^{-8} cm)
$\mathscr{a}$	energy-dependent correction term in semiclassical expressions (6-56) and (6-57) for density of states
$\mathscr{a}_k$ $(k = 0, 1)$	$\mathscr{a}$ in Eq. (6-63)
$\tilde{A}$, ^1A, ^3A	spectroscopic designation of molecular quantum state
A	rotational constant $= h/8\pi^2 I_a$
A	preexponential factor in thermal rate constant expression ($\sec^{-1}$); factor in Eq. (2-2)
A_D, A_H	designation of A in hydrogen–deuterium isotope effect: subscript D for A of deuterated species, subscript H for A of normal unsubstituted species
A_∞	high-pressure preexponential factor
A_ϕ	vibrational amplitude for bending vibration
$\mathscr{A}$	fragment, generally polyatomic, of quasidiatomic molecule $\mathscr{AB}$
A	general symbol for reactant subject to unimolecular decomposition
$[\mathsf{A}]$	total concentration of A of all energies and angular momenta at time t
$[\mathsf{A}]^0$	value of $[\mathsf{A}]$ at $t = 0$
$[\mathsf{A}]_e$	value of $[\mathsf{A}]$ at equilibrium
A_{ij}^+	ith ionic fragment of the jth parallel sequence
$\mathsf{A}(E)$	A of specified internal energy E
$[\mathsf{A}(E)]$	concentration of $\mathsf{A}(E)$ at time t
$\mathsf{A}(E, \mathscr{J})$	A of specified E and $\mathscr{J}$
$\mathsf{A}_{ij}^+(E_{ij})$	A_{ij}^+ of specified energy E_{ij}

A(i), A_i	A in quantum state i
A(X)	A in internal state(s) specified by parameter X
$\mathscr{AB}$	quasidiatomic model of polyatomic A
AB	diatomic molecule
ABC	triatomic molecule
AP	appearance potential
b, B	constants defined locally
b_{xy}	constant in Eq. (6-121)
$\mathscr{b}$	impact parameter
$\tilde{B}$, ^1B, ^3B	spectroscopic designation of molecular quantum state
B	rotational constant $= h/8\pi^2 I_b$
$\mathscr{B}$	fragment part, generally polyatomic, of $\mathscr{AB}$
B, B_{ij}	neutral fragment produced in $M^+ \rightarrow A^+ + B$; $A^+_{i-1,j} \rightarrow A^+_{ij} + B_{ij}$
B_m	Bernoulli number
c_i, c_i', C, C_1, C_2	constants defined locally
c	speed of light
c	contour of inversion integral (6-65)
c	$= (\ln \theta^{-1})^{-1}$ constant in Eq. (6-97)
$\tilde{C}$	spectroscopic designation of molecular quantum state
C	rotational constant $= h/8\pi^2 I_c$
C_s, C_{2v}, C_{3v}	designation of symmetry groups
$\mathscr{C}$	constant in attractive part of potential, $-\mathscr{C}/r^n$
$C_k(E)$ ($k = 0, 1$)	anharmonic correction factor at energy E
C_p	specific heat at constant pressure
d	parameter in Eq. (4-32)
$\hat{d}$	electric dipole moment operator
$\hat{d}_{el}$	electronic part of $\hat{d}$
$\hat{d}_N$	nuclear part of $\hat{d}$
$\mathscr{d}$	collision diameter
D	parameter in Eq. (4-32)
D	$= \langle D_i^{-1}\rangle^{-1}$, i.e. harmonic mean of D_i
D_e	potential energy of dissociation (Fig. 3-1)
D_i	D_e of ith oscillator
$D_i^{(v)}$	coefficient defined by Eq. (6-72)
D_J	potential energy of dissociation of rotational state $\mathscr{J}$ (Fig. 7-1)
D_m	coefficient defined by Eq. (6-74)
DEN	denominator of expression defined locally
D_0	quantum mechanical dissociation energy (Fig. 3-1)
D_{2d}, D_{2h}, D_{3d}	designation of symmetry groups
D	total amount of decomposition products present at time t in chemical activation
$\mathbf{D}_1$	amount of **D** formed if deactivation occurs in one step
$\mathbf{D}_N$	amount of **D** formed when deactivation occurs in N steps
$\mathscr{D}(y)$	probability per unit energy that ionic excitation process shall deposit energy y in M^+ = energy deposition function for M^+
$\mathfrak{D}_1(E_1, t)$	probability of fragment $A_1^+(E_1)$ decomposing (not surviving) in $(0, t)$
e	superscript indicating electronic excitation
e	electronic charge (4.8×10^{-10} esu)
E	total internal energy, in excess of zero-point energy, of a specified collection of degrees of freedom. No distinction is made in E per molecule and E per mole, the difference being obvious from context
$\langle E\rangle$	thermal average of E
$E' = E - E_J$	
E_a	experimental (Arrhenius) activation energy in thermal reactions
$E_{a(D)}$, $E_{a(H)}$	designation of E_a in hydrogen deuterium isotope effect (D = deuterated species, H = normal, unsubstituted species)
E_{a0}	E_a at the low-pressure limit of a thermal reaction
$E^{\dagger}_{a0}$	E_{a0} calculated under nonequilibrium conditions
$E_{a\infty}$	E_a at the high-pressure limit of a thermal reaction

$E_d = \sum_i D_i - E_z$ energy required to dissociate all oscillators in molecule ("total dissociation energy")

E_i energy of quantum state i

E_J critical energy when $\mathcal{J} \neq 0$

E_{ar}^J critical energy for the reverse of the activating reaction in chemical activation when $\mathcal{J} \neq 0$

E_m lower limit of integral in Eq. (10-39)

E_n energy of quantum level n

E_r total rotational energy

$\langle E_r \rangle$ thermal average of E_r

$E_r(\mathcal{J})$ $\mathcal{J}$-dependent part of E_r in a symmetric top

$E_r(\mathcal{K})$ $\mathcal{K}$-dependent part of E_r in a symmetric top

E_t translational energy in 3-dimensional coordinate space

E_v vibrational energy

$\langle E_v \rangle$ thermal average of E_v

E_z total zero-point energy

$E_{z(D)}$, $E_{z(H)}$ designation of E_z in hydrogen deuterium isotope effect (D = deuterated species, H = normal, unsubstituted species)

E_0 critical energy for decomposition when $\mathcal{J} = 0$

E_0^{ar} critical energy for the reverse of the activating reaction in chemical activation when $\mathcal{J} = 0$

$E_{0(D)}$, $E_{0(H)}$ designation of E_0 in hydrogen–deuterium isotope effect (D = deuterated species, H = normal, unsubstituted species)

E_{00} energy of the 0–0 transition in photochemistry [Fig. (9-7)]

$E_{0(ij)}$ critical energy for decomposition of A_{ij}^+

E_λ energy of radiation of wavelength λ

E_π π-bond energy

$E_{max,F}$ energy at which is located the maximum of $F(E)$

$E_{max,P}$ energy at which is located the maximum of $P(E)_e$

E_{max} energy above which $\mathcal{F}(E)$ is zero

E_{mono} energy of monoenergetic reactant

$\mathcal{E}$ total energy available for partition between two fragments

$f(\cdots)$ function, not otherwise specified, of variable(s) in parentheses

f_0 low-pressure centrifugal correction factor in thermal reactions

f_∞ high-pressure centrifugal correction factor in thermal reactions

f_2 quadratic force constant

f_3 cubic force constant

f_ϕ bending force constant

$f(E)$ optical oscillator strength

$\not f$ arbitary function

$\mathfrak{f}(\ldots, t)$ time-dependent function, not otherwise specified, of variables in parentheses

F first derivative of ln $Q(z)$, Eq. (6-96)

F thermodynamic Gibbs free energy

F_c Franck–Condon factor, Eq. (10-49)

$F_i^0 - E_{i0}^{\circ}$ free energy function of ith oscillator

F_j fraction of A's of energy E_j

$F(E)$, $F(E, \mathcal{J})$ in chemical activation systems: probability per unit energy range (and per unit $\mathcal{J}$) of forming reactant with internal energy in the range E, $E + dE$ (and angular momentum quantum number in the range $\mathcal{J}$, $\mathcal{J} + d\mathcal{J}$): energy (and angular momentum) distribution function for chemically activated reactant

$F(E)_{ss}$, $F(E, \mathcal{J})_{ss}$ $F(E)$ and $F(E, \mathcal{J})$ at steady state, respectively

$\mathcal{F}(E)$ probability per unit energy range of forming M^+ with internal energy in the range E, $E + dE$; energy distribution function for M^+

$\mathfrak{F}_{i\star}(E, t)$ fractional abundance of metastable i at t when excitation energy in M^+ is E

$\mathfrak{F}_{ij}(E, t)$ fractional abundance of A_{ij}^+ at time t when excitation energy of M^+ is E

$\mathfrak{F}(\ldots, t)$ fractional abundance of product at time t

g_i degeneracy of states in quantum level n_i
$g(s)$ arbitrary function of transform parameter s

$\mathscr{g}$ arbitrary function
$G(E)$ integrated density at energy E for any collection of degrees of freedom: total number of quantum states at E
$G_r(E)$, $G_v(E)$, $G_{vr}(E)$ $G(E)$ for rotational, vibrational, and vibrational–rotational states, respectively
$G'(E) = G(E)/Q_r'$
$\mathscr{G}^*(E)$ $G(E)$ for the transition state of the activating reaction in chemical activation

h Planck's constant
$\hbar = h/2\pi$
H thermodynamic enthalpy
$H_i^0 - E_{0i}^0$ enthalpy function of the ith oscillator
$H(E)$ function defined in Eq. (9-45) as $k(E)F(E)/k_\infty$
$\mathscr{H}$, $\mathscr{H}(p_i; q_i)$ Hamiltonian of A
$\mathscr{H}'(p_i; q_i)$ Hamiltonian of the critical configuration of A
$\mathscr{H}^*(p_2, \ldots, p_n; q_2, \ldots, q_n)$ Hamiltonian of the transition state of A
$\mathscr{H}_1$, $\mathscr{H}_1(p_1; q_1^0)$ one-dimensional Hamiltonian for motion in the reaction coordinate
$\nabla\mathscr{H}' =$ grad $\mathscr{H}'$, gradient of $\mathscr{H}'$, defined as $|\nabla\mathscr{H}'| = [\sum_i (\partial\mathscr{H}'/\partial p_i)^2 + \sum_i (\partial\mathscr{H}'/\partial q_i)^2]^{1/2}$
$\mathfrak{H}(x)$ Heaviside function $= 1$ for $x > 0$ $= 0$ for $x < 0$
$\mathsf{H}(E_k)$ ($k = i$ or j) thermal distribution function for fractional deviation from the equilibrium distribution, defined in Eq. (8-132) as $\mathsf{S}(E_k)/P(E_k)_e$

i running index (as superscript or subscript)
$i(z)$, $i[m/e]$ ion current
i imaginary number, $\sqrt{-1}$
I moment of inertia
I_a, I_b, I_c moment of inertia about axes a, b, c
I_i moment of inertia of ith rotor
$\mathscr{I}$, IP ionization potential
$\mathfrak{I}_a$ number of light quanta absorbed per cubic centimeter per second

j running index (as superscript or subscript)
$\mathscr{J}$ rotational quantum number for two-dimensional rotor
$\langle\mathscr{J}(\mathscr{J}+1)\rangle$ thermal average of $\mathscr{J}(\mathscr{J}+1)$ i.e., average of $\mathscr{J}(\mathscr{J}+1)$ over $P(\mathscr{J})_e$
$\mathscr{J}_{max}$ highest value of $\mathscr{J}$ for which $\mathscr{AB}$ is rotationally stable, ($\mathscr{AB}$ is rotationally dissociated for all $\mathscr{J} > \mathscr{J}_{max}$)

k index (as subscript)
k, k', k_1, k_2 general designation of rate constant (not necessarily unimolecular) for processes not otherwise specified in detail
k_a observable overall unimolecular rate constant in chemical activation
k_{a0} k_a at the low-pressure limit
$k_{a\infty}$ k_a at the high-pressure limit
k_{bi} second-order rate constant at the low-pressure limit of a thermal reaction, defined as $k_0'/[\mathsf{M}]$
$k_D(E)$, $k_H(E)$ designation of $k(E)$ in hydrogen–deuterium isotope effect (D = deuterated species, H = normal, unsubstituted species)
$k_{0j}(E)$ rate constant for decomposition of M^+ of energy E into the jth parallel channel
k_{ij} in Chapter 1, rate constant for transition from state i to state j;
$k_{ij}(E_{ij})$ in Chapter 10, rate constant for decomposition of A_{ij}^+ of energy E_{ij}
$\langle k_{ss}\rangle$ observable apparent steady-state rate constant
$\langle k_t\rangle$ observable apparent nonsteady-state rate constant
k_{uni} general-pressure thermal rate constant defined as $\langle k(E, \mathscr{J})\rangle_{ss}$

k_0 thermal unimolecular rate constant at the low-pressure limit

$k_0' = f_0 k_0$

$k_0^{\dagger}$ k_0 calculated for nonequilibrium conditions

$k_{0(D)}$, $k_{0(H)}$ designation of k_0 in hydrogen–deuterium isotope effect (D = deuterated species, H = normal, unsubstituted species)

k_∞ thermal unimolecular rate constant at the high-pressure limit

$k_\infty' = f_\infty k_\infty$

$k'_{\infty(D)}$, $k'_{\infty(H)}$ designation of k_∞' in hydrogen–deuterium isotope effect (D = deuterated species, H = normal, unsubstitued species)

k^{af} second-order rate constant for the activating reaction forming the chemically activated A

k^{ar} rate constant for the reverse of the activating reaction in chemical activation

$k(E)$ microcanonical rate constant for decomposition of A(E) into all channels

$k^1(E)$ rate constant for decomposition of A(E) into one channel [Eq. (4-29)]

$k'(E) = (f_\infty/f_0)k(E)$ Eq. (8-52)

$\langle k(E)\rangle_e = \langle k(E, \mathscr{J} = 0)\rangle_e$ average of $k(E)$ over $P(E)_e$

$k(E, \mathscr{J})$ $k(E)$ for species with specified value of $\mathscr{J}$

$k(E, \mathscr{J})_{app}$ apparent, or observable, value of $k(E, \mathscr{J})$

$k(E, \mathscr{J})_{app,ss}$ steady-state value of $k(E, \mathscr{J})_{app}$

$\langle k(E, \mathscr{J})\rangle_e$ average of $k(E, \mathscr{J})$ over $P(E, \mathscr{J})_e$

$\langle k(E', \mathscr{J})\rangle_J = (f_\infty/f_0)k(E')$ Eq. (8-47)

$\langle k(E, \mathscr{J})_{ss}\rangle_P$ average of f over $P(E, \mathscr{J})$

$\langle k(E, \mathscr{J})\rangle_{ss}$ average of $k(E, \mathscr{J})$ over $P(\cdots)_{ss}$

$\langle k(E, \mathscr{J})\rangle_t$ average of $k(E, \mathscr{J})$ over $\mathfrak{P}(\ldots, t)$

$\langle k(E, \mathscr{J})_t\rangle$ average of $\mathfrak{f}$ over $P(E, \mathscr{J})$

$k(E, \varepsilon_t)$ rate constant for decomposition of A(E) of specified energy ε_t in the reaction coordinate

$k(X)$ rate constant for decomposition of A(X) into all channels

$k(X)_{ss}$ steady-state value of $k(X)$

$\boldsymbol{k}$ unit vector

$\mathscr{k}$ Boltzmann constant (0.695 cm^{-1} deg^{-1} $molecule^{-1}$ = 1.38×10^{-16} erg deg^{-1} $molecule^{-1}$)

K equilibrium constant for specified reaction

K^* equilibrium constant for the formation of the transition state

$\mathscr{K}$ quantum number for component of angular momentum along symmetry axis of symmetric top

K_i constant of Eq. (6-124)

log or $\log_{10}$ logarithm base 10

ln logarithm base e

$\ell(s)$ function of transform parameter s

$L(s) = \ell(s)/s$

$\mathscr{L}$ orbital angular momentum

$\boldsymbol{L}$ orbital angular momentum vector

$\mathfrak{L}^{-1}\{\cdots\}$ Laplace transform of function in braces

m/e mass to charge ratio in mass spectrometry

m_0/e m/e for parent M^+

m_1/e m/e for fragment A_1^+

$m_\star/e$ m/e for metastable in $M^+ \rightarrow A_1^+$

m_e mass of electron

m_i mass of atom i

$m_1, m_2, \ldots$ integers (Chapter 6)

$\mathscr{m}$ rotational quantum number for one-dimensional rotor

m exponent of r in the repulsive part of intermolecular potential B/r^{m}

M slope of k_{uni} versus 1/[M] defined by Eq. (8-76)

M mass of molecule

$\boldsymbol{M}$ intrinsic angular momentum vector

M_2, M_3 coefficients in Eq. (6-106)

$M(x - z)$ instrument function in Eq. (10-69)

$M_{\mathscr{A}\mathscr{B}}$, $M_{\mathscr{A}}$, $M_{\mathscr{B}}$ $\boldsymbol{M}$ of quasidiatom $\mathscr{A}\mathscr{B}$ and its fragments $\mathscr{A}$ and $\mathscr{B}$, respectively

$\mathscr{M}$ intrinsic angular momentum

$\mathscr{M}_c$ component of $\mathscr{M}$ along axis c

n number of elements, locally defined; Chapter 1, number of time intervals; Chapter 4, number of degrees of freedom; Chapter 9, number of levels above E_0

$\mathscr{n}$ Chapter 6 and 10, number of vibrational quanta; vibrational quantum number (except in Figs. 10–4 and 10–7, where the symbol $\mathscr{v}$ is used)

$\mathscr{n}' = (1 - \mathscr{x})\mathscr{n} - \mathscr{n}^2\mathscr{x}$

$\tilde{n}$ translational quantum number

n exponent in the attractive part of intermolecular potential $-\mathscr{C}/r^{\mathrm{n}}$

$\mathrm{n}' = \mathrm{n}/(\mathrm{n} - 2)$

$\mathsf{n} = \mathsf{v} + k - 1 + \frac{1}{2}\mathsf{r}$: for a specified k, the number of vibrational degrees of freedom, plus one-half the number of rotational degrees of freedom, plus $k - 1$

$\mathsf{n}^* = \mathsf{v}^* + k - 1 + \frac{1}{2}\mathsf{r}^*$, n for the transition state

N slope of $k_{\mathrm{uni}}/[\mathrm{M}]$ versus M defined by Eq. (8-79)

N number of elements, locally defined: Chapter 4, Section 5, number of transition states; Chapters 7 and 10, number of electrons in molecule; Chapters 11 and 5, total number of atoms in molecule; Chapter 9, number of transitions in cascade deactivation

N_f fractional number of representative points per unit volume of phase space (Chapter 4)

$N_r(E)$, $N_v(E)$, $N_{vr}(E)$ $N(E)$ for rotational, vibrational, and vibrational–rotational states, respectively

$N'_{vr}(E) = N_{vr}(E)/Q_r'$

$N(E)$ density of quantum states at energy E for any collection of degrees of freedom (number of quantum states per unit energy interval)

$\mathscr{N}^*(E)$ $N(E)$ for transition state of the activating reaction in chemical activation

$\mathfrak{N}$ total number of ions of all energies at time t

$\mathfrak{N}_0$ total number of ions of all energies at time $t = 0$

$\mathfrak{N}(E)$ total number of ions of energy E at time t (per unit energy range)

$\mathfrak{N}_0(E)$ total number of ions of energy E at time $t = 0$ (per unit energy range)

$\mathscr{p}$ pressure

p, p_i generalized momentum

$^3\mathrm{P}$ spectroscopic designation of quantum state of atom

P general symbol for probability, usually specified by parameter in parentheses

$\hat{P}$ perturbation operator for intersystem crossing from quantum state A to quantum state B

$P(\cdots)_{ss}$ probability per unit E and $\mathscr{J}$ of finding A with specified E and $\mathscr{J}$ at steady state: steady-state distribution function

$P(E)_e$ probability per unit energy range of finding A with internal energy in the range E, $E + dE$ at thermal equilibrium: Boltzmann distribution

$P(E, \mathscr{J})_e = P(E)_e \times P(\mathscr{J})_e$

$P(E_0, E)_{e,\mathrm{norm}}$ portion of $P(E)_e$ above E_0 normalized to unity between E_0 and infinity

$P(i)$ probability of finding A in quantum state i

$P(\mathscr{J})_e$ probability per unit range of $\mathscr{J}$ of finding A with rotational quantum number in the range $\mathscr{J}$, $\mathscr{J} + d\mathscr{J}$ at thermal equilibrium: Boltzmann distribution for two-dimensional rotation

$P(p_i; q_i)$ probability per unit phase space volume of finding representative point with specified p_i; q_i in the ranges $p_i, p_i + dp_i$ and $q_i, q_i + dq_i$

$P(t)$, $P(\tau)$ probability of lifetime t (or τ), probability of survival in $(0, t)$ (or, $0, \tau$)

$P_{\mathrm{norm}}(t)$ $P(t)$ normalized to unity in $(0, \infty)$

$_{u}P_{\mathrm{norm}}(t)$ general form of lifetime probability with specified parameter u, normalized to unity in $(0, \infty)$

$P(n_a, n)$ (discrete) probability of finding n_a vibrational quanta in one of two degenerate oscillators if the number of vibrational quanta available to both oscillators is n

$P(x, \mathscr{E})$ (continuous) probability per unit energy range of finding energy x in one fragment when the energy available to both fragments is $\mathscr{E}$

$P(\varepsilon_t, E - E_0)$ probability per unit energy range of finding energy ε_t in the reaction coordinate if the total energy available to the transition state is $E - E_0$

$P(\omega)$ probability per unit time that an intact A shall suffer a collision in $t, t + dt$

$P(\omega)_{\mathrm{abs}} = \int_0^\infty P(\omega)\, dt$ absolute probability of an intact A suffering a collision

$\mathscr{P}$ matrix element of perturbation

$\mathfrak{P}(\ldots, t)$ probability per unit E and $\mathscr{J}$ of finding A at time t with specified E and $\mathscr{J}$ (nonsteady-state distribution function)

q, q_i generalized coordinate

q_1 critical coordinate

q_1^0 value of q_1 specifying location of critical surface

$\mathscr{q}_{ij}$ transition probability per collision from state i to state j

$\mathscr{q}(E_j, E_i)$ transition probability per collision for transition from state(s) of energy E_j to state(s) of energy E_i

q oscillator degeneracy (number of oscillators of the same frequency)

Q thermodynamic partition function for any collection of degrees of freedom

Q_D, Q_H designation of Q in hydrogen–deuterium isotope effect (D = deuterated species, H = normal, unsubstituted species)

Q_r, Q_v, Q_{vr} Q for rotational, vibrational, and vibrational–rotational degrees of freedom, respectively

$Q_r' = Q_r/(kT)^{r/2}$

Q_v^{class} classical vibrational partition function for v vibrational degrees of freedom

$Q(s)$ Q written as a function of transform parameter $s = e^{-1/kT}$

$Q(z)$ Q written as a function of $z = e^{-s}$

$Q(\theta)$ Q written as a function of $\theta = \exp(-s^\star)$

$\mathscr{Q}^*$ Q for the transition state of the activating reaction in chemical activation

r internuclear distance (except in Table 10-3, where the symbol R is used)

r_e equilibrium value of r

r_m r at the maximum of effective potential

$\langle r_m \rangle$ thermal average of r_m

r_s r for a single bond

r_x r for bond of bond order x

r_0 r at the maximum of type 2 or 3 potential

r_∞ r at large separation

r number of rotational degrees of freedom participating in the randomization of internal energy

RE recombination energy of ion

R gas constant (1.987 cal deg^{-1} mole^{-1})

$R(E)$ function defined in Eq. (8-71) as $k(E)P(E)_e/k_\infty$

$R^{\mathrm{ar}}(E)$ $R(E)$ for the reverse of the activating reaction in chemical activation

$\mathscr{R}$ electric dipole transition moment

$\mathfrak{R}$ total rate of formation of A by chemical activation

R_j constant in Eq. (6-123)

$s = e^{-1/kT}$ transform parameter

$s^\star$ value of s when $\phi'(s) = 0$

$\mathscr{s}$ Kassel parameter [Eq. (2-2)]; effective number of oscillators

s step size in cascade deactivation

S second derivative of ln $Q(z)$ in Eq. (6-96)

^1S, ^3S spectroscopic designation of quantum state of an atom
S total amount of stabilization product present at time t in chemical activation
$\mathbf{S}_1$ amount of **S** formed if deactivation occurs in one step
$\mathbf{S}_N$ amount of **S** formed if deactivation occurs in N steps
S thermodynamic entropy
S_r, S_v S for rotational and vibrational degrees of freedom, respectively
$\mathscr{S}$ critical (hyper)surface in phase space
$\mathsf{S}(E_k)$ ($k = i$ or j) nonequilibrium distribution function in thermal systems at low pressures, defined by Eq. (8-128)

t time
T temperature
T third derivative of ln $Q(z)$ in Eq. (6-96)
$T^\star = kT/\epsilon$ reduced temperature used in calculating transport properties
T_d designation of symmetry group

$u = h\nu/kT$
u scalar velocity in phase space
$\boldsymbol{u}$ vector velocity in phase space
U_i constant in Eq. (6-123)

v superscript indicating vibrational excitation
$\boldsymbol{v}$ vector velocity in coordinate space representing translation of center of mass of A
v vibrational level
v number of vibrational degrees of freedom involved in randomization of internal energy
v′ number of degenerate oscillators
V potential energy
V_{disp} dispersion contribution to V_W
V_{eff} effective potential (vibrational plus rotational)
$\langle V_{eff} \rangle$ thermal average of V_{eff}
V_{ind} inductive contribution to V_W
V_r rotational potential
V_{vib} vibrational potential
V_W van der Waals potential

$\mathscr{V}$ energy of incident particle in electron volts
w energy-dependent parameter in Eq. (6-58)
$w = \mathscr{J}(\mathscr{J} + 1)$
$W(n)$ number of quantum states for a collection of degrees of freedom containing a total of n quanta of energy
$W(E)$ number of quantum states at energy E for any collection of degrees of freedom
$W_r(E)$, $W_v(E)$, $W_{vr}(E)$ $W(E)$ for rotational, vibrational, and vibrational–rotational degrees of freedom, respectively

x variable defined locally
x anharmonicity constant
x Cartesian coordinate
X parameter specifying collection of initial states from which A is dissociating
$\tilde{X}$, X spectroscopic designation of molecular ground state
X reactant, usually polyatomic, in bimolecular reaction X + Y →

y variable defined locally
y Cartesian coordinate
Y reactant, usually polyatomic, in bimolecular reaction X + Y →

z variable defined locally; in particular, $z = e^{-s}$ in Chapter 6
z_0 solution of Eq. (8-69)
z Cartesian coordinate
Z collision number under hard-sphere potential
$Z_{L-J} = Z\Omega^{(2,2)}$ collision number under Lennard-Jones potential
$Z_{rel} = Z_{L-J}$ for M as colliding species relative for Z_{L-J} for A itself as colliding species

α reaction path degeneracy
α_a, α_b polarizabilities of $\mathscr{A}$ and $\mathscr{B}$
α_D, α_H designation of α in hydrogen–deuterium isotope effect (D = deuterated species, H = normal unsubstituted species)

α_{ij} reaction path degeneracy in the reaction $A_{ij}^{+} \rightarrow A_{i+1,j}^{+} + B_{i+1,j}$

α^{ar} α for the reverse of the activating reaction in chemical activation

$\beta = [(v - 1)/v]\nu_d$ [Eq. (6-59)]

β_c collision-per-collision efficiency of foreign gas relative to reactant in thermal reactions

β_k general formulation of β [Eq. (6-64)]

β_p pressure-for-pressure efficiency of foreign gas relative to reactant in thermal reactions

$\gamma = \langle\Delta E\rangle$ average amount of energy transferred per deactivating collision in thermal reactions

$\gamma' = \gamma^{-1} + (kT)^{-1}$

Γ uncertainty in energy of quantum state due to finite lifetime: energy "width" of state

$d\Gamma$ element of volume in coordinate space

$\Gamma(a, z)$ incomplete gamma function

$\Gamma(x)$ gamma function of x

δE energy interval

δq_1 small distance in the vicinity of $q_1{}^0$ within which the reaction coordinate is assumed to be separable from the other degrees of freedom

$\delta\varepsilon_t$ interval of ε_t

Δ[A] change of [A] in Δt

$\Delta E = E_i - E_{i-1}$

$\Delta E'$ excess internal energy of ion at ionization threshold

$\langle\Delta E\rangle$ average amount of energy transferred per deactivating collision in thermal reactions

$\langle\Delta E\rangle_a$ average amount of energy transferred per activating collision in thermal reactions

$\langle\Delta E\rangle_{Ex}$ $\langle\Delta E\rangle$ calculated on the exponential model

$\langle\Delta E\rangle_{Sl}$ $\langle\Delta E\rangle$ calculated on the step-ladder model

$\Delta E_z = E_{z(H)} - E_{z(D)}$

$\Delta E_r(\mathscr{J}) = E_r(\mathscr{J}) - E_r{}^*(\mathscr{J})$

$\Delta E_0 = E_{0(H)} - E_{0(D)}$

ΔS° standard entropy change in specified reaction at specified temperature

$\Delta S^* = R \ln(A_\infty h/\alpha f_\infty e kT)$ entropy change (per mole) in the formation of the transition state of a thermal reaction [Eq. (8-114)]

$\Delta S_* = \Delta S^* + R \ln(\alpha f_\infty)$ (per mole)

ΔH° standard enthalpy change in specified reaction at specified temperature

$\Delta H_f{}^\circ$ standard enthalpy of formation at 0 °K

ΔF° standard free energy change in specified reaction at specified temperature

Δt time interval

ε general symbol for energy assigned to one degree of freedom; also, in Chapter 6, equal to $E + E_z$, for one or several oscillators

ε_t translational energy in the reaction coordinate (one-dimensional)

$\langle\varepsilon_t\rangle$ average translational energy in the reaction coordinate when $\mathscr{J} = 0$ and total internal energy of transition state is $E - E_0$

$\langle\langle\varepsilon_t\rangle\rangle$ $\langle\varepsilon_t\rangle$ averaged over $P(E)_e$

$\langle\varepsilon_t(\mathscr{J})\rangle$ average translational energy in the reaction coordinate when $\mathscr{J} \neq 0$ and total internal energy of transition state is $E - E_0$

$\langle\langle\varepsilon_t(\mathscr{J})\rangle\rangle$ $\langle\varepsilon_t(\mathscr{J})\rangle$ averaged over $P(E, \mathscr{J})_e$

ε_z one-oscillator zero-point energy

ϵ symbol used for D_e in Eq. (8-121)

$\eta = E/E_z$

θ angle in radians; in Chapter 6 only: $\theta = \exp(-s^\star)$

$\kappa(E)$ transmission coefficient for barrier penetration by particle of energy E [Eq. (4-32)]

λ wavelength of exciting radiation

$\lambdabar$ de Broglie wavelength

Λ parameter in Eq. (10-68)

μ — reduced mass of A or AB
μ^* — effective mass in reaction coordinate
μ_a, μ_b — dipole moments of $\mathscr{A}$ and $\mathscr{B}$
μ_{AB} — reduced mass of $\mathscr{A}\mathscr{B}$
μ_D, μ_H — designation of μ in hydrogen–deuterium isotope effect (D = deuterated species, H = normal, unsubstituted species)
μsec — microsecond

ν_i — general symbol for vibrational frequency of ith oscillator, understood to be fundamental frequency
$\langle\nu\rangle$ — average frequency $(1/\mathrm{v}) \sum_i \nu_i$
ν_d — dispersion parameter defined as $\langle\nu^2\rangle/\langle\nu\rangle^2$
$\langle\nu\rangle_g$ — geometric mean of frequencies $(\Pi_i\ \nu_i)^{1/\mathrm{v}}$
ν_θ — frequency of torsional vibration

ξ — probability of elastic collision

ζ — exponent in the plot of k_{uni}^{-1} versus $[\mathrm{M}]^{-\zeta}$ for extrapolation to k_∞

$\rho = r_e^2/(r_0^2 - r_e^2)$

σ — symmetry number for rotation
σ_F^2 — variance of $F(E)$
σ_i — σ of ith rotor
σ_P^2 — variance of $P(E)_e$
σ^2 — variance
$\sigma(E)$ — total cross section for photoionization at energy E

τ — time; lifetime with respect to decomposition; also flight time in Table 10-1
$\langle\tau\rangle$ — average lifetime
$\tau^1(E)$ — lifetime of A(E) decomposing into one channel

Υ — scaling factor in Eq. (10-68); in Chapter 6: inversion integral [Eq. (6-65)]

ϕ — angle
$\phi(\cdots)$ — specified function of argument in parenthesis
$\phi(J) = E_J + E_r(\mathscr{J}) - E_0$
$\phi^a(J)$ — $\phi(\mathscr{J})$ for the activating reaction in chemical activation
$\phi(\theta)$ — function defined by Eq. (6-65) with $\theta = e^{-s}$
$\phi'(\theta)$, $\phi''(\theta)$, $\phi'''(\theta)$ — first, second, and third derivatives of $\phi(\theta)$ with respect to θ
$\langle\Phi\rangle$ — average value of $\Phi(p_1)$
Φ_D — photochemical quantum yield for **D**
$\Phi(p_i; q_i)$ — flux in phase space through critical surface

χ — total energy absorbed by all molecules

ψ — time-independent wave function
ψ_{el}, ψ_{rot}, ψ_{vib} — electronic, rotational, and vibrational wave function
$\psi^\star$ — complex conjugate of ψ

$\omega = Z[\mathrm{M}]$ collision frequency
ω_i — normal (calculated) frequency of ith oscillator

Ω — hypervolume in phase space
$\Omega^{(2,2)}$ — collision integral used in calculation of transport properties

Ξ — constant in Eq. (10-9)

Author Index

Numbers in parentheses are reference numbers and indicate that an author's work is referred to although his name is not cited in the text. Numbers in italics show the page on which the complete reference is listed.

G

H

Q

R

S

T

Y

Z

Subject Index

Page numbers in boldface refer to definitions or to a more thorough discussion of the subjects; page numbers in italics refer to figures or tables.

C

D

F

G

H

I

N

O

Q

R

S

Physical Chemistry

A Series of Monographs

Ernest M. Loebl, Editor

Department of Chemistry, Polytechnic Institute of Brooklyn, Brooklyn, New York

1 W. Jost: Diffusion in Solids, Liquids, Gases, 1952
2 S. Mizushima: Structure of Molecules and Internal Rotation, 1954
3 H. H. G. Jellinek: Degradation of Vinyl Polymers, 1955
4 M. E. L. McBain and E. Hutchinson: Solubilization and Related Phenomena, 1955
5 C. H. Bamford, A. Elliott, and W. E. Hanby: Synthetic Polypeptides, 1956
6 George J. Janz: Thermodynamic Properties of Organic Compounds — Estimation Methods, Principles and Practice, Revised Edition, 1967
7 G. K. T. Conn and D. G. Avery: Infrared Methods, 1960
8 C. B. Monk: Electrolytic Dissociation, 1961
9 P. Leighton: Photochemistry of Air Pollution, 1961
10 P. J. Holmes: Electrochemistry of Semiconductors, 1962
11 H. Fujita: The Mathematical Theory of Sedimentation Analysis, 1962
12 K. Shinoda, T. Nakagawa, B. Tamamushi, and T. Isemura: Colloidal Surfactants, 1963
13 J. E. Wollrab: Rotational Spectra and Molecular Structure, 1967
14 A. Nelson Wright and C. A. Winkler: Active Nitrogen, 1968
15 R. B. Anderson: Experimental Methods in Catalytic Research, 1968
16 Milton Kerker: The Scattering of Light and Other Electromagnetic Radiation, 1969
17 Oleg V. Krylov: Catalysis by Nonmetals — Rules for Catalyst Selection, 1970
18 Alfred Clark: The Theory of Adsorption and Catalysis, 1970

Physical Chemistry

A Series of Monographs

19 Arnold Reisman: Phase Equilibria: Basic Principles, Applications, Experimental Techniques, 1970
20 J. J. Bikerman: Physical Surfaces, 1970
21 R. T. Sanderson: Chemical Bonds and Bond Energy, 1970
22 S. Petrucci, ed.: Ionic Interactions: From Dilute Solutions to Fused Salts (In Two Volumes), 1971
23 A. B. F. Duncan: Rydberg Series in Atoms and Molecules, 1971
24 J. R. Anderson: Chemisorption and Reactions on Metallic Films, 1971
25 E. A. Moelwyn-Hughes: Chemical Statics and Kinetics of Solution, 1971
26 Ivan Draganic and Zorica Draganic: The Radiation Chemistry of Water, 1971
27 M. B. Huglin: Light Scattering from Polymer Solutions, 1972
28 M. J. Blandamer: Introduction to Chemical Ultrasonics, 1973
29 A. I. Kitaigorodsky: Molecular Crystals and Molecules, 1973
30 Wendell Forst: Theory of Unimolecular Reactions, 1973

In Preparation

Jerry Goodisman: Diatomic Interaction Potential Theory. Volume 1, Fundamentals; Volume 2, Applications

I. G. Kaplan: Symmetry of Many-Electron Systems